The Essential Aristotle

The Essential Aristotle

by Aristotle

Table of Contents

Rhetoric

Table of Contents

Book I

1

RHETORIC the counterpart of Dialectic. Both alike are concerned with such things as come, more or less, within the general ken of all men and belong to no definite science. Accordingly all men make use, more or less, of both; for to a certain extent all men attempt to discuss statements and to maintain them, to defend themselves and to attack others. Ordinary people do this either at random or through practice and from acquired habit. Both ways being possible, the subject can plainly be handled systematically, for it is possible to inquire the reason why some speakers succeed through practice and others spontaneously; and every one will at once agree that such an inquiry is the function of an art.

Now, the framers of the current treatises on rhetoric have constructed but a small portion of that art. The modes of persuasion are the only true constituents of the art: everything else is merely accessory. These writers, however, say nothing about enthymemes, which are the substance of rhetorical persuasion, but deal mainly with non-essentials. The arousing of prejudice, pity, anger, and similar emotions has nothing to do with the essential facts, but is merely a personal appeal to the man who is judging the case. Consequently if the rules for trials which are now laid down some states—especially in well-governed states—were applied everywhere, such people would have nothing to say. All men, no doubt, think that the laws should prescribe such rules, but some, as in the court of Areopagus, give practical effect to their thoughts and forbid talk about non-essentials. This is sound law and custom. It is not right to pervert the judge by moving him to anger or envy or pity—one might as well warp a carpenter's rule before using it. Again, a litigant has clearly nothing to do but to show that the alleged fact is so or is not so, that it has or has not happened. As to whether a thing is important or unimportant, just or unjust, the judge must surely refuse to take his instructions from the litigants: he must decide for himself all such points as the law-giver has not already defined for him.

Now, it is of great moment that well-drawn laws should themselves define all the points they possibly can and leave as few as may be to the decision of the judges; and this for several reasons. First, to find one man, or a few men, who are sensible persons and capable of legislating and administering justice is easier than to find a large number. Next, laws are made after long consideration, whereas decisions in the courts are given at short notice, which makes it hard for those who try the case to satisfy the claims of justice and expediency. The weightiest reason of all is that the decision of the lawgiver is not particular but prospective and general, whereas members of the assembly and the jury find it their duty to decide on definite cases brought before them. They will often have allowed themselves to be so much influenced by feelings of friendship or hatred or self-interest that they lose any clear vision of the truth and have their judgement obscured by considerations of personal pleasure or pain. In general, then, the judge should, we say, be allowed to decide as few things as possible. But questions as to whether something has happened or has not happened, will be or will not be, is or is not, must of necessity be left to the judge, since the lawgiver cannot foresee them. If this is so, it is evident that any one who lays down rules about other matters, such as what must be the contents of the 'introduction' or the 'narration' or any of the other divisions of a speech, is theorizing about non-essentials as if they belonged to the art. The only question with which these writers here deal is how to put the judge into a given frame of mind. About the orator's proper modes of persuasion they have nothing to tell us; nothing, that is, about how to gain skill in enthymemes.

Hence it comes that, although the same systematic principles apply to political as to forensic oratory, and although the former is a nobler business, and fitter for a citizen, than that which concerns the relations of private individuals, these authors say nothing about political oratory, but try, one and all, to write treatises on the way to plead in court. The reason for this is that in political oratory there is less inducement to talk about nonessentials. Political oratory is less given to unscrupulous practices than forensic, because it treats of wider issues. In a political debate the man who is forming a judgement is making a decision about his own vital interests. There is no need, therefore, to prove anything except that the facts are what the supporter of a measure maintains they are. In forensic oratory this is not enough; to conciliate the listener is what pays here. It is other people's affairs that are to be decided, so that the judges, intent on their own satisfaction and listening with partiality, surrender themselves to the disputants instead of judging between them. Hence in many places, as we have said already, irrelevant speaking is forbidden in the law-courts: in the public assembly those who have to form a judgement are themselves well able to guard against that.

It is clear, then, that rhetorical study, in its strict sense, is concerned with the modes of persuasion. Persuasion is clearly a sort of demonstration, since we are most fully persuaded when we consider a thing to have been demonstrated. The orator's demonstration is an enthymeme, and this is, in general, the most effective of the modes of persuasion. The enthymeme is a sort of syllogism, and the consideration of syllogisms of all kinds, without distinction, is the business of dialectic, either of dialectic as a whole or of one of its branches. It follows plainly, therefore, that he who is best able to see how and from what elements a syllogism is produced will also be best skilled in the enthymeme, when he has further learnt what its subject-matter is and in what respects it differs from the syllogism of strict logic. The true and the approximately true are apprehended by the same faculty; it may also be noted that men have a sufficient natural instinct for what is true, and usually do arrive at the truth. Hence the man who makes a good guess at truth is likely to make a good guess at probabilities.

It has now been shown that the ordinary writers on rhetoric treat of non-essentials; it has also been shown why they have inclined more towards the forensic branch of oratory.

Rhetoric is useful (1) because things that are true and things that are just have a natural tendency to prevail over their opposites, so that if the decisions of judges are not what they ought to be, the defeat must be due to the speakers themselves, and they must be blamed accordingly. Moreover, (2) before some audiences not even the possession of the exactest knowledge will make it easy for what we say to produce conviction. For argument based on knowledge implies instruction, and there are people whom one cannot instruct. Here, then, we must use, as our modes of persuasion and argument, notions possessed by everybody, as we observed in the Topics when dealing with the way to handle a popular audience. Further, (3) we must be able to employ persuasion, just as strict reasoning can be employed, on opposite sides of a question, not in order that we may in practice employ it in both ways (for we must not make people believe what is wrong), but in order that we may see clearly what the facts are, and that, if another man argues unfairly, we on our part may be able to confute him. No other of the arts draws opposite conclusions: dialectic and rhetoric alone do this. Both these arts draw opposite conclusions impartially. Nevertheless, the underlying facts do not lend themselves equally well to the contrary views. No; things that are true and things that are better are, by their nature, practically always easier to prove and easier to believe in. Again, (4) it is absurd to hold that a man ought to be ashamed of being unable to defend himself with his limbs, but not of being unable to defend himself with speech and reason, when the use of rational speech is more distinctive of a human being than the use of his limbs. And if it be objected that one who uses such power of speech unjustly might do great harm, that is a charge which may be made in common against all good things except virtue, and above all against the things that are most useful, as strength, health, wealth, generalship. A man can confer the greatest of benefits by a right use of these, and inflict the greatest of injuries by using them wrongly.

It is clear, then, that rhetoric is not bound up with a single definite class of subjects, but is as universal as dialectic; it is clear, also, that it is useful. It is clear, further, that its function is not simply to succeed in persuading, but rather to discover the means of coming as near such success as the circumstances of each particular case allow. In this it resembles all other arts. For example, it is not the function of medicine simply to make a man quite healthy, but to put him as far as may be on the road to health; it is possible to give excellent treatment even to those who can never enjoy sound health. Furthermore, it is plain that it is the function of one and the same art to discern the real and the apparent means of persuasion, just as it is the function of dialectic to discern the real and the apparent syllogism. What makes a man a 'sophist' is not his faculty, but his moral purpose. In rhetoric, however, the term 'rhetorician' may describe either the speaker's knowledge of the art, or his moral purpose. In dialectic it is different: a man is a 'sophist' because he has a certain kind of moral purpose, a 'dialectician' in respect, not of his moral purpose, but of his faculty.

Let us now try to give some account of the systematic principles of Rhetoric itself—of the right method and means of succeeding in the object we set before us. We must make as it were a fresh start, and before going further define what rhetoric is.

2

Rhetoric may be defined as the faculty of observing in any given case the available means of persuasion. This is not a function of any other art. Every other art can instruct or persuade about its own particular subject-matter; for instance, medicine about what is healthy and unhealthy, geometry about the properties of magnitudes, arithmetic about numbers, and the same is true of the other arts and sciences. But rhetoric we look upon as the power of observing the means of persuasion on almost any subject presented to us; and that is why we say that, in its technical character, it is not concerned with any special or definite class of subjects.

Of the modes of persuasion some belong strictly to the art of rhetoric and some do not. By the latter I mean such things as are not supplied by the speaker but are there at the outset—witnesses, evidence given under torture, written contracts, and so on. By the former I mean such as we can ourselves construct by means of the principles of rhetoric. The one kind has merely to be used, the other has to be invented.

Of the modes of persuasion furnished by the spoken word there are three kinds. The first kind depends on the personal character of the speaker; the second on putting the audience into a certain frame of mind; the third on the proof, or apparent proof, provided by the words of the speech itself. Persuasion is achieved by the speaker's personal character when the speech is so spoken as to make us think him credible. We believe good men more fully and more readily than others: this is true generally whatever the question is, and absolutely true where exact certainty is impossible and opinions are divided. This kind of persuasion, like the others, should be achieved by what the speaker says, not by what people think of his character before he begins to speak. It is not true, as some writers assume in their treatises on rhetoric, that the personal goodness revealed by the speaker contributes nothing to his power of persuasion; on the contrary, his character may almost be called the most effective means of persuasion he possesses. Secondly, persuasion may come through the hearers, when the speech stirs their emotions. Our judgements when we are pleased and friendly are not the same as when we are pained and hostile. It is towards producing these effects, as we maintain, that present-day writers on rhetoric direct the whole of their efforts. This subject shall be treated in detail when we come to speak of the emotions. Thirdly, persuasion is effected through the speech itself when we have proved a truth or an apparent truth by means of the persuasive arguments suitable to the case in question.

There are, then, these three means of effecting persuasion. The man who is to be in command of them must, it is clear, be able (1) to reason logically, (2) to understand human character and goodness in their various forms, and (3) to understand the emotions—that is, to name them and describe them, to know their causes and the way in which they are excited. It thus appears that rhetoric is an offshoot of dialectic and also of ethical studies. Ethical studies may fairly be called political; and for this reason rhetoric masquerades as political science, and the professors of it as political experts—sometimes from want of education, sometimes from ostentation, sometimes owing to other human failings. As a matter of fact, it is a branch of dialectic and similar to it, as we said at the outset. Neither rhetoric nor dialectic is the scientific study of any one separate subject: both are faculties for providing arguments. This is perhaps a sufficient account of their scope and of how they are related to each other.

With regard to the persuasion achieved by proof or apparent proof: just as in dialectic there is induction on the one hand and syllogism or apparent syllogism on the other, so it is in rhetoric. The example is an induction, the enthymeme is a syllogism, and the apparent enthymeme is an apparent syllogism. I call the enthymeme a rhetorical syllogism, and the example a rhetorical induction. Every one who effects persuasion through proof does in fact use either enthymemes or examples: there is no other way. And since every one who proves anything at all is bound to use either syllogisms or inductions (and this is clear to us from the Analytics), it must follow that enthymemes are syllogisms and examples are inductions. The difference between example and enthymeme is made plain by the passages in the Topics where induction and syllogism have already been discussed. When we base the proof of a proposition on a number of similar cases, this is induction in dialectic, example in rhetoric; when it is shown that, certain propositions being true, a further and quite distinct proposition must also be true in consequence, whether invariably or usually, this is called syllogism in dialectic, enthymeme in rhetoric. It is plain also that each of these types of oratory has its advantages. Types of oratory, I say: for what has been said in the Methodics applies equally well here; in some oratorical styles examples prevail, in others enthymemes; and in like manner, some orators are better at the former and some at the latter. Speeches that rely on examples are as persuasive as the other kind, but those which rely on enthymemes excite the louder applause. The sources of examples and enthymemes, and their proper uses, we will discuss later. Our next step is to define the processes themselves more clearly.

A statement is persuasive and credible either because it is directly self-evident or because it appears to be proved from other statements that are so. In either case it is persuasive because there is somebody whom it persuades. But none of the arts theorize about individual cases. Medicine, for instance, does not theorize about what will help to cure Socrates or Callias, but only about what will help to cure any or all of a given class of patients: this alone is business: individual cases are so infinitely various that no systematic knowledge of them is possible. In the same way the theory of rhetoric is concerned not with what seems probable to a given individual like Socrates or Hippias, but with what seems probable to men of a given type; and this is true of dialectic also. Dialectic does not construct its syllogisms out of any haphazard materials, such as the fancies of crazy people, but out of materials that call for discussion; and rhetoric, too, draws upon the regular subjects of debate. The duty of rhetoric is to deal with such matters as we deliberate upon without arts or systems to guide us, in the hearing of persons who cannot take in at a glance a complicated argument, or follow a long chain of reasoning. The subjects of our deliberation are such as seem to present us with alternative possibilities: about things that could not have been, and cannot now or in the future be, other than they are, nobody who takes them to be of this nature wastes his time in deliberation.

It is possible to form syllogisms and draw conclusions from the results of previous syllogisms; or, on the other hand, from premises which have not been thus proved, and at the same time are so little accepted that they call for proof. Reasonings of the former kind will necessarily be hard to follow owing to their length, for we assume an audience of untrained thinkers; those of the latter kind will fail to win assent, because they are based on premises that are not generally admitted or believed.

The enthymeme and the example must, then, deal with what is in the main contingent, the example being an induction, and the enthymeme a syllogism, about such matters. The enthymeme must consist of few propositions, fewer often than those which make up the normal syllogism. For if any of these propositions is a familiar fact, there is no need even to mention it; the hearer adds it himself. Thus, to show that Dorieus has been victor in a contest for which the prize is a crown, it is enough to say 'For he has been victor in the Olympic games', without adding 'And in the Olympic games the prize is a crown', a fact which everybody knows.

There are few facts of the 'necessary' type that can form the basis of rhetorical syllogisms. Most of the things about which we make decisions, and into which therefore we inquire, present us with alternative possibilities. For it is about our actions that we deliberate and inquire, and all our actions have a contingent character; hardly any of them are determined by necessity. Again, conclusions that state what is merely usual or possible must be drawn from premises that do the same, just as 'necessary' conclusions must be drawn from 'necessary' premises; this too is clear to us from the Analytics. It is evident, therefore, that the propositions forming the basis of enthymemes, though some of them may be 'necessary', will most of them be only usually true. Now the materials of enthymemes are Probabilities and Signs, which we can see must correspond respectively with the propositions that are generally and those that are necessarily true. A Probability is a thing that usually happens; not, however, as some definitions would suggest, anything whatever that usually happens, but only if it belongs to the class of the 'contingent' or 'variable'. It bears the same relation to that in respect of which it is probable as the universal bears to the particular. Of Signs, one kind bears the same relation to the statement it supports as the particular bears to the universal, the other the same as the universal bears to the particular. The infallible kind is a 'complete proof' (tekmerhiou); the fallible kind has no specific name. By infallible signs I mean those on which syllogisms proper may be based: and this shows us why this kind of Sign is called 'complete proof': when people think that what they have said cannot be refuted, they then think that they are bringing forward a 'complete proof', meaning that the matter has now been demonstrated and completed (peperhasmeuou); for the word 'perhas' has the same meaning (of 'end' or 'boundary') as the word 'tekmarh' in the ancient tongue. Now the one kind of Sign (that which bears to the proposition it supports the relation of particular to universal) may be illustrated thus. Suppose it were said, 'The fact that Socrates was wise and just is a sign that the wise are just'. Here we certainly have a Sign; but even though the proposition be true, the argument is refutable, since it does not form a syllogism. Suppose, on the other hand, it were said, 'The fact that he has a fever is a sign that he is ill', or, 'The fact that she is giving milk is a sign that she has lately borne a child'. Here we have the infallible kind of Sign, the only kind that constitutes a complete proof, since it is the only kind that, if the particular statement is true, is irrefutable. The other kind of Sign, that which bears to the proposition it supports the relation of universal to particular, might be illustrated by saying, 'The fact that he breathes fast is a sign that he has a fever'. This argument also is refutable, even if the statement about the fast breathing be true, since a man may breathe hard without having a fever.

It has, then, been stated above what is the nature of a Probability, of a Sign, and of a complete proof, and what are the differences between them. In the Analytics a more explicit description has been given of these points; it is there shown why some of these reasonings can be put into syllogisms and some cannot.

The 'example' has already been described as one kind of induction; and the special nature of the subject-matter that distinguishes it from the other kinds has also been stated above. Its relation to the proposition it supports is not that of part to whole, nor whole to part, nor whole to whole, but of part to part, or like to like. When two statements are of the same order, but one is more familiar than the other, the former is an 'example'. The argument may, for instance, be that Dionysius, in asking as he does for a bodyguard, is scheming to make himself a despot. For in the past Peisistratus kept asking for a bodyguard in order to carry out such a scheme, and did make himself a despot as soon as he got it; and so did Theagenes at Megara; and in the same way all other instances known to the speaker are made into examples, in order to show what is not yet known, that Dionysius has the same purpose in making the same request: all these being instances of the one general principle, that a man who asks for a bodyguard is scheming to make himself a despot. We have now described the sources of those means of persuasion which are popularly supposed to be demonstrative.

There is an important distinction between two sorts of enthymemes that has been wholly overlooked by almost everybody—one that also subsists between the syllogisms treated of in dialectic. One sort of enthymeme really belongs to rhetoric, as one sort of syllogism really belongs to dialectic; but the other sort really belongs to other arts and faculties, whether to those we already exercise or to those we have not yet acquired. Missing this distinction, people fail to notice that the more correctly they handle their particular subject the further they are getting away from pure rhetoric or dialectic. This statement will be clearer if expressed more fully. I mean that the proper subjects of dialectical and rhetorical syllogisms are the things with which we say the regular or universal Lines of Argument are concerned, that is to say those lines of argument that apply equally to questions of right conduct, natural science, politics, and many other things that have nothing to do with one another. Take, for instance, the line of argument concerned with 'the more or less'. On this line of argument it is equally easy to base a syllogism or enthymeme about any of what nevertheless are essentially disconnected subjects—right conduct, natural science, or anything else whatever. But there are also those special Lines of Argument which are based on such propositions as apply only to particular groups or classes of things. Thus there are propositions about natural science on which it is impossible to base any enthymeme or syllogism about ethics, and other propositions about ethics on which nothing can be based about natural science. The same principle applies throughout. The general Lines of Argument have no special subject-matter, and therefore will not increase our understanding of any particular class of things. On the other hand, the better the selection one makes of propositions suitable for special Lines of Argument, the nearer one comes, unconsciously, to setting up a science that is distinct from dialectic and rhetoric. One may succeed in stating the required principles, but one's science will be no longer dialectic or rhetoric, but the science to which the principles thus discovered belong. Most enthymemes are in fact based upon these particular or special Lines of Argument; comparatively few on the common or general kind. As in the therefore, so in this work, we must distinguish, in dealing with enthymemes, the special and the general Lines of Argument on which they are to be founded. By special Lines of Argument I mean the propositions peculiar to each several class of things, by general those common to all classes alike. We may begin with the special Lines of Argument. But, first of all, let us classify rhetoric into its varieties. Having distinguished these we may deal with them one by one, and try to discover the elements of which each is composed, and the propositions each must employ.

3

Rhetoric falls into three divisions, determined by the three classes of listeners to speeches. For of the three elements in speech-making—speaker, subject, and person addressed—it is the last one, the hearer, that determines the speech's end and object. The hearer must be either a judge, with a decision to make about things past or future, or an observer. A member of the assembly decides about future events, a juryman about past events: while those who merely decide on the orator's skill are observers. From this it follows that there are three divisions of oratory—(1) political, (2) forensic, and (3) the ceremonial oratory of display.

Political speaking urges us either to do or not to do something: one of these two courses is always taken by private counsellors, as well as by men who address public assemblies. Forensic speaking either attacks or defends somebody: one or other of these two things must always be done by the parties in a case. The ceremonial oratory of display either praises or censures somebody. These three kinds of rhetoric refer to three different kinds of time. The political orator is concerned with the future: it is about things to be done hereafter that he advises, for or against. The party in a case at law is concerned with the past; one man accuses the other, and the other defends himself, with reference to things already done. The ceremonial orator is, properly speaking, concerned with the present, since all men praise or blame in view of the state of things existing at the time, though they often find it useful also to recall the past and to make guesses at the future.

Rhetoric has three distinct ends in view, one for each of its three kinds. The political orator aims at establishing the expediency or the harmfulness of a proposed course of action; if he urges its acceptance, he does so on the ground that it will do good; if he urges its rejection, he does so on the ground that it will do harm; and all other points, such as whether the proposal is just or unjust, honourable or dishonourable, he brings in as subsidiary and relative to this main consideration. Parties in a law-case aim at establishing the justice or injustice of some action, and they too bring in all other points as subsidiary and relative to this one. Those who praise or attack a man aim at proving him worthy of honour or the reverse, and they too treat all other considerations with reference to this one.

That the three kinds of rhetoric do aim respectively at the three ends we have mentioned is shown by the fact that speakers will sometimes not try to establish anything else. Thus, the litigant will sometimes not deny that a thing has happened or that he has done harm. But that he is guilty of injustice he will never admit; otherwise there would be no need of a trial. So too, political orators often make any concession short of admitting that they are recommending their hearers to take an inexpedient course or not to take an expedient one. The question whether it is not unjust for a city to enslave its innocent neighbours often does not trouble them at all. In like manner those who praise or censure a man do not consider whether his acts have been expedient or not, but often make it a ground of actual praise that he has neglected his own interest to do what was honourable. Thus, they praise Achilles because he championed his fallen friend Patroclus, though he knew that this meant death, and that otherwise he need not die: yet while to die thus was the nobler thing for him to do, the expedient thing was to live on.

It is evident from what has been said that it is these three subjects, more than any others, about which the orator must be able to have propositions at his command. Now the propositions of Rhetoric are Complete Proofs, Probabilities, and Signs. Every kind of syllogism is composed of propositions, and the enthymeme is a particular kind of syllogism composed of the aforesaid propositions.

Since only possible actions, and not impossible ones, can ever have been done in the past or the present, and since things which have not occurred, or will not occur, also cannot have been done or be going to be done, it is necessary for the political, the forensic, and the ceremonial speaker alike to be able to have at their command propositions about the possible and the impossible, and about whether a thing has or has not occurred, will or will not occur. Further, all men, in giving praise or blame, in urging us to accept or reject proposals for action, in accusing others or defending themselves, attempt not only to prove the points mentioned but also to show that the good or the harm, the honour or disgrace, the justice or injustice, is great or small, either absolutely or relatively; and therefore it is plain that we must also have at our command propositions about greatness or smallness and the greater or the lesser-propositions both universal and particular. Thus, we must be able to say which is the greater or lesser good, the greater or lesser act of justice or injustice; and so on.

Such, then, are the subjects regarding which we are inevitably bound to master the propositions relevant to them. We must now discuss each particular class of these subjects in turn, namely those dealt with in political, in ceremonial, and lastly in legal, oratory.

4

First, then, we must ascertain what are the kinds of things, good or bad, about which the political orator offers counsel. For he does not deal with all things, but only with such as may or may not take place. Concerning things which exist or will exist inevitably, or which cannot possibly exist or take place, no counsel can be given. Nor, again, can counsel be given about the whole class of things which may or may not take place; for this class includes some good things that occur naturally, and some that occur by accident; and about these it is useless to offer counsel. Clearly counsel can only be given on matters about which people deliberate; matters, namely, that ultimately depend on ourselves, and which we have it in our power to set going. For we turn a thing over in our mind until we have reached the point of seeing whether we can do it or not.

Now to enumerate and classify accurately the usual subjects of public business, and further to frame, as far as possible, true definitions of them is a task which we must not attempt on the present occasion. For it does not belong to the art of rhetoric, but to a more instructive art and a more real branch of knowledge; and as it is, rhetoric has been given a far wider subject-matter than strictly belongs to it. The truth is, as indeed we have said already, that rhetoric is a combination of the science of logic and of the ethical branch of politics; and it is partly like dialectic, partly like sophistical reasoning. But the more we try to make either dialectic rhetoric not, what they really are, practical faculties, but sciences, the more we shall inadvertently be destroying their true nature; for we shall be re-fashioning them and shall be passing into the region of sciences dealing with definite subjects rather than simply with words and forms of reasoning. Even here, however, we will mention those points which it is of practical importance to distinguish, their fuller treatment falling naturally to political science.

The main matters on which all men deliberate and on which political speakers make speeches are some five in number: ways and means, war and peace, national defence, imports and exports, and legislation.

As to Ways and Means, then, the intending speaker will need to know the number and extent of the country's sources of revenue, so that, if any is being overlooked, it may be added, and, if any is defective, it may be increased. Further, he should know all the expenditure of the country, in order that, if any part of it is superfluous, it may be abolished, or, if any is too large, it may be reduced. For men become richer not only by increasing their existing wealth but also by reducing their expenditure. A comprehensive view of these questions cannot be gained solely from experience in home affairs; in order to advise on such matters a man must be keenly interested in the methods worked out in other lands.

As to Peace and War, he must know the extent of the military strength of his country, both actual and potential, and also the mature of that actual and potential strength; and further, what wars his country has waged, and how it has waged them. He must know these facts not only about his own country, but also about neighbouring countries; and also about countries with which war is likely, in order that peace may be maintained with those stronger than his own, and that his own may have power to make war or not against those that are weaker. He should know, too, whether the military power of another country is like or unlike that of his own; for this is a matter that may affect their relative strength. With the same end in view he must, besides, have studied the wars of other countries as well as those of his own, and the way they ended; similar causes are likely to have similar results.

With regard to National Defence: he ought to know all about the methods of defence in actual use, such as the strength and character of the defensive force and the positions of the forts—this last means that he must be well acquainted with the lie of the country—in order that a garrison may be increased if it is too small or removed if it is not wanted, and that the strategic points may be guarded with special care.

With regard to the Food Supply: he must know what outlay will meet the needs of his country; what kinds of food are produced at home and what imported; and what articles must be exported or imported. This last he must know in order that agreements and commercial treaties may be made with the countries concerned. There are, indeed, two sorts of state to which he must see that his countrymen give no cause for offence, states stronger than his own, and states with which it is advantageous to trade.

But while he must, for security's sake, be able to take all this into account, he must before all things understand the subject of legislation; for it is on a country's laws that its whole welfare depends. He must, therefore, know how many different forms of constitution there are; under what conditions each of these will prosper and by what internal developments or external attacks each of them tends to be destroyed. When I speak of destruction through internal developments I refer to the fact that all constitutions, except the best one of all, are destroyed both by not being pushed far enough and by being pushed too far. Thus, democracy loses its vigour, and finally passes into oligarchy, not only when it is not pushed far enough, but also when it is pushed a great deal too far; just as the aquiline and the snub nose not only turn into normal noses by not being aquiline or snub enough, but also by being too violently aquiline or snub arrive at a condition in which they no longer look like noses at all. It is useful, in framing laws, not only to study the past history of one's own country, in order to understand which constitution is desirable for it now, but also to have a knowledge of the constitutions of other nations, and so to learn for what kinds of nation the various kinds of constitution are suited. From this we can see that books of travel are useful aids to legislation, since from these we may learn the laws and customs of different races. The political speaker will also find the researches of historians useful. But all this is the business of political science and not of rhetoric.

These, then, are the most important kinds of information which the political speaker must possess. Let us now go back and state the premises from which he will have to argue in favour of adopting or rejecting measures regarding these and other matters.

5

It may be said that every individual man and all men in common aim at a certain end which determines what they choose and what they avoid. This end, to sum it up briefly, is happiness and its constituents. Let us, then, by way of illustration only, ascertain what is in general the nature of happiness, and what are the elements of its constituent parts. For all advice to do things or not to do them is concerned with happiness and with the things that make for or against it; whatever creates or increases happiness or some part of happiness, we ought to do; whatever destroys or hampers happiness, or gives rise to its opposite, we ought not to do.

We may define happiness as prosperity combined with virtue; or as independence of life; or as the secure enjoyment of the maximum of pleasure; or as a good condition of property and body, together with the power of guarding one's property and body and making use of them. That happiness is one or more of these things, pretty well everybody agrees.

From this definition of happiness it follows that its constituent parts are:—good birth, plenty of friends, good friends, wealth, good children, plenty of children, a happy old age, also such bodily excellences as health, beauty, strength, large stature, athletic powers, together with fame, honour, good luck, and virtue. A man cannot fail to be completely independent if he possesses these internal and these external goods; for besides these there are no others to have. (Goods of the soul and of the body are internal. Good birth, friends, money, and honour are external.) Further, we think that he should possess resources and luck, in order to make his life really secure. As we have already ascertained what happiness in general is, so now let us try to ascertain what of these parts of it is.

Now good birth in a race or a state means that its members are indigenous or ancient: that its earliest leaders were distinguished men, and that from them have sprung many who were distinguished for qualities that we admire.

The good birth of an individual, which may come either from the male or the female side, implies that both parents are free citizens, and that, as in the case of the state, the founders of the line have been notable for virtue or wealth or something else which is highly prized, and that many distinguished persons belong to the family, men and women, young and old.

The phrases 'possession of good children' and 'of many children' bear a quite clear meaning. Applied to a community, they mean that its young men are numerous and of good a quality: good in regard to bodily excellences, such as stature, beauty, strength, athletic powers; and also in regard to the excellences of the soul, which in a young man are temperance and courage. Applied to an individual, they mean that his own children are numerous and have the good qualities we have described. Both male and female are here included; the excellences of the latter are, in body, beauty and stature; in soul, self-command and an industry that is not sordid. Communities as well as individuals should lack none of these perfections, in their women as well as in their men. Where, as among the Lacedaemonians, the state of women is bad, almost half of human life is spoilt.

The constituents of wealth are: plenty of coined money and territory; the ownership of numerous, large, and beautiful estates; also the ownership of numerous and beautiful implements, live stock, and slaves. All these kinds of property are our own, are secure, gentlemanly, and useful. The useful kinds are those that are productive, the gentlemanly kinds are those that provide enjoyment. By 'productive' I mean those from which we get our income; by 'enjoyable', those from which we get nothing worth mentioning except the use of them. The criterion of 'security' is the ownership of property in such places and under such Conditions that the use of it is in our power; and it is 'our own' if it is in our own power to dispose of it or keep it. By 'disposing of it' I mean giving it away or selling it. Wealth as a whole consists in using things rather than in owning them; it is really the activity—that is, the use-of property that constitutes wealth.

Fame means being respected by everybody, or having some quality that is desired by all men, or by most, or by the good, or by the wise.

Honour is the token of a man's being famous for doing good. it is chiefly and most properly paid to those who have already done good; but also to the man who can do good in future. Doing good refers either to the preservation of life and the means of life, or to wealth, or to some other of the good things which it is hard to get either always or at that particular place or time—for many gain honour for things which seem small, but the place and the occasion account for it. The constituents of honour are: sacrifices; commemoration, in verse or prose; privileges; grants of land; front seats at civic celebrations; state burial; statues; public maintenance; among foreigners, obeisances and giving place; and such presents as are among various bodies of men regarded as marks of honour. For a present is not only the bestowal of a piece of property, but also a token of honour; which explains why honour-loving as well as money-loving persons desire it. The present brings to both what they want; it is a piece of property, which is what the lovers of money desire; and it brings honour, which is what the lovers of honour desire.

The excellence of the body is health; that is, a condition which allows us, while keeping free from disease, to have the use of our bodies; for many people are 'healthy' as we are told Herodicus was; and these no one can congratulate on their 'health', for they have to abstain from everything or nearly everything that men do.—Beauty varies with the time of life. In a young man beauty is the possession of a body fit to endure the exertion of running and of contests of strength; which means that he is pleasant to look at; and therefore all-round athletes are the most beautiful, being naturally adapted both for contests of strength and for speed also. For a man in his prime, beauty is fitness for the exertion of warfare, together with a pleasant but at the same time formidable appearance. For an old man, it is to be strong enough for such exertion as is necessary, and to be free from all those deformities of old age which cause pain to others. Strength is the power of moving some one else at will; to do this, you must either pull, push, lift, pin, or grip him; thus you must be strong in all of those ways or at least in some. Excellence in size is to surpass ordinary people in height, thickness, and breadth by just as much as will not make one's movements slower in consequence. Athletic excellence of the body consists in size, strength, and swiftness; swiftness implying strength. He who can fling forward his legs in a certain way, and move them fast and far, is good at running; he who can grip and hold down is good at wrestling; he who can drive an adversary from his ground with the right blow is a good boxer: he who can do both the last is a good pancratiast, while he who can do all is an 'all-round' athlete.

Happiness in old age is the coming of old age slowly and painlessly; for a man has not this happiness if he grows old either quickly, or tardily but painfully. It arises both from the excellences of the body and from good luck. If a man is not free from disease, or if he is strong, he will not be free from suffering; nor can he continue to live a long and painless life unless he has good luck. There is, indeed, a capacity for long life that is quite independent of health or strength; for many people live long who lack the excellences of the body; but for our present purpose there is no use in going into the details of this.

The terms 'possession of many friends' and 'possession of good friends' need no explanation; for we define a 'friend' as one who will always try, for your sake, to do what he takes to be good for you. The man towards whom many feel thus has many friends; if these are worthy men, he has good friends.

'Good luck' means the acquisition or possession of all or most, or the most important, of those good things which are due to luck. Some of the things that are due to luck may also be due to artificial contrivance; but many are independent of art, as for example those which are due to nature—though, to be sure, things due to luck may actually be contrary to nature. Thus health may be due to artificial contrivance, but beauty and stature are due to nature. All such good things as excite envy are, as a class, the outcome of good luck. Luck is also the cause of good things that happen contrary to reasonable expectation: as when, for instance, all your brothers are ugly, but you are handsome yourself; or when you find a treasure that everybody else has overlooked; or when a missile hits the next man and misses you; or when you are the only man not to go to a place you have gone to regularly, while the others go there for the first time and are killed. All such things are reckoned pieces of good luck.

As to virtue, it is most closely connected with the subject of Eulogy, and therefore we will wait to define it until we come to discuss that subject.

6

It is now plain what our aims, future or actual, should be in urging, and what in depreciating, a proposal; the latter being the opposite of the former. Now the political or deliberative orator's aim is utility: deliberation seeks to determine not ends but the means to ends, i.e. what it is most useful to do. Further, utility is a good thing. We ought therefore to assure ourselves of the main facts about Goodness and Utility in general.

We may define a good thing as that which ought to be chosen for its own sake; or as that for the sake of which we choose something else; or as that which is sought after by all things, or by all things that have sensation or reason, or which will be sought after by any things that acquire reason; or as that which must be prescribed for a given individual by reason generally, or is prescribed for him by his individual reason, this being his individual good; or as that whose presence brings anything into a satisfactory and self-sufficing condition; or as self-sufficiency; or as what produces, maintains, or entails characteristics of this kind, while preventing and destroying their opposites. One thing may entail another in either of two ways—(1) simultaneously, (2) subsequently. Thus learning entails knowledge subsequently, health entails life simultaneously. Things are productive of other things in three senses: first as being healthy produces health; secondly, as food produces health; and thirdly, as exercise does—i.e. it does so usually. All this being settled, we now see that both the acquisition of good things and the removal of bad things must be good; the latter entails freedom from the evil things simultaneously, while the former entails possession of the good things subsequently. The acquisition of a greater in place of a lesser good, or of a lesser in place of a greater evil, is also good, for in proportion as the greater exceeds the lesser there is acquisition of good or removal of evil. The virtues, too, must be something good; for it is by possessing these that we are in a good condition, and they tend to produce good works and good actions. They must be severally named and described elsewhere. Pleasure, again, must be a good thing, since it is the nature of all animals to aim at it. Consequently both pleasant and beautiful things must be good things, since the former are productive of pleasure, while of the beautiful things some are pleasant and some desirable in and for themselves.

The following is a more detailed list of things that must be good. Happiness, as being desirable in itself and sufficient by itself, and as being that for whose sake we choose many other things. Also justice, courage, temperance, magnanimity, magnificence, and all such qualities, as being excellences of the soul. Further, health, beauty, and the like, as being bodily excellences and productive of many other good things: for instance, health is productive both of pleasure and of life, and therefore is thought the greatest of goods, since these two things which it causes, pleasure and life, are two of the things most highly prized by ordinary people. Wealth, again: for it is the excellence of possession, and also productive of many other good things. Friends and friendship: for a friend is desirable in himself and also productive of many other good things. So, too, honour and reputation, as being pleasant, and productive of many other good things, and usually accompanied by the presence of the good things that cause them to be bestowed. The faculty of speech and action; since all such qualities are productive of what is good. Further—good parts, strong memory, receptiveness, quickness of intuition, and the like, for all such faculties are productive of what is good. Similarly, all the sciences and arts. And life: since, even if no other good were the result of life, it is desirable in itself. And justice, as the cause of good to the community.

The above are pretty well all the things admittedly good. In dealing with things whose goodness is disputed, we may argue in the following ways:—That is good of which the contrary is bad. That is good the contrary of which is to the advantage of our enemies; for example, if it is to the particular advantage of our enemies that we should be cowards, clearly courage is of particular value to our countrymen. And generally, the contrary of that which our enemies desire, or of that at which they rejoice, is evidently valuable. Hence the passage beginning:

Surely would Priam exult.

This principle usually holds good, but not always, since it may well be that our interest is sometimes the same as that of our enemies. Hence it is said that 'evils draw men together'; that is, when the same thing is hurtful to them both.

Further: that which is not in excess is good, and that which is greater than it should be is bad. That also is good on which much labour or money has been spent; the mere fact of this makes it seem good, and such a good is assumed to be an end—an end reached through a long chain of means; and any end is a good. Hence the lines beginning:

And for Priam (and Troy-town's folk) should
they leave behind them a boast;

and

Oh, it were shame
To have tarried so long and return empty-handed
as erst we came;

and there is also the proverb about 'breaking the pitcher at the door'.

That which most people seek after, and which is obviously an object of contention, is also a good; for, as has been shown, that is good which is sought after by everybody, and 'most people' is taken to be equivalent to 'everybody'. That which is praised is good, since no one praises what is not good. So, again, that which is praised by our enemies [or by the worthless] for when even those who have a grievance think a thing good, it is at once felt that every one must agree with them; our enemies can admit the fact only because it is evident, just as those must be worthless whom their friends censure and their enemies do not. (For this reason the Corinthians conceived themselves to be insulted by Simonides when he wrote:

Against the Corinthians hath Ilium no complaint.)

Again, that is good which has been distinguished by the favour of a discerning or virtuous man or woman, as Odysseus was distinguished by Athena, Helen by Theseus, Paris by the goddesses, and Achilles by Homer. And, generally speaking, all things are good which men deliberately choose to do; this will include the things already mentioned, and also whatever may be bad for their enemies or good for their friends, and at the same time practicable. Things are 'practicable' in two senses: (1) it is possible to do them, (2) it is easy to do them. Things are done 'easily' when they are done either without pain or quickly: the 'difficulty' of an act lies either in its painfulness or in the long time it takes. Again, a thing is good if it is as men wish; and they wish to have either no evil at an or at least a balance of good over evil. This last will happen where the penalty is either imperceptible or slight. Good, too, are things that are a man's very own, possessed by no one else, exceptional; for this increases the credit of having them. So are things which befit the possessors, such as whatever is appropriate to their birth or capacity, and whatever they feel they ought to have but lack—such things may indeed be trifling, but none the less men deliberately make them the goal of their action. And things easily effected; for these are practicable (in the sense of being easy); such things are those in which every one, or most people, or one's equals, or one's inferiors have succeeded. Good also are the things by which we shall gratify our friends or annoy our enemies; and the things chosen by those whom we admire: and the things for which we are fitted by nature or experience, since we think we shall succeed more easily in these: and those in which no worthless man can succeed, for such things bring greater praise: and those which we do in fact desire, for what we desire is taken to be not only pleasant but also better. Further, a man of a given disposition makes chiefly for the corresponding things: lovers of victory make for victory, lovers of honour for honour, money-loving men for money, and so with the rest. These, then, are the sources from which we must derive our means of persuasion about Good and Utility.

7

Since, however, it often happens that people agree that two things are both useful but do not agree about which is the more so, the next step will be to treat of relative goodness and relative utility.

A thing which surpasses another may be regarded as being that other thing plus something more, and that other thing which is surpassed as being what is contained in the first thing. Now to call a thing 'greater' or 'more' always implies a comparison of it with one that is 'smaller' or 'less', while 'great' and 'small', 'much' and 'little', are terms used in comparison with normal magnitude. The 'great' is that which surpasses the normal, the 'small' is that which is surpassed by the normal; and so with 'many' and 'few'.

Now we are applying the term 'good' to what is desirable for its own sake and not for the sake of something else; to that at which all things aim; to what they would choose if they could acquire understanding and practical wisdom; and to that which tends to produce or preserve such goods, or is always accompanied by them. Moreover, that for the sake of which things are done is the end (an end being that for the sake of which all else is done), and for each individual that thing is a good which fulfils these conditions in regard to himself. It follows, then, that a greater number of goods is a greater good than one or than a smaller number, if that one or that smaller number is included in the count; for then the larger number surpasses the smaller, and the smaller quantity is surpassed as being contained in the larger.

Again, if the largest member of one class surpasses the largest member of another, then the one class surpasses the other; and if one class surpasses another, then the largest member of the one surpasses the largest member of the other. Thus, if the tallest man is taller than the tallest woman, then men in general are taller than women. Conversely, if men in general are taller than women, then the tallest man is taller than the tallest woman. For the superiority of class over class is proportionate to the superiority possessed by their largest specimens. Again, where one good is always accompanied by another, but does not always accompany it, it is greater than the other, for the use of the second thing is implied in the use of the first. A thing may be accompanied by another in three ways, either simultaneously, subsequently, or potentially. Life accompanies health simultaneously (but not health life), knowledge accompanies the act of learning subsequently, cheating accompanies sacrilege potentially, since a man who has committed sacrilege is always capable of cheating. Again, when two things each surpass a third, that which does so by the greater amount is the greater of the two; for it must surpass the greater as well as the less of the other two. A thing productive of a greater good than another is productive of is itself a greater good than that other. For this conception of 'productive of a greater' has been implied in our argument. Likewise, that which is produced by a greater good is itself a greater good; thus, if what is wholesome is more desirable and a greater good than what gives pleasure, health too must be a greater good than pleasure. Again, a thing which is desirable in itself is a greater good than a thing which is not desirable in itself, as for example bodily strength than what is wholesome, since the latter is not pursued for its own sake, whereas the former is; and this was our definition of the good. Again, if one of two things is an end, and the other is not, the former is the greater good, as being chosen for its own sake and not for the sake of something else; as, for example, exercise is chosen for the sake of physical well-being. And of two things that which stands less in need of the other, or of other things, is the greater good, since it is more self-sufficing. (That which stands 'less' in need of others is that which needs either fewer or easier things.) So when one thing does not exist or cannot come into existence without a second, while the second can exist without the first, the second is the better. That which does not need something else is more self-sufficing than that which does, and presents itself as a greater good for that reason. Again, that which is a beginning of other things is a greater good than that which is not, and that which is a cause is a greater good than that which is not; the reason being the same in each case, namely that without a cause and a beginning nothing can exist or come into existence. Again, where there are two sets of consequences arising from two different beginnings or causes, the consequences of the more important beginning or cause are themselves the more important; and conversely, that beginning or cause is itself the more important which has the more important consequences. Now it is plain, from all that has been said, that one thing may be shown to be more important than another from two opposite points of view: it may appear the more important (1) because it is a beginning and the other thing is not, and also (2) because it is not a beginning and the other thing is—on the ground that the end is more important and is not a beginning. So Leodamas, when accusing Callistratus, said that the man who prompted the deed was more guilty than the doer, since it would not have been done if he had not planned it. On the other hand, when accusing Chabrias he said that the doer was worse than the prompter, since there would have been no deed without some one to do it; men, said he, plot a thing only in order to carry it out.

Further, what is rare is a greater good than what is plentiful. Thus, gold is a better thing than iron, though less useful: it is harder to get, and therefore better worth getting. Reversely, it may be argued that the plentiful is a better thing than the rare, because we can make more use of it. For what is often useful surpasses what is seldom useful, whence the saying:

The best of things is water.

More generally: the hard thing is better than the easy, because it is rarer: and reversely, the easy thing is better than the hard, for it is as we wish it to be. That is the greater good whose contrary is the greater evil, and whose loss affects us more. Positive goodness and badness are more important than the mere absence of goodness and badness: for positive goodness and badness are ends, which the mere absence of them cannot be. Further, in proportion as the functions of things are noble or base, the things themselves are good or bad: conversely, in proportion as the things themselves are good or bad, their functions also are good or bad; for the nature of results corresponds with that of their causes and beginnings, and conversely the nature of causes and beginnings corresponds with that of their results. Moreover, those things are greater goods, superiority in which is more desirable or more honourable. Thus, keenness of sight is more desirable than keenness of smell, sight generally being more desirable than smell generally; and similarly, unusually great love of friends being more honourable than unusually great love of money, ordinary love of friends is more honourable than ordinary love of money. Conversely, if one of two normal things is better or nobler than the other, an unusual degree of that thing is better or nobler than an unusual degree of the other. Again, one thing is more honourable or better than another if it is more honourable or better to desire it; the importance of the object of a given instinct corresponds to the importance of the instinct itself; and for the same reason, if one thing is more honourable or better than another, it is more honourable and better to desire it. Again, if one science is more honourable and valuable than another, the activity with which it deals is also more honourable and valuable; as is the science, so is the reality that is its object, each science being authoritative in its own sphere. So, also, the more valuable and honourable the object of a science, the more valuable and honourable the science itself is—in consequence. Again, that which would be judged, or which has been judged, a good thing, or a better thing than something else, by all or most people of understanding, or by the majority of men, or by the ablest, must be so; either without qualification, or in so far as they use their understanding to form their judgement. This is indeed a general principle, applicable to all other judgements also; not only the goodness of things, but their essence, magnitude, and general nature are in fact just what knowledge and understanding will declare them to be. Here the principle is applied to judgements of goodness, since one definition of 'good' was 'what beings that acquire understanding will choose in any given case': from which it clearly follows that that thing is hetter which understanding declares to be so. That, again, is a better thing which attaches to better men, either absolutely, or in virtue of their being better; as courage is better than strength. And that is a greater good which would be chosen by a better man, either absolutely, or in virtue of his being better: for instance, to suffer wrong rather than to do wrong, for that would be the choice of the juster man. Again, the pleasanter of two things is the better, since all things pursue pleasure, and things instinctively desire pleasurable sensation for its own sake; and these are two of the characteristics by which the 'good' and the 'end' have been defined. One pleasure is greater than another if it is more unmixed with pain, or more lasting. Again, the nobler thing is better than the less noble, since the noble is either what is pleasant or what is desirable in itself. And those things also are greater goods which men desire more earnestly to bring about for themselves or for their friends, whereas those things which they least desire to bring about are greater evils. And those things which are more lasting are better than those which are more fleeting, and the more secure than the less; the enjoyment of the lasting has the advantage of being longer, and that of the secure has the advantage of suiting our wishes, being there for us whenever we like. Further, in accordance with the rule of co-ordinate terms and inflexions of the same stem, what is true of one such related word is true of all. Thus if the action qualified by the term 'brave' is more noble and desirable than the action qualified by the term 'temperate', then 'bravery' is more desirable than 'temperance' and 'being brave' than 'being temperate'. That, again, which is chosen by all is a greater good than that which is not, and that chosen by the majority than that chosen by the minority. For that which all desire is good, as we have said;' and so, the more a thing is desired, the better it is. Further, that is the better thing which is considered so by competitors or enemies, or, again, by authorized judges or those whom they select to represent them. In the first two cases the decision is virtually that of every one, in the last two that of authorities and experts. And sometimes it may be argued that what all share is the better thing, since it is a dishonour not to share in it; at other times, that what none or few share is better, since it is rarer. The more praiseworthy things are, the nobler and therefore the better they are. So with the things that earn greater honours than others—honour is, as it were, a measure of value; and the things whose absence involves comparatively heavy penalties; and the things that are better than others admitted or believed to be good. Moreover, things look better merely by being divided into their parts, since they then seem to surpass a greater number of things than before. Hence Homer says that Meleager was roused to battle by the thought of

> All horrors that light on a folk whose city
> is ta'en of their foes,
> When they slaughter the men, when the burg is
> wasted with ravening flame,
> When strangers are haling young children to thraldom,
> (fair women to shame.)

The same effect is produced by piling up facts in a climax after the manner of Epicharmus. The reason is partly the same as in the case of division (for combination too makes the impression of great superiority), and partly that the original thing appears to be the cause and origin of important results. And since a thing is better when it is harder or rarer than other things, its superiority may be due to seasons, ages, places, times, or one's natural powers. When a man accomplishes something beyond his natural power, or beyond his years, or beyond the measure of people like him, or in a special way, or at a special place or time, his deed will have a high degree of nobleness, goodness, and justice, or of their opposites. Hence the epigram on the victor at the Olympic games:

In time past, heaving a Yoke on my shoulders,
of wood unshaven,
I carried my loads of fish from, Argos to Tegea town.

So Iphicrates used to extol himself by describing the low estate from which he had risen. Again, what is natural is better than what is acquired, since it is harder to come by. Hence the words of Homer:

I have learnt from none but mysell.

And the best part of a good thing is particularly good; as when Pericles in his funeral oration said that the country's loss of its young men in battle was 'as if the spring were taken out of the year'. So with those things which are of service when the need is pressing; for example, in old age and times of sickness. And of two things that which leads more directly to the end in view is the better. So too is that which is better for people generally as well as for a particular individual. Again, what can be got is better than what cannot, for it is good in a given case and the other thing is not. And what is at the end of life is better than what is not, since those things are ends in a greater degree which are nearer the end. What aims at reality is better than what aims at appearance. We may define what aims at appearance as what a man will not choose if nobody is to know of his having it. This would seem to show that to receive benefits is more desirable than to confer them, since a man will choose the former even if nobody is to know of it, but it is not the general view that he will choose the latter if nobody knows of it. What a man wants to be is better than what a man wants to seem, for in aiming at that he is aiming more at reality. Hence men say that justice is of small value, since it is more desirable to seem just than to be just, whereas with health it is not so. That is better than other things which is more useful than they are for a number of different purposes; for example, that which promotes life, good life, pleasure, and noble conduct. For this reason wealth and health are commonly thought to be of the highest value, as possessing all these advantages. Again, that is better than other things which is accompanied both with less pain and with actual pleasure; for here there is more than one advantage; and so here we have the good of feeling pleasure and also the good of not feeling pain. And of two good things that is the better whose addition to a third thing makes a better whole than the addition of the other to the same thing will make. Again, those things which we are seen to possess are better than those which we are not seen to possess, since the former have the air of reality. Hence wealth may be regarded as a greater good if its existence is known to others. That which is dearly prized is better than what is not—the sort of thing that some people have only one of, though others have more like it. Accordingly, blinding a one-eyed man inflicts worse injury than half-blinding a man with two eyes; for the one-eyed man has been robbed of what he dearly prized.

The grounds on which we must base our arguments, when we are speaking for or against a proposal, have now been set forth more or less completely.

8

The most important and effective qualification for success in persuading audiences and speaking well on public affairs is to understand all the forms of government and to discriminate their respective customs, institutions, and interests. For all men are persuaded by considerations of their interest, and their interest lies in the maintenance of the established order. Further, it rests with the supreme authority to give authoritative decisions, and this varies with each form of government; there are as many different supreme authorities as there are different forms of government. The forms of government are four—democracy, oligarchy, aristocracy, monarchy. The supreme right to judge and decide always rests, therefore, with either a part or the whole of one or other of these governing powers.

A Democracy is a form of government under which the citizens distribute the offices of state among themselves by lot, whereas under oligarchy there is a property qualification, under aristocracy one of education. By education I mean that education which is laid down by the law; for it is those who have been loyal to the national institutions that hold office under an aristocracy. These are bound to be looked upon as 'the best men', and it is from this fact that this form of government has derived its name ('the rule of the best'). Monarchy, as the word implies, is the constitution a in which one man has authority over all. There are two forms of monarchy: kingship, which is limited by prescribed conditions, and 'tyranny', which is not limited by anything.

We must also notice the ends which the various forms of government pursue, since people choose in practice such actions as will lead to the realization of their ends. The end of democracy is freedom; of oligarchy, wealth; of aristocracy, the maintenance of education and national institutions; of tyranny, the protection of the tyrant. It is clear, then, that we must distinguish those particular customs, institutions, and interests which tend to realize the ideal of each constitution, since men choose their means with reference to their ends. But rhetorical persuasion is effected not only by demonstrative but by ethical argument; it helps a speaker to convince us, if we believe that he has certain qualities himself, namely, goodness, or goodwill towards us, or both together. Similarly, we should know the moral qualities characteristic of each form of government, for the special moral character of each is bound to provide us with our most effective means of persuasion in dealing with it. We shall learn the qualities of governments in the same way as we learn the qualities of individuals, since they are revealed in their deliberate acts of choice; and these are determined by the end that inspires them.

We have now considered the objects, immediate or distant, at which we are to aim when urging any proposal, and the grounds on which we are to base our arguments in favour of its utility. We have also briefly considered the means and methods by which we shall gain a good knowledge of the moral qualities and institutions peculiar to the various forms of government—only, however, to the extent demanded by the present occasion; a detailed account of the subject has been given in the Politics.

9

We have now to consider Virtue and Vice, the Noble and the Base, since these are the objects of praise and blame. In doing so, we shall at the same time be finding out how to make our hearers take the required view of our own characters—our second method of persuasion. The ways in which to make them trust the goodness of other people are also the ways in which to make them trust our own. Praise, again, may be serious or frivolous; nor is it always of a human or divine being but often of inanimate things, or of the humblest of the lower animals. Here too we must know on what grounds to argue, and must, therefore, now discuss the subject, though by way of illustration only.

The Noble is that which is both desirable for its own sake and also worthy of praise; or that which is both good and also pleasant because good. If this is a true definition of the Noble, it follows that virtue must be noble, since it is both a good thing and also praiseworthy. Virtue is, according to the usual view, a faculty of providing and preserving good things; or a faculty of conferring many great benefits, and benefits of all kinds on all occasions. The forms of Virtue are justice, courage, temperance, magnificence, magnanimity, liberality, gentleness, prudence, wisdom. If virtue is a faculty of beneficence, the highest kinds of it must be those which are most useful to others, and for this reason men honour most the just and the courageous, since courage is useful to others in war, justice both in war and in peace. Next comes liberality; liberal people let their money go instead of fighting for it, whereas other people care more for money than for anything else. Justice is the virtue through which everybody enjoys his own possessions in accordance with the law; its opposite is injustice, through which men enjoy the possessions of others in defiance of the law. Courage is the virtue that disposes men to do noble deeds in situations of danger, in accordance with the law and in obedience to its commands; cowardice is the opposite. Temperance is the virtue that disposes us to obey the law where physical pleasures are concerned; incontinence is the opposite. Liberality disposes us to spend money for others' good; illiberality is the opposite. Magnanimity is the virtue that disposes us to do good to others on a large scale; [its opposite is meanness of spirit]. Magnificence is a virtue productive of greatness in matters involving the spending of money. The opposites of these two are smallness of spirit and meanness respectively. Prudence is that virtue of the understanding which enables men to come to wise decisions about the relation to happiness of the goods and evils that have been previously mentioned.

The above is a sufficient account, for our present purpose, of virtue and vice in general, and of their various forms. As to further aspects of the subject, it is not difficult to discern the facts; it is evident that things productive of virtue are noble, as tending towards virtue; and also the effects of virtue, that is, the signs of its presence and the acts to which it leads. And since the signs of virtue, and such acts as it is the mark of a virtuous man to do or have done to him, are noble, it follows that all deeds or signs of courage, and everything done courageously, must be noble things; and so with what is just and actions done justly. (Not, however, actions justly done to us; here justice is unlike the other virtues; 'justly' does not always mean 'nobly'; when a man is punished, it is more shameful that this should be justly than unjustly done to him). The same is true of the other virtues. Again, those actions are noble for which the reward is simply honour, or honour more than money. So are those in which a man aims at something desirable for some one else's sake; actions good absolutely, such as those a man does for his country without thinking of himself; actions good in their own nature; actions that are not good simply for the individual, since individual interests are selfish. Noble also are those actions whose advantage may be enjoyed after death, as opposed to those whose advantage is enjoyed during one's lifetime: for the latter are more likely to be for one's own sake only. Also, all actions done for the sake of others, since less than other actions are done for one's own sake; and all successes which benefit others and not oneself; and services done to one's benefactors, for this is just; and good deeds generally, since they are not directed to one's own profit. And the opposites of those things of which men feel ashamed, for men are ashamed of saying, doing, or intending to do shameful things. So when Alcacus said

Something I fain would say to thee,
Only shame restraineth me,
Sappho wrote
If for things good and noble thou wert yearning,
If to speak baseness were thy tongue not burning,
No load of shame would on thine eyelids weigh;
What thou with honour wishest thou wouldst say.

Those things, also, are noble for which men strive anxiously, without feeling fear; for they feel thus about the good things which lead to fair fame. Again, one quality or action is nobler than another if it is that of a naturally finer being: thus a man's will be nobler than a woman's. And those qualities are noble which give more pleasure to other people than to their possessors; hence the nobleness of justice and just actions. It is noble to avenge oneself on one's enemies and not to come to terms with them; for requital is just, and the just is noble; and not to surrender is a sign of courage. Victory, too, and honour belong to the class of noble things, since they are desirable even when they yield no fruits, and they prove our superiority in good qualities. Things that deserve to be remembered are noble, and the more they deserve this, the nobler they are. So are the things that continue even after death; those which are always attended by honour; those which are exceptional; and those which are possessed by one person alone—these last are more readily remembered than others. So again are possessions that bring no profit, since they are more fitting than others for a gentleman. So are the distinctive qualities of a particular people, and the symbols of what it specially admires, like long hair in Sparta, where this is a mark of a free man, as it is not easy to perform any menial task when one's hair is long. Again, it is noble not to practise any sordid craft, since it is the mark of a free man not to live at another's beck and call. We are also to assume when we wish either to praise a man or blame him that qualities closely allied to those which he actually has are identical with them; for instance, that the cautious man is cold-blooded and treacherous, and that the stupid man is an honest fellow or the thick-skinned man a good-tempered one. We can always idealize any given man by drawing on the virtues akin to his actual qualities; thus we may say that the passionate and excitable man is 'outspoken'; or that the arrogant man is 'superb' or 'impressive'. Those who run to extremes will be said to possess the corresponding good qualities; rashness will be called courage, and extravagance generosity. That will be what most people think; and at the same time this method enables an advocate to draw a misleading inference from the motive, arguing that if a man runs into danger needlessly, much more will he do so in a noble cause; and if a man is open-handed to any one and every one, he will be so to his friends also, since it is the extreme form of goodness to be good to everybody.

We must also take into account the nature of our particular audience when making a speech of praise; for, as Socrates used to say, 'it is not difficult to praise the Athenians to an Athenian audience.' If the audience esteems a given quality, we must say that our hero has that quality, no matter whether we are addressing Scythians or Spartans or philosophers. Everything, in fact, that is esteemed we are to represent as noble. After all, people regard the two things as much the same.

All actions are noble that are appropriate to the man who does them: if, for instance, they are worthy of his ancestors or of his own past career. For it makes for happiness, and is a noble thing, that he should add to the honour he already has. Even inappropriate actions are noble if they are better and nobler than the appropriate ones would be; for instance, if one who was just an average person when all went well becomes a hero in adversity, or if he becomes better and easier to get on with the higher he rises. Compare the saying of lphicrates, 'Think what I was and what I am'; and the epigram on the victor at the Olympic games,

In time past, bearing a yoke on my shoulders,
of wood unshaven,
and the encomium of Simonides,

A woman whose father, whose husband, whose
brethren were princes all.

Since we praise a man for what he has actually done, and fine actions are distinguished from others by being intentionally good, we must try to prove that our hero's noble acts are intentional. This is all the easier if we can make out that he has often acted so before, and therefore we must assert coincidences and accidents to have been intended. Produce a number of good actions, all of the same kind, and people will think that they must have been intended, and that they prove the good qualities of the man who did them.

Praise is the expression in words of the eminence of a man's good qualities, and therefore we must display his actions as the product of such qualities. Encomium refers to what he has actually done; the mention of accessories, such as good birth and education, merely helps to make our story credible—good fathers are likely to have good sons, and good training is likely to produce good character. Hence it is only when a man has already done something that we bestow encomiums upon him. Yet the actual deeds are evidence of the doer's character: even if a man has not actually done a given good thing, we shall bestow praise on him, if we are sure that he is the sort of man who would do it. To call any one blest is, it may be added, the same thing as to call him happy; but these are not the same thing as to bestow praise and encomium upon him; the two latter are a part of 'calling happy', just as goodness is a part of happiness.

To praise a man is in one respect akin to urging a course of action. The suggestions which would be made in the latter case become encomiums when differently expressed. When we know what action or character is required, then, in order to express these facts as suggestions for action, we have to change and reverse our form of words. Thus the statement 'A man should be proud not of what he owes to fortune but of what he owes to himself', if put like this, amounts to a suggestion; to make it into praise we must put it thus, 'Since he is proud not of what he owes to fortune but of what he owes to himself.' Consequently, whenever you want to praise any one, think what you would urge people to do; and when you want to urge the doing of anything, think what you would praise a man for having done. Since suggestion may or may not forbid an action, the praise into which we convert it must have one or other of two opposite forms of expression accordingly.

There are, also, many useful ways of heightening the effect of praise. We must, for instance, point out that a man is the only one, or the first, or almost the only one who has done something, or that he has done it better than any one else; all these distinctions are honourable. And we must, further, make much of the particular season and occasion of an action, arguing that we could hardly have looked for it just then. If a man has often achieved the same success, we must mention this; that is a strong point; he himself, and not luck, will then be given the credit. So, too, if it is on his account that observances have been devised and instituted to encourage or honour such achievements as his own: thus we may praise Hippolochus because the first encomium ever made was for him, or Harmodius and Aristogeiton because their statues were the first to be put up in the market-place. And we may censure bad men for the opposite reason.

Again, if you cannot find enough to say of a man himself, you may pit him against others, which is what Isocrates used to do owing to his want of familiarity with forensic pleading. The comparison should be with famous men; that will strengthen your case; it is a noble thing to surpass men who are themselves great. It is only natural that methods of 'heightening the effect' should be attached particularly to speeches of praise; they aim at proving superiority over others, and any such superiority is a form of nobleness. Hence if you cannot compare your hero with famous men, you should at least compare him with other people generally, since any superiority is held to reveal excellence. And, in general, of the lines of argument which are common to all speeches, this 'heightening of effect' is most suitable for declamations, where we take our hero's actions as admitted facts, and our business is simply to invest these with dignity and nobility. 'Examples' are most suitable to deliberative speeches; for we judge of future events by divination from past events. Enthymemes are most suitable to forensic speeches; it is our doubts about past events that most admit of arguments showing why a thing must have happened or proving that it did happen.

The above are the general lines on which all, or nearly all, speeches of praise or blame are constructed. We have seen the sort of thing we must bear in mind in making such speeches, and the materials out of which encomiums and censures are made. No special treatment of censure and vituperation is needed. Knowing the above facts, we know their contraries; and it is out of these that speeches of censure are made.

10

We have next to treat of Accusation and Defence, and to enumerate and describe the ingredients of the syllogisms used therein. There are three things we must ascertain first, the nature and number of the incentives to wrong-doing; second, the state of mind of wrongdoers; third, the kind of persons who are wronged, and their condition. We will deal with these questions in order. But before that let us define the act of 'wrong-doing'.

We may describe 'wrong-doing' as injury voluntarily inflicted contrary to law. 'Law' is either special or general. By special law I mean that written law which regulates the life of a particular community; by general law, all those unwritten principles which are supposed to be acknowledged everywhere. We do things 'voluntarily' when we do them consciously and without constraint. (Not all voluntary acts are deliberate, but all deliberate acts are conscious—no one is ignorant of what he deliberately intends.) The causes of our deliberately intending harmful and wicked acts contrary to law are (1) vice, (2) lack of self-control. For the wrongs a man does to others will correspond to the bad quality or qualities that he himself possesses. Thus it is the mean man who will wrong others about money, the profligate in matters of physical pleasure, the effeminate in matters of comfort, and the coward where danger is concerned—his terror makes him abandon those who are involved in the same danger. The ambitious man does wrong for sake of honour, the quick-tempered from anger, the lover of victory for the sake of victory, the embittered man for the sake of revenge, the stupid man because he has misguided notions of right and wrong, the shameless man because he does not mind what people think of him; and so with the rest—any wrong that any one does to others corresponds to his particular faults of character.

However, this subject has already been cleared up in part in our discussion of the virtues and will be further explained later when we treat of the emotions. We have now to consider the motives and states of mind of wrongdoers, and to whom they do wrong.

Let us first decide what sort of things people are trying to get or avoid when they set about doing wrong to others. For it is plain that the prosecutor must consider, out of all the aims that can ever induce us to do wrong to our neighbours, how many, and which, affect his adversary; while the defendant must consider how many, and which, do not affect him. Now every action of every person either is or is not due to that person himself. Of those not due to himself some are due to chance, the others to necessity; of these latter, again, some are due to compulsion, the others to nature. Consequently all actions that are not due to a man himself are due either to chance or to nature or to compulsion. All actions that are due to a man himself and caused by himself are due either to habit or to rational or irrational craving. Rational craving is a craving for good, i.e. a wish—nobody wishes for anything unless he thinks it good. Irrational craving is twofold, viz. anger and appetite.

Thus every action must be due to one or other of seven causes: chance, nature, compulsion, habit, reasoning, anger, or appetite. It is superfluous further to distinguish actions according to the doers' ages, moral states, or the like; it is of course true that, for instance, young men do have hot tempers and strong appetites; still, it is not through youth that they act accordingly, but through anger or appetite. Nor, again, is action due to wealth or poverty; it is of course true that poor men, being short of money, do have an appetite for it, and that rich men, being able to command needless pleasures, do have an appetite for such pleasures: but here, again, their actions will be due not to wealth or poverty but to appetite. Similarly, with just men, and unjust men, and all others who are said to act in accordance with their moral qualities, their actions will really be due to one of the causes mentioned—either reasoning or emotion: due, indeed, sometimes to good dispositions and good emotions, and sometimes to bad; but that good qualities should be followed by good emotions, and bad by bad, is merely an accessory fact—it is no doubt true that the temperate man, for instance, because he is temperate, is always and at once attended by healthy opinions and appetites in regard to pleasant things, and the intemperate man by unhealthy ones. So we must ignore such distinctions. Still we must consider what kinds of actions and of people usually go together; for while there are no definite kinds of action associated with the fact that a man is fair or dark, tall or short, it does make a difference if he is young or old, just or unjust. And, generally speaking, all those accessory qualities that cause distinctions of human character are important: e.g. the sense of wealth or poverty, of being lucky or unlucky. This shall be dealt with later—let us now deal first with the rest of the subject before us.

The things that happen by chance are all those whose cause cannot be determined, that have no purpose, and that happen neither always nor usually nor in any fixed way. The definition of chance shows just what they are. Those things happen by nature which have a fixed and internal cause; they take place uniformly, either always or usually. There is no need to discuss in exact detail the things that happen contrary to nature, nor to ask whether they happen in some sense naturally or from some other cause; it would seem that chance is at least partly the cause of such events. Those things happen through compulsion which take place contrary to the desire or reason of the doer, yet through his own agency. Acts are done from habit which men do because they have often done them before. Actions are due to reasoning when, in view of any of the goods already mentioned, they appear useful either as ends or as means to an end, and are performed for that reason: 'for that reason,' since even licentious persons perform a certain number of useful actions, but because they are pleasant and not because they are useful. To passion and anger are due all acts of revenge. Revenge and punishment are different things. Punishment is inflicted for the sake of the person punished; revenge for that of the punisher, to satisfy his feelings. (What anger is will be made clear when we come to discuss the emotions.) Appetite is the cause of all actions that appear pleasant. Habit, whether acquired by mere familiarity or by effort, belongs to the class of pleasant things, for there are many actions not naturally pleasant which men perform with pleasure, once they have become used to them. To sum up then, all actions due to ourselves either are or seem to be either good or pleasant. Moreover, as all actions due to ourselves are done voluntarily and actions not due to ourselves are done involuntarily, it follows that all voluntary actions must either be or seem to be either good or pleasant; for I reckon among goods escape from evils or apparent evils and the exchange of a greater evil for a less (since these things are in a sense positively desirable), and likewise I count among pleasures escape from painful or apparently painful things and the exchange of a greater pain for a less. We must ascertain, then, the number and nature of the things that are useful and pleasant. The useful has been previously examined in connexion with political oratory; let us now proceed to examine the pleasant. Our various definitions must be regarded as adequate, even if they are not exact, provided they are clear.

11

We may lay it down that Pleasure is a movement, a movement by which the soul as a whole is consciously brought into its normal state of being; and that Pain is the opposite. If this is what pleasure is, it is clear that the pleasant is what tends to produce this condition, while that which tends to destroy it, or to cause the soul to be brought into the opposite state, is painful. It must therefore be pleasant as a rule to move towards a natural state of being, particularly when a natural process has achieved the complete recovery of that natural state. Habits also are pleasant; for as soon as a thing has become habitual, it is virtually natural; habit is a thing not unlike nature; what happens often is akin to what happens always, natural events happening always, habitual events often. Again, that is pleasant which is not forced on us; for force is unnatural, and that is why what is compulsory, painful, and it has been rightly said

All that is done on compulsion is bitterness unto the soul.

So all acts of concentration, strong effort, and strain are necessarily painful; they all involve compulsion and force, unless we are accustomed to them, in which case it is custom that makes them pleasant. The opposites to these are pleasant; and hence ease, freedom from toil, relaxation, amusement, rest, and sleep belong to the class of pleasant things; for these are all free from any element of compulsion. Everything, too, is pleasant for which we have the desire within us, since desire is the craving for pleasure. Of the desires some are irrational, some associated with reason. By irrational I mean those which do not arise from any opinion held by the mind. Of this kind are those known as 'natural'; for instance, those originating in the body, such as the desire for nourishment, namely hunger and thirst, and a separate kind of desire answering to each kind of nourishment; and the desires connected with taste and sex and sensations of touch in general; and those of smell, hearing, and vision. Rational desires are those which we are induced to have; there are many things we desire to see or get because we have been told of them and induced to believe them good. Further, pleasure is the consciousness through the senses of a certain kind of emotion; but imagination is a feeble sort of sensation, and there will always be in the mind of a man who remembers or expects something an image or picture of what he remembers or expects. If this is so, it is clear that memory and expectation also, being accompanied by sensation, may be accompanied by pleasure. It follows that anything pleasant is either present and perceived, past and remembered, or future and expected, since we perceive present pleasures, remember past ones, and expect future ones. Now the things that are pleasant to remember are not only those that, when actually perceived as present, were pleasant, but also some things that were not, provided that their results have subsequently proved noble and good. Hence the words

Sweet 'tis when rescued to remember pain,

and

Even his griefs are a joy long after to one that remembers
All that he wrought and endured.

The reason of this is that it is pleasant even to be merely free from evil. The things it is pleasant to expect are those that when present are felt to afford us either great delight or great but not painful benefit. And in general, all the things that delight us when they are present also do so, as a rule, when we merely remember or expect them. Hence even being angry is pleasant—Homer said of wrath that

Sweeter it is by far than the honeycomb dripping with sweetness—

for no one grows angry with a person on whom there is no prospect of taking vengeance, and we feel comparatively little anger, or none at all, with those who are much our superiors in power. Some pleasant feeling is associated with most of our appetites we are enjoying either the memory of a past pleasure or the expectation of a future one, just as persons down with fever, during their attacks of thirst, enjoy remembering the drinks they have had and looking forward to having more. So also a lover enjoys talking or writing about his loved one, or doing any little thing connected with him; all these things recall him to memory and make him actually present to the eye of imagination. Indeed, it is always the first sign of love, that besides enjoying some one's presence, we remember him when he is gone, and feel pain as well as pleasure, because he is there no longer. Similarly there is an element of pleasure even in mourning and lamentation for the departed. There is grief, indeed, at his loss, but pleasure in remembering him and as it were seeing him before us in his deeds and in his life. We can well believe the poet when he says

He spake, and in each man's heart he awakened
the love of lament.

Revenge, too, is pleasant; it is pleasant to get anything that it is painful to fail to get, and angry people suffer extreme pain when they fail to get their revenge; but they enjoy the prospect of getting it. Victory also is pleasant, and not merely to 'bad losers', but to every one; the winner sees himself in the light of a champion, and everybody has a more or less keen appetite for being that. The pleasantness of victory implies of course that combative sports and intellectual contests are pleasant (since in these it often happens that some one wins) and also games like knuckle-bones, ball, dice, and draughts. And similarly with the serious sports; some of these become pleasant when one is accustomed to them; while others are pleasant from the first, like hunting with hounds, or indeed any kind of hunting. For where there is competition, there is victory. That is why forensic pleading and debating contests are pleasant to those who are accustomed to them and have the capacity for them. Honour and good repute are among the most pleasant things of all; they make a man see himself in the character of a fine fellow, especially when he is credited with it by people whom he thinks good judges. His neighbours are better judges than people at a distance; his associates and fellow-countrymen better than strangers; his contemporaries better than posterity; sensible persons better than foolish ones; a large number of people better than a small number: those of the former class, in each case, are the more likely to be good judges of him. Honour and credit bestowed by those whom you think much inferior to yourself—e.g. children or animals—you do not value: not for its own sake, anyhow: if you do value it, it is for some other reason. Friends belong to the class of pleasant things;

it is pleasant to love—if you love wine, you certainly find it delightful: and it is pleasant to be loved, for this too makes a man see himself as the possessor of goodness, a thing that every being that has a feeling for it desires to possess: to be loved means to be valued for one's own personal qualities. To be admired is also pleasant, simply because of the honour implied. Flattery and flatterers are pleasant: the flatterer is a man who, you believe, admires and likes To do the same thing often is pleasant, since, as we saw, anything habitual is pleasant. And to change is also pleasant: change means an approach to nature, whereas invariable repetition of anything causes the excessive prolongation of a settled condition: therefore, says the poet,

Change is in all things sweet.

That is why what comes to us only at long intervals is pleasant, whether it be a person or a thing; for it is a change from what we had before, and, besides, what comes only at long intervals has the value of rarity. Learning things and wondering at things are also pleasant as a rule; wondering implies the desire of learning, so that the object of wonder is an object of desire; while in learning one is brought into one's natural condition. Conferring and receiving benefits belong to the class of pleasant things; to receive a benefit is to get what one desires; to confer a benefit implies both posses sion and superiority, both of which are things we try to attain. It is because beneficent acts are pleasant that people find it pleasant to put their neighbours straight again and to supply what they lack. Again, since learning and wondering are pleasant, it follows that such things as acts of imitation must be pleasant—for instance, painting, sculpture, poetry and every product of skilful imitation; this latter, even if the object imitated is not itself pleasant; for it is not the object itself which here gives delight; the spectator draws inferences ('That is a so-and-so') and thus learns something fresh. Dramatic turns of fortune and hairbreadth escapes from perils are pleasant, because we feel all such things are wonderful.

And since what is natural is pleasant, and things akin to each other seem natural to each other, therefore all kindred and similar things are usually pleasant to each other; for instance, one man, horse, or young person is pleasant to another man, horse, or young person. Hence the proverbs 'mate delights mate', 'like to like', 'beast knows beast', 'jackdaw to jackdaw', and the rest of them. But since everything like and akin to oneself is pleasant, and since every man is himself more like and akin to himself than any one else is, it follows that all of us must be more or less fond of ourselves. For all this resemblance and kinship is present particularly in the relation of an individual to himself. And because we are all fond of ourselves, it follows that what is our own is pleasant to all of us, as for instance our own deeds and words. That is why we are usually fond of our flatterers, [our lovers,] and honour; also of our children, for our children are our own work. It is also pleasant to complete what is defective, for the whole thing thereupon becomes our own work. And since power over others is very pleasant, it is pleasant to be thought wise, for practical wisdom secures us power over others. (Scientific wisdom is also pleasant, because it is the knowledge of many wonderful things.) Again, since most of us are ambitious, it must be pleasant to disparage our neighbours as well as to have power over them. It is pleasant for a man to spend his time over what he feels he can do best; just as the poet says,

To that he bends himself,
To that each day allots most time, wherein
He is indeed the best part of himself.

Similarly, since amusement and every kind of relaxation and laughter too belong to the class of pleasant things, it follows that ludicrous things are pleasant, whether men, words, or deeds. We have discussed the ludicrous separately in the treatise on the Art of Poetry.

So much for the subject of pleasant things: by considering their opposites we can easily see what things are unpleasant.

12

The above are the motives that make men do wrong to others; we are next to consider the states of mind in which they do it, and the persons to whom they do it.

They must themselves suppose that the thing can be done, and done by them: either that they can do it without being found out, or that if they are found out they can escape being punished, or that if they are punished the disadvantage will be less than the gain for themselves or those they care for. The general subject of apparent possibility and impossibility will be handled later on, since it is relevant not only to forensic but to all kinds of speaking. But it may here be said that people think that they can themselves most easily do wrong to others without being punished for it if they possess eloquence, or practical ability, or much legal experience, or a large body of friends, or a great deal of money. Their confidence is greatest if they personally possess the advantages mentioned: but even without them they are satisfied if they have friends or supporters or partners who do possess them: they can thus both commit their crimes and escape being found out and punished for committing them. They are also safe, they think, if they are on good terms with their victims or with the judges who try them. Their victims will in that case not be on their guard against being wronged, and will make some arrangement with them instead of prosecuting; while their judges will favour them because they like them, either letting them off altogether or imposing light sentences. They are not likely to be found out if their appearance contradicts the charges that might be brought against them: for instance, a weakling is unlikely to be charged with violent assault, or a poor and ugly man with adultery. Public and open injuries are the easiest to do, because nobody could at all suppose them possible, and therefore no precautions are taken. The same is true of crimes so great and terrible that no man living could be suspected of them: here too no precautions are taken. For all men guard against ordinary offences, just as they guard against ordinary diseases; but no one takes precautions against a disease that nobody has ever had. You feel safe, too, if you have either no enemies or a great many; if you have none, you expect not to be watched and therefore not to be detected; if you have a great many, you will be watched, and therefore people will think you can never risk an attempt on them, and you can defend your innocence by pointing out that you could never have taken such a risk. You may also trust to hide your crime by the way you do it or the place you do it in, or by some convenient means of disposal.

You may feel that even if you are found out you can stave off a trial, or have it postponed, or corrupt your judges: or that even if you are sentenced you can avoid paying damages, or can at least postpone doing so for a long time: or that you are so badly off that you will have nothing to lose. You may feel that the gain to be got by wrong-doing is great or certain or immediate, and that the penalty is small or uncertain or distant. It may be that the advantage to be gained is greater than any possible retribution: as in the case of despotic power, according to the popular view. You may consider your crimes as bringing you solid profit, while their punishment is nothing more than being called bad names. Or the opposite argument may appeal to you: your crimes may bring you some credit (thus you may, incidentally, be avenging your father or mother, like Zeno), whereas the punishment may amount to a fine, or banishment, or something of that sort. People may be led on to wrong others by either of these motives or feelings; but no man by both—they will affect people of quite opposite characters. You may be encouraged by having often escaped detection or punishment already; or by having often tried and failed; for in crime, as in war, there are men who will always refuse to give up the struggle. You may get your pleasure on the spot and the pain later, or the gain on the spot and the loss later. That is what appeals to weak-willed persons—and weakness of will may be shown with regard to all the objects of desire. It may on the contrary appeal to you as it does appeal to self-controlled and sensible people—that the pain and loss are immediate, while the pleasure and profit come later and last longer. You may feel able to make it appear that your crime was due to chance, or to necessity, or to natural causes, or to habit: in fact, to put it generally, as if you had failed to do right rather than actually done wrong. You may be able to trust other people to judge you equitably. You may be stimulated by being in want: which may mean that you want necessaries, as poor people do, or that you want luxuries, as rich people do. You may be encouraged by having a particularly good reputation, because that will save you from being suspected: or by having a particularly bad one, because nothing you are likely to do will make it worse.

The above, then, are the various states of mind in which a man sets about doing wrong to others. The kind of people to whom he does wrong, and the ways in which he does it, must be considered next. The people to whom he does it are those who have what he wants himself, whether this means necessities or luxuries and materials for enjoyment. His victims may be far off or near at hand. If they are near, he gets his profit quickly; if they are far off, vengeance is slow, as those think who plunder the Carthaginians. They may be those who are trustful instead of being cautious and watchful, since all such people are easy to elude. Or those who are too easy-going to have enough energy to prosecute an offender. Or sensitive people, who are not apt to show fight over questions of money. Or those who have been wronged already by many people, and yet have not prosecuted; such men must surely be the proverbial 'Mysian prey'. Or those who have either never or often been wronged before; in neither case will they take precautions; if they have never been wronged they think they never will, and if they have often been wronged they feel that surely it cannot happen again. Or those whose character has been attacked in the past, or is exposed to attack in the future: they will be too much frightened of the judges to make up their minds to prosecute, nor can they win their case if they do: this is true of those who are hated or unpopular. Another likely class of victim is those who their injurer can pretend have, themselves or through their ancestors or friends, treated badly, or intended to treat badly, the man himself, or his ancestors, or those he cares for; as the proverb says, 'wickedness needs but a pretext'. A man may wrong his enemies, because that is pleasant: he may equally wrong his friends, because that is easy. Then there are those who have no friends, and those who lack eloquence and practical capacity; these will either not attempt to prosecute, or they will come to terms, or failing that they will lose their case. There are those whom it does not pay to waste time in waiting for trial or damages, such as foreigners and small farmers; they will settle for a trifle, and always be ready to leave off. Also those who have themselves wronged others, either often, or in the same way as they are now being wronged themselves—for it is felt that next to no wrong is done to people when it is the same wrong as they have often themselves done to others: if, for instance, you assault a man who has been accustomed to behave with violence to others. So too with those who have done wrong to others, or have meant to, or mean to, or are likely to do so; there is something fine and pleasant in wronging such persons, it seems as though almost no wrong were done. Also those by doing wrong to whom we shall be gratifying our friends, or those we admire or love, or our masters, or in general the people by reference to whom we mould our lives. Also those whom we may wrong and yet be sure of equitable treatment. Also those against whom we have had any grievance, or any previous differences with them, as Callippus had when he behaved as he did to Dion· here too it seems as if almost no wrong were being done. Also those who are on the point of being wronged by others if we fail to wrong them ourselves, since here we feel we have no time left for thinking the matter over. So Aenesidemus is said to have sent the 'cottabus' prize to Gelon, who had just reduced a town to slavery, because Gelon had got there first and forestalled his own attempt. Also those by wronging whom we shall be able to do many righteous acts; for we feel that we can then easily cure the harm done. Thus Jason the Thessalian said that it is a duty to do some unjust acts in order to be able to do many just ones.

Among the kinds of wrong done to others are those that are done universally, or at least commonly: one expects to be forgiven for doing these. Also those that can easily be kept dark, as where things that can rapidly be consumed like eatables are concerned, or things that can easily be changed in shape, colour, or combination, or things that can easily be stowed away almost anywhere—portable objects that you can stow away in small corners, or things so like others of which you have plenty already that nobody can tell the difference. There are also wrongs of a kind that shame prevents the victim speaking about, such as outrages done to the women in his household or to himself or to his sons. Also those for which you would be thought very litigious to prosecute any one-trifling wrongs, or wrongs for which people are usually excused.

The above is a fairly complete account of the circumstances under which men do wrong to others, of the sort of wrongs they do, of the sort of persons to whom they do them, and of their reasons for doing them.

13

It will now be well to make a complete classification of just and unjust actions. We may begin by observing that they have been defined relatively to two kinds of law, and also relatively to two classes of persons. By the two kinds of law I mean particular law and universal law. Particular law is that which each community lays down and applies to its own members: this is partly written and partly unwritten. Universal law is the law of Nature. For there really is, as every one to some extent divines, a natural justice and injustice that is binding on all men, even on those who have no association or covenant with each other. It is this that Sophocles' Antigone clearly means when she says that the burial of Polyneices was a just act in spite of the prohibition: she means that it was just by nature.

Not of to-day or yesterday it is,
But lives eternal: none can date its birth.

And so Empedocles, when he bids us kill no living creature, says that doing this is not just for some people while unjust for others,

Nay, but, an all-embracing law, through the realms of the sky
Unbroken it stretcheth, and over the earth's immensity.

And as Alcidamas says in his Messeniac Oration....

The actions that we ought to do or not to do have also been divided into two classes as affecting either the whole community or some one of its members. From this point of view we can perform just or unjust acts in either of two ways—towards one definite person, or towards the community. The man who is guilty of adultery or assault is doing wrong to some definite person; the man who avoids service in the army is doing wrong to the community.

Thus the whole class of unjust actions may be divided into two classes, those affecting the community, and those affecting one or more other persons. We will next, before going further, remind ourselves of what 'being wronged' means. Since it has already been settled that 'doing a wrong' must be intentional, 'being wronged' must consist in having an injury done to you by some one who intends to do it. In order to be wronged, a man must (1) suffer actual harm, (2) suffer it against his will. The various possible forms of harm are clearly explained by our previous, separate discussion of goods and evils. We have also seen that a voluntary action is one where the doer knows what he is doing. We now see that every accusation must be of an action affecting either the community or some individual. The doer of the action must either understand and intend the action, or not understand and intend it. In the former case, he must be acting either from deliberate choice or from passion. (Anger will be discussed when we speak of the passions the motives for crime and the state of mind of the criminal have already been discussed.) Now it often happens that a man will admit an act, but will not admit the prosecutor's label for the act nor the facts which that label implies. He will admit that he took a thing but not that he 'stole' it; that he struck some one first, but not that he committed 'outrage'; that he had intercourse with a woman, but not that he committed 'adultery'; that he is guilty of theft, but not that he is guilty of 'sacrilege', the object stolen not being consecrated; that he has encroached, but not that he has 'encroached on State lands'; that he has been in communication with the enemy, but not that he has been guilty of 'treason'. Here therefore we must be able to distinguish what is theft, outrage, or adultery, from what is not, if we are to be able to make the justice of our case clear, no matter whether our aim is to establish a man's guilt or to establish his innocence. Wherever such charges are brought against a man, the question is whether he is or is not guilty of a criminal offence. It is deliberate purpose that constitutes wickedness and criminal guilt, and such names as 'outrage' or 'theft' imply deliberate purpose as well as the mere action. A blow does not always amount to 'outrage', but only if it is struck with some such purpose as to insult the man struck or gratify the striker himself. Nor does taking a thing without the owner's knowledge always amount to 'theft', but only if it is taken with the intention of keeping it and injuring the owner. And as with these charges, so with all the others.

We saw that there are two kinds of right and wrong conduct towards others, one provided for by written ordinances, the other by unwritten. We have now discussed the kind about which the laws have something to say. The other kind has itself two varieties. First, there is the conduct that springs from exceptional goodness or badness, and is visited accordingly with censure and loss of honour, or with praise and increase of honour and decorations: for instance, gratitude to, or requital of, our benefactors, readiness to help our friends, and the like. The second kind makes up for the defects of a community's written code of law. This is what we call equity; people regard it as just; it is, in fact, the sort of justice which goes beyond the written law. Its existence partly is and partly is not intended by legislators; not intended, where they have noticed no defect in the law; intended, where find themselves unable to define things exactly, and are obliged to legislate as if that held good always which in fact only holds good usually; or where it is not easy to be complete owing to the endless possible cases presented, such as the kinds and sizes of weapons that may be used to inflict wounds—a lifetime would be too short to make out a complete list of these. If, then, a precise statement is impossible and yet legislation is necessary, the law must be expressed in wide terms; and so, if a man has no more than a finger-ring on his hand when he lifts it to strike or actually strikes another man, he is guilty of a criminal act according to the unwritten words of the law; but he is innocent really, and it is equity that declares him to be so. From this definition of equity it is plain what sort of actions, and what sort of persons, are equitable or the reverse. Equity must be applied to forgivable actions; and it must make us distinguish between criminal acts on the one hand, and errors of judgement, or misfortunes, on the other. (A 'misfortune' is an act, not due to moral badness, that has unexpected results: an 'error of judgement' is an act, also not due to moral badness, that has results that might have been expected: a 'criminal act' has results that might have been expected, but is due to moral badness, for that is the source of all actions inspired by our appetites.) Equity bids us be merciful to the weakness of human nature; to think less about the laws than about the man who framed them, and less about what he said than about what he meant; not to consider the actions of the accused so much as his intentions, nor this or that detail so much as the whole story; to ask not what a man is now but what he has always or usually been. It bids us remember benefits rather than injuries, and benefits received rather than benefits conferred; to be patient when we are wronged; to settle a dispute by negotiation and not by force; to prefer arbitration to motion—for an arbitrator goes by the equity of a case, a judge by the strict law, and arbitration was invented with the express purpose of securing full power for equity.

The above may be taken as a sufficient account of the nature of equity.

14

The worse of two acts of wrong done to others is that which is prompted by the worse disposition. Hence the most trifling acts may be the worst ones; as when Callistratus charged Melanopus with having cheated the temple—builders of three consecrated half—obols. The converse is true of just acts. This is because the greater is here potentially contained in the less: there is no crime that a man who has stolen three consecrated half—obols would shrink from committing. Sometimes, however, the worse act is reckoned not in this way but by the greater harm that it does. Or it may be because no punishment for it is severe enough to be adequate; or the harm done may be incurable—a difficult and even hopeless crime to defend; or the sufferer may not be able to get his injurer legally punished, a fact that makes the harm incurable, since legal punishment and chastisement are the proper cure. Or again, the man who has suffered wrong may have inflicted some fearful punishment on himself; then the doer of the wrong ought in justice to receive a still more fearful punishment. Thus Sophocles, when pleading for retribution to Euctemon, who had cut his own throat because of the outrage done to him, said he would not fix a penalty less than the victim had fixed for himself.

Again, a man's crime is worse if he has been the first man, or the only man, or almost the only man, to commit it: or if it is by no means the first time he has gone seriously wrong in the same way: or if his crime has led to the thinking-out and invention of measures to prevent and punish similar crimes—thus in Argos a penalty is inflicted on a man on whose account a law is passed, and also on those on whose account the prison was built: or if a crime is specially brutal, or specially deliberate: or if the report of it awakes more terror than pity. There are also such rhetorically effective ways of putting it as the following: That the accused has disregarded and broken not one but many solemn obligations like oaths, promises, pledges, or rights of intermarriage between states—here the crime is worse because it consists of many crimes; and that the crime was committed in the very place where criminals are punished, as for example perjurers do—it is argued that a man who will commit a crime in a law-court would commit it anywhere. Further, the worse deed is that which involves the doer in special shame; that whereby a man wrongs his benefactors—for he does more than one wrong, by not merely doing them harm but failing to do them good; that which breaks the unwritten laws of justice—the better sort of man will be just without being forced to be so, and the written laws depend on force while the unwritten ones do not. It may however be argued otherwise, that the crime is worse which breaks the written laws: for the man who commits crimes for which terrible penalties are provided will not hesitate over crimes for which no penalty is provided at all.—So much, then, for the comparative badness of criminal actions.

15

There are also the so-called 'non-technical' means of persuasion; and we must now take a cursory view of these, since they are specially characteristic of forensic oratory. They are five in number: laws, witnesses, contracts, tortures, oaths.

First, then, let us take laws and see how they are to be used in persuasion and dissuasion, in accusation and defence. If the written law tells against our case, clearly we must appeal to the universal law, and insist on its greater equity and justice. We must argue that the juror's oath 'I will give my verdict according to honest opinion' means that one will not simply follow the letter of the written law. We must urge that the principles of equity are permanent and changeless, and that the universal law does not change either, for it is the law of nature, whereas written laws often do change. This is the bearing the lines in Sophocles' Antigone, where Antigone pleads that in burying her brother she had broken Creon's law, but not the unwritten law:

> Not of to-day or yesterday they are,
> But live eternal: (none can date their birth.)
> Not I would fear the wrath of any man
> (And brave God's vengeance) for defying these.

We shall argue that justice indeed is true and profitable, but that sham justice is not, and that consequently the written law is not, because it does not fulfil the true purpose of law. Or that justice is like silver, and must be assayed by the judges, if the genuine is to be distinguished from the counterfeit. Or that the better a man is, the more he will follow and abide by the unwritten law in preference to the written. Or perhaps that the law in question contradicts some other highly-esteemed law, or even contradicts itself. Thus it may be that one law will enact that all contracts must be held binding, while another forbids us ever to make illegal contracts. Or if a law is ambiguous, we shall turn it about and consider which construction best fits the interests of justice or utility, and then follow that way of looking at it. Or if, though the law still exists, the situation to meet which it was passed exists no longer, we must do our best to prove this and to combat the law thereby. If however the written law supports our case, we must urge that the oath 'to give my verdict according to my honest opinion' not meant to make the judges give a verdict that is contrary to the law, but to save them from the guilt of perjury if they misunderstand what the law really means. Or that no one chooses what is absolutely good, but every one what is good for himself. Or that not to use the laws is as ahas to have no laws at all. Or that, as in the other arts, it does not pay to try to be cleverer than the doctor: for less harm comes from the doctor's mistakes than from the growing habit of disobeying authority. Or that trying to be cleverer than the laws is just what is forbidden by those codes of law that are accounted best.—So far as the laws are concerned, the above discussion is probably sufficient.

As to witnesses, they are of two kinds, the ancient and the recent; and these latter, again, either do or do not share in the risks of the trial. By 'ancient' witnesses I mean the poets and all other notable persons whose judgements are known to all. Thus the Athenians appealed to Homer as a witness about Salamis; and the men of Tenedos not long ago appealed to Periander of Corinth in their dispute with the people of Sigeum; and Cleophon supported his accusation of Critias by quoting the elegiac verse of Solon, maintaining that discipline had long been slack in the family of Critias, or Solon would never have written,

Pray thee, bid the red-haired Critias do what
his father commands him.

These witnesses are concerned with past events. As to future events we shall also appeal to soothsayers: thus Themistocles quoted the oracle about 'the wooden wall' as a reason for engaging the enemy's fleet. Further, proverbs are, as has been said, one form of evidence. Thus if you are urging somebody not to make a friend of an old man, you will appeal to the proverb,

Never show an old man kindness.

Or if you are urging that he who has made away with fathers should also make away with their sons, quote,

Fool, who slayeth the father and leaveth his sons to avenge him.

'Recent' witnesses are well-known people who have expressed their opinions about some disputed matter: such opinions will be useful support for subsequent disputants on the same oints: thus Eubulus used in the law-courts against the reply Plato had made to Archibius, 'It has become the regular custom in this country to admit that one is a scoundrel'. There are also those witnesses who share the risk of punishment if their evidence is pronounced false. These are valid witnesses to the fact that an action was or was not done, that something is or is not the case; they are not valid witnesses to the quality of an action, to its being just or unjust, useful or harmful. On such questions of quality the opinion of detached persons is highly trustworthy. Most trustworthy of all are the 'ancient' witnesses, since they cannot be corrupted.

In dealing with the evidence of witnesses, the following are useful arguments. If you have no witnesses on your side, you will argue that the judges must decide from what is probable; that this is meant by 'giving a verdict in accordance with one's honest opinion'; that probabilities cannot be bribed to mislead the court; and that probabilities are never convicted of perjury. If you have witnesses, and the other man has not, you will argue that probabilities cannot be put on their trial, and that we could do without the evidence of witnesses altogether if we need do no more than balance the pleas advanced on either side.

The evidence of witnesses may refer either to ourselves or to our opponent; and either to questions of fact or to questions of personal character: so, clearly, we need never be at a loss for useful evidence. For if we have no evidence of fact supporting our own case or telling against that of our opponent, at least we can always find evidence to prove our own worth or our opponent's worthlessness. Other arguments about a witness—that he is a friend or an enemy or neutral, or has a good, bad, or indifferent reputation, and any other such distinctions—we must construct upon the same general lines as we use for the regular rhetorical proofs.

Concerning contracts argument can be so far employed as to increase or diminish their importance and their credibility; we shall try to increase both if they tell in our favour, and to diminish both if they tell in favour of our opponent. Now for confirming or upsetting the credibility of contracts the procedure is just the same as for dealing with witnesses, for the credit to be attached to contracts depends upon the character of those who have signed them or have the custody of them. The contract being once admitted genuine, we must insist on its importance, if it supports our case. We may argue that a contract is a law, though of a special and limited kind; and that, while contracts do not of course make the law binding, the law does make any lawful contract binding, and that the law itself as a whole is a of contract, so that any one who disregards or repudiates any contract is repudiating the law itself. Further, most business relations—those, namely, that are voluntary—are regulated by contracts, and if these lose their binding force, human intercourse ceases to exist. We need not go very deep to discover the other appropriate arguments of this kind. If, however, the contract tells against us and for our opponents, in the first place those arguments are suitable which we can use to fight a law that tells against us. We do not regard ourselves as bound to observe a bad law which it was a mistake ever to pass: and it is ridiculous to suppose that we are bound to observe a bad and mistaken contract. Again, we may argue that the duty of the judge as umpire is to decide what is just, and therefore he must ask where justice lies, and not what this or that document means. And that it is impossible to pervert justice by fraud or by force, since it is founded on nature, but a party to a contract may be the victim of either fraud or force. Moreover, we must see if the contract contravenes either universal law or any written law of our own or another country; and also if it contradicts any other previous or subsequent contract; arguing that the subsequent is the binding contract, or else that the previous one was right and the subsequent one fraudulent—whichever way suits us. Further, we must consider the question of utility, noting whether the contract is against the interest of the judges or not; and so on—these arguments are as obvious as the others.

Examination by torture is one form of evidence, to which great weight is often attached because it is in a sense compulsory. Here again it is not hard to point out the available grounds for magnifying its value, if it happens to tell in our favour, and arguing that it is the only form of evidence that is infallible; or, on the other hand, for refuting it if it tells against us and for our opponent, when we may say what is true of torture of every kind alike, that people under its compulsion tell lies quite as often as they tell the truth, sometimes persistently refusing to tell the truth, sometimes recklessly making a false charge in order to be let off sooner. We ought to be able to quote cases, familiar to the judges, in which this sort of thing has actually happened. [We must say that evidence under torture is not trustworthy, the fact being that many men whether thick-witted, tough-skinned, or stout of heart endure their ordeal nobly, while cowards and timid men are full of boldness till they see the ordeal of these others: so that no trust can be placed in evidence under torture.]

In regard to oaths, a fourfold division can be made. A man may either both offer and accept an oath, or neither, or one without the other—that is, he may offer an oath but not accept one, or accept an oath but not offer one. There is also the situation that arises when an oath has already been sworn either by himself or by his opponent.

If you refuse to offer an oath, you may argue that men do not hesitate to perjure themselves; and that if your opponent does swear, you lose your money, whereas, if he does not, you think the judges will decide against him; and that the risk of an unfavourable verdict is prefer, able, since you trust the judges and do not trust him.

If you refuse to accept an oath, you may argue that an oath is always paid for; that you would of course have taken it if you had been a rascal, since if you are a rascal you had better make something by it, and you would in that case have to swear in order to succeed. Thus your refusal, you argue, must be due to high principle, not to fear of perjury: and you may aptly quote the saying of Xenophanes,

'Tis not fair that he who fears not God

should challenge him who doth.

It is as if a strong man were to challenge a weakling to strike, or be struck by, him.

If you agree to accept an oath, you may argue that you trust yourself but not your opponent; and that (to invert the remark of Xenophanes) the fair thing is for the impious man to offer the oath and for the pious man to accept it; and that it would be monstrous if you yourself were unwilling to accept an oath in a case where you demand that the judges should do so before giving their verdict. If you wish to offer an oath, you may argue that piety disposes you to commit the issue to the gods; and that your opponent ought not to want other judges than himself, since you leave the decision with him; and that it is outrageous for your opponents to refuse to swear about this question, when they insist that others should do so.

Now that we see how we are to argue in each case separately, we see also how we are to argue when they occur in pairs, namely, when you are willing to accept the oath but not to offer it; to offer it but not to accept it; both to accept and to offer it; or to do neither. These are of course combinations of the cases already mentioned, and so your arguments also must be combinations of the arguments already mentioned.

If you have already sworn an oath that contradicts your present one, you must argue that it is not perjury, since perjury is a crime, and a crime must be a voluntary action, whereas actions due to the force or fraud of others are involuntary. You must further reason from this that perjury depends on the intention and not on the spoken words. But if it is your opponent who has already sworn an oath that contradicts his present one, you must say that if he does not abide by his oaths he is the enemy of society, and that this is the reason why men take an oath before administering the laws. 'My opponents insist that you, the judges, must abide by the oath you have sworn, and yet they are not abiding by their own oaths.' And there are other arguments which may be used to magnify the importance of the oath. [So much, then, for the 'non-technical' modes of persuasion.]

Book II

1

WE have now considered the materials to be used in supporting or opposing a political measure, in pronouncing eulogies or censures, and for prosecution and defence in the law courts. We have considered the received opinions on which we may best base our arguments so as to convince our hearers—those opinions with which our enthymemes deal, and out of which they are built, in each of the three kinds of oratory, according to what may be called the special needs of each.

But since rhetoric exists to affect the giving of decisions—the hearers decide between one political speaker and another, and a legal verdict is a decision—the orator must not only try to make the argument of his speech demonstrative and worthy of belief; he must also make his own character look right and put his hearers, who are to decide, into the right frame of mind. Particularly in political oratory, but also in lawsuits, it adds much to an orator's influence that his own character should look right and that he should be thought to entertain the right feelings towards his hearers; and also that his hearers themselves should be in just the right frame of mind. That the orator's own character should look right is particularly important in political speaking: that the audience should be in the right frame of mind, in lawsuits. When people are feeling friendly and placable, they think one sort of thing; when they are feeling angry or hostile, they think either something totally different or the same thing with a different intensity: when they feel friendly to the man who comes before them for judgement, they regard him as having done little wrong, if any; when they feel hostile, they take the opposite view. Again, if they are eager for, and have good hopes of, a thing that will be pleasant if it happens, they think that it certainly will happen and be good for them: whereas if they are indifferent or annoyed, they do not think so.

There are three things which inspire confidence in the orator's own character—the three, namely, that induce us to believe a thing apart from any proof of it: good sense, good moral character, and goodwill. False statements and bad advice are due to one or more of the following three causes. Men either form a false opinion through want of good sense; or they form a true opinion, but because of their moral badness do not say what they really think; or finally, they are both sensible and upright, but not well disposed to their hearers, and may fail in consequence to recommend what they know to be the best course. These are the only possible cases. It follows that any one who is thought to have all three of these good qualities will inspire trust in his audience. The way to make ourselves thought to be sensible and morally good must be gathered from the analysis of goodness already given: the way to establish your own goodness is the same as the way to establish that of others. Good will and friendliness of disposition will form part of our discussion of the emotions, to which we must now turn.

The Emotions are all those feelings that so change men as to affect their judgements, and that are also attended by pain or pleasure. Such are anger, pity, fear and the like, with their opposites. We must arrange what we have to say about each of them under three heads. Take, for instance, the emotion of anger: here we must discover (1) what the state of mind of angry people is, (2) who the people are with whom they usually get angry, and (3) on what grounds they get angry with them. It is not enough to know one or even two of these points; unless we know all three, we shall be unable to arouse anger in any one. The same is true of the other emotions. So just as earlier in this work we drew up a list of useful propositions for the orator, let us now proceed in the same way to analyse the subject before us.

2

Anger may be defined as an impulse, accompanied by pain, to a conspicuous revenge for a conspicuous slight directed without justification towards what concerns oneself or towards what concerns one's friends. If this is a proper definition of anger, it must always be felt towards some particular individual, e.g. Cleon, and not 'man' in general. It must be felt because the other has done or intended to do something to him or one of his friends. It must always be attended by a certain pleasure—that which arises from the expectation of revenge. For since nobody aims at what he thinks he cannot attain, the angry man is aiming at what he can attain, and the belief that you will attain your aim is pleasant. Hence it has been well said about wrath,

Sweeter it is by far than the honeycomb
dripping with sweetness,
And spreads through the hearts of men.

It is also attended by a certain pleasure because the thoughts dwell upon the act of vengeance, and the images then called up cause pleasure, like the images called up in dreams.

Now slighting is the actively entertained opinion of something as obviously of no importance. We think bad things, as well as good ones, have serious importance; and we think the same of anything that tends to produce such things, while those which have little or no such tendency we consider unimportant. There are three kinds of slighting—contempt, spite, and insolence. (1) Contempt is one kind of slighting: you feel contempt for what you consider unimportant, and it is just such things that you slight. (2) Spite is another kind; it is a thwarting another man's wishes, not to get something yourself but to prevent his getting it. The slight arises just from the fact that you do not aim at something for yourself: clearly you do not think that he can do you harm, for then you would be afraid of him instead of slighting him, nor yet that he can do you any good worth mentioning, for then you would be anxious to make friends with him. (3) Insolence is also a form of slighting, since it consists in doing and saying things that cause shame to the victim, not in order that anything may happen to yourself, or because anything has happened to yourself, but simply for the pleasure involved. (Retaliation is not 'insolence', but vengeance.) The cause of the pleasure thus enjoyed by the insolent man is that he thinks himself greatly superior to others when ill-treating them. That is why youths and rich men are insolent; they think themselves superior when they show insolence. One sort of insolence is to rob people of the honour due to them; you certainly slight them thus; for it is the unimportant, for good or evil, that has no honour paid to it. So Achilles says in anger:

He hath taken my prize for himself
and hath done me dishonour,
and
Like an alien honoured by none,

meaning that this is why he is angry. A man expects to be specially respected by his inferiors in birth, in capacity, in goodness, and generally in anything in which he is much their superior: as where money is concerned a wealthy man looks for respect from a poor man; where speaking is concerned, the man with a turn for oratory looks for respect from one who cannot speak; the ruler demands the respect of the ruled, and the man who thinks he ought to be a ruler demands the respect of the man whom he thinks he ought to be ruling. Hence it has been said

Great is the wrath of kings, whose father is Zeus almighty,

and

Yea, but his rancour abideth long afterward also,

their great resentment being due to their great superiority. Then again a man looks for respect from those who he thinks owe him good treatment, and these are the people whom he has treated or is treating well, or means or has meant to treat well, either himself, or through his friends, or through others at his request.

It will be plain by now, from what has been said, (1) in what frame of mind, (2) with what persons, and (3) on what grounds people grow angry. (1) The frame of mind is that of one in which any pain is being felt. In that condition, a man is always aiming at something. Whether, then, another man opposes him either directly in any way, as by preventing him from drinking when he is thirsty, or indirectly, the act appears to him just the same; whether some one works against him, or fails to work with him, or otherwise vexes him while he is in this mood, he is equally angry in all these cases. Hence people who are afflicted by sickness or poverty or love or thirst or any other unsatisfied desires are prone to anger and easily roused: especially against those who slight their present distress. Thus a sick man is angered by disregard of his illness, a poor man by disregard of his poverty, a man aging war by disregard of the war he is waging, a lover by disregard of his love, and so throughout, any other sort of slight being enough if special slights are wanting. Each man is predisposed, by the emotion now controlling him, to his own particular anger. Further, we are angered if we happen to be expecting a contrary result: for a quite unexpected evil is specially painful, just as the quite unexpected fulfilment of our wishes is specially pleasant. Hence it is plain what seasons, times, conditions, and periods of life tend to stir men easily to anger, and where and when this will happen; and it is plain that the more we are under these conditions the more easily we are stirred.

These, then, are the frames of mind in which men are easily stirred to anger. The persons with whom we get angry are those who laugh, mock, or jeer at us, for such conduct is insolent. Also those who inflict injuries upon us that are marks of insolence. These injuries must be such as are neither retaliatory nor profitable to the doers: for only then will they be felt to be due to insolence. Also those who speak ill of us, and show contempt for us, in connexion with the things we ourselves most care about: thus those who are eager to win fame as philosophers get angry with those who show contempt for their philosophy; those who pride themselves upon their appearance get angry with those who show contempt for their appearance and so on in other cases. We feel particularly angry on this account if we suspect that we are in fact, or that people think we are, lacking completely or to any effective extent in the qualities in question. For when we are convinced that we excel in the qualities for which we are jeered at, we can ignore the jeering. Again, we are angrier with our friends than with other people, since we feel that our friends ought to treat us well and not badly. We are angry with those who have usually treated us with honour or regard, if a change comes and they behave to us otherwise: for we think that they feel contempt for us, or they would still be behaving as they did before. And with those who do not return our kindnesses or fail to return them adequately, and with those who oppose us though they are our inferiors: for all such persons seem to feel contempt for us; those who oppose us seem to think us inferior to themselves, and those who do not return our kindnesses seem to think that those kindnesses were conferred by inferiors. And we feel particularly angry with men of no account at all, if they slight us. For, by our hypothesis, the anger caused by the slight is felt towards people who are not justified in slighting us, and our inferiors are not thus justified. Again, we feel angry with friends if they do not speak well of us or treat us well; and still more, if they do the contrary; or if they do not perceive our needs, which is why Plexippus is angry with Meleager in Antiphon's play; for this want of perception shows that they are slighting us—we do not fail to perceive the needs of those for whom we care. Again we are angry with those who rejoice at our misfortunes or simply keep cheerful in the midst of our misfortunes, since this shows that they either hate us or are slighting us. Also with those who are indifferent to the pain they give us: this is why we get angry with bringers of bad news. And with those who listen to stories about us or keep on looking at our weaknesses; this seems like either slighting us or hating us; for those who love us share in all our distresses and it must distress any one to keep on looking at his own weaknesses. Further, with those who slight us before five classes of people: namely, (1) our rivals, (2) those whom we admire, (3) those whom we wish to admire us, (4) those for whom we feel reverence, (5) those who feel reverence for us: if any one slights us before such persons, we feel particularly angry. Again, we feel angry with those who slight us in connexion with what we are as honourable men bound to champion   our parents, children, wives, or subjects. And with those who do not return a favour, since such a slight is unjustifiable. Also with those who reply with humorous levity when we are speaking seriously, for such behaviour indicates contempt. And with those who treat us less well than they treat everybody else; it is another mark of contempt that they should think we do not deserve what every one else deserves. Forgetfulness, too, causes anger, as when our own names are forgotten, trifling as this may be; since forgetfulness is felt to be another sign that we are being slighted; it is due to negligence, and to neglect us is to slight us.

The persons with whom we feel anger, the frame of mind in which we feel it, and the reasons why we feel it, have now all been set forth. Clearly the orator will have to speak so as to bring his hearers into a frame of mind that will dispose them to anger, and to represent his adversaries as open to such charges and possessed of such qualities as do make people angry.

3

Since growing calm is the opposite of growing angry, and calmness the opposite of anger, we must ascertain in what frames of mind men are calm, towards whom they feel calm, and by what means they are made so. Growing calm may be defined as a settling down or quieting of anger. Now we get angry with those who slight us; and since slighting is a voluntary act, it is plain that we feel calm towards those who do nothing of the kind, or who do or seem to do it involuntarily. Also towards those who intended to do the opposite of what they did do. Also towards those who treat themselves as they have treated us: since no one can be supposed to slight himself. Also towards those who admit their fault and are sorry: since we accept their grief at what they have done as satisfaction, and cease to be angry. The punishment of servants shows this: those who contradict us and deny their offence we punish all the more, but we cease to be incensed against those who agree that they deserved their punishment. The reason is that it is shameless to deny what is obvious, and those who are shameless towards us slight us and show contempt for us: anyhow, we do not feel shame before those of whom we are thoroughly contemptuous. Also we feel calm towards those who humble themselves before us and do not gainsay us; we feel that they thus admit themselves our inferiors, and inferiors feel fear, and nobody can slight any one so long as he feels afraid of him. That our anger ceases towards those who humble themselves before us is shown even by dogs, who do not bite people when they sit down. We also feel calm towards those who are serious when we are serious, because then we feel that we are treated seriously and not contemptuously. Also towards those who have done us more kindnesses than we have done them. Also towards those who pray to us and beg for mercy, since they humble themselves by doing so. Also towards those who do not insult or mock at or slight any one at all, or not any worthy person or any one like ourselves. In general, the things that make us calm may be inferred by seeing what the opposites are of those that make us angry. We are not angry with people we fear or respect, as long as we fear or respect them; you cannot be afraid of a person and also at the same time angry with him. Again, we feel no anger, or comparatively little, with those who have done what they did through anger: we do not feel that they have done it from a wish to slight us, for no one slights people when angry with them, since slighting is painless, and anger is painful. Nor do we grow angry with those who reverence us.

As to the frame of mind that makes people calm, it is plainly the opposite to that which makes them angry, as when they are amusing themselves or laughing or feasting; when they are feeling prosperous or successful or satisfied; when, in fine, they are enjoying freedom from pain, or inoffensive pleasure, or justifiable hope. Also when time has passed and their anger is no longer fresh, for time puts an end to anger. And vengeance previously taken on one person puts an end to even greater anger felt against another person. Hence Philocrates, being asked by some one, at a time when the public was angry with him, 'Why don't you defend yourself?' did right to reply, 'The time is not yet.' 'Why, when is the time?' 'When I see someone else calumniated.' For men become calm when they have spent their anger on somebody else. This happened in the case of Ergophilus: though the people were more irritated against him than against Callisthenes, they acquitted him because they had condemned Callisthenes to death the day before. Again, men become calm if they have convicted the offender; or if he has already suffered worse things than they in their anger would have themselves inflicted upon him; for they feel as if they were already avenged. Or if they feel that they themselves are in the wrong and are suffering justly (for anger is not excited by what is just), since men no longer think then that they are suffering without justification; and anger, as we have seen, means this. Hence we ought always to inflict a preliminary punishment in words: if that is done, even slaves are less aggrieved by the actual punishment. We also feel calm if we think that the offender will not see that he is punished on our account and because of the way he has treated us. For anger has to do with individuals. This is plain from the definition. Hence the poet has well written:

Say that it was Odysseus, sacker of cities,

implying that Odysseus would not have considered himself avenged unless the Cyclops perceived both by whom and for what he had been blinded. Consequently we do not get angry with any one who cannot be aware of our anger, and in particular we cease to be angry with people once they are dead, for we feel that the worst has been done to them, and that they will neither feel pain nor anything else that we in our anger aim at making them feel. And therefore the poet has well made Apollo say, in order to put a stop to the anger of Achilles against the dead Hector,

For behold in his fury he doeth despite to the senseless clay.

It is now plain that when you wish to calm others you must draw upon these lines of argument; you must put your hearers into the corresponding frame of mind, and represent those with whom they are angry as formidable, or as worthy of reverence, or as benefactors, or as involuntary agents, or as much distressed at what they have done.

4

Let us now turn to Friendship and Enmity, and ask towards whom these feelings are entertained, and why. We will begin by defining and friendly feeling. We may describe friendly feeling towards any one as wishing for him what you believe to be good things, not for your own sake but for his, and being inclined, so far as you can, to bring these things about. A friend is one who feels thus and excites these feelings in return: those who think they feel thus towards each other think themselves friends. This being assumed, it follows that your friend is the sort of man who shares your pleasure in what is good and your pain in what is unpleasant, for your sake and for no other reason. This pleasure and pain of his will be the token of his good wishes for you, since we all feel glad at getting what we wish for, and pained at getting what we do not. Those, then, are friends to whom the same things are good and evil; and those who are, moreover, friendly or unfriendly to the same people; for in that case they must have the same wishes, and thus by wishing for each other what they wish for themselves, they show themselves each other's friends. Again, we feel friendly to those who have treated us well, either ourselves or those we care for, whether on a large scale, or readily, or at some particular crisis; provided it was for our own sake. And also to those who we think wish to treat us well. And also to our friends' friends, and to those who like, or are liked by, those whom we like ourselves. And also to those who are enemies to those whose enemies we are, and dislike, or are disliked by, those whom we dislike. For all such persons think the things good which we think good, so that they wish what is good for us; and this, as we saw, is what friends must do. And also to those who are willing to treat us well where money or our personal safety is concerned: and therefore we value those who are liberal, brave, or just. The just we consider to be those who do not live on others; which means those who work for their living, especially farmers and others who work with their own hands. We also like temperate men, because they are not unjust to others; and, for the same reason, those who mind their own business. And also those whose friends we wish to be, if it is plain that they wish to be our friends: such are the morally good, and those well thought of by every one, by the best men, or by those whom we admire or who admire us. And also those with whom it is pleasant to live and spend our days: such are the good-tempered, and those who are not too ready to show us our mistakes, and those who are not cantankerous or quarrelsome—such people are always wanting to fight us, and those who fight us we feel wish for the opposite of what we wish for ourselves—and those who have the tact to make and take a joke; here both parties have the same object in view, when they can stand being made fun of as well as do it prettily themselves. And we also feel friendly towards those who praise such good qualities as we possess, and especially if they praise the good qualities that we are not too sure we do possess. And towards those who are cleanly in their person, their dress, and all their way of life. And towards those who do not reproach us with what we have done amiss to them or they have done to help us, for both actions show a tendency to criticize us. And towards those who do not nurse grudges or store up grievances, but are always ready to make friends again; for we take it that they will behave to us just as we find them behaving to every one else. And towards those who are not evil speakers and who are aware of neither their neighbours' bad points nor our own, but of our good ones only, as a good man always will be. And towards those who do not try to thwart us when we are angry or in earnest, which would mean being ready to fight us. And towards

those who have some serious feeling towards us, such as admiration for us, or belief
in our goodness, or pleasure in our company; especially if they feel like this about
qualities in us for which we especially wish to be admired, esteemed, or liked. And
towards those who are like ourselves in character and occupation, provided they
do not get in our way or gain their living from the same source as we do—for then
it will be a case of 'potter against potter':

Potter to potter and builder to builder begrudge their reward.

And those who desire the same things as we desire, if it is possible for us both
to share them together; otherwise the same trouble arises here too. And towards
those with whom we are on such terms that, while we respect their opinions, we
need not blush before them for doing what is conventionally wrong: as well as
towards those before whom we should be ashamed to do anything really wrong.
Again, our rivals, and those whom we should like to envy us—though without
ill-feeling—either we like these people or at least we wish them to like us. And we
feel friendly towards those whom we help to secure good for themselves, provided
we are not likely to suffer heavily by it ourselves. And those who feel as friendly to
us when we are not with them as when we are—which is why all men feel friendly
towards those who are faithful to their dead friends. And, speaking generally,
towards those who are really fond of their friends and do not desert them in
trouble; of all good men, we feel most friendly to those who show their goodness
as friends. Also towards those who are honest with us, including those who will tell
us of their own weak points: it has just said that with our friends we are not
ashamed of what is conventionally wrong, and if we do have this feeling, we do not
love them; if therefore we do not have it, it looks as if we did love them. We also
like those with whom we do not feel frightened or uncomfortable—nobody can like
a man of whom he feels frightened. Friendship has various forms—comradeship,
intimacy, kinship, and so on.

Things that cause friendship are: doing kindnesses; doing them unasked; and
not proclaiming the fact when they are done, which shows that they were done for
our own sake and not for some other reason.

Enmity and Hatred should clearly be studied by reference to their opposites.
Enmity may be produced by anger or spite or calumny. Now whereas anger arises
from offences against oneself, enmity may arise even without that; we may hate
people merely because of what we take to be their character. Anger is always
concerned with individuals—a Callias or a Socrates—whereas hatred is directed
also against classes: we all hate any thief and any informer. Moreover, anger can be
cured by time; but hatred cannot. The one aims at giving pain to its object, the
other at doing him harm; the angry man wants his victims to feel; the hater does
not mind whether they feel or not. All painful things are felt; but the greatest evils,
injustice and folly, are the least felt, since their presence causes no pain. And anger
is accompanied by pain, hatred is not; the angry man feels pain, but the hater does
not. Much may happen to make the angry man pity those who offend him, but the
hater under no circumstances wishes to pity a man whom he has once hated: for

the one would have the offenders suffer for what they have done; the other would have them cease to exist.

It is plain from all this that we can prove people to be friends or enemies; if they are not, we can make them out to be so; if they claim to be so, we can refute their claim; and if it is disputed whether an action was due to anger or to hatred, we can attribute it to whichever of these we prefer.

5

To turn next to Fear, what follows will show things and persons of which, and the states of mind in which, we feel afraid. Fear may be defined as a pain or disturbance due to a mental picture of some destructive or painful evil in the future. Of destructive or painful evils only; for there are some evils, e.g. wickedness or stupidity, the prospect of which does not frighten us: I mean only such as amount to great pains or losses. And even these only if they appear not remote but so near as to be imminent: we do not fear things that are a very long way off: for instance, we all know we shall die, but we are not troubled thereby, because death is not close at hand. From this definition it will follow that fear is caused by whatever we feel has great power of destroying or of harming us in ways that tend to cause us great pain. Hence the very indications of such things are terrible, making us feel that the terrible thing itself is close at hand; the approach of what is terrible is just what we mean by 'danger'. Such indications are the enmity and anger of people who have power to do something to us; for it is plain that they have the will to do it, and so they are on the point of doing it. Also injustice in possession of power; for it is the unjust man's will to do evil that makes him unjust. Also outraged virtue in possession of power; for it is plain that, when outraged, it always has the will to retaliate, and now it has the power to do so. Also fear felt by those who have the power to do something to us, since such persons are sure to be ready to do it. And since most men tend to be bad—slaves to greed, and cowards in danger—it is, as a rule, a terrible thing to be at another man's mercy; and therefore, if we have done anything horrible, those in the secret terrify us with the thought that they may betray or desert us. And those who can do us wrong are terrible to us when we are liable to be wronged; for as a rule men do wrong to others whenever they have the power to do it. And those who have been wronged, or believe themselves to be wronged, are terrible; for they are always looking out for their opportunity. Also those who have done people wrong, if they possess power, since they stand in fear of retaliation: we have already said that wickedness possessing power is terrible. Again, our rivals for a thing cause us fear when we cannot both have it at once; for we are always at war with such men. We also fear those who are to be feared by stronger people than ourselves: if they can hurt those stronger people, still more can they hurt us; and, for the same reason, we fear those whom those stronger people are actually afraid of. Also those who have destroyed people stronger than we are. Also those who are attacking people weaker than we are: either they are already formidable, or they will be so when they have thus grown stronger. Of those we have wronged, and of our enemies or rivals, it is not the passionate and outspoken whom we have to fear, but the quiet, dissembling, unscrupulous; since we never know when they are upon us, we can never be sure they are at a safe distance. All terrible things are more terrible if they give us no chance of retrieving a blunder either no chance at all, or only one that depends on our enemies and not ourselves. Those things are also worse which we cannot, or cannot easily, help. Speaking generally, anything causes us to feel fear that when it happens to, or threatens, others cause us to feel pity.

The above are, roughly, the chief things that are terrible and are feared. Let us now describe the conditions under which we ourselves feel fear. If fear is associated with the expectation that something destructive will happen to us, plainly nobody will be afraid who believes nothing can happen to him; we shall not fear things that we believe cannot happen to us, nor people who we believe cannot inflict them upon us; nor shall we be afraid at times when we think ourselves safe from them. It follows therefore that fear is felt by those who believe something to be likely to happen to them, at the hands of particular persons, in a particular form, and at a particular time. People do not believe this when they are, or think they are, in the midst of great prosperity, and are in consequence insolent, contemptuous, and reckless—the kind of character produced by wealth, physical strength, abundance of friends, power: nor yet when they feel they have experienced every kind of horror already and have grown callous about the future, like men who are being flogged and are already nearly dead—if they are to feel the anguish of uncertainty, there must be some faint expectation of escape. This appears from the fact that fear sets us thinking what can be done, which of course nobody does when things are hopeless. Consequently, when it is advisable that the audience should be frightened, the orator must make them feel that they really are in danger of something, pointing out that it has happened to others who were stronger than they are, and is happening, or has happened, to people like themselves, at the hands of unexpected people, in an unexpected form, and at an unexpected time.

Having now seen the nature of fear, and of the things that cause it, and the various states of mind in which it is felt, we can also see what Confidence is, about what things we feel it, and under what conditions. It is the opposite of fear, and what causes it is the opposite of what causes fear; it is, therefore, the expectation associated with a mental picture of the nearness of what keeps us safe and the absence or remoteness of what is terrible: it may be due either to the near presence of what inspires confidence or to the absence of what causes alarm. We feel it if we can take steps—many, or important, or both—to cure or prevent trouble; if we have neither wronged others nor been wronged by them; if we have either no rivals at all or no strong ones; if our rivals who are strong are our friends or have treated us well or been treated well by us; or if those whose interest is the same as ours are the more numerous party, or the stronger, or both.

As for our own state of mind, we feel confidence if we believe we have often succeeded and never suffered reverses, or have often met danger and escaped it safely. For there are two reasons why human beings face danger calmly: they may have no experience of it, or they may have means to deal with it: thus when in danger at sea people may feel confident about what will happen either because they have no experience of bad weather, or because their experience gives them the means of dealing with it. We also feel confident whenever there is nothing to terrify other people like ourselves, or people weaker than ourselves, or people than whom we believe ourselves to be stronger—and we believe this if we have conquered them, or conquered others who are as strong as they are, or stronger. Also if we believe ourselves superior to our rivals in the number and importance of the advantages that make men formidable—wealth, physical strength, strong bodies of supporters, extensive territory, and the possession of all, or the most important, appliances of war. Also if we have wronged no one, or not many, or not those of whom we are afraid; and generally, if our relations with the gods are satisfactory, as will be shown especially by signs and oracles. The fact is that anger makes us confident—that anger is excited by our knowledge that we are not the wrongers but the wronged, and that the divine power is always supposed to be on the side of the wronged. Also when, at the outset of an enterprise, we believe that we cannot and shall not fail, or that we shall succeed completely.—So much for the causes of fear and confidence.

6

We now turn to Shame and Shamelessness; what follows will explain the things that cause these feelings, and the persons before whom, and the states of mind under which, they are felt. Shame may be defined as pain or disturbance in regard to bad things, whether present, past, or future, which seem likely to involve us in discredit; and shamelessness as contempt or indifference in regard to these same bad things. If this definition be granted, it follows that we feel shame at such bad things as we think are disgraceful to ourselves or to those we care for. These evils are, in the first place, those due to moral badness. Such are throwing away one's shield or taking to flight; for these bad things are due to cowardice. Also, withholding a deposit or otherwise wronging people about money; for these acts are due to injustice. Also, having carnal intercourse with forbidden persons, at wrong times, or in wrong places; for these things are due to licentiousness. Also, making profit in petty or disgraceful ways, or out of helpless persons, e.g. the poor, or the dead—whence the proverb 'He would pick a corpse's pocket'; for all this is due to low greed and meanness. Also, in money matters, giving less help than you might, or none at all, or accepting help from those worse off than yourself; so also borrowing when it will seem like begging; begging when it will seem like asking the return of a favour; asking such a return when it will seem like begging; praising a man in order that it may seem like begging; and going on begging in spite of failure: all such actions are tokens of meanness. Also, praising people to their face, and praising extravagantly a man's good points and glozing over his weaknesses, and showing extravagant sympathy with his grief when you are in his presence, and all that sort of thing; all this shows the disposition of a flatterer. Also, refusing to endure hardships that are endured by people who are older, more delicately brought up, of higher rank, or generally less capable of endurance than ourselves: for all this shows effeminacy. Also, accepting benefits, especially accepting them often, from another man, and then abusing him for conferring them: all this shows a mean, ignoble disposition. Also, talking incessantly about yourself, making loud professions, and appropriating the merits of others; for this is due to boastfulness. The same is true of the actions due to any of the other forms of badness of moral character, of the tokens of such badness, &c.: they are all disgraceful and shameless. Another sort of bad thing at which we feel shame is, lacking a share in the honourable things shared by every one else, or by all or nearly all who are like ourselves. By 'those like ourselves' I mean those of our own race or country or age or family, and generally those who are on our own level. Once we are on a level with others, it is a disgrace to be, say, less well educated than they are; and so with other advantages: all the more so, in each case, if it is seen to be our own fault: wherever we are ourselves to blame for our present, past, or future circumstances, it follows at once that this is to a greater extent due to our moral badness. We are moreover ashamed of having done to us, having had done, or being about to have done to us acts that involve us in dishonour and reproach; as when we surrender our persons, or lend ourselves to vile deeds, e.g. when we submit to outrage. And acts of yielding to the lust of others are shameful whether willing or unwilling (yielding to force being an instance of unwillingness), since unresisting submission to them is due to unmanliness or cowardice.

These things, and others like them, are what cause the feeling of shame. Now since shame is a mental picture of disgrace, in which we shrink from the disgrace itself and not from its consequences, and we only care what opinion is held of us because of the people who form that opinion, it follows that the people before whom we feel shame are those whose opinion of us matters to us. Such persons are: those who admire us, those whom we admire, those by whom we wish to be admired, those with whom we are competing, and those whose opinion of us we respect. We admire those, and wish those to admire us, who possess any good thing that is highly esteemed; or from whom we are very anxious to get something that they are able to give us—as a lover feels. We compete with our equals. We respect, as true, the views of sensible people, such as our elders and those who have been well educated. And we feel more shame about a thing if it is done openly, before all men's eyes. Hence the proverb, 'shame dwells in the eyes'. For this reason we feel most shame before those who will always be with us and those who notice what we do, since in both cases eyes are upon us. We also feel it before those not open to the same imputation as ourselves: for it is plain that their opinions about it are the opposite of ours. Also before those who are hard on any one whose conduct they think wrong; for what a man does himself, he is said not to resent when his neighbours do it: so that of course he does resent their doing what he does not do himself. And before those who are likely to tell everybody about you; not telling others is as good as not be lieving you wrong. People are likely to tell others about you if you have wronged them, since they are on the look out to harm you; or if they speak evil of everybody, for those who attack the innocent will be still more ready to attack the guilty. And before those whose main occupation is with their neighbours' failings—people like satirists and writers of comedy; these are really a kind of evil-speakers and tell-tales. And before those who have never yet known us come to grief, since their attitude to us has amounted to admiration so far: that is why we feel ashamed to refuse those a favour who ask one for the first time—we have not as yet lost credit with them. Such are those who are just beginning to wish to be our friends; for they have seen our best side only (hence the appropriateness of Euripides' reply to the Syracusans): and such also are those among our old acquaintances who know nothing to our discredit. And we are ashamed not merely of the actual shameful conduct mentioned, but also of the evidences of it: not merely, for example, of actual sexual intercourse, but also of its evidences; and not merely of disgraceful acts but also of disgraceful talk. Similarly we feel shame not merely in presence of the persons mentioned but also of those who will tell them what we have done, such as their servants or friends. And, generally, we feel no shame before those upon whose opinions we quite look down as untrustworthy (no one feels shame before small children or animals); nor are we ashamed of the same things before intimates as before strangers, but before the former of what seem genuine faults, before the latter of what seem conventional ones.

The conditions under which we shall feel shame are these: first, having people related to us like those before whom, as has been said, we feel shame. These are, as was stated, persons whom we admire, or who admire us, or by whom we wish to be admired, or from whom we desire some service that we shall not obtain if we forfeit their good opinion. These persons may be actually looking on (as Cydias represented them in his speech on land assignments in Samos, when he told the Athenians to imagine the Greeks to be standing all around them, actually seeing the way they voted and not merely going to hear about it afterwards): or again they may be near at hand, or may be likely to find out about what we do. This is why in misfortune we do not wish to be seen by those who once wished themselves like us; for such a feeling implies admiration. And men feel shame when they have acts or exploits to their credit on which they are bringing dishonour, whether these are their own, or those of their ancestors, or those of other persons with whom they have some close connexion. Generally, we feel shame before those for whose own misconduct we should also feel it—those already mentioned; those who take us as their models; those whose teachers or advisers we have been; or other people, it may be, like ourselves, whose rivals we are. For there are many things that shame before such people makes us do or leave undone. And we feel more shame when we are likely to be continually seen by, and go about under the eyes of, those who know of our disgrace. Hence, when Antiphon the poet was to be cudgelled to death by order of Dionysius, and saw those who were to perish with him covering their faces as they went through the gates, he said, 'Why do you cover your faces? Is it lest some of these spectators should see you to-morrow?'

So much for Shame; to understand Shamelessness, we need only consider the converse cases, and plainly we shall have all we need.

7

To take Kindness next: the definition of it will show us towards whom it is felt, why, and in what frames of mind. Kindness—under the influence of which a man is said to 'be kind' may be defined as helpfulness towards some one in need, not in return for anything, nor for the advantage of the helper himself, but for that of the person helped. Kindness is great if shown to one who is in great need, or who needs what is important and hard to get, or who needs it at an important and difficult crisis; or if the helper is the only, the first, or the chief person to give the help. Natural cravings constitute such needs; and in particular cravings, accompanied by pain, for what is not being attained. The appetites are cravings for this kind: sexual desire, for instance, and those which arise during bodily injuries and in dangers; for appetite is active both in danger and in pain. Hence those who stand by us in poverty or in banishment, even if they do not help us much, are yet really kind to us, because our need is great and the occasion pressing; for instance, the man who gave the mat in the Lyceum. The helpfulness must therefore meet, preferably, just this kind of need; and failing just this kind, some other kind as great or greater. We now see to whom, why, and under what conditions kindness is shown; and these facts must form the basis of our arguments. We must show that the persons helped are, or have been, in such pain and need as has been described, and that their helpers gave, or are giving, the kind of help described, in the kind of need described. We can also see how to eliminate the idea of kindness and make our opponents appear unkind: we may maintain that they are being or have been helpful simply to promote their own interest—this, as has been stated, is not kindness; or that their action was accidental, or was forced upon them; or that they were not doing a favour, but merely returning one, whether they know this or not—in either case the action is a mere return, and is therefore not a kindness even if the doer does not know how the case stands. In considering this subject we must look at all the categories: an act may be an act of kindness because (1) it is a particular thing, (2) it has a particular magnitude or (3) quality, or (4) is done at a particular time or (5) place. As evidence of the want of kindness, we may point out that a smaller service had been refused to the man in need; or that the same service, or an equal or greater one, has been given to his enemies; these facts show that the service in question was not done for the sake of the person helped. Or we may point out that the thing desired was worthless and that the helper knew it: no one will admit that he is in need of what is worthless.

8

So much for Kindness and Unkindness. Let us now consider Pity, asking ourselves what things excite pity, and for what persons, and in what states of our mind pity is felt. Pity may be defined as a feeling of pain caused by the sight of some evil, destructive or painful, which befalls one who does not deserve it, and which we might expect to befall ourselves or some friend of ours, and moreover to befall us soon. In order to feel pity, we must obviously be capable of supposing that some evil may happen to us or some friend of ours, and moreover some such evil as is stated in our definition or is more or less of that kind. It is therefore not felt by those completely ruined, who suppose that no further evil can befall them, since the worst has befallen them already; nor by those who imagine themselves immensely fortunate—their feeling is rather presumptuous insolence, for when they think they possess all the good things of life, it is clear that the impossibility of evil befalling them will be included, this being one of the good things in question. Those who think evil may befall them are such as have already had it befall them and have safely escaped from it; elderly men, owing to their good sense and their experience; weak men, especially men inclined to cowardice; and also educated people, since these can take long views. Also those who have parents living, or children, or wives; for these are our own, and the evils mentioned above may easily befall them. And those who neither moved by any courageous emotion such as anger or confidence (these emotions take no account of the future), nor by a disposition to presumptuous insolence (insolent men, too, take no account of the possibility that something evil will happen to them), nor yet by great fear (panic-stricken people do not feel pity, because they are taken up with what is happening to themselves); only those feel pity who are between these two extremes. In order to feel pity we must also believe in the goodness of at least some people; if you think nobody good, you will believe that everybody deserves evil fortune. And, generally, we feel pity whenever we are in the condition of remembering that similar misfortunes have happened to us or ours, or expecting them to happen in the future.

So much for the mental conditions under which we feel pity. What we pity is stated clearly in the definition. All unpleasant and painful things excite pity if they tend to destroy pain and annihilate; and all such evils as are due to chance, if they are serious. The painful and destructive evils are: death in its various forms, bodily injuries and afflictions, old age, diseases, lack of food. The evils due to chance are: friendlessness, scarcity of friends (it is a pitiful thing to be torn away from friends and companions), deformity, weakness, mutilation; evil coming from a source from which good ought to have come; and the frequent repetition of such misfortunes. Also the coming of good when the worst has happened: e.g. the arrival of the Great King's gifts for Diopeithes after his death. Also that either no good should have befallen a man at all, or that he should not be able to enjoy it when it has.

The grounds, then, on which we feel pity are these or like these. The people we pity are: those whom we know, if only they are not very closely related to us—in that case we feel about them as if we were in danger ourselves. For this reason Amasis did not weep, they say, at the sight of his son being led to death, but did weep when he saw his friend begging: the latter sight was pitiful, the former terrible, and the terrible is different from the pitiful; it tends to cast out pity, and often helps to produce the opposite of pity. Again, we feel pity when the danger is near ourselves. Also we pity those who are like us in age, character, disposition, social standing, or birth; for in all these cases it appears more likely that the same misfortune may befall us also. Here too we have to remember the general principle that what we fear for ourselves excites our pity when it happens to others. Further, since it is when the sufferings of others are close to us that they excite our pity (we cannot remember what disasters happened a hundred centuries ago, nor look forward to what will happen a hundred centuries hereafter, and therefore feel little pity, if any, for such things): it follows that those who heighten the effect of their words with suitable gestures, tones, dress, and dramatic action generally, are especially successful in exciting pity: they thus put the disasters before our eyes, and make them seem close to us, just coming or just past. Anything that has just happened, or is going to happen soon, is particularly piteous: so too therefore are the tokens and the actions of sufferers—the garments and the like of those who have already suffered; the words and the like of those actually suffering—of those, for instance, who are on the point of death. Most piteous of all is it when, in such times of trial, the victims are persons of noble character: whenever they are so, our pity is especially excited, because their innocence, as well as the setting of their misfortunes before our eyes, makes their misfortunes seem close to ourselves.

9

Most directly opposed to pity is the feeling called Indignation. Pain at unmerited good fortune is, in one sense, opposite to pain at unmerited bad fortune, and is due to the same moral qualities. Both feelings are associated with good moral character; it is our duty both to feel sympathy and pity for unmerited distress, and to feel indignation at unmerited prosperity; for whatever is undeserved is unjust, and that is why we ascribe indignation even to the gods. It might indeed be thought that envy is similarly opposed to pity, on the ground that envy it closely akin to indignation, or even the same thing. But it is not the same. It is true that it also is a disturbing pain excited by the prosperity of others. But it is excited not by the prosperity of the undeserving but by that of people who are like us or equal with us. The two feelings have this in common, that they must be due not to some untoward thing being likely to befall ourselves, but only to what is happening to our neighbour. The feeling ceases to be envy in the one case and indignation in the other, and becomes fear, if the pain and disturbance are due to the prospect of something bad for ourselves as the result of the other man's good fortune. The feelings of pity and indignation will obviously be attended by the converse feelings of satisfaction. If you are pained by the unmerited distress of others, you will be pleased, or at least not pained, by their merited distress. Thus no good man can be pained by the punishment of parricides or murderers. These are things we are bound to rejoice at, as we must at the prosperity of the deserving; both these things are just, and both give pleasure to any honest man, since he cannot help expecting that what has happened to a man like him will happen to him too. All these feelings are associated with the same type of moral character. And their contraries are associated with the contrary type; the man who is delighted by others' misfortunes is identical with the man who envies others' prosperity. For any one who is pained by the occurrence or existence of a given thing must be pleased by that thing's non-existence or destruction. We can now see that all these feelings tend to prevent pity (though they differ among themselves, for the reasons given), so that all are equally useful for neutralizing an appeal to pity.

We will first consider Indignation—reserving the other emotions for subsequent discussion—and ask with whom, on what grounds, and in what states of mind we may be indignant. These questions are really answered by what has been said already. Indignation is pain caused by the sight of undeserved good fortune. It is, then, plain to begin with that there are some forms of good the sight of which cannot cause it. Thus a man may be just or brave, or acquire moral goodness: but we shall not be indignant with him for that reason, any more than we shall pity him for the contrary reason. Indignation is roused by the sight of wealth, power, and the like—by all those things, roughly speaking, which are deserved by good men and by those who possess the goods of nature-noble birth, beauty, and so on. Again, what is long established seems akin to what exists by nature; and therefore we feel more indignation at those possessing a given good if they have as a matter of fact only just got it and the prosperity it brings with it. The newly rich give more offence than those whose wealth is of long standing and inherited. The same is true of those who have office or power, plenty of friends, a fine family, &c. We feel the same when these advantages of theirs secure them others. For here again, the newly rich give us more offence by obtaining office through their riches than do those whose wealth is of long standing; and so in all other cases. The reason is that what the latter have is felt to be really their own, but what the others have is not; what appears to have been always what it is is regarded as real, and so the possessions of the newly rich do not seem to be really their own. Further, it is not any and every man that deserves any given kind of good; there is a certain correspondence and appropriateness in such things; thus it is appropriate for brave men, not for just men, to have fine weapons, and for men of family, not for parvenus, to make distinguished marriages. Indignation may therefore properly be felt when any one gets what is not appropriate for him, though he may be a good man enough. It may also be felt when any one sets himself up against his superior, especially against his superior in some particular respect—whence the lines

Only from battle he shrank with Aias Telamon's son;

Zeus had been angered with him,
had he fought with a mightier one;

    but also, even apart from that, when the inferior in any sense contends with his superior; a musician, for instance, with a just man, for justice is a finer thing than music.

    Enough has been said to make clear the grounds on which, and the persons against whom, Indignation is felt—they are those mentioned, and others like him. As for the people who feel it; we feel it if we do ourselves deserve the greatest possible goods and moreover have them, for it is an injustice that those who are not our equals should have been held to deserve as much as we have. Or, secondly, we feel it if we are really good and honest people; our judgement is then sound, and we loathe any kind of injustice. Also if we are ambitious and eager to gain particular ends, especially if we are ambitious for what others are getting without deserving to get it. And, generally, if we think that we ourselves deserve a thing and that others do not, we are disposed to be indignant with those others so far as that thing is concerned. Hence servile, worthless, unambitious persons are not inclined to Indignation, since there is nothing they can believe themselves to deserve.

    From all this it is plain what sort of men those are at whose misfortunes, distresses, or failures we ought to feel pleased, or at least not pained: by considering the facts described we see at once what their contraries are. If therefore our speech puts the judges in such a frame of mind as that indicated and shows that those who claim pity on certain definite grounds do not deserve to secure pity but do deserve not to secure it, it will be impossible for the judges to feel pity.

<h1 style="text-align:center">10</h1>

    To take Envy next: we can see on what grounds, against what persons, and in what states of mind we feel it. Envy is pain at the sight of such good fortune as consists of the good things already mentioned; we feel it towards our equals; not with the idea of getting something for ourselves, but because the other people have it. We shall feel it if we have, or think we have, equals; and by 'equals' I mean equals in birth, relationship, age, disposition, distinction, or wealth. We feel envy also if we fall but a little short of having everything; which is why people in high place and prosperity feel it—they think every one else is taking what belongs to themselves. Also if we are exceptionally distinguished for some particular thing, and especially if that thing is wisdom or good fortune. Ambitious men are more envious than those who are not. So also those who profess wisdom; they are ambitious to be thought wise. Indeed, generally, those who aim at a reputation for anything are envious on this particular point. And small-minded men are envious, for everything seems great to them. The good things which excite envy have already been mentioned. The deeds or possessions which arouse the love of reputation and honour and the desire for fame, and the various gifts of fortune, are almost all subject to envy; and particularly if we desire the thing ourselves, or think we are entitled to it, or if having it puts us a little above others, or not having it a little below them. It is clear also what kind of people we envy; that was included in what has been said already: we envy those who are near us in time, place, age, or reputation. Hence the line:

Ay, kin can even be jealous of their kin.

Also our fellow-competitors, who are indeed the people just mentioned—we do not compete with men who lived a hundred centuries ago, or those not yet born, or the dead, or those who dwell near the Pillars of Hercules, or those whom, in our opinion or that of others, we take to be far below us or far above us. So too we compete with those who follow the same ends as ourselves: we compete with our rivals in sport or in love, and generally with those who are after the same things; and it is therefore these whom we are bound to envy beyond all others. Hence the saying:

Potter against potter.

We also envy those whose possession of or success in a thing is a reproach to us: these are our neighbours and equals; for it is clear that it is our own fault we have missed the good thing in question; this annoys us, and excites envy in us. We also envy those who have what we ought to have, or have got what we did have once. Hence old men envy younger men, and those who have spent much envy those who have spent little on the same thing. And men who have not got a thing, or not got it yet, envy those who have got it quickly. We can also see what things and what persons give pleasure to envious people, and in what states of mind they feel it: the states of mind in which they feel pain are those under which they will feel pleasure in the contrary things. If therefore we ourselves with whom the decision rests are put into an envious state of mind, and those for whom our pity, or the award of something desirable, is claimed are such as have been described, it is obvious that they will win no pity from us.

11

We will next consider Emulation, showing in what follows its causes and objects, and the state of mind in which it is felt. Emulation is pain caused by seeing the presence, in persons whose nature is like our own, of good things that are highly valued and are possible for ourselves to acquire; but it is felt not because others have these goods, but because we have not got them ourselves. It is therefore a good feeling felt by good persons, whereas envy is a bad feeling felt by bad persons. Emulation makes us take steps to secure the good things in question, envy makes us take steps to stop our neighbour having them. Emulation must therefore tend to be felt by persons who believe themselves to deserve certain good things that they have not got, it being understood that no one aspires to things which appear impossible. It is accordingly felt by the young and by persons of lofty disposition. Also by those who possess such good things as are deserved by men held in honour—these are wealth, abundance of friends, public office, and the like;

on the assumption that they ought to be good men, they are emulous to gain such goods because they ought, in their belief, to belong to men whose state of mind is good. Also by those whom all others think deserving. We also feel it about anything for which our ancestors, relatives, personal friends, race, or country are specially honoured, looking upon that thing as really our own, and therefore feeling that we deserve to have it. Further, since all good things that are highly honoured are objects of emulation, moral goodness in its various forms must be such an object, and also all those good things that are useful and serviceable to others: for men honour those who are morally good, and also those who do them service. So with those good things our possession of which can give enjoyment to our neighbours—wealth and beauty rather than health. We can see, too, what persons are the objects of the feeling. They are those who have these and similar things—those already mentioned, as courage, wisdom, public office. Holders of public office—generals, orators, and all who possess such powers—can do many people a good turn. Also those whom many people wish to be like; those who have many acquaintances or friends; those whom admire, or whom we ourselves admire; and those who have been praised and eulogized by poets or prose-writers. Persons of the contrary sort are objects of contempt: for the feeling and notion of contempt are opposite to those of emulation. Those who are such as to emulate or be emulated by others are inevitably disposed to be contemptuous of all such persons as are subject to those bad things which are contrary to the good things that are the objects of emulation: despising them for just that reason. Hence we often despise the fortunate, when luck comes to them without their having those good things which are held in honour.

This completes our discussion of the means by which the several emotions may be produced or dissipated, and upon which depend the persuasive arguments connected with the emotions.

12

Let us now consider the various types of human character, in relation to the emotions and moral qualities, showing how they correspond to our various ages and fortunes. By emotions I mean anger, desire, and the like; these we have discussed already. By moral qualities I mean virtues and vices; these also have been discussed already, as well as the various things that various types of men tend to will and to do. By ages I mean youth, the prime of life, and old age. By fortune I mean birth, wealth, power, and their opposites—in fact, good fortune and ill fortune.

To begin with the Youthful type of character. Young men have strong passions, and tend to gratify them indiscriminately. Of the bodily desires, it is the sexual by which they are most swayed and in which they show absence of self-control. They are changeable and fickle in their desires, which are violent while they last, but quickly over: their impulses are keen but not deep-rooted, and are like sick people's attacks of hunger and thirst. They are hot-tempered, and quick-tempered, and apt to give way to their anger; bad temper often gets the better of them, for owing to their love of honour they cannot bear being slighted, and are indignant if they imagine themselves unfairly treated. While they love honour, they love victory still more; for youth is eager for superiority over others, and victory is one form of this. They love both more than they love money, which indeed they love very little, not having yet learnt what it means to be without it—this is the point of Pittacus' remark about Amphiaraus. They look at the good side rather than the bad, not having yet witnessed many instances of wickedness. They trust others readily, because they have not yet often been cheated. They are sanguine; nature warms their blood as though with excess of wine; and besides that, they have as yet met with few disappointments. Their lives are mainly spent not in memory but in expectation; for expectation refers to the future, memory to the past, and youth has a long future before it and a short past behind it: on the first day of one's life one has nothing at all to remember, and can only look forward. They are easily cheated, owing to the sanguine disposition just mentioned. Their hot tempers and hopeful dispositions make them more courageous than older men are; the hot temper prevents fear, and the hopeful disposition creates confidence; we cannot feel fear so long as we are feeling angry, and any expectation of good makes us confident. They are shy, accepting the rules of society in which they have been trained, and not yet believing in any other standard of honour. They have exalted notions, because they have not yet been humbled by life or learnt its necessary limitations; moreover, their hopeful disposition makes them think themselves equal to great things—and that means having exalted notions. They would always rather do noble deeds than useful ones: their lives are regulated more by moral feeling than by reasoning; and whereas reasoning leads us to choose what is useful, moral goodness leads us to choose what is noble. They are fonder of their friends, intimates, and companions than older men are, because they like spending their days in the company of others, and have not yet come to value either their friends or anything else by their usefulness to themselves. All their mistakes are in the direction of doing things excessively and vehemently. They disobey Chilon's precept by overdoing everything, they love too much and hate too much, and the same thing with everything else. They think they know everything, and are always quite sure about it; this, in fact, is why they overdo everything. If they do wrong to others, it is because they mean to insult them, not to do them actual harm. They are ready to pity others, because they think every one an honest man, or anyhow better than he is: they judge their neighbour by their own harmless natures, and so cannot think he deserves to be treated in that way. They are fond of fun and therefore witty, wit being well-bred insolence.

13

Such, then is the character of the Young. The character of Elderly Men—men who are past their prime—may be said to be formed for the most part of elements that are the contrary of all these. They have lived many years; they have often been taken in, and often made mistakes; and life on the whole is a bad business. The result is that they are sure about nothing and under-do everything. They 'think', but they never 'know'; and because of their hesitation they always add a 'possibly' or a 'perhaps', putting everything this way and nothing positively. They are cynical; that is, they tend to put the worse construction on everything. Further, their experience makes them distrustful and therefore suspicious of evil. Consequently they neither love warmly nor hate bitterly, but following the hint of Bias they love as though they will some day hate and hate as though they will some day love. They are small-minded, because they have been humbled by life: their desires are set upon nothing more exalted or unusual than what will help them to keep alive. They are not generous, because money is one of the things they must have, and at the same time their experience has taught them how hard it is to get and how easy to lose. They are cowardly, and are always anticipating danger; unlike that of the young, who are warm-blooded, their temperament is chilly; old age has paved the way for cowardice; fear is, in fact, a form of chill. They love life; and all the more when their last day has come, because the object of all desire is something we have not got, and also because we desire most strongly that which we need most urgently. They are too fond of themselves; this is one form that small-mindedness takes. Because of this, they guide their lives too much by considerations of what is useful and too little by what is noble—for the useful is what is good for oneself, and the noble what is good absolutely. They are not shy, but shameless rather; caring less for what is noble than for what is useful, they feel contempt for what people may think of them. They lack confidence in the future; partly through experience—for most things go wrong, or anyhow turn out worse than one expects; and partly because of their cowardice. They live by memory rather than by hope; for what is left to them of life is but little as compared with the long past; and hope is of the future, memory of the past. This, again, is the cause of their loquacity; they are continually talking of the past, because they enjoy remembering it. Their fits of anger are sudden but feeble. Their sensual passions have either altogether gone or have lost their vigour: consequently they do not feel their passions much, and their actions are inspired less by what they do feel than by the love of gain. Hence men at this time of life are often supposed to have a self-controlled character; the fact is that their passions have slackened, and they are slaves to the love of gain. They guide their lives by reasoning more than by moral feeling; reasoning being directed to utility and moral feeling to moral goodness. If they wrong others, they mean to injure them, not to insult them. Old men may feel pity, as well as young men, but not for the same reason. Young men feel it out of kindness; old men out of weakness, imagining that anything that befalls any one else might easily happen to them, which, as we saw, is a thought that excites pity. Hence they are querulous, and not disposed to jesting or laughter—the love of laughter being the very opposite of querulousness.

Such are the characters of Young Men and Elderly Men. People always think well of speeches adapted to, and reflecting, their own character: and we can now see how to compose our speeches so as to adapt both them and ourselves to our audiences.

14

As for Men in their Prime, clearly we shall find that they have a character between that of the young and that of the old, free from the extremes of either. They have neither that excess of confidence which amounts to rashness, nor too much timidity, but the right amount of each. They neither trust everybody nor distrust everybody, but judge people correctly. Their lives will be guided not by the sole consideration either of what is noble or of what is useful, but by both; neither by parsimony nor by prodigality, but by what is fit and proper. So, too, in regard to anger and desire; they will be brave as well as temperate, and temperate as well as brave; these virtues are divided between the young and the old; the young are brave but intemperate, the old temperate but cowardly. To put it generally, all the valuable qualities that youth and age divide between them are united in the prime of life, while all their excesses or defects are replaced by moderation and fitness. The body is in its prime from thirty to five-and-thirty; the mind about forty-nine.

15

So much for the types of character that distinguish youth, old age, and the prime of life. We will now turn to those Gifts of Fortune by which human character is affected. First let us consider Good Birth. Its effect on character is to make those who have it more ambitious; it is the way of all men who have something to start with to add to the pile, and good birth implies ancestral distinction. The well-born man will look down even on those who are as good as his own ancestors, because any far-off distinction is greater than the same thing close to us, and better to boast about. Being well-born, which means coming of a fine stock, must be distinguished from nobility, which means being true to the family nature—a quality not usually found in the well-born, most of whom are poor creatures. In the generations of men as in the fruits of the earth, there is a varying yield; now and then, where the stock is good, exceptional men are produced for a while, and then decadence sets in. A clever stock will degenerate towards the insane type of character, like the descendants of Alcibiades or of the elder Dionysius; a steady stock towards the fatuous and torpid type, like the descendants of Cimon, Pericles, and Socrates.

16

The type of character produced by Wealth lies on the surface for all to see. Wealthy men are insolent and arrogant; their possession of wealth affects their understanding; they feel as if they had every good thing that exists; wealth becomes a sort of standard of value for everything else, and therefore they imagine there is nothing it cannot buy. They are luxurious and ostentatious; luxurious, because of the luxury in which they live and the prosperity which they display; ostentatious and vulgar, because, like other people's, their minds are regularly occupied with the object of their love and admiration, and also because they think that other people's idea of happiness is the same as their own. It is indeed quite natural that they should be affected thus; for if you have money, there are always plenty of people who come begging from you. Hence the saying of Simonides about wise men and rich men, in answer to Hiero's wife, who asked him whether it was better to grow rich or wise. 'Why, rich,' he said; 'for I see the wise men spending their days at the rich men's doors.' Rich men also consider themselves worthy to hold public office; for they consider they already have the things that give a claim to office. In a word, the type of character produced by wealth is that of a prosperous fool. There is indeed one difference between the type of the newly-enriched and those who have long been rich: the newly-enriched have all the bad qualities mentioned in an exaggerated and worse form—to be newly-enriched means, so to speak, no education in riches. The wrongs they do others are not meant to injure their victims, but spring from insolence or self-indulgence, e.g. those that end in assault or in adultery.

17

As to Power: here too it may fairly be said that the type of character it produces is mostly obvious enough. Some elements in this type it shares with the wealthy type, others are better. Those in power are more ambitious and more manly in character than the wealthy, because they aspire to do the great deeds that their power permits them to do. Responsibility makes them more serious: they have to keep paying attention to the duties their position involves. They are dignified rather than arrogant, for the respect in which they are held inspires them with dignity and therefore with moderation—dignity being a mild and becoming form of arrogance. If they wrong others, they wrong them not on a small but on a great scale.

Good fortune in certain of its branches produces the types of character belonging to the conditions just described, since these conditions are in fact more or less the kinds of good fortune that are regarded as most important. It may be added that good fortune leads us to gain all we can in the way of family happiness and bodily advantages. It does indeed make men more supercilious and more reckless; but there is one excellent quality that goes with it—piety, and respect for the divine power, in which they believe because of events which are really the result of chance.

This account of the types of character that correspond to differences of age or fortune may end here; for to arrive at the opposite types to those described, namely, those of the poor, the unfortunate, and the powerless, we have only to ask what the opposite qualities are.

18

The use of persuasive speech is to lead to decisions. (When we know a thing, and have decided about it, there is no further use in speaking about it.) This is so even if one is addressing a single person and urging him to do or not to do something, as when we scold a man for his conduct or try to change his views: the single person is as much your 'judge' as if he were one of many; we may say, without qualification, that any one is your judge whom you have to persuade. Nor does it matter whether we are arguing against an actual opponent or against a mere proposition; in the latter case we still have to use speech and overthrow the opposing arguments, and we attack these as we should attack an actual opponent. Our principle holds good of ceremonial speeches also; the 'onlookers' for whom such a speech is put together are treated as the judges of it. Broadly speaking, however, the only sort of person who can strictly be called a judge is the man who decides the issue in some matter of public controversy; that is, in law suits and in political debates, in both of which there are issues to be decided. In the section on political oratory an account has already been given of the types of character that mark the different constitutions.

The manner and means of investing speeches with moral character may now be regarded as fully set forth.

Each of the main divisions of oratory has, we have seen, its own distinct purpose. With regard to each division, we have noted the accepted views and propositions upon which we may base our arguments—for political, for ceremonial, and for forensic speaking. We have further determined completely by what means speeches may be invested with the required moral character. We are now to proceed to discuss the arguments common to all oratory. All orators, besides their special lines of argument, are bound to use, for instance, the topic of the Possible and Impossible; and to try to show that a thing has happened, or will happen in future. Again, the topic of Size is common to all oratory; all of us have to argue that things are bigger or smaller than they seem, whether we are making political speeches, speeches of eulogy or attack, or prosecuting or defending in the law-courts. Having analysed these subjects, we will try to say what we can about the general principles of arguing by 'enthymeme' and 'example', by the addition of which we may hope to complete the project with which we set out. Of the above—mentioned general lines of argument, that concerned with Amplification is—as has been already said—most appropriate to ceremonial speeches; that concerned with the Past, to forensic speeches, where the required decision is always about the past; that concerned with Possibility and the Future, to political speeches.

19

Let us first speak of the Possible and Impossible. It may plausibly be argued: That if it is possible for one of a pair of contraries to be or happen, then it is possible for the other: e.g. if a man can be cured, he can also fall ill; for any two contraries are equally possible, in so far as they are contraries. That if of two similar things one is possible, so is the other. That if the harder of two things is possible, so is the easier. That if a thing can come into existence in a good and beautiful form, then it can come into existence generally; thus a house can exist more easily than a beautiful house. That if the beginning of a thing can occur, so can the end; for nothing impossible occurs or begins to occur; thus the commensurability of the diagonal of a square with its side neither occurs nor can begin to occur. That if the end is possible, so is the beginning; for all things that occur have a beginning. That if that which is posterior in essence or in order of generation can come into being, so can that which is prior: thus if a man can come into being, so can a boy, since the boy comes first in order of generation; and if a boy can, so can a man, for the man also is first. That those things are possible of which the love or desire is natural; for no one, as a rule, loves or desires impossibilities. That things which are the object of any kind of science or art are possible and exist or come into existence. That anything is possible the first step in whose production depends on men or things which we can compel or persuade to produce it, by our greater strength, our control of them, or our friendship with them. That where the parts are possible, the whole is possible; and where the whole is possible, the parts are usually possible. For if the slit in front, the toe-piece, and the upper leather can be made, then shoes can be made; and if shoes, then also the front slit and toe-piece. That if a whole genus is a thing that can occur, so can the species; and if the species can occur, so can the genus: thus, if a sailing vessel can be made, so also can a trireme; and if a trireme, then a sailing vessel also. That if one of two things whose existence depends on each other is possible, so is the other; for instance, if 'double', then 'half', and if 'half', then 'double'. That if a thing can be produced without art or preparation, it can be produced still more certainly by the careful application of art to it. Hence Agathon has said:

To some things we by art must needs attain,
Others by destiny or luck we gain.

That if anything is possible to inferior, weaker, and stupider people, it is more so for their opposites; thus Isocrates said that it would be a strange thing if he could not discover a thing that Euthynus had found out. As for Impossibility, we can clearly get what we want by taking the contraries of the arguments stated above.

Questions of Past Fact may be looked at in the following ways: First, that if the less likely of two things has occurred, the more likely must have occurred also. That if one thing that usually follows another has happened, then that other thing has happened; that, for instance, if a man has forgotten a thing, he has also once learnt it. That if a man had the power and the wish to do a thing, he has done it; for every one does do whatever he intends to do whenever he can do it, there being nothing to stop him. That, further, he has done the thing in question either if he intended it and nothing external prevented him; or if he had the power to do it and was angry at the time; or if he had the power to do it and his heart was set upon it—for people as a rule do what they long to do, if they can; bad people through lack of self-control; good people, because their hearts are set upon good things. Again, that if a thing was 'going to happen', it has happened; if a man was 'going to do something', he has done it, for it is likely that the intention was carried out. That if one thing has happened which naturally happens before another or with a view to it, the other has happened; for instance, if it has lightened, it has also thundered; and if an action has been attempted, it has been done. That if one thing has happened which naturally happens after another, or with a view to which that other happens, then that other (that which happens first, or happens with a view to this thing) has also happened; thus, if it has thundered it has lightened, and if an action has been done it has been attempted. Of all these sequences some are inevitable and some merely usual. The arguments for the non-occurrence of anything can obviously be found by considering the opposites of those that have been mentioned.

How questions of Future Fact should be argued is clear from the same considerations: That a thing will be done if there is both the power and the wish to do it; or if along with the power to do it there is a craving for the result, or anger, or calculation, prompting it. That the thing will be done, in these cases, if the man is actually setting about it, or even if he means to do it later—for usually what we mean to do happens rather than what we do not mean to do. That a thing will happen if another thing which naturally happens before it has already happened; thus, if it is clouding over, it is likely to rain. That if the means to an end have occurred, then the end is likely to occur; thus, if there is a foundation, there will be a house.

For arguments about the Greatness and Smallness of things, the greater and the lesser, and generally great things and small, what we have already said will show the line to take. In discussing deliberative oratory we have spoken about the relative greatness of various goods, and about the greater and lesser in general. Since therefore in each type oratory the object under discussion is some kind of good—whether it is utility, nobleness, or justice—it is clear that every orator must obtain the materials of amplification through these channels. To go further than this, and try to establish abstract laws of greatness and superiority, is to argue without an object; in practical life, particular facts count more than generalizations.

Enough has now been said about these questions of possibility and the reverse, of past or future fact, and of the relative greatness or smallness of things.

20

The special forms of oratorical argument having now been discussed, we have next to treat of those which are common to all kinds of oratory. These are of two main kinds, 'Example' and 'Enthymeme'; for the 'Maxim' is part of an enthymeme.

We will first treat of argument by Example, for it has the nature of induction, which is the foundation of reasoning. This form of argument has two varieties; one consisting in the mention of actual past facts, the other in the invention of facts by the speaker. Of the latter, again, there are two varieties, the illustrative parallel and the fable (e.g. the fables of Aesop, those from Libya). As an instance of the mention of actual facts, take the following. The speaker may argue thus: 'We must prepare for war against the king of Persia and not let him subdue Egypt. For Darius of old did not cross the Aegean until he had seized Egypt; but once he had seized it, he did cross. And Xerxes, again, did not attack us until he had seized Egypt; but once he had seized it, he did cross. If therefore the present king seizes Egypt, he also will cross, and therefore we must not let him.'

The illustrative parallel is the sort of argument Socrates used: e.g. 'Public officials ought not to be selected by lot. That is like using the lot to select athletes, instead of choosing those who are fit for the contest; or using the lot to select a steersman from among a ship's crew, as if we ought to take the man on whom the lot falls, and not the man who knows most about it.'

Instances of the fable are that of Stesichorus about Phalaris, and that of Aesop in defence of the popular leader. When the people of Himera had made Phalaris military dictator, and were going to give him a bodyguard, Stesichorus wound up a long talk by telling them the fable of the horse who had a field all to himself. Presently there came a stag and began to spoil his pasturage. The horse, wishing to revenge himself on the stag, asked a man if he could help him to do so. The man said, 'Yes, if you will let me bridle you and get on to your back with javelins in my hand'. The horse agreed, and the man mounted; but instead of getting his revenge on the stag, the horse found himself the slave of the man. 'You too', said Stesichorus, 'take care lest your desire for revenge on your enemies, you meet the same fate as the horse. By making Phalaris military dictator, you have already let yourselves be bridled. If you let him get on to your backs by giving him a bodyguard, from that moment you will be his slaves.'

Aesop, defending before the assembly at Samos a poular leader who was being tried for his life, told this story: A fox, in crossing a river, was swept into a hole in the rocks; and, not being able to get out, suffered miseries for a long time through the swarms of fleas that fastened on her. A hedgehog, while roaming around, noticed the fox; and feeling sorry for her asked if he might remove the fleas. But the fox declined the offer; and when the hedgehog asked why, she replied, 'These fleas are by this time full of me and not sucking much blood; if you take them away, others will come with fresh appetites and drink up all the blood I have left.' 'So, men of Samos', said Aesop, 'my client will do you no further harm; he is wealthy already. But if you put him to death, others will come along who are not rich, and their peculations will empty your treasury completely.'

Fables are suitable for addresses to popular assemblies; and they have one advantage—they are comparatively easy to invent, whereas it is hard to find parallels among actual past events. You will in fact frame them just as you frame illustrative parallels: all you require is the power of thinking out your analogy, a power developed by intellectual training. But while it is easier to supply parallels by inventing fables, it is more valuable for the political speaker to supply them by quoting what has actually happened, since in most respects the future will be like what the past has been.

Where we are unable to argue by Enthymeme, we must try to demonstrate our point by this method of Example, and to convince our hearers thereby. If we can argue by Enthymeme, we should use our Examples as subsequent supplementary evidence. They should not precede the Enthymemes: that will give the argument an inductive air, which only rarely suits the conditions of speech-making. If they follow the enthymemes, they have the effect of witnesses giving evidence, and this alway tells. For the same reason, if you put your examples first you must give a large number of them; if you put them last, a single one is sufficient; even a single witness will serve if he is a good one. It has now been stated how many varieties of argument by Example there are, and how and when they are to be employed.

21

We now turn to the use of Maxims, in order to see upon what subjects and occasions, and for what kind of speaker, they will appropriately form part of a speech. This will appear most clearly when we have defined a maxim. It is a statement; not a particular fact, such as the character of lphicrates, but of a general kind; nor is it about any and every subject—e.g. 'straight is the contrary of curved' is not a maxim—but only about questions of practical conduct, courses of conduct to be chosen or avoided. Now an Enthymeme is a syllogism dealing with such practical subjects. It is therefore roughly true that the premisses or conclusions of Enthymemes, considered apart from the rest of the argument, are Maxims: e.g.

Never should any man whose wits are sound
Have his sons taught more wisdom than their fellows.

Here we have a Maxim; add the reason or explanation, and the whole thing is an Enthymeme; thus—

It makes them idle; and therewith they earn
Ill-will and jealousy throughout the city.

Again,

There is no man in all things prosperous,

and

There is no man among us all is free,

are maxims; but the latter, taken with what follows it, is an Enthymeme—

For all are slaves of money or of chance.

From this definition of a maxim it follows that there are four kinds of maxims. In the first Place, the maxim may or may not have a supplement. Proof is needed where the statement is paradoxical or disputable; no supplement is wanted where the statement contains nothing paradoxical, either because the view expressed is already a known truth, e.g.

Chiefest of blessings is health for a man, as it seemeth to me,

this being the general opinion: or because, as soon as the view is stated, it is clear at a glance, e.g.

No love is true save that which loves for ever.

Of the Maxims that do have a supplement attached, some are part of an Enthymeme, e.g.

Never should any man whose wits are sound, &c.

Others have the essential character of Enthymemes, but are not stated as parts of Enthymemes; these latter are reckoned the best; they are those in which the reason for the view expressed is simply implied, e.g.

O mortal man, nurse not immortal wrath.

To say 'it is not right to nurse immortal wrath' is a maxim; the added words 'mortal man' give the reason. Similarly, with the words Mortal creatures ought to cherish mortal, not immortal thoughts.

What has been said has shown us how many kinds of Maxims there are, and to what subjects the various kinds are appropriate. They must not be given without supplement if they express disputed or paradoxical views: we must, in that case, either put the supplement first and make a maxim of the conclusion, e.g. you might say, 'For my part, since both unpopularity and idleness are undesirable, I hold that it is better not to be educated'; or you may say this first, and then add the previous clause. Where a statement, without being paradoxical, is not obviously true, the reason should be added as concisely as possible. In such cases both laconic and enigmatic sayings are suitable: thus one might say what Stesichorus said to the Locrians, 'Insolence is better avoided, lest the cicalas chirp on the ground'.

The use of Maxims is appropriate only to elderly men, and in handling subjects in which the speaker is experienced. For a young man to use them is—like telling stories— unbecoming; to use them in handling things in which one has no experience is silly and ill-bred: a fact sufficiently proved by the special fondness of country fellows for striking out maxims, and their readiness to air them.

To declare a thing to be universally true when it is not is most appropriate when working up feelings of horror and indignation in our hearers; especially by way of preface, or after the facts have been proved. Even hackneyed and commonplace maxims are to be used, if they suit one's purpose: just because they are commonplace, every one seems to agree with them, and therefore they are taken for truth. Thus, any one who is calling on his men to risk an engagement without obtaining favourable omens may quote

One omen of all is hest, that we fight for our fatherland.

Or, if he is calling on them to attack a stronger force—
  The War-God showeth no favour.
Or, if he is urging people to destroy the innocent children of their enemies—
  Fool, who slayeth the father and leaveth his sons to avenge him.
    Some proverbs are also maxims, e.g. the proverb 'An Attic neighbour'. You are not to avoid uttering maxims that contradict such sayings as have become public property (I mean such sayings as 'know thyself' and 'nothing in excess') if doing so will raise your hearers' opinion of your character, or convey an effect of strong emotion—e.g. an angry speaker might well say, 'It is not true that we ought to know ourselves: anyhow, if this man had known himself, he would never have thought himself fit for an army command.' It will raise people's opinion of our character to say, for instance, 'We ought not to follow the saying that bids us treat our friends as future enemies: much better to treat our enemies as future friends.' The moral purpose should be implied partly by the very wording of our maxim. Failing this, we should add our reason: e.g. having said 'We should treat our friends, not as the saying advises, but as if they were going to be our friends always', we should add 'for the other behaviour is that of a traitor': or we might put it, I disapprove of that saying. A true friend will treat his friend as if he were going to be his friend for ever'; and again, 'Nor do I approve of the saying "nothing in excess": we are bound to hate bad men excessively.' One great advantage of Maxims to a speaker is due to the want of intelligence in his hearers, who love to hear him succeed in expressing as a universal truth the opinions which they hold themselves about particular cases. I will explain what I mean by this, indicating at the same time how we are to hunt down the maxims required. The maxim, as has been already said, a general statement and people love to hear stated in general terms what they already believe in some particular connexion: e.g. if a man happens to have bad neighbours or bad children, he will agree with any one who tells him, 'Nothing is more annoying than having neighbours', or, 'Nothing is more foolish than to be the parent of children.' The orator has therefore to guess the subjects on which his hearers really hold views already, and what those views are, and then must express, as general truths, these same views on these same subjects. This is one advantage of using maxims. There is another which is more important—it invests a speech with moral character. There is moral character in every speech in which the moral purpose is conspicuous: and maxims always produce this effect, because the utterance of them amounts to a general declaration of moral principles: so that, if the maxims are sound, they display the speaker as a man of sound moral character. So much for the Maxim—its nature, varieties, proper use, and advantages.

22

    We now come to the Enthymemes, and will begin the subject with some general consideration of the proper way of looking for them, and then proceed to what is a distinct question, the lines of argument to be embodied in them. It has already been pointed out that the Enthymeme is a syllogism, and in what sense it is so. We have also noted the differences between it and the syllogism of dialectic. Thus we must not carry its reasoning too far back, or the length of our argument will cause obscurity: nor must we put in all the steps that lead to our conclusion, or we shall waste words in saying what is manifest. It is this simplicity that makes the uneducated more effective than the educated when addressing popular audiences—makes them, as the poets tell us, 'charm the crowd's ears more finely'. Educated men lay down broad general principles; uneducated men argue from common knowledge and draw obvious conclusions. We must not, therefore, start from any and every accepted opinion, but only from those we have defined—those accepted by our judges or by those whose authority they recognize: and there must, moreover, be no doubt in the minds of most, if not all, of our judges that the opinions put forward really are of this sort. We should also base our arguments upon probabilities as well as upon certainties.

The first thing we have to remember is this. Whether our argument concerns public affairs or some other subject, we must know some, if not all, of the facts about the subject on which we are to speak and argue. Otherwise we can have no materials out of which to construct arguments. I mean, for instance, how could we advise the Athenians whether they should go to war or not, if we did not know their strength, whether it was naval or military or both, and how great it is; what their revenues amount to; who their friends and enemies are; what wars, too, they have waged, and with what success; and so on? Or how could we eulogize them if we knew nothing about the sea-fight at Salamis, or the battle of Marathon, or what they did for the Heracleidae, or any other facts like that? All eulogy is based upon the noble deeds—real or imaginary—that stand to the credit of those eulogized. On the same principle, invectives are based on facts of the opposite kind: the orator looks to see what base deeds—real or imaginary—stand to the discredit of those he is attacking, such as treachery to the cause of Hellenic freedom, or the enslavement of their gallant allies against the barbarians (Aegina, Potidaea, &c.), or any other misdeeds of this kind that are recorded against them. So, too, in a court of law: whether we are prosecuting or defending, we must pay attention to the existing facts of the case. It makes no difference whether the subject is the Lacedaemonians or the Athenians, a man or a god; we must do the same thing. Suppose it to be Achilles whom we are to advise, to praise or blame, to accuse or defend; here too we must take the facts, real or imaginary; these must be our material, whether we are to praise or blame him for the noble or base deeds he has done, to accuse or defend him for his just or unjust treatment of others, or to advise him about what is or is not to his interest. The same thing applies to any subject whatever. Thus, in handling the question whether justice is or is not a good, we must start with the real facts about justice and goodness. We see, then, that this is the only way in which any one ever proves anything, whether his arguments are strictly cogent or not: not all facts can form his basis, but only those that bear on the matter in hand: nor, plainly, can proof be effected otherwise by means of the speech. Consequently, as appears in the Topics, we must first of all have by us a selection of arguments about questions that may arise and are suitable for us to handle; and then we must try to think out arguments of the same type for special needs as they emerge; not vaguely and indefinitely, but by keeping our eyes on the actual facts of the subject we have to speak on, and gathering in as many of them as we can that bear closely upon it: for the more actual facts we have at our command, the more easily we prove our case; and the more closely they bear on the subject, the more they will seem to belong to that speech only instead of being commonplaces. By 'commonplaces' I mean, for example, eulogy of Achilles because he is a human being or a demi-god, or because he joined the expedition against Troy: these things are true of many others, so that this kind of eulogy applies no better to Achilles than to Diomede. The special facts here needed are those that are true of Achilles alone; such facts as that he slew Hector, the bravest of the Trojans, and Cycnus the invulnerable, who prevented all the Greeks from landing, and again that he was the youngest man who joined the expedition, and was not bound by oath to join it, and so on.

Here, again, we have our first principle of selection of Enthymemes—that which refers to the lines of argument selected. We will now consider the various elementary classes of enthymemes. (By an 'elementary class' of enthymeme I mean the same thing as a 'line of argument'.) We will begin, as we must begin, by observing that there are two kinds of enthymemes. One kind proves some affirmative or negative proposition; the other kind disproves one. The difference between the two kinds is the same as that between syllogistic proof and disproof in dialectic. The demonstrative enthymeme is formed by the conjunction of compatible propositions; the refutative, by the conjunction of incompatible propositions.

We may now be said to have in our hands the lines of argument for the various special subjects that it is useful or necessary to handle, having selected the propositions suitable in various cases. We have, in fact, already ascertained the lines of argument applicable to enthymemes about good and evil, the noble and the base, justice and injustice, and also to those about types of character, emotions, and moral qualities. Let us now lay hold of certain facts about the whole subject, considered from a different and more general point of view. In the course of our discussion we will take note of the distinction between lines of proof and lines of disproof: and also of those lines of argument used in what seems to be enthymemes, but are not, since they do not represent valid syllogisms. Having made all this clear, we will proceed to classify Objections and Refutations, showing how they can be brought to bear upon enthymemes.

23

1. One line of positive proof is based upon consideration of the opposite of the thing in question. Observe whether that opposite has the opposite quality. If it has not, you refute the original proposition; if it has, you establish it. E.g. 'Temperance is beneficial; for licentiousness is hurtful'. Or, as in the Messenian speech, 'If war is the cause of our present troubles, peace is what we need to put things right again'. Or—

For if not even evil-doers should
Anger us if they meant not what they did,
Then can we owe no gratitude to such
As were constrained to do the good they did us.
    Or—
Since in this world liars may win belief,
Be sure of the opposite likewise—that this world
Hears many a true word and believes it not.

2. Another line of proof is got by considering some modification of the key-word, and arguing that what can or cannot be said of the one, can or cannot be said of the other: e.g. 'just' does not always mean 'beneficial', or 'justly' would always mean 'beneficially', whereas it is not desirable to be justly put to death.

3. Another line of proof is based upon correlative ideas. If it is true that one man noble or just treatment to another, you argue that the other must have received noble or just treatment; or that where it is right to command obedience, it must have been right to obey the command. Thus Diomedon, the tax-farmer, said of the taxes: 'If it is no disgrace for you to sell them, it is no disgrace for us to buy them'. Further, if 'well' or 'justly' is true of the person to whom a thing is done, you argue that it is true of the doer. But it is possible to draw a false conclusion here. It may be just that A should be treated in a certain way, and yet not just that he should be so treated by B. Hence you must ask yourself two distinct questions: (1) Is it right that A should be thus treated? (2) Is it right that B should thus treat him? and apply your results properly, according as your answers are Yes or No. Sometimes in such a case the two answers differ: you may quite easily have a position like that in the Alcmaeon of Theodectes:

And was there none to loathe thy mother's crime?

to which question Alcmaeon in reply says,

Why, there are two things to examine here.

And when Alphesiboea asks what he means, he rejoins:

They judged her fit to die, not me to slay her.

Again there is the lawsuit about Demosthenes and the men who killed Nicanor; as they were judged to have killed him justly, it was thought that he was killed justly. And in the case of the man who was killed at Thebes, the judges were requested to decide whether it was unjust that he should be killed, since if it was not, it was argued that it could not have been unjust to kill him.

4. Another line of proof is the 'a fortiori'. Thus it may be argued that if even the gods are not omniscient, certainly human beings are not. The principle here is that, if a quality does not in fact exist where it is more likely to exist, it clearly does not exist where it is less likely. Again, the argument that a man who strikes his father also strikes his neighbours follows from the principle that, if the less likely thing is true, the more likely thing is true also; for a man is less likely to strike his father than to strike his neighbours. The argument, then, may run thus. Or it may be urged that, if a thing is not true where it is more likely, it is not true where it is less likely; or that, if it is true where it is less likely, it is true where it is more likely: according as we have to show that a thing is or is not true. This argument might also be used in a case of parity, as in the lines:

Thou hast pity for thy sire, who has lost his sons:
Hast none for Oeneus, whose brave son is dead?

And, again, 'if Theseus did no wrong, neither did Paris'; or 'the sons of Tyndareus did no wrong, neither did Paris'; or 'if Hector did well to slay Patroclus, Paris did well to slay Achilles'. And 'if other followers of an art are not bad men, neither are philosophers'. And 'if generals are not bad men because it often happens that they are condemned to death, neither are sophists'. And the remark that 'if each individual among you ought to think of his own city's reputation, you ought all to think of the reputation of Greece as a whole'.

5. Another line of argument is based on considerations of time. Thus Iphicrates, in the case against Harmodius, said, 'if before doing the deed I had bargained that, if I did it, I should have a statue, you would have given me one. Will you not give me one now that I have done the deed? You must not make promises when you are expecting a thing to be done for you, and refuse to fulfil them when the thing has been done.' And, again, to induce the Thebans to let Philip pass through their territory into Attica, it was argued that 'if he had insisted on this before he helped them against the Phocians, they would have promised to do it. It is monstrous, therefore, that just because he threw away his advantage then, and trusted their honour, they should not let him pass through now'.

6. Another line is to apply to the other speaker what he has said against yourself. It is an excellent turn to give to a debate, as may be seen in the Teucer. It was employed by Iphicrates in his reply to Aristophon. 'Would you', he asked, 'take a bribe to betray the fleet?' 'No', said Aristophon; and Iphicrates replied, 'Very good: if you, who are Aristophon, would not betray the fleet, would I, who am Iphicrates?' Only, it must be recognized beforehand that the other man is more likely than you are to commit the crime in question. Otherwise you will make yourself ridiculous; it is Aristeides who is prosecuting, you cannot say that sort of thing to him. The purpose is to discredit the prosecutor, who as a rule would have it appear that his character is better than that of the defendant, a pretension which it is desirable to upset. But the use of such an argument is in all cases ridiculous if you are attacking others for what you do or would do yourself, or are urging others to do what you neither do nor would do yourself.

7. Another line of proof is secured by defining your terms. Thus, 'What is the supernatural? Surely it is either a god or the work of a god. Well, any one who believes that the work of a god exists, cannot help also believing that gods exist.' Or take the argument of Iphicrates, 'Goodness is true nobility; neither Harmodius nor Aristogeiton had any nobility before they did a noble deed'. He also argued that he himself was more akin to Harmodius and Aristogeiton than his opponent was. 'At any rate, my deeds are more akin to those of Harmodius and Aristogeiton than yours are'. Another example may be found in the Alexander. 'Every one will agree that by incontinent people we mean those who are not satisfied with the enjoyment of one love.' A further example is to be found in the reason given by Socrates for not going to the court of Archelaus. He said that 'one is insulted by being unable to requite benefits, as well as by being unable to requite injuries'. All the persons mentioned define their term and get at its essential meaning, and then use the result when reasoning on the point at issue.

8. Another line of argument is founded upon the various senses of a word. Such a word is 'rightly', as has been explained in the Topics. Another line is based upon logical division. Thus, 'All men do wrong from one of three motives, A, B, or C: in my case A and B are out of the question, and even the accusers do not allege C'.

10. Another line is based upon induction. Thus from the case of the woman of Peparethus it might be argued that women everywhere can settle correctly the facts about their children. Another example of this occurred at Athens in the case between the orator Mantias and his son, when the boy's mother revealed the true facts: and yet another at Thebes, in the case between Ismenias and Stilbon, when Dodonis proved that it was Ismenias who was the father of her son Thettaliscus, and he was in consequence always regarded as being so. A further instance of induction may be taken from the Law of Theodectes: 'If we do not hand over our horses to the care of men who have mishandled other people's horses, nor ships to those who have wrecked other people's ships, and if this is true of everything else alike, then men who have failed to secure other people's safety are not to be employed to secure our own.' Another instance is the argument of Alcidamas: 'Every one honours the wise'. Thus the Parians have honoured Archilochus, in spite of his bitter tongue; the Chians Homer, though he was not their countryman; the Mytilenaeans Sappho, though she was a woman; the Lacedaemonians actually made Chilon a member of their senate, though they are the least literary of men; the Italian Greeks honoured Pythagoras; the inhabitants of Lampsacus gave public burial to Anaxagoras, though he was an alien, and honour him even to this day. (It may be argued that peoples for whom philosophers legislate are always prosperous) on the ground that the Athenians became prosperous under Solon's laws and the Lacedaemonians under those of Lycurgus, while at Thebes no sooner did the leading men become philosophers than the country began to prosper.

11. Another line of argument is founded upon some decision already pronounced, whether on the same subject or on one like it or contrary to it. Such

a proof is most effective if every one has always decided thus; but if not every one, then at any rate most people; or if all, or most, wise or good men have thus decided, or the actual judges of the present question, or those whose authority they accept, or any one whose decision they cannot gainsay because he has complete control over them, or those whom it is not seemly to gainsay, as the gods, or one's father, or one's teachers. Thus Autocles said, when attacking Mixidemides, that it was a strange thing that the Dread Goddesses could without loss of dignity submit to the judgement of the Areopagus, and yet Mixidemides could not. Or as Sappho said, 'Death is an evil thing; the gods have so judged it, or they would die'. Or again as Aristippus said in reply to Plato when he spoke somewhat too dogmatically, as Aristippus thought: 'Well, anyhow, our friend', meaning Socrates, 'never spoke like that'. And Hegesippus, having previously consulted Zeus at Olympia, asked Apollo at Delphi 'whether his opinion was the same as his father's', implying that it would be shameful for him to contradict his father. Thus too Isocrates argued that Helen must have been a good woman, because Theseus decided that she was; and Paris a good man, because the goddesses chose him before all others; and Evagoras also, says Isocrates, was good, since when Conon met with his misfortune he betook himself to Evagoras without trying any one else on the way.

12. Another line of argument consists in taking separately the parts of a subject. Such is that given in the Topics: 'What sort of motion is the soul? for it must be this or that.' The Socrates of Theodectes provides an example: 'What temple has he profaned? What gods recognized by the state has he not honoured?'

13. Since it happens that any given thing usually has both good and bad consequences, another line of argument consists in using those consequences as a reason for urging that a thing should or should not be done, for prosecuting or defending any one, for eulogy or censure. E.g. education leads both to unpopularity, which is bad, and to wisdom, which is good. Hence you either argue, 'It is therefore not well to be educated, since it is not well to be unpopular': or you answer, 'No, it is well to be educated, since it is well to be wise'. The Art of Rhetoric of Callippus is made up of this line of argument, with the addition of those of Possibility and the others of that kind already described.

14. Another line of argument is used when we have to urge or discourage a course of action that may be done in either of two opposite ways, and have to apply the method just mentioned to both. The difference between this one and the last is that, whereas in the last any two things are contrasted, here the things contrasted are opposites. For instance, the priestess enjoined upon her son not to take to public speaking: 'For', she said, 'if you say what is right, men will hate you; if you say what is wrong, the gods will hate you.' The reply might be, 'On the contrary, you ought to take to public speaking: for if you say what is right the gods will love you; if you say what is wrong, men will love you.' This amounts to the proverbial 'buying the marsh with the salt'. It is just this situation, viz. when each of two opposites has both a good and a bad consequence opposite respectively to

each other, that has been termed divarication.

15. Another line of argument is this: The things people approve of openly are not those which they approve of secretly: openly, their chief praise is given to justice and nobleness; but in their hearts they prefer their own advantage. Try, in face of this, to establish the point of view which your opponent has not adopted. This is the most effective of the forms of argument that contradict common opinion.

16. Another line is that of rational correspondence. E.g. Iphicrates, when they were trying to compel his son, a youth under the prescribed age, to perform one of the state duties because he was tall, said 'If you count tall boys men, you will next be voting short men boys'. And Theodectes in his Law said, 'You make citizens of such mercenaries as Strabax and Charidemus, as a reward of their merits; will you not make exiles of such citizens as those who have done irreparable harm among the mercenaries?'

17. Another line is the argument that if two results are the same their antecedents are also the same. For instance, it was a saying of Xenophanes that to assert that the gods had birth is as impious as to say that they die; the consequence of both statements is that there is a time when the gods do not exist. This line of proof assumes generally that the result of any given thing is always the same: e.g. 'you are going to decide not about Isocrates, but about the value of the whole profession of philosophy.' Or, 'to give earth and water' means slavery; or, 'to share in the Common Peace' means obeying orders. We are to make either such assumptions or their opposite, as suits us best.

18. Another line of argument is based on the fact that men do not always make the same choice on a later as on an earlier occasion, but reverse their previous choice. E.g. the following enthymeme: 'When we were exiles, we fought in order to return; now we have returned, it would be strange to choose exile in order not to have to fight.' one occasion, that is, they chose to be true to their homes at the cost of fighting, and on the other to avoid fighting at the cost of deserting their homes.

19. Another line of argument is the assertion that some possible motive for an event or state of things is the real one: e.g. that a gift was given in order to cause pain by its withdrawal. This notion underlies the lines:

God gives to many great prosperity,
Not of good God towards them, but to make
The ruin of them more conspicuous.

Or take the passage from the Meleager of Antiphon:

To slay no boar, but to be witnesses
Of Meleager's prowess unto Greece.

Or the argument in the Ajax of Theodectes, that Diomede chose out Odysseus not to do him honour, but in order that his companion might be a lesser man than himself—such a motive for doing so is quite possible.

20. Another line of argument is common to forensic and deliberative oratory,

namely, to consider inducements and deterrents, and the motives people have for doing or avoiding the actions in question. These are the conditions which make us bound to act if they are for us, and to refrain from action if they are against us: that is, we are bound to act if the action is possible, easy, and useful to ourselves or our friends or hurtful to our enemies; this is true even if the action entails loss, provided the loss is outweighed by the solid advantage. A speaker will urge action by pointing to such conditions, and discourage it by pointing to the opposite. These same arguments also form the materials for accusation or defence—the deterrents being pointed out by the defence, and the inducements by the prosecution. As for the defence,...This topic forms the whole Art of Rhetoric both of Pamphilus and of Callippus.

21. Another line of argument refers to things which are supposed to happen and yet seem incredible. We may argue that people could not have believed them, if they had not been true or nearly true: even that they are the more likely to be true because they are incredible. For the things which men believe are either facts or probabilities: if, therefore, a thing that is believed is improbable and even incredible, it must be true, since it is certainly not believed because it is at all probable or credible. An example is what Androcles of the deme Pitthus said in his well-known arraignment of the law. The audience tried to shout him down when he observed that the laws required a law to set them right. 'Why', he went on, 'fish need salt, improbable and incredible as this might seem for creatures reared in salt water; and olive-cakes need oil, incredible as it is that what produces oil should need it.'

22. Another line of argument is to refute our opponent's case by noting any contrasts or contradictions of dates, acts, or words that it anywhere displays; and this in any of the three following connexions. (1) Referring to our opponent's conduct, e.g. 'He says he is devoted to you, yet he conspired with the Thirty.' (2) Referring to our own conduct, e.g. 'He says I am litigious, and yet he cannot prove that I have been engaged in a single lawsuit.' (3) Referring to both of us together, e.g. 'He has never even lent any one a penny, but I have ransomed quite a number of you.'

23. Another line that is useful for men and causes that have been really or seemingly slandered, is to show why the facts are not as supposed; pointing out that there is a reason for the false impression given. Thus a woman, who had palmed off her son on another woman, was thought to be the lad's mistress because she embraced him; but when her action was explained the charge was shown to be groundless. Another example is from the Ajax of Theodectes, where Odysseus tells Ajax the reason why, though he is really braver than Ajax, he is not thought so.

24. Another line of argument is to show that if the cause is present, the effect is present, and if absent, absent. For by proving the cause you at once prove the effect, and conversely nothing can exist without its cause. Thus Thrasybulus accused Leodamas of having had his name recorded as a criminal on the slab in

the Acropolis, and of erasing the record in the time of the Thirty Tyrants: to which Leodamas replied, 'Impossible: for the Thirty would have trusted me all the more if my quarrel with the commons had been inscribed on the slab.'

25. Another line is to consider whether the accused person can take or could have taken a better course than that which he is recommending or taking, or has taken. If he has not taken this better course, it is clear that he is not guilty, since no one deliberately and consciously chooses what is bad. This argument is, however, fallacious, for it often becomes clear after the event how the action could have been done better, though before the event this was far from clear.

26. Another line is, when a contemplated action is inconsistent with any past action, to examine them both together. Thus, when the people of Elea asked Xenophanes if they should or should not sacrifice to Leucothea and mourn for her, he advised them not to mourn for her if they thought her a goddess, and not to sacrifice to her if they thought her a mortal woman.

27. Another line is to make previous mistakes the grounds of accusation or defence. Thus, in the Medea of Carcinus the accusers allege that Medea has slain her children; 'at all events', they say, 'they are not to be seen'—Medea having made the mistake of sending her children away. In defence she argues that it is not her children, but Jason, whom she would have slain; for it would have been a mistake on her part not to do this if she had done the other. This special line of argument for enthymeme forms the whole of the Art of Rhetoric in use before Theodorus.

Another line is to draw meanings from names. Sophocles, for instance, says,

O steel in heart as thou art steel in name.

This line of argument is common in praises of the gods. Thus, too, Conon called Thrasybulus rash in counsel. And Herodicus said of Thrasymachus, 'You are always bold in battle'; of Polus, 'you are always a colt'; and of the legislator Draco that his laws were those not of a human being but of a dragon, so savage were they. And, in Euripides, Hecuba says of Aphrodite,

Her name and Folly's (aphrosuns) lightly begin alike,

and Chaeremon writes

Pentheus a name foreshadowing grief (penthos) to come.

The Refutative Enthymeme has a greater reputation than the Demonstrative, because within a small space it works out two opposing arguments, and arguments put side by side are clearer to the audience. But of all syllogisms, whether refutative or demonstrative, those are most applauded of which we foresee the conclusions from the beginning, so long as they are not obvious at first sight—for part of the pleasure we feel is at our own intelligent anticipation; or those which we follow well enough to see the point of them as soon as the last word has been uttered.

24

Besides genuine syllogisms, there may be syllogisms that look genuine but are not; and since an enthymeme is merely a syllogism of a particular kind, it follows that, besides genuine enthymemes, there may be those that look genuine but are not.

1. Among the lines of argument that form the Spurious Enthymeme the first is that which arises from the particular words employed.

(a) One variety of this is when—as in dialectic, without having gone through any reasoning process, we make a final statement as if it were the conclusion of such a process, 'Therefore so-and-so is not true', 'Therefore also so-and-so must be true'—so too in rhetoric a compact and antithetical utterance passes for an enthymeme, such language being the proper province of enthymeme, so that it is seemingly the form of wording here that causes the illusion mentioned. In order to produce the effect of genuine reasoning by our form of wording it is useful to summarize the results of a number of previous reasonings: as 'some he saved—others he avenged—the Greeks he freed'. Each of these statements has been previously proved from other facts; but the mere collocation of them gives the impression of establishing some fresh conclusion.

(b) Another variety is based on the use of similar words for different things; e.g. the argument that the mouse must be a noble creature, since it gives its name to the most august of all religious rites—for such the Mysteries are. Or one may introduce, into a eulogy of the dog, the dog-star; or Pan, because Pindar said:

O thou blessed one!
Thou whom they of Olympus call
The hound of manifold shape
That follows the Mother of Heaven:

or we may argue that, because there is much disgrace in there not being a dog about, there is honour in being a dog. Or that Hermes is readier than any other god to go shares, since we never say 'shares all round' except of him. Or that speech is a very excellent thing, since good men are not said to be worth money but to be worthy of esteem—the phrase 'worthy of esteem' also having the meaning of 'worth speech'.

2. Another line is to assert of the whole what is true of the parts, or of the parts what is true of the whole. A whole and its parts are supposed to be identical, though often they are not. You have therefore to adopt whichever of these two lines better suits your purpose. That is how Euthydemus argues: e.g. that any one knows that there is a trireme in the Peiraeus, since he knows the separate details that make up this statement. There is also the argument that one who knows the letters knows the whole word, since the word is the same thing as the letters which compose it; or that, if a double portion of a certain thing is harmful to health, then a single portion must not be called wholesome, since it is absurd that two good things should make one bad thing. Put thus, the enthymeme is refutative; put as

follows; demonstrative: 'For one good thing cannot be made up of two bad things.' The whole line of argument is fallacious. Again, there is Polycrates' saying that Thrasybulus put down thirty tyrants, where the speaker adds them up one by one. Or the argument in the Orestes of Theodectes, where the argument is from part to whole:

'Tis right that she who slays her lord should die.

'It is right, too, that the son should avenge his father. Very good: these two things are what Orestes has done.' Still, perhaps the two things, once they are put together, do not form a right act. The fallacy might also be said to be due to omission, since the speaker fails to say by whose hand a husband-slayer should die.

3. Another line is the use of indignant language, whether to support your own case or to overthrow your opponent's. We do this when we paint a highly-coloured picture of the situation without having proved the facts of it: if the defendant does so, he produces an impression of his innocence; and if the prosecutor goes into a passion, he produces an impression of the defendant's guilt. Here there is no genuine enthymeme: the hearer infers guilt or innocence, but no proof is given, and the inference is fallacious accordingly.

4. Another line is to use a 'Sign', or single instance, as certain evidence; which, again, yields no valid proof. Thus, it might be said that lovers are useful to their countries, since the love of Harmodius and Aristogeiton caused the downfall of the tyrant Hipparchus. Or, again, that Dionysius is a thief, since he is a vicious man—there is, of course, no valid proof here; not every vicious man is a thief, though every thief is a vicious man.

5. Another line represents the accidental as essential. An instance is what Polycrates says of the mice, that they 'came to the rescue' because they gnawed through the bowstrings. Or it might be maintained that an invitation to dinner is a great honour, for it was because he was not invited that Achilles was 'angered' with the Greeks at Tenedos? As a fact, what angered him was the insult involved; it was a mere accident that this was the particular form that the insult took.

6. Another is the argument from consequence. In the Alexander, for instance, it is argued that Paris must have had a lofty disposition, since he despised society and lived by himself on Mount Ida: because lofty people do this kind of thing, therefore Paris too, we are to suppose, had a lofty soul. Or, if a man dresses fashionably and roams around at night, he is a rake, since that is the way rakes behave. Another similar argument points out that beggars sing and dance in temples, and that exiles can live wherever they please, and that such privileges are at the disposal of those we account happy and therefore every one might be regarded as happy if only he has those privileges. What matters, however, is the circumstances under which the privileges are enjoyed. Hence this line too falls under the head of fallacies by omission.

7. Another line consists in representing as causes things which are not causes, on the ground that they happened along with or before the event in question. They assume that, because B happens after A, it happens because of A. Politicians

are especially fond of taking this line. Thus Demades said that the policy of Demosthenes was the cause of all the mischief, 'for after it the war occurred'.

8. Another line consists in leaving out any mention of time and circumstances. E.g. the argument that Paris was justified in taking Helen, since her father left her free to choose: here the freedom was presumably not perpetual; it could only refer to her first choice, beyond which her father's authority could not go. Or again, one might say that to strike a free man is an act of wanton outrage; but it is not so in every case—only when it is unprovoked.

9. Again, a spurious syllogism may, as in 'eristical' discussions, be based on the confusion of the absolute with that which is not absolute but particular. As, in dialectic, for instance, it may be argued that what-is-not is, on the ground that what-is-not is what-is-not: or that the unknown can be known, on the ground that it can be known to he unknown: so also in rhetoric a spurious enthymeme may be based on the confusion of some particular probability with absolute probability. Now no particular probability is universally probable: as Agathon says,

One might perchance say that was probable—
That things improbable oft will hap to men.

For what is improbable does happen, and therefore it is probable that improbable things will happen. Granted this, one might argue that 'what is improbable is probable'. But this is not true absolutely. As, in eristic, the imposture comes from not adding any clause specifying relationship or reference or manner; so here it arises because the probability in question is not general but specific. It is of this line of argument that Corax's Art of Rhetoric is composed. If the accused is not open to the charge—for instance if a weakling be tried for violent assault—the defence is that he was not likely to do such a thing. But if he is open to the charge—i.e. if he is a strong man—the defence is still that he was not likely to do such a thing, since he could be sure that people would think he was likely to do it. And so with any other charge: the accused must be either open or not open to it: there is in either case an appearance of probable innocence, but whereas in the latter case the probability is genuine, in the former it can only be asserted in the special sense mentioned. This sort of argument illustrates what is meant by making the worse argument seem the better. Hence people were right in objecting to the training Protagoras undertook to give them. It was a fraud; the probability it handled was not genuine but spurious, and has a place in no art except Rhetoric and Eristic.

25

Enthymemes, genuine and apparent, have now been described; the next subject is their Refutation.

An argument may be refuted either by a counter-syllogism or by bringing an objection. It is clear that counter-syllogisms can be built up from the same lines of arguments as the original syllogisms: for the materials of syllogisms are the ordinary opinions of men, and such opinions often contradict each other. Objections, as appears in the Topics, may be raised in four ways—either by directly attacking your opponent's own statement, or by putting forward another statement like it, or by putting forward a statement contrary to it, or by quoting previous decisions.

1. By 'attacking your opponent's own statement' I mean, for instance, this: if his enthymeme should assert that love is always good, the objection can be brought in two ways, either by making the general statement that 'all want is an evil', or by making the particular one that there would be no talk of 'Caunian love' if there were not evil loves as well as good ones.

2. An objection 'from a contrary statement' is raised when, for instance, the opponent's enthymeme having concluded that a good man does good to all his friends, you object, 'That proves nothing, for a bad man does not do evil to all his friends'.

3. An example of an objection 'from a like statement' is, the enthymeme having shown that ill-used men always hate their ill-users, to reply, 'That proves nothing, for well-used men do not always love those who used them well'.

4. The 'decisions' mentioned are those proceeding from well-known men; for instance, if the enthymeme employed has concluded that 'that allowance ought to be made for drunken offenders, since they did not know what they were doing', the objection will be, 'Pittacus, then, deserves no approval, or he would not have prescribed specially severe penalties for offences due to drunkenness'.

Enthymemes are based upon one or other of four kinds of alleged fact: (1) Probabilities, (2) Examples, (3) Infallible Signs, (4) Ordinary Signs. (1) Enthymemes based upon Probabilities are those which argue from what is, or is supposed to be, usually true. (2) Enthymemes based upon Example are those which proceed by induction from one or more similar cases, arrive at a general proposition, and then argue deductively to a particular inference. (3) Enthymemes based upon Infallible Signs are those which argue from the inevitable and invariable. (4) Enthymemes based upon ordinary Signs are those which argue from some universal or particular proposition, true or false.

Now (1) as a Probability is that which happens usually but not always, Enthymemes founded upon Probabilities can, it is clear, always be refuted by raising some objection. The refutation is not always genuine: it may be spurious: for it consists in showing not that your opponent's premiss is not probable, but only in showing that it is not inevitably true. Hence it is always in defence rather than in accusation that it is possible to gain an advantage by using this fallacy. For the accuser uses probabilities to prove his case: and to refute a conclusion as improbable is not the same thing as to refute it as not inevitable. Any argument based upon what usually happens is always open to objection: otherwise it would not be a probability but an invariable and necessary truth. But the judges think, if the refutation takes this form, either that the accuser's case is not probable or that they must not decide it; which, as we said, is a false piece of reasoning. For they ought to decide by considering not merely what must be true but also what is likely to be true: this is, indeed, the meaning of 'giving a verdict in accordance with one's honest opinion'. Therefore it is not enough for the defendant to refute the accusation by proving that the charge is not hound to be true: he must do so by showing that it is not likely to be true. For this purpose his objection must state what is more usually true than the statement attacked. It may do so in either of two ways: either in respect of frequency or in respect of exactness. It will be most convincing if it does so in both respects; for if the thing in question both happens oftener as we represent it and happens more as we represent it, the probability is particularly great.

(2) Fallible Signs, and Enthymemes based upon them, can be refuted even if the facts are correct, as was said at the outset. For we have shown in the Analytics that no Fallible Sign can form part of a valid logical proof.

(3) Enthymemes depending on examples may be refuted in the same way as probabilities. If we have a negative instance, the argument is refuted, in so far as it is proved not inevitable, even though the positive examples are more similar and more frequent. And if the positive examples are more numerous and more frequent, we must contend that the present case is dissimilar, or that its conditions are dissimilar, or that it is different in some way or other.

(4) It will be impossible to refute Infallible Signs, and Enthymemes resting on them, by showing in any way that they do not form a valid logical proof: this, too, we see from the Analytics. All we can do is to show that the fact alleged does not exist. If there is no doubt that it does, and that it is an Infallible Sign, refutation now becomes impossible: for this is equivalent to a demonstration which is clear in every respect.

26

Amplification and Depreciation are not an element of enthymeme. By 'an element of enthymeme' I mean the same thing as a line of enthymematic argument—a general class embracing a large number of particular kinds of enthymeme. Amplification and Depreciation are one kind of enthymeme, viz. the kind used to show that a thing is great or small; just as there are other kinds used to show that a thing is good or bad, just or unjust, and anything else of the sort. All these things are the subject—matter of syllogisms and enthymemes; none of these is the line of argument of an enthymeme; no more, therefore, are Amplification and Depreciation. Nor are Refutative Enthymemes a different species from Constructive. For it is clear that refutation consists either in offering positive proof or in raising an objection. In the first case we prove the opposite of our adversary's statements. Thus, if he shows that a thing has happened, we show that it has not; if he shows that it has not happened, we show that it has. This, then, could not be the distinction if there were one, since the same means are employed by both parties, enthymemes being adduced to show that the fact is or is not so-and-so. An objection, on the other hand, is not an enthymeme at all, as was said in the Topics, consists in stating some accepted opinion from which it will be clear that our opponent has not reasoned correctly or has made a false assumption.

Three points must be studied in making a speech; and we have now completed the account of (1) Examples, Maxims, Enthymemes, and in general the thought—element the way to invent and refute arguments. We have next to discuss (2) Style, and (3) Arrangement.

Book III

1

IN making a speech one must study three points: first, the means of producing persuasion; second, the style, or language, to be used; third, the proper arrangement of the various parts of the speech. We have already specified the sources of persuasion. We have shown that these are three in number; what they are; and why there are only these three: for we have shown that persuasion must in every case be effected either (1) by working on the emotions of the judges themselves, (2) by giving them the right impression of the speakers' character, or (3) by proving the truth of the statements made.

Enthymemes also have been described, and the sources from which they should be derived; there being both special and general lines of argument for enthymemes.

Our next subject will be the style of expression. For it is not enough to know what we ought to say; we must also say it as we ought; much help is thus afforded towards producing the right impression of a speech. The first question to receive attention was naturally the one that comes first naturally—how persuasion can be produced from the facts themselves. The second is how to set these facts out in language. A third would be the proper method of delivery; this is a thing that affects the success of a speech greatly; but hitherto the subject has been neglected. Indeed, it was long before it found a way into the arts of tragic drama and epic recitation: at first poets acted their tragedies themselves. It is plain that delivery has just as much to do with oratory as with poetry. (In connexion with poetry, it has been studied by Glaucon of Teos among others.) It is, essentially, a matter of the right management of the voice to express the various emotions—of speaking loudly, softly, or between the two; of high, low, or intermediate pitch; of the various rhythms that suit various subjects. These are the three things—volume of sound, modulation of pitch, and rhythm—that a speaker bears in mind. It is those who do bear them in mind who usually win prizes in the dramatic contests; and just as in drama the actors now count for more than the poets, so it is in the contests of public life, owing to the defects of our political institutions. No systematic treatise upon the rules of delivery has yet been composed; indeed, even the study of language made no progress till late in the day. Besides, delivery is—very properly—not regarded as an elevated subject of inquiry. Still, the whole business of rhetoric being concerned with appearances, we must pay attention to the subject of delivery, unworthy though it is, because we cannot do without it. The right thing in speaking really is that we should be satisfied not to annoy our hearers, without trying to delight them: we ought in fairness to fight our case with no help beyond the bare facts: nothing, therefore, should matter except the proof of those facts. Still, as has been already said, other things affect the result considerably, owing to the defects of our hearers. The arts of language cannot help having a small but real importance, whatever it is we have to expound to others: the way in which a thing is said does affect its intelligibility. Not, however, so much importance as people think. All such arts are fanciful and meant to charm the hearer. Nobody uses fine language when teaching geometry.

When the principles of delivery have been worked out, they will produce the same effect as on the stage. But only very slight attempts to deal with them have been made and by a few people, as by Thrasymachus in his 'Appeals to Pity'. Dramatic ability is a natural gift, and can hardly be systematically taught. The principles of good diction can be so taught, and therefore we have men of ability in this direction too, who win prizes in their turn, as well as those speakers who excel in delivery—speeches of the written or literary kind owe more of their effect to their direction than to their thought.

It was naturally the poets who first set the movement going; for words represent things, and they had also the human voice at their disposal, which of all our organs can best represent other things. Thus the arts of recitation and acting were formed, and others as well. Now it was because poets seemed to win fame through their fine language when their thoughts were simple enough, that the language of oratorical prose at first took a poetical colour, e.g. that of Gorgias. Even now most uneducated people think that poetical language makes the finest discourses. That is not true: the language of prose is distinct from that of poetry. This is shown by the state of things to-day, when even the language of tragedy has altered its character. Just as iambics were adopted, instead of tetrameters, because they are the most prose-like of all metres, so tragedy has given up all those words, not used in ordinary talk, which decorated the early drama and are still used by the writers of hexameter poems. It is therefore ridiculous to imitate a poetical manner which the poets themselves have dropped; and it is now plain that we have not to treat in detail the whole question of style, but may confine ourselves to that part of it which concerns our present subject, rhetoric. The other—the poetical—part of it has been discussed in the treatise on the Art of Poetry.

2

We may, then, start from the observations there made, including the definition of style. Style to be good must be clear, as is proved by the fact that speech which fails to convey a plain meaning will fail to do just what speech has to do. It must also be appropriate, avoiding both meanness and undue elevation; poetical language is certainly free from meanness, but it is not appropriate to prose.

Clearness is secured by using the words (nouns and verbs alike) that are current and ordinary. Freedom from meanness, and positive adornment too, are secured by using the other words mentioned in the Art of Poetry. Such variation from what is usual makes the language appear more stately. People do not feel towards strangers as they do towards their own countrymen, and the same thing is true of their feeling for language. It is therefore well to give to everyday speech an unfamiliar air: people like what strikes them, and are struck by what is out of the way. In verse such effects are common, and there they are fitting: the persons and things there spoken of are comparatively remote from ordinary life. In prose passages they are far less often fitting because the subject-matter is less exalted. Even in poetry, it is not quite appropriate that fine language should be used by a slave or a very young man, or about very trivial subjects: even in poetry the style, to be appropriate, must sometimes be toned down, though at other times heightened. We can now see that a writer must disguise his art and give the impression of speaking naturally and not artificially. Naturalness is persuasive, artificiality is the contrary; for our hearers are prejudiced and think we have some design against them, as if we were mixing their wines for them. It is like the difference between the quality of Theodorus' voice and the voices of all other actors: his really seems to be that of the character who is speaking, theirs do not. We can hide our purpose successfully by taking the single words of our composition from the speech of ordinary life. This is done in poetry by Euripides, who was the first to show the way to his successors.

Language is composed of nouns and verbs. Nouns are of the various kinds considered in the treatise on Poetry. Strange words, compound words, and invented words must be used sparingly and on few occasions: on what occasions we shall state later. The reason for this restriction has been already indicated: they depart from what is suitable, in the direction of excess. In the language of prose, besides the regular and proper terms for things, metaphorical terms only can be used with advantage. This we gather from the fact that these two classes of terms, the proper or regular and the metaphorical—these and no others—are used by everybody in conversation. We can now see that a good writer can produce a style that is distinguished without being obtrusive, and is at the same time clear, thus satisfying our definition of good oratorical prose. Words of ambiguous meaning are chiefly useful to enable the sophist to mislead his hearers. Synonyms are useful to the poet, by which I mean words whose ordinary meaning is the same, e.g. 'porheueseai' (advancing) and 'badizein' (proceeding); these two are ordinary words and have the same meaning.

In the Art of Poetry, as we have already said, will be found definitions of these kinds of words; a classification of Metaphors; and mention of the fact that metaphor is of great value both in poetry and in prose. Prose-writers must, however, pay specially careful attention to metaphor, because their other resources are scantier than those of poets. Metaphor, moreover, gives style clearness, charm, and distinction as nothing else can: and it is not a thing whose use can be taught by one man to another. Metaphors, like epithets, must be fitting, which means that they must fairly correspond to the thing signified: failing this, their inappropriateness will be conspicuous: the want of harmony between two things is emphasized by their being placed side by side. It is like having to ask ourselves what dress will suit an old man; certainly not the crimson cloak that suits a young man. And if you wish to pay a compliment, you must take your metaphor from something better in the same line; if to disparage, from something worse. To illustrate my meaning: since opposites are in the same class, you do what I have suggested if you say that a man who begs 'prays', and a man who prays 'begs'; for praying and begging are both varieties of asking. So Iphicrates called Callias a 'mendicant priest' instead of a 'torch-bearer', and Callias replied that Iphicrates must be uninitiated or he would have called him not a 'mendicant priest' but a 'torch-bearer'. Both are religious titles, but one is honourable and the other is not. Again, somebody calls actors 'hangers-on of Dionysus', but they call themselves 'artists': each of these terms is a metaphor, the one intended to throw dirt at the actor, the other to dignify him. And pirates now call themselves 'purveyors'. We can thus call a crime a mistake, or a mistake a crime. We can say that a thief 'took' a thing, or that he 'plundered' his victim. An expression like that of Euripides' Telephus,

King of the oar, on Mysia's coast he landed,

is inappropriate; the word 'king' goes beyond the dignity of the subject, and so the art is not concealed. A metaphor may be amiss because the very syllables of the words conveying it fail to indicate sweetness of vocal utterance. Thus Dionysius the Brazen in his elegies calls poetry 'Calliope's screech'. Poetry and screeching are both, to be sure, vocal utterances. But the metaphor is bad, because the sounds of 'screeching', unlike those of poetry, are discordant and unmeaning. Further, in using metaphors to give names to nameless things, we must draw them not from remote but from kindred and similar things, so that the kinship is clearly perceived as soon as the words are said. Thus in the celebrated riddle

I marked how a man glued bronze with fire to another man's body,

the process is nameless; but both it and gluing are a kind of application, and that is why the application of the cupping-glass is here called a 'gluing'. Good riddles do, in general, provide us with satisfactory metaphors: for metaphors imply riddles, and therefore a good riddle can furnish a good metaphor. Further, the materials of metaphors must be beautiful; and the beauty, like the ugliness, of all words may, as Licymnius says, lie in their sound or in their meaning. Further, there is a third consideration—one that upsets the fallacious argument of the sophist Bryson, that there is no such thing as foul language, because in whatever words you put a given thing your meaning is the same. This is untrue. One term may describe a thing more truly than another, may be more like it, and set it more intimately before our eyes. Besides, two different words will represent a thing in two different lights; so on this ground also one term must be held fairer or fouler than another. For both of two terms will indicate what is fair, or what is foul, but not simply their fairness or their foulness, or if so, at any rate not in an equal degree. The materials of metaphor must be beautiful to the ear, to the understanding, to the eye or some other physical sense. It is better, for instance, to say 'rosy-fingered morn', than 'crimson-fingered' or, worse still, 'red-fingered morn'. The epithets that we apply, too, may have a bad and ugly aspect, as when Orestes is called a 'mother-slayer'; or a better one, as when he is called his 'father's avenger'. Simonides, when the victor in the mule-race offered him a small fee, refused to write him an ode, because, he said, it was so unpleasant to write odes to half-asses: but on receiving an adequate fee, he wrote

> Hail to you, daughters of storm-footed steeds?

though of course they were daughters of asses too. The same effect is attained by the use of diminutives, which make a bad thing less bad and a good thing less good. Take, for instance, the banter of Aristophanes in the Babylonians where he uses 'goldlet' for 'gold', 'cloaklet' for 'cloak', 'scoffiet' for 'scoff', and 'plaguelet'. But alike in using epithets and in using diminutives we must be wary and must observe the mean.

3

Bad taste in language may take any of four forms:

(1) The misuse of compound words. Lycophron, for instance, talks of the 'many visaged heaven' above the 'giant-crested earth', and again the 'strait-pathed shore'; and Gorgias of the 'pauper-poet flatterer' and 'oath-breaking and over-oath-keeping'. Alcidamas uses such expressions as 'the soul filling with rage and face becoming flame-flushed', and 'he thought their enthusiasm would be issue—fraught' and 'issue-fraught he made the persuasion of his words', and 'sombre-hued is the floor of the sea'.The way all these words are compounded makes them, we feel, fit for verse only. This, then, is one form in which bad taste is shown.

(2) Another is the employment of strange words. For instance, Lycophron talks of 'the prodigious Xerxes' and 'spoliative Sciron'; Alcidamas of 'a toy for poetry' and 'the witlessness of nature', and says 'whetted with the unmitigated temper of his spirit'.

(3) A third form is the use of long, unseasonable, or frequent epithets. It is appropriate enough for a poet to talk of 'white milk', in prose such epithets are sometimes lacking in appropriateness or, when spread too thickly, plainly reveal the author turning his prose into poetry. Of course we must use some epithets, since they lift our style above the usual level and give it an air of distinction. But we must aim at the due mean, or the result will be worse than if we took no trouble at all; we shall get something actually bad instead of something merely not good. That is why the epithets of Alcidamas seem so tasteless; he does not use them as the seasoning of the meat, but as the meat itself, so numerous and swollen and aggressive are they. For instance, he does not say 'sweat', but 'the moist sweat'; not

'to the Isthmian games', but 'to the world-concourse of the Isthmian games'; not 'laws', but 'the laws that are monarchs of states'; not 'at a run', but 'his heart impelling him to speed of foot'; not 'a school of the Muses', but 'Nature's school of the Muses had he inherited'; and so 'frowning care of heart', and 'achiever' not of 'popularity' but of 'universal popularity', and 'dispenser of pleasure to his audience', and 'he concealed it' not 'with boughs' but 'with boughs of the forest trees', and 'he clothed' not 'his body' but 'his body's nakedness', and 'his soul's desire was counter imitative' (this's at one and the same time a compound and an epithet, so that it seems a poet's effort), and 'so extravagant the excess of his wickedness'. We thus see how the inappropriateness of such poetical language imports absurdity and tastelessness into speeches, as well as the obscurity that comes from all this verbosity—for when the sense is plain, you only obscure and spoil its clearness by piling up words.

The ordinary use of compound words is where there is no term for a thing and some compound can be easily formed, like 'pastime' (chronotribein); but if this is much done, the prose character disappears entirely. We now see why the language of compounds is just the thing for writers of dithyrambs, who love sonorous noises; strange words for writers of epic poetry, which is a proud and stately affair; and metaphor for iambic verse, the metre which (as has been already' said) is widely used to-day.

(4) There remains the fourth region in which bad taste may be shown, metaphor. Metaphors like other things may be inappropriate. Some are so because they are ridiculous; they are indeed used by comic as well as tragic poets. Others are too grand and theatrical; and these, if they are far-fetched, may also be obscure. For instance, Gorgias talks of 'events that are green and full of sap', and says 'foul was the deed you sowed and evil the harvest you reaped'. That is too much like poetry. Alcidamas, again, called philosophy 'a fortress that threatens the power of law', and the Odyssey 'a goodly looking-glass of human life',' talked about 'offering no such toy to poetry': all these expressions fail, for the reasons given, to carry the hearer with them. The address of Gorgias to the swallow, when she had let her droppings fall on him as she flew overhead, is in the best tragic manner. He said, 'Nay, shame, O Philomela'. Considering her as a bird, you could not call her act shameful; considering her as a girl, you could; and so it was a good gibe to address her as what she was once and not as what she is.

<h1 style="text-align:center">4</h1>

The Simile also is a metaphor; the difference is but slight. When the poet says of Achilles that he

    Leapt on the foe as a lion,

this is a simile; when he says of him 'the lion leapt', it is a metaphor—here, since both are courageous, he has transferred to Achilles the name of 'lion'. Similes are useful in prose as well as in verse; but not often, since they are of the nature of poetry. They are to be employed just as metaphors are employed, since they are really the same thing except for the difference mentioned.

The following are examples of similes. Androtion said of Idrieus that he was like a terrier let off the chain, that flies at you and bites you—Idrieus too was savage now that he was let out of his chains. Theodamas compared Archidamus to an Euxenus who could not do geometry—a proportional simile, implying that Euxenus is an Archidamus who can do geometry. In Plato's Republic those who strip the dead are compared to curs which bite the stones thrown at them but do not touch the thrower, and there is the simile about the Athenian people, who are compared to a ship's captain who is strong but a little deaf; and the one about poets' verses, which are likened to persons who lack beauty but possess youthful freshness—when the freshness has faded the charm perishes, and so with verses when broken up into prose. Pericles compared the Samians to children who take their pap but go on crying; and the Boeotians to holm-oaks, because they were ruining one another by civil wars just as one oak causes another oak's fall. Demosthenes said that the Athenian people were like sea-sick men on board ship. Again, Demosthenes compared the political orators to nurses who swallow the bit of food themselves and then smear the children's lips with the spittle. Antisthenes compared the lean Cephisodotus to frankincense, because it was his consumption that gave one pleasure. All these ideas may be expressed either as similes or as metaphors; those which succeed as metaphors will obviously do well also as similes, and similes, with the explanation omitted, will appear as metaphors. But the proportional metaphor must always apply reciprocally to either of its co-ordinate terms. For instance, if a drinking-bowl is the shield of Dionysus, a shield may fittingly be called the drinking-bowl of Ares.

5

Such, then, are the ingredients of which speech is composed. The foundation of good style is correctness of language, which falls under five heads. (1) First, the proper use of connecting words, and the arrangement of them in the natural sequence which some of them require. For instance, the connective 'men' (e.g. ego men) requires the correlative de (e.g. o de). The answering word must be brought in before the first has been forgotten, and not be widely separated from it; nor, except in the few cases where this is appropriate, is another connective to be introduced before the one required. Consider the sentence, 'But as soon as he told me (for Cleon had come begging and praying), took them along and set out.' In this sentence many connecting words are inserted in front of the one required to complete the sense; and if there is a long interval before 'set out', the result is obscurity. One merit, then, of good style lies in the right use of connecting words. (2) The second lies in calling things by their own special names and not by vague general ones. (3) The third is to avoid ambiguities; unless, indeed, you definitely desire to be ambiguous, as those do who have nothing to say but are pretending to mean something. Such people are apt to put that sort of thing into verse. Empedocles, for instance, by his long circumlocutions imposes on his hearers; these are affected in the same way as most people are when they listen to diviners, whose ambiguous utterances are received with nods of acquiescence—

Croesus by crossing the Halys will ruin a mighty realm.

Diviners use these vague generalities about the matter in hand because their predictions are thus, as a rule, less likely to be falsified. We are more likely to be right, in the game of 'odd and even', if we simply guess 'even' or 'odd' than if we guess at the actual number; and the oracle-monger is more likely to be right if he simply says that a thing will happen than if he says when it will happen, and therefore he refuses to add a definite date. All these ambiguities have the same sort of effect, and are to be avoided unless we have some such object as that mentioned. (4) A fourth rule is to observe Protagoras' classification of nouns into male, female, and inanimate; for these distinctions also must be correctly given. 'Upon her arrival she said her say and departed (e d elthousa kai dialechtheisa ocheto).' (5) A fifth rule is to express plurality, fewness, and unity by the correct wording, e.g. 'Having come, they struck me (oi d elthontes etupton me).'

It is a general rule that a written composition should be easy to read and therefore easy to deliver. This cannot be so where there are many connecting words or clauses, or where punctuation is hard, as in the writings of Heracleitus. To punctuate Heracleitus is no easy task, because we often cannot tell whether a particular word belongs to what precedes or what follows it. Thus, at the outset of his treatise he says, 'Though this truth is always men understand it not', where it is not clear with which of the two clauses the word 'always' should be joined by the punctuation. Further, the following fact leads to solecism, viz. that the sentence does not work out properly if you annex to two terms a third which does not suit them both. Thus either 'sound' or 'colour' will fail to work out properly with some verbs: 'perceive' will apply to both, 'see' will not. Obscurity is also caused if, when you intend to insert a number of details, you do not first make your meaning clear; for instance, if you say, 'I meant, after telling him this, that and the other thing, to set out', rather than something of this kind 'I meant to set out after telling him; then this, that, and the other thing occurred.'

6

The following suggestions will help to give your language impressiveness. (1) Describe a thing instead of naming it: do not say 'circle', but 'that surface which extends equally from the middle every way'. To achieve conciseness, do the opposite—put the name instead of the description. When mentioning anything ugly or unseemly, use its name if it is the description that is ugly, and describe it if it is the name that is ugly. (2) Represent things with the help of metaphors and epithets, being careful to avoid poetical effects. (3) Use plural for singular, as in poetry, where one finds

Unto havens Achaean,
though only one haven is meant, and
Here are my letter's many-leaved folds.

(4) Do not bracket two words under one article, but put one article with each; e.g. 'that wife of ours.' The reverse to secure conciseness; e.g. 'our wife.' Use plenty of connecting words; conversely, to secure conciseness, dispense with connectives, while still preserving connexion; e.g. 'having gone and spoken', and 'having gone, I spoke', respectively. (6) And the practice of Antimachus, too, is useful—to describe a thing by mentioning attributes it does not possess; as he does in talking of Teumessus

There is a little wind-swept knoll...

A subject can be developed indefinitely along these lines. You may apply this method of treatment by negation either to good or to bad qualities, according to which your subject requires. It is from this source that the poets draw expressions such as the 'stringless' or 'lyreless' melody, thus forming epithets out of negations. This device is popular in proportional metaphors, as when the trumpet's note is called 'a lyreless melody'.

7

Your language will be appropriate if it expresses emotion and character, and if it corresponds to its subject. 'Correspondence to subject' means that we must neither speak casually about weighty matters, nor solemnly about trivial ones; nor must we add ornamental epithets to commonplace nouns, or the effect will be comic, as in the works of Cleophon, who can use phrases as absurd as 'O queenly fig-tree'. To express emotion, you will employ the language of anger in speaking of outrage; the language of disgust and discreet reluctance to utter a word when speaking of impiety or foulness; the language of exultation for a tale of glory, and that of humiliation for a tale of and so in all other cases.

This aptness of language is one thing that makes people believe in the truth of your story: their minds draw the false conclusion that you are to be trusted from the fact that others behave as you do when things are as you describe them; and therefore they take your story to be true, whether it is so or not. Besides, an emotional speaker always makes his audience feel with him, even when there is nothing in his arguments; which is why many speakers try to overwhelm their audience by mere noise.

Furthermore, this way of proving your story by displaying these signs of its genuineness expresses your personal character. Each class of men, each type of disposition, will have its own appropriate way of letting the truth appear. Under 'class' I include differences of age, as boy, man, or old man; of sex, as man or woman; of nationality, as Spartan or Thessalian. By 'dispositions' I here mean those dispositions only which determine the character of a man's for it is not every disposition that does this. If, then, a speaker uses the very words which are in keeping with a particular disposition, he will reproduce the corresponding character; for a rustic and an educated man will not say the same things nor speak in the same way. Again, some impression is made upon an audience by a device which speech-writers employ to nauseous excess, when they say 'Who does not know this?' or 'It is known to everybody.' The hearer is ashamed of his ignorance, and agrees with the speaker, so as to have a share of the knowledge that everybody else possesses.

All the variations of oratorical style are capable of being used in season or out of season. The best way to counteract any exaggeration is the well-worn device by which the speaker puts in some criticism of himself; for then people feel it must be all right for him to talk thus, since he certainly knows what he is doing. Further, it is better not to have everything always just corresponding to everything else—your hearers will see through you less easily thus. I mean for instance, if your words are harsh, you should not extend this harshness to your voice and your countenance and have everything else in keeping. If you do, the artificial character of each detail becomes apparent; whereas if you adopt one device and not another, you are using art all the same and yet nobody notices it. (To be sure, if mild sentiments are expressed in harsh tones and harsh sentiments in mild tones, you become comparatively unconvincing.) Compound words, fairly plentiful epithets, and strange words best suit an emotional speech. We forgive an angry man for talking about a wrong as 'heaven-high' or 'colossal'; and we excuse such language when the speaker has his hearers already in his hands and has stirred them deeply either by praise or blame or anger or affection, as Isocrates, for instance, does at the end of his Panegyric, with his 'name and fame' and 'in that they brooked'. Men do speak in this strain when they are deeply stirred, and so, once the audience is in a like state of feeling, approval of course follows. This is why such language is fitting in poetry, which is an inspired thing. This language, then, should be used either under stress of emotion, or ironically, after the manner of Gorgias and of the passages in the Phaedrus.

8

The form of a prose composition should be neither metrical nor destitute of rhythm. The metrical form destroys the hearer's trust by its artificial appearance, and at the same time it diverts his attention, making him watch for metrical recurrences, just as children catch up the herald's question, 'Whom does the freedman choose as his advocate?', with the answer 'Cleon!' On the other hand, unrhythmical language is too unlimited; we do not want the limitations of metre, but some limitation we must have, or the effect will be vague and unsatisfactory. Now it is number that limits all things; and it is the numerical limitation of the forms of a composition that constitutes rhythm, of which metres are definite sections. Prose, then, is to be rhythmical, but not metrical, or it will become not prose but verse. It should not even have too precise a prose rhythm, and therefore should only be rhythmical to a certain extent.

Of the various rhythms, the heroic has dignity, but lacks the tones of the spoken language. The iambic is the very language of ordinary people, so that in common talk iambic lines occur oftener than any others: but in a speech we need dignity and the power of taking the hearer out of his ordinary self. The trochee is too much akin to wild dancing: we can see this in tetrameter verse, which is one of the trochaic rhythms.

There remains the paean, which speakers began to use in the time of Thrasymachus, though they had then no name to give it. The paean is a third class of rhythm, closely akin to both the two already mentioned; it has in it the ratio of three to two, whereas the other two kinds have the ratio of one to one, and two to one respectively. Between the two last ratios comes the ratio of one-and-a-half to one, which is that of the paean.

Now the other two kinds of rhythm must be rejected in writing prose, partly for the reasons given, and partly because they are too metrical; and the paean must be adopted, since from this alone of the rhythms mentioned no definite metre arises, and therefore it is the least obtrusive of them. At present the same form of paean is employed at the beginning a at the end of sentences, whereas the end should differ from the beginning. There are two opposite kinds of paean, one of which is suitable to the beginning of a sentence, where it is indeed actually used; this is the kind that begins with a long syllable and ends with three short ones, as

Dalogenes | eite Luki | an,

and

Chruseokom | a Ekate | pai Dios.

The other paean begins, conversely, with three short syllables and ends with a long one, as

meta de lan | udata t ok | eanon e | oanise nux.

This kind of paean makes a real close: a short syllable can give no effect of finality, and therefore makes the rhythm appear truncated. A sentence should break off with the long syllable: the fact that it is over should be indicated not by the scribe, or by his period-mark in the margin, but by the rhythm itself.

We have now seen that our language must be rhythmical and not destitute of rhythm, and what rhythms, in what particular shape, make it so.

9

The language of prose must be either free-running, with its parts united by nothing except the connecting words, like the preludes in dithyrambs; or compact and antithetical, like the strophes of the old poets. The free-running style is the ancient one, e.g. 'Herein is set forth the inquiry of Herodotus the Thurian.' Every one used this method formerly; not many do so now. By 'free-running' style I mean the kind that has no natural stopping-places, and comes to a stop only because there is no more to say of that subject. This style is unsatisfying just because it goes on indefinitely—one always likes to sight a stopping-place in front of one: it is only at the goal that men in a race faint and collapse; while they see the end of the course before them, they can keep on going. Such, then, is the free-running kind of style; the compact is that which is in periods. By a period I mean a portion of speech that has in itself a beginning and an end, being at the same time not too big to be taken in at a glance. Language of this kind is satisfying and easy to follow. It is satisfying, because it is just the reverse of indefinite; and moreover, the hearer always feels that he is grasping something and has reached some definite conclusion; whereas it is unsatisfactory to see nothing in front of you and get nowhere. It is easy to follow, because it can easily be remembered; and this because language when in periodic form can be numbered, and number is the easiest of all things to remember. That is why verse, which is measured, is always more easily remembered than prose, which is not: the measures of verse can be numbered. The period must, further, not be completed until the sense is complete: it must not be capable of breaking off abruptly, as may happen with the following iambic lines of Sophocles—

Calydon's soil is this; of Pelops' land
(The smiling plains face us across the strait.)

By a wrong division of the words the hearer may take the meaning to be the reverse of what it is: for instance, in the passage quoted, one might imagine that Calydon is in the Peloponnesus.

A Period may be either divided into several members or simple. The period of several members is a portion of speech (1) complete in itself, (2) divided into parts, and (3) easily delivered at a single breath—as a whole, that is; not by fresh breath being taken at the division. A member is one of the two parts of such a period. By a 'simple' period, I mean that which has only one member. The members, and the whole periods, should be neither curt nor long. A member which is too short often makes the listener stumble; he is still expecting the rhythm to go on to the limit his mind has fixed for it; and if meanwhile he is pulled back by the speaker's stopping, the shock is bound to make him, so to speak, stumble. If, on the other hand, you go on too long, you make him feel left behind, just as people who when walking pass beyond the boundary before turning back leave their companions behind So too if a period is too long you turn it into a speech, or something like a dithyrambic prelude. The result is much like the preludes that Democritus of Chios jeered at Melanippides for writing instead of antistrophic stanzas—

He that sets traps for another man's feet
Is like to fall into them first;
And long-winded preludes do harm to us all,
But the preluder catches it worst.

Which applies likewise to long-membered orators. Periods whose members are altogether too short are not periods at all; and the result is to bring the hearer down with a crash.

The periodic style which is divided into members is of two kinds. It is either simply divided, as in 'I have often wondered at the conveners of national gatherings and the founders of athletic contests'; or it is antithetical, where, in each of the two members, one of one pair of opposites is put along with one of another pair, or the same word is used to bracket two opposites, as 'They aided both parties—not only those who stayed behind but those who accompanied them: for the latter they acquired new territory larger than that at home, and to the former they left territory at home that was large enough'. Here the contrasted words are 'staying behind' and 'accompanying', 'enough' and 'larger'. So in the example, 'Both to those who want to get property and to those who desire to enjoy it' where 'enjoyment' is contrasted with 'getting'. Again, 'it often happens in such enterprises that the wise men fail and the fools succeed'; 'they were awarded the prize of valour immediately, and won the command of the sea not long afterwards'; 'to sail through the mainland and march through the sea, by bridging the Hellespont and cutting through Athos'; 'nature gave them their country and law took it away again'; 'of them perished in misery, others were saved in disgrace'; 'Athenian citizens keep foreigners in their houses as servants, while the city of Athens allows her allies by thousands to live as the foreigner's slaves'; and 'to possess in life or to bequeath at death'. There is also what some one said about Peitholaus and Lycophron in a law-court, 'These men used to sell you when they were at home, and now they have come to you here and bought you'. All these passages have the structure described above. Such a form of speech is satisfying, because the significance of contrasted ideas is easily felt, especially when they are thus put side by side, and also because it has the effect of a logical argument; it is by putting two opposing conclusions side by side that you prove one of them false.

Such, then, is the nature of antithesis. Parisosis is making the two members of a period equal in length. Paromoeosis is making the extreme words of both members like each other. This must happen either at the beginning or at the end of each member. If at the beginning, the resemblance must always be between whole words; at the end, between final syllables or inflexions of the same word or the same word repeated. Thus, at the beginning

    agron gar elaben arlon par' autou
and
    dorhetoi t epelonto pararretoi t epeessin
At the end
    ouk wethesan auton paidion tetokenai,
all autou aitlon lelonenai,
and
    en pleiotals de opontisi kai en elachistais elpisin
An example of inflexions of the same word is
    axios de staoenai chalkous ouk axios on chalkou;
Of the same word repeated,
    su d' auton kai zonta eleges kakos kai nun grafeis kakos.
Of one syllable,

ti d' an epaoes deinon, ei andrh' eides arhgon;

It is possible for the same sentence to have all these features together—antithesis, parison, and homoeoteleuton. (The possible beginnings of periods have been pretty fully enumerated in the Theodectea.) There are also spurious antitheses, like that of Epicharmus—

There one time I as their guest did stay,
And they were my hosts on another day.

10

We may now consider the above points settled, and pass on to say something about the way to devise lively and taking sayings. Their actual invention can only come through natural talent or long practice; but this treatise may indicate the way it is done. We may deal with them by enumerating the different kinds of them. We will begin by remarking that we all naturally find it agreeable to get hold of new ideas easily: words express ideas, and therefore those words are the most agreeable that enable us to get hold of new ideas. Now strange words simply puzzle us; ordinary words convey only what we know already; it is from metaphor that we can best get hold of something fresh. When the poet calls 'old age a withered stalk', he conveys a new idea, a new fact, to us by means of the general notion of bloom, which is common to both things. The similes of the poets do the same, and therefore, if they are good similes, give an effect of brilliance. The simile, as has been said before, is a metaphor, differing from it only in the way it is put; and just because it is longer it is less attractive. Besides, it does not say outright that 'this' is 'that', and therefore the hearer is less interested in the idea. We see, then, that both speech and reasoning are lively in proportion as they make us seize a new idea promptly. For this reason people are not much taken either by obvious arguments (using the word 'obvious' to mean what is plain to everybody and needs no investigation), nor by those which puzzle us when we hear them stated, but only by those which convey their information to us as soon as we hear them, provided we had not the information already; or which the mind only just fails to keep up with. These two kinds do convey to us a sort of information: but the obvious and the obscure kinds convey nothing, either at once or later on. It is these qualities, then, that, so far as the meaning of what is said is concerned, make an argument acceptable. So far as the style is concerned, it is the antithetical form that appeals to us, e.g. 'judging that the peace common to all the rest was a war upon their own private interests', where there is an antithesis between war and peace. It is also good to use metaphorical words; but the metaphors must not be far-fetched, or they will be difficult to grasp, nor obvious, or they will have no effect. The words, too, ought to set the scene before our eyes; for events ought to be seen in progress rather than in prospect. So we must aim at these three points: Antithesis, Metaphor, and Actuality.

Of the four kinds of Metaphor the most taking is the proportional kind. Thus

Pericles, for instance, said that the vanishing from their country of the young men who had fallen in the war was 'as if the spring were taken out of the year'. Leptines, speaking of the Lacedaemonians, said that he would not have the Athenians let Greece 'lose one of her two eyes'. When Chares was pressing for leave to be examined upon his share in the Olynthiac war, Cephisodotus was indignant, saying that he wanted his examination to take place 'while he had his fingers upon the people's throat'. The same speaker once urged the Athenians to march to Euboea, 'with Miltiades' decree as their rations'. Iphicrates, indignant at the truce made by the Athenians with Epidaurus and the neighbouring sea-board, said that they had stripped themselves of their travelling money for the journey of war. Peitholaus called the state-galley 'the people's big stick', and Sestos 'the corn-bin of the Peiraeus'. Pericles bade his countrymen remove Aegina, 'that eyesore of the Peiraeus.' And Moerocles said he was no more a rascal than was a certain respectable citizen he named, 'whose rascality was worth over thirty per cent per annum to him, instead of a mere ten like his own'. There is also the iambic line of Anaxandrides about the way his daughters put off marrying—

My daughters' marriage-bonds are overdue.

Polyeuctus said of a paralytic man named Speusippus that he could not keep quiet, 'though fortune had fastened him in the pillory of disease'. Cephisodotus called warships 'painted millstones'. Diogenes the Dog called taverns 'the mess-rooms of Attica'. Aesion said that the Athenians had 'emptied' their town into Sicily: this is a graphic metaphor. 'Till all Hellas shouted aloud' may be regarded as a metaphor, and a graphic one again. Cephisodotus bade the Athenians take care not to hold too many 'parades'. Isocrates used the same word of those who 'parade at the national festivals.' Another example occurs in the Funeral Speech: 'It is fitting that Greece should cut off her hair beside the tomb of those who fell at Salamis, since her freedom and their valour are buried in the same grave.' Even if the speaker here had only said that it was right to weep when valour was being buried in their grave, it would have been a metaphor, and a graphic one; but the coupling of 'their valour' and 'her freedom' presents a kind of antithesis as well. 'The course of my words', said Iphicrates, 'lies straight through the middle of Chares' deeds': this is a proportional metaphor, and the phrase 'straight through the middle' makes it graphic. The expression 'to call in one danger to rescue us from another' is a graphic metaphor. Lycoleon said, defending Chabrias, 'They did not respect even that bronze statue of his that intercedes for him yonder'. This was a metaphor for the moment, though it would not always apply; a vivid metaphor, however; Chabrias is in danger, and his statue intercedes for him—that lifeless yet living thing which records his services to his country. 'Practising in every way littleness of mind' is metaphorical, for practising a quality implies increasing it. So is 'God kindled our reason to be a lamp within our soul', for both reason and light reveal things. So is 'we are not putting an end to our wars, but only postponing them', for both literal postponement and the making of such a peace as this apply to future action. So is such a saying as 'This treaty is a far nobler trophy than those we set up on fields of battle; they celebrate small gains and single successes; it celebrates our triumph in the war as a whole'; for both trophy and treaty are signs of victory. So is 'A country pays a heavy reckoning in being condemned by the judgement of mankind', for a reckoning is damage deservedly incurred.

11

It has already been mentioned that liveliness is got by using the proportional type of metaphor and being making (ie. making your hearers see things). We have still to explain what we mean by their 'seeing things', and what must be done to effect this. By 'making them see things' I mean using expressions that represent things as in a state of activity. Thus, to say that a good man is 'four-square' is certainly a metaphor; both the good man and the square are perfect; but the metaphor does not suggest activity. On the other hand, in the expression 'with his vigour in full bloom' there is a notion of activity; and so in 'But you must roam as free as a sacred victim'; and in

Thereas up sprang the Hellenes to their feet,

where 'up sprang' gives us activity as well as metaphor, for it at once suggests swiftness. So with Homer's common practice of giving metaphorical life to lifeless things: all such passages are distinguished by the effect of activity they convey. Thus,

Downward anon to the valley rebounded the boulder remorseless;
and

The (bitter) arrow flew;
and

Flying on eagerly;
and

Stuck in the earth, still panting to feed on the flesh of the heroes;
and

And the point of the spear in its fury drove
full through his breastbone.

In all these examples the things have the effect of being active because they are made into living beings; shameless behaviour and fury and so on are all forms of activity. And the poet has attached these ideas to the things by means of proportional metaphors: as the stone is to Sisyphus, so is the shameless man to his victim. In his famous similes, too, he treats inanimate things in the same way:

Curving and crested with white, host following
host without ceasing.

Here he represents everything as moving and living; and activity is movement.

Metaphors must be drawn, as has been said already, from things that are related to the original thing, and yet not obviously so related—just as in philosophy also an acute mind will perceive resemblances even in things far apart. Thus Archytas said that an arbitrator and an altar were the same, since the injured fly to both for refuge. Or you might say that an anchor and an overhead hook were the same, since both are in a way the same, only the one secures things from below and the other from above. And to speak of states as 'levelled' is to identify two widely different things, the equality of a physical surface and the equality of political powers.

Liveliness is specially conveyed by metaphor, and by the further power of surprising the hearer; because the hearer expected something different, his acquisition of the new idea impresses him all the more. His mind seems to say, 'Yes, to be sure; I never thought of that'. The liveliness of epigrammatic remarks is due to the meaning not being just what the words say: as in the saying of Stesichorus that 'the cicalas will chirp to themselves on the ground'. Well-constructed riddles are attractive for the same reason; a new idea is conveyed, and there is metaphorical expression. So with the 'novelties' of Theodorus. In these the thought is startling, and, as Theodorus puts it, does not fit in with the ideas you already have. They are like the burlesque words that one finds in the comic writers. The effect is produced even by jokes depending upon changes of the letters of a word; this too is a surprise. You find this in verse as well as in prose. The word which comes is not what the hearer imagined: thus

Onward he came, and his feet were shod with his—chilblains,

where one imagined the word would be 'sandals'. But the point should be clear the moment the words are uttered. Jokes made by altering the letters of a word consist in meaning, not just what you say, but something that gives a twist to the word used; e.g. the remark of Theodorus about Nicon the harpist Thratt' ei su ('you Thracian slavey'), where he pretends to mean Thratteis su ('you harpplayer'), and surprises us when we find he means something else. So you enjoy the point when you see it, though the remark will fall flat unless you are aware that Nicon is Thracian. Or again: Boulei auton persai. In both these cases the saying must fit the facts. This is also true of such lively remarks as the one to the effect that to the Athenians their empire (arche) of the sea was not the beginning (arche) of their troubles, since they gained by it. Or the opposite one of Isocrates, that their empire (arche) was the beginning (arche) of their troubles. Either way, the speaker says something unexpected, the soundness of which is thereupon recognized. There would be nothing clever in saying 'empire is empire'. Isocrates means more than that, and uses the word with a new meaning. So too with the former saying, which denies that arche in one sense was arche in another sense. In all these jokes, whether a word is used in a second sense or metaphorically, the joke is good if it fits the facts. For instance, Anaschetos (proper name) ouk anaschetos: where you say that what is so-and-so in one sense is not so-and-so in another; well, if the man is unpleasant, the joke fits the facts. Again, take—

Thou must not be a stranger stranger than Thou should'st.

Do not the words 'thou must not be', &c., amount to saying that the stranger must not always be strange? Here again is the use of one word in different senses. Of the same kind also is the much-praised verse of Anaxandrides:

Death is most fit before you do
Deeds that would make death fit for you.

This amounts to saying 'it is a fit thing to die when you are not fit to die', or 'it is a fit thing to die when death is not fit for you', i.e. when death is not the fit return for what you are doing. The type of language employed—is the same in all these examples; but the more briefly and antithetically such sayings can be expressed, the more taking they are, for antithesis impresses the new idea more firmly and brevity more quickly. They should always have either some personal application or some merit of expression, if they are to be true without being commonplace—two requirements not always satisfied simultaneously. Thus 'a man should die having done no wrong' is true but dull: 'the right man should marry the right woman' is also true but dull. No, there must be both good qualities together, as in 'it is fitting to die when you are not fit for death'. The more a saying has these qualitis, the livelier it appears: if, for instance, its wording is metaphorical, metaphorical in the right way, antithetical, and balanced, and at the same time it gives an idea of activity.

Successful similes also, as has been said above, are in a sense metaphors, since they always involve two relations like the proportional metaphor. Thus: a shield, we say, is the 'drinking-bowl of Ares', and a bow is the 'chordless lyre'. This way of putting a metaphor is not 'simple', as it would be if we called the bow a lyre or the shield a drinking-bowl. There are 'simple' similes also: we may say that a flute-player is like a monkey, or that a short-sighted man's eyes are like a lamp-flame with water dropping on it, since both eyes and flame keep winking. A simile succeeds best when it is a converted metaphor, for it is possible to say that a shield is like the drinking-bowl of Ares, or that a ruin is like a house in rags, and to say that Niceratus is like a Philoctetes stung by Pratys—the simile made by Thrasyniachus when he saw Niceratus, who had been beaten by Pratys in a recitation competition, still going about unkempt and unwashed. It is in these respects that poets fail worst when they fail, and succeed best when they succeed, i.e. when they give the resemblance pat, as in

Those legs of his curl just like parsley leaves;

and

Just like Philammon struggling with his punchball.

These are all similes; and that similes are metaphors has been stated often already.

Proverbs, again, are metaphors from one species to another. Suppose, for instance, a man to start some undertaking in hope of gain and then to lose by it later on, 'Here we have once more the man of Carpathus and his hare', says he. For both alike went through the said experience.

It has now been explained fairly completely how liveliness is secured and why it has the effect it has. Successful hyperboles are also metaphors, e.g. the one about the man with a black eye, 'you would have thought he was a basket of mulberries'; here the 'black eye' is compared to a mulberry because of its colour, the exaggeration lying in the quantity of mulberries suggested. The phrase 'like so-and-so' may introduce a hyperbole under the form of a simile. Thus

Just like Philammon struggling with his punchball

is equivalent to 'you would have thought he was Philammon struggling with his punchball'; and

Those legs of his curl just like parsley leaves

is equivalent to 'his legs are so curly that you would have thought they were not legs but parsley leaves'. Hyperboles are for young men to use; they show vehemence of character; and this is why angry people use them more than other people.

Not though he gave me as much as the dust
or the sands of the sea...
But her, the daughter of Atreus' son, I never will marry,
Nay, not though she were fairer than Aphrodite the Golden,
Defter of hand than Athene...

(The Attic orators are particularly fond of this method of speech.) Consequently it does not suit an elderly speaker.

12

It should be observed that each kind of rhetoric has its own appropriate style. The style of written prose is not that of spoken oratory, nor are those of political and forensic speaking the same. Both written and spoken have to be known. To know the latter is to know how to speak good Greek. To know the former means that you are not obliged, as otherwise you are, to hold your tongue when you wish to communicate something to the general public.

The written style is the more finished: the spoken better admits of dramatic delivery—like the kind of oratory that reflects character and the kind that reflects emotion. Hence actors look out for plays written in the latter style, and poets for actors competent to act in such plays. Yet poets whose plays are meant to be read are read and circulated: Chaeremon, for instance, who is as finished as a professional speech-writer; and Licymnius among the dithyrambic poets. Compared with those of others, the speeches of professional writers sound thin in actual contests. Those of the orators, on the other hand, are good to hear spoken, but look amateurish enough when they pass into the hands of a reader. This is just because they are so well suited for an actual tussle, and therefore contain many dramatic touches, which, being robbed of all dramatic rendering, fail to do their own proper work, and consequently look silly. Thus strings of unconnected words, and constant repetitions of words and phrases, are very properly condemned in written speeches: but not in spoken speeches-speakers use them freely, for they have a dramatic effect. In this repetition there must be variety of tone, paving the way, as it were, to dramatic effect; e.g. 'This is the villain among you who deceived you, who cheated you, who meant to betray you completely'. This is the sort of thing that Philemon the actor used to do in the Old Men's Madness of Anaxandrides whenever he spoke the words 'Rhadamanthus and Palamedes', and also in the prologue to the Saints whenever he pronounced the pronoun 'I'. If one does not deliver such things cleverly, it becomes a case of 'the man who swallowed a poker'. So too with strings of unconnected words, e.g.'I came to him; I met him; I besought him'. Such passages must be acted, not delivered with the same quality and pitch of voice, as though they had only one idea in them. They have the further peculiarity of suggesting that a number of separate statements have been made in the time usually occupied by one. Just as the use of conjunctions makes many statements into a single one, so the omission of conjunctions acts in the reverse way and makes a single one into many. It thus makes everything more important: e.g. 'I came to him; I talked to him; I entreated him'—what a lot of facts! the hearer thinks—'he paid no attention to anything I said'. This is the effect which Homer seeks when he writes,

Nireus likewise from Syme (three well-fashioned ships did bring),
Nireus, the son of Aglaia (and Charopus, bright-faced king),
Nireus, the comeliest man (of all that to Ilium's strand).

If many things are said about a man, his name must be mentioned many times; and therefore people think that, if his name is mentioned many times, many things have been said about him. So that Homer, by means of this illusion, has made a great deal of though he has mentioned him only in this one passage, and has preserved his memory, though he nowhere says a word about him afterwards.

Now the style of oratory addressed to public assemblies is really just like scene-painting. The bigger the throng, the more distant is the point of view: so that, in the one and the other, high finish in detail is superfluous and seems better away. The forensic style is more highly finished; still more so is the style of language addressed to a single judge, with whom there is very little room for rhetorical artifices, since he can take the whole thing in better, and judge of what is to the point and what is not; the struggle is less intense and so the judgement is undisturbed. This is why the same speakers do not distinguish themselves in all these branches at once; high finish is wanted least where dramatic delivery is wanted most, and here the speaker must have a good voice, and above all, a strong one. It is ceremonial oratory that is most literary, for it is meant to be read; and next to it forensic oratory.

To analyse style still further, and add that it must be agreeable or magnificent, is useless; for why should it have these traits any more than 'restraint', 'liberality', or any other moral excellence? Obviously agreeableness will be produced by the qualities already mentioned, if our definition of excellence of style has been correct. For what other reason should style be 'clear', and 'not mean' but 'appropriate'? If it is prolix, it is not clear; nor yet if it is curt. Plainly the middle way suits best. Again, style will be made agreeable by the elements mentioned, namely by a good blending of ordinary and unusual words, by the rhythm, and by—the persuasiveness that springs from appropriateness.

This concludes our discussion of style, both in its general aspects and in its special applications to the various branches of rhetoric. We have now to deal with Arrangement.

13

A speech has two parts. You must state your case, and you must prove it. You cannot either state your case and omit to prove it, or prove it without having first stated it; since any proof must be a proof of something, and the only use of a preliminary statement is the proof that follows it. Of these two parts the first part is called the Statement of the case, the second part the Argument, just as we distinguish between Enunciation and Demonstration. The current division is absurd. For 'narration' surely is part of a forensic speech only: how in a political speech or a speech of display can there be 'narration' in the technical sense? or a reply to a forensic opponent? or an epilogue in closely-reasoned speeches? Again, introduction, comparison of conflicting arguments, and recapitulation are only found in political speeches when there is a struggle between two policies. They may occur then; so may even accusation and defence, often enough; but they form no essential part of a political speech. Even forensic speeches do not always need epilogues; not, for instance, a short speech, nor one in which the facts are easy to remember, the effect of an epilogue being always a reduction in the apparent length. It follows, then, that the only necessary parts of a speech are the Statement and the Argument. These are the essential features of a speech; and it cannot in any case have more than Introduction, Statement, Argument, and Epilogue. 'Refutation of the Opponent' is part of the arguments: so is 'Comparison' of the opponent's case with your own, for that process is a magnifying of your own case and therefore a part of the arguments, since one who does this proves something. The Introduction does nothing like this; nor does the Epilogue—it merely reminds us of what has been said already. If we make such distinctions we shall end, like Theodorus and his followers, by distinguishing 'narration' proper from 'post-narration' and 'pre-narration', and 'refutation' from 'final refutation'. But we ought only to bring in a new name if it indicates a real species with distinct specific qualities; otherwise the practice is pointless and silly, like the way Licymnius invented names in his Art of Rhetoric—'Secundation', 'Divagation', 'Ramification'.

14

The Introduction is the beginning of a speech, corresponding to the prologue in poetry and the prelude in flute—music; they are all beginnings, paving the way, as it were, for what is to follow. The musical prelude resembles the introduction to speeches of display; as flute players play first some brilliant passage they know well and then fit it on to the opening notes of the piece itself, so in speeches of display the writer should proceed in the same way; he should begin with what best takes his fancy, and then strike up his theme and lead into it; which is indeed what is always done. (Take as an example the introduction to the Helen of Isocrates—there is nothing in common between the 'eristics' and Helen.) And here, even if you travel far from your subject, it is fitting, rather than that there should be sameness in the entire speech.

The usual subject for the introductions to speeches of display is some piece of praise or censure. Thus Gorgias writes in his Olympic Speech, 'You deserve widespread admiration, men of Greece', praising thus those who start,ed the festival gatherings.' Isocrates, on the other hand, censures them for awarding distinctions to fine athletes but giving no prize for intellectual ability. Or one may begin with a piece of advice, thus: 'We ought to honour good men and so I myself am praising Aristeides' or 'We ought to honour those who are unpopular but not bad men, men whose good qualities have never been noticed, like Alexander son of Priam.' Here the orator gives advice. Or we may begin as speakers do in the law-courts; that is to say, with appeals to the audience to excuse us if our speech is about something paradoxical, difficult, or hackneyed; like Choerilus in the lines—

But now when allotment of all has been made...

Introductions to speeches of display, then, may be composed of some piece of praise or censure, of advice to do or not to do something, or of appeals to the audience; and you must choose between making these preliminary passages connected or disconnected with the speech itself.

Introductions to forensic speeches, it must be observed, have the same value as the prologues of dramas and the introductions to epic poems; the dithyrambic prelude resembling the introduction to a speech of display, as

For thee, and thy gilts, and thy battle-spoils....

In prologues, and in epic poetry, a foretaste of the theme is given, intended to inform the hearers of it in advance instead of keeping their minds in suspense. Anything vague puzzles them: so give them a grasp of the beginning, and they can hold fast to it and follow the argument. So we find—

Sing, O goddess of song, of the Wrath...

Tell me, O Muse, of the hero...

Lead me to tell a new tale, how there came great warfare to Europe
Out of the Asian land...

The tragic poets, too, let us know the pivot of their play; if not at the outset like Euripides, at least somewhere in the preface to a speech like Sophocles—

Polybus was my father...;

and so in Comedy. This, then, is the most essential function and distinctive property of the introduction, to show what the aim of the speech is; and therefore no introduction ought to be employed where the subject is not long or intricate.

The other kinds of introduction employed are remedial in purpose, and may be used in any type of speech. They are concerned with the speaker, the hearer, the subject, or the speaker's opponent. Those concerned with the speaker himself or with his opponent are directed to removing or exciting prejudice. But whereas the defendant will begin by dealing with this sort of thing, the prosecutor will take quite another line and deal with such matters in the closing part of his speech. The reason for this is not far to seek. The defendant, when he is going to bring himself on the stage, must clear away any obstacles, and therefore must begin by removing any prejudice felt against him. But if you are to excite prejudice, you must do so at the close, so that the judges may more easily remember what you have said.

The appeal to the hearer aims at securing his goodwill, or at arousing his resentment, or sometimes at gaining his serious attention to the case, or even at distracting it—for gaining it is not always an advantage, and speakers will often for that reason try to make him laugh.

You may use any means you choose to make your hearer receptive; among others, giving him a good impression of your character, which always helps to secure his attention. He will be ready to attend to anything that touches himself and to anything that is important, surprising, or agreeable; and you should accordingly convey to him the impression that what you have to say is of this nature. If you wish to distract his attention, you should imply that the subject does not affect him, or is trivial or disagreeable. But observe, all this has nothing to do with the speech itself. It merely has to do with the weak-minded tendency of the hearer to listen to what is beside the point. Where this tendency is absent, no introduction wanted beyond a summary statement of your subject, to put a sort of head on the main body of your speech. Moreover, calls for attention, when required, may come equally well in any part of a speech; in fact, the beginning of it is just where there is least slackness of interest; it is therefore ridiculous to put this kind of thing at the beginning, when every one is listening with most attention. Choose therefore any point in the speech where such an appeal is needed, and then say 'Now I beg you to note this point—it concerns you quite as much as myself'; or

I will tell you that whose like you have never yet

heard for terror, or for wonder. This is what Prodicus called 'slipping in a bit of the fifty-drachma show-lecture for the audience whenever they began to nod'. It is plain that such introductions are addressed not to ideal hearers, but to hearers as we find them. The use of introductions to excite prejudice or to dispel misgivings is universal—

My lord, I will not say that eagerly...

or

Why all this preface?

Introductions are popular with those whose case is weak, or looks weak; it pays them to dwell on anything rather than the actual facts of it. That is why slaves, instead of answering the questions put to them, make indirect replies with long preambles. The means of exciting in your hearers goodwill and various other feelings of the same kind have already been described. The poet finely says May I find in Phaeacian hearts, at my coming, goodwill and compassion; and these are the two things we should aim at. In speeches of display we must make the hearer

feel that the eulogy includes either himself or his family or his way of life or
something or other of the kind. For it is true, as Socrates says in the Funeral
Speech, that 'the difficulty is not to praise the Athenians at Athens but at Sparta'.

The introductions of political oratory will be made out of the same materials as
those of the forensic kind, though the nature of political oratory makes them very
rare. The subject is known already, and therefore the facts of the case need no
introduction; but you may have to say something on account of yourself or to your
opponents; or those present may be inclined to treat the matter either more or less
seriously than you wish them to. You may accordingly have to excite or dispel some
prejudice, or to make the matter under discussion seem more or less important
than before: for either of which purposes you will want an introduction. You may
also want one to add elegance to your remarks, feeling that otherwise they will have
a casual air, like Gorgias' eulogy of the Eleans, in which, without any preliminary
sparring or fencing, he begins straight off with 'Happy city of Elis!'

15

In dealing with prejudice, one class of argument is that whereby you can dispel
objectionable suppositions about yourself. It makes no practical difference whether
such a supposition has been put into words or not, so that this distinction may be
ignored. Another way is to meet any of the issues directly: to deny the alleged fact;
or to say that you have done no harm, or none to him, or not as much as he says;
or that you have done him no injustice, or not much; or that you have done
nothing disgraceful, or nothing disgraceful enough to matter: these are the sort of
questions on which the dispute hinges. Thus Iphicrates replying to Nausicrates,
admitted that he had done the deed alleged, and that he had done Nausicrates
harm, but not that he had done him wrong. Or you may admit the wrong, but
balance it with other facts, and say that, if the deed harmed him, at any rate it was
honourable; or that, if it gave him pain, at least it did him good; or something else
like that. Another way is to allege that your action was due to mistake, or bad luck,
or necessity as Sophocles said he was not trembling, as his traducer maintained, in
order to make people think him an old man, but because he could not help it; he
would rather not be eighty years old. You may balance your motive against your
actual deed; saying, for instance, that you did not mean to injure him but to do
so-and-so; that you did not do what you are falsely charged with doing—the
damage was accidental—'I should indeed be a detestable person if I had
deliberately intended this result.' Another way is open when your calumniator, or
any of his connexions, is or has been subject to the same grounds for suspicion. Yet
another, when others are subject to the same grounds for suspicion but are
admitted to be in fact innocent of the charge: e.g. 'Must I be a profligate because
I am well-groomed? Then so-and-so must be one too.' Another, if other people
have been calumniated by the same man or some one else, or, without being
calumniated, have been suspected, like yourself now, and yet have been proved

innocent. Another way is to return calumny for calumny and say, 'It is monstrous to trust the man's statements when you cannot trust the man himself.' Another is when the question has been already decided. So with Euripides' reply to Hygiaenon, who, in the action for an exchange of properties, accused him of impiety in having written a line encouraging perjury—

My tongue hath sworn: no oath is on my soul.

Euripides said that his opponent himself was guilty in bringing into the law-courts cases whose decision belonged to the Dionysiac contests. 'If I have not already answered for my words there, I am ready to do so if you choose to prosecute me there.' Another method is to denounce calumny, showing what an enormity it is, and in particular that it raises false issues, and that it means a lack of confidence in the merits of his case. The argument from evidential circumstances is available for both parties: thus in the Teucer Odysseus says that Teucer is closely bound to Priam, since his mother Hesione was Priam's sister. Teucer replies that Telamon his father was Priam's enemy, and that he himself did not betray the spies to Priam. Another method, suitable for the calumniator, is to praise some trifling merit at great length, and then attack some important failing concisely; or after mentioning a number of good qualities to attack one bad one that really bears on the question. This is the method of thoroughly skilful and unscrupulous prosecutors. By mixing up the man's merits with what is bad, they do their best to make use of them to damage him.

There is another method open to both calumniator and apologist. Since a given action can be done from many motives, the former must try to disparage it by selecting the worse motive of two, the latter to put the better construction on it. Thus one might argue that Diomedes chose Odysseus as his companion because he supposed Odysseus to be the best man for the purpose; and you might reply to this that it was, on the contrary, because he was the only hero so worthless that Diomedes need not fear his rivalry.

16

We may now pass from the subject of calumny to that of Narration.

Narration in ceremonial oratory is not continuous but intermittent. There must, of course, be some survey of the actions that form the subject-matter of the speech. The speech is a composition containing two parts. One of these is not provided by the orator's art, viz. the actions themselves, of which the orator is in no sense author. The other part is provided by his namely, the proof (where proof is needed) that the actions were done, the description of their quality or of their extent, or even all these three things together. Now the reason why sometimes it is not desirable to make the whole narrative continuous is that the case thus expounded is hard to keep in mind. Show, therefore, from one set of facts that your hero is, e.g. brave, and from other sets of facts that he is able, just, &c. A speech thus arranged is comparatively simple, instead of being complicated and elaborate. You will have to recall well-known deeds among others; and because they are well-known, the hearer usually needs no narration of them; none, for instance, if your object is the praise of Achilles; we all know the facts of his life—what you have to do is to apply those facts. But if your object is the praise of Critias, you must narrate his deeds, which not many people know of...

Nowadays it is said, absurdly enough, that the narration should be rapid. Remember what the man said to the baker who asked whether he was to make the cake hard or soft: 'What, can't you make it right?' Just so here. We are not to make long narrations, just as we are not to make long introductions or long arguments. Here, again, rightness does not consist either in rapidity or in conciseness, but in the happy mean; that is, in saying just so much as will make the facts plain, or will lead the hearer to believe that the thing has happened, or that the man has caused injury or wrong to some one, or that the facts are really as important as you wish them to be thought: or the opposite facts to establish the opposite arguments.

You may also narrate as you go anything that does credit to yourself, e.g. 'I kept telling him to do his duty and not abandon his children'; or discredit to your adversary, e.g. 'But he answered me that, wherever he might find himself, there he would find other children', the answer Herodotus' records of the Egyptian mutineers. Slip in anything else that the judges will enjoy.

The defendant will make less of the narration. He has to maintain that the thing has not happened, or did no harm, or was not unjust, or not so bad as is alleged. He must therefor snot waste time about what is admitted fact, unless this bears on his own contention; e.g. that the thing was done, but was not wrong. Further, we must speak of events as past and gone, except where they excite pity or indignation by being represented as present. The Story told to Alcinous is an example of a brief chronicle, when it is repeated to Penelope in sixty lines. Another instance is the Epic Cycle as treated by Phayllus, and the prologue to the Oeneus.

The narration should depict character; to which end you must know what makes it do so. One such thing is the indication of moral purpose; the quality of purpose indicated determines the quality of character depicted and is itself determined by the end pursued. Thus it is that mathematical discourses depict no character; they have nothing to do with moral purpose, for they represent nobody as pursuing any end. On the other hand, the Socratic dialogues do depict character, being concerned with moral questions. This end will also be gained by describing the manifestations of various types of character, e.g. 'he kept walking along as he talked', which shows the man's recklessness and rough manners. Do not let your words seem inspired so much by intelligence, in the manner now current, as by moral purpose: e.g. 'I willed this; aye, it was my moral purpose; true, I gained nothing by it, still it is better thus.' For the other way shows good sense, but this shows good character; good sense making us go after what is useful, and good character after what is noble. Where any detail may appear incredible, then add the cause of it; of this Sophocles provides an example in the Antigone, where

Antigone says she had cared more for her brother than for husband or children, since if the latter perished they might be replaced,

> But since my father and mother in their graves
> Lie dead, no brother can be born to me.

If you have no such cause to suggest, just say that you are aware that no one will believe your words, but the fact remains that such is our nature, however hard the world may find it to believe that a man deliberately does anything except what pays him.

Again, you must make use of the emotions. Relate the familiar manifestations of them, and those that distinguish yourself and your opponent; for instance, 'he went away scowling at me'. So Aeschines described Cratylus as 'hissing with fury and shaking his fists'. These details carry conviction: the audience take the truth of what they know as so much evidence for the truth of what they do not. Plenty of such details may be found in Homer:

> Thus did she say: but the old woman buried her face in her hands:

a true touch—people beginning to cry do put their hands over their eyes.

Bring yourself on the stage from the first in the right character, that people may regard you in that light; and the same with your adversary; but do not let them see what you are about. How easily such impressions may be conveyed we can see from the way in which we get some inkling of things we know nothing of by the mere look of the messenger bringing news of them. Have some narrative in many different parts of your speech; and sometimes let there be none at the beginning of it.

In political oratory there is very little opening for narration; nobody can 'narrate' what has not yet happened. If there is narration at all, it will be of past events, the recollection of which is to help the hearers to make better plans for the future. Or it may be employed to attack some one's character, or to eulogize him—only then you will not be doing what the political speaker, as such, has to do.

If any statement you make is hard to believe, you must guarantee its truth, and at once offer an explanation, and then furnish it with such particulars as will be expected. Thus Carcinus' Jocasta, in his Oedipus, keeps guaranteeing the truth of her answers to the inquiries of the man who is seeking her son; and so with Haemon in Sophocles.

17

The duty of the Arguments is to attempt demonstrative proofs. These proofs must bear directly upon the question in dispute, which must fall under one of four heads. (1) If you maintain that the act was not committed, your main task in court is to prove this. (2) If you maintain that the act did no harm, prove this. If you maintain that (3) the act was less than is alleged, or (4) justified, prove these facts, just as you would prove the act not to have been committed if you were maintaining that.

It should be noted that only where the question in dispute falls under the first of these heads can it be true that one of the two parties is necessarily a rogue. Here ignorance cannot be pleaded, as it might if the dispute were whether the act was justified or not. This argument must therefore be used in this case only, not in the others.

In ceremonial speeches you will develop your case mainly by arguing that what has been done is, e.g., noble and useful. The facts themselves are to be taken on trust; proof of them is only submitted on those rare occasions when they are not easily credible or when they have been set down to some one else.

In political speeches you may maintain that a proposal is impracticable; or that, though practicable, it is unjust, or will do no good, or is not so important as its proposer thinks. Note any falsehoods about irrelevant matters—they will look like proof that his other statements also are false. Argument by 'example' is highly suitable for political oratory, argument by 'enthymeme' better suits forensic. Political oratory deals with future events, of which it can do no more than quote past events as examples. Forensic oratory deals with what is or is not now true, which can better be demonstrated, because not contingent—there is no contingency in what has now already happened. Do not use a continuous succession of enthymemes: intersperse them with other matter, or they will spoil one another's effect. There are limits to their number—

Friend, you have spoken as much as a sensible man would have spoken. ,as much' says Homer, not 'as well'. Nor should you try to make enthymemes on every point; if you do, you will be acting just like some students of philosophy, whose conclusions are more familiar and believable than the premises from which they draw them. And avoid the enthymeme form when you are trying to rouse feeling; for it will either kill the feeling or will itself fall flat: all simultaneous motions tend to cancel each other either completely or partially. Nor should you go after the enthymeme form in a passage where you are depicting character—the process of demonstration can express neither moral character nor moral purpose. Maxims should be employed in the Arguments—and in the Narration too—since these do express character: 'I have given him this, though I am quite aware that one should "Trust no man".' Or if you are appealing to the emotions: 'I do not regret it, though I have been wronged; if he has the profit on his side, I have justice on mine.'

Political oratory is a more difficult task than forensic; and naturally so, since it deals with the future, whereas the pleader deals with the past, which, as Epimenides of Crete said, even the diviners already know. (Epimenides did not practise divination about the future; only about the obscurities of the past.) Besides, in forensic oratory you have a basis in the law; and once you have a starting-point, you can prove anything with comparative ease. Then again, political oratory affords few chances for those leisurely digressions in which you may attack your adversary, talk about yourself, or work on your hearers' emotions; fewer chances indeed, than any other affords, unless your set purpose is to divert your hearers' attention. Accordingly, if you find yourself in difficulties, follow the lead of the Athenian speakers, and that of Isocrates, who makes regular attacks upon people in the course of a political speech, e.g. upon the Lacedaemonians in the Panegyricus, and upon Chares in the speech about the allies. In ceremonial oratory, intersperse your speech with bits of episodic eulogy, like Isocrates, who is always bringing some one forward for this purpose. And this is what Gorgias meant by saying that he always found something to talk about. For if he speaks of Achilles, he praises Peleus, then Aeacus, then Zeus; and in like manner the virtue of valour, describing its good results, and saying what it is like.

Now if you have proofs to bring forward, bring them forward, and your moral discourse as well; if you have no enthymemes, then fall back upon moral discourse: after all, it is more fitting for a good man to display himself as an honest fellow than as a subtle reasoner. Refutative enthymemes are more popular than demonstrative ones: their logical cogency is more striking: the facts about two opposites always stand out clearly when the two are nut side by side.

The 'Reply to the Opponent' is not a separate division of the speech; it is part of the Arguments to break down the opponent's case, whether by objection or by counter-syllogism. Both in political speaking and when pleading in court, if you are the first speaker you should put your own arguments forward first, and then meet the arguments on the other side by refuting them and pulling them to pieces beforehand. If, however, the case for the other side contains a great variety of arguments, begin with these, like Callistratus in the Messenian assembly, when he demolished the arguments likely to be used against him before giving his own. If you speak later, you must first, by means of refutation and counter-syllogism, attempt some answer to your opponent's speech, especially if his arguments have been well received. For just as our minds refuse a favourable reception to a person against whom they are prejudiced, so they refuse it to a speech when they have been favourably impressed by the speech on the other side. You should, therefore, make room in the minds of the audience for your coming speech; and this will be done by getting your opponent's speech out of the way. So attack that first—either the whole of it, or the most important, successful, or vulnerable points in it, and thus inspire confidence in what you have to say yourself—

First, champion will I be of Goddesses...

Never, I ween, would Hera...

where the speaker has attacked the silliest argument first. So much for the Arguments.

With regard to the element of moral character: there are assertions which, if made about yourself, may excite dislike, appear tedious, or expose you to the risk of contradiction; and other things which you cannot say about your opponent without seeming abusive or ill-bred. Put such remarks, therefore, into the mouth of some third person. This is what Isocrates does in the Philippus and in the Antidosis, and Archilochus in his satires. The latter represents the father himself as attacking his daughter in the lampoon

Think nought impossible at all,

Nor swear that it shall not befall...

and puts into the mouth of Charon the carpenter the lampoon which begins
    Not for the wealth of Gyes...
So too Sophocles makes Haemon appeal to his father on behalf of Antigone as if it were others who were speaking.

Again, sometimes you should restate your enthymemes in the form of maxims; e.g. 'Wise men will come to terms in the hour of success; for they will gain most if they do'. Expressed as an enthymeme, this would run, 'If we ought to come to terms when doing so will enable us to gain the greatest advantage, then we ought to come to terms in the hour of success.'

18

Next as to Interrogation. The best moment to a employ this is when your opponent has so answered one question that the putting of just one more lands him in absurdity. Thus Pericles questioned Lampon about the way of celebrating the rites of the Saviour Goddess. Lampon declared that no uninitiated person could be told of them. Pericles then asked, 'Do you know them yourself?' 'Yes', answered Lampon. 'Why,' said Pericles, 'how can that be, when you are uninitiated?'

Another good moment is when one premiss of an argument is obviously true, and you can see that your opponent must say 'yes' if you ask him whether the other is true. Having first got this answer about the other, do not go on to ask him about the obviously true one, but just state the conclusion yourself. Thus, when Meletus denied that Socrates believed in the existence of gods but admitted that he talked about a supernatural power, Socrates proceeded to to ask whether 'supernatural beings were not either children of the gods or in some way divine?' 'Yes', said Meletus. 'Then', replied Socrates, 'is there any one who believes in the existence of children of the gods and yet not in the existence of the gods themselves?' Another good occasion is when you expect to show that your opponent is contradicting either his own words or what every one believes. A fourth is when it is impossible for him to meet your question except by an evasive answer. If he answers 'True, and yet not true', or 'Partly true and partly not true', or 'True in one sense but not in another', the audience thinks he is in difficulties, and applauds his discomfiture. In other cases do not attempt interrogation; for if your opponent gets in an objection, you are felt to have been worsted. You cannot ask a series of questions owing to the incapacity of the audience to follow them; and for this reason you should also make your enthymemes as compact as possible.

In replying, you must meet ambiguous questions by drawing reasonable distinctions, not by a curt answer. In meeting questions that seem to involve you in a contradiction, offer the explanation at the outset of your answer, before your opponent asks the next question or draws his conclusion. For it is not difficult to see the drift of his argument in advance. This point, however, as well as the various means of refutation, may be regarded as known to us from the Topics.

When your opponent in drawing his conclusion puts it in the form of a question, you must justify your answer. Thus when Sophocles was asked by

Peisander whether he had, like the other members of the Board of Safety, voted for setting up the Four Hundred, he said 'Yes.'—'Why, did you not think it wicked?'—'Yes.'—'So you committed this wickedness?' 'Yes', said Sophocles, 'for there was nothing better to do.' Again, the Lacedaemonian, when he was being examined on his conduct as ephor, was asked whether he thought that the other ephors had been justly put to death. 'Yes', he said. 'Well then', asked his opponent, 'did not you propose the same measures as they?'—'Yes.'—'Well then, would not you too be justly put to death?'—'Not at all', said he; 'they were bribed to do it, and I did it from conviction'. Hence you should not ask any further questions after drawing the conclusion, nor put the conclusion itself in the form of a further question, unless there is a large balance of truth on your side.

As to jests. These are supposed to be of some service in controversy. Gorgias said that you should kill your opponents' earnestness with jesting and their jesting with earnestness; in which he was right. jests have been classified in the Poetics. Some are becoming to a gentleman, others are not; see that you choose such as become you. Irony better befits a gentleman than buffoonery; the ironical man jokes to amuse himself, the buffoon to amuse other people.

19

The Epilogue has four parts. You must (1) make the audience well-disposed towards yourself and ill-disposed towards your opponent (2) magnify or minimize the leading facts, (3) excite the required state of emotion in your hearers, and (4) refresh their memories.

(1) Having shown your own truthfulness and the untruthfulness of your opponent, the natural thing is to commend yourself, censure him, and hammer in your points. You must aim at one of two objects—you must make yourself out a good man and him a bad one either in yourselves or in relation to your hearers. How this is to be managed—by what lines of argument you are to represent people as good or bad—this has been already explained.

(2) The facts having been proved, the natural thing to do next is to magnify or minimize their importance. The facts must be admitted before you can discuss how important they are; just as the body cannot grow except from something already present. The proper lines of argument to be used for this purpose of amplification and depreciation have already been set forth.

(3) Next, when the facts and their importance are clearly understood, you must excite your hearers' emotions. These emotions are pity, indignation, anger, hatred, envy, emulation, pugnacity. The lines of argument to be used for these purposes also have been previously mentioned.

(4) Finally you have to review what you have already said. Here you may properly do what some wrongly recommend doing in the introduction—repeat your points frequently so as to make them easily understood. What you should do in your introduction is to state your subject, in order that the point to be judged

may be quite plain; in the epilogue you should summarize the arguments by which your case has been proved. The first step in this reviewing process is to observe that you have done what you undertook to do. You must, then, state what you have said and why you have said it. Your method may be a comparison of your own case with that of your opponent; and you may compare either the ways you have both handled the same point or make your comparison less direct: 'My opponent said so-and-so on this point; I said so-and-so, and this is why I said it'. Or with modest irony, e.g. 'He certainly said so-and-so, but I said so-and-so'. Or 'How vain he would have been if he had proved all this instead of that!' Or put it in the form of a question. 'What has not been proved by me?' or 'What has my opponent proved?' You may proceed then, either in this way by setting point against point, or by following the natural order of the arguments as spoken, first giving your own, and then separately, if you wish, those of your opponent.

For the conclusion, the disconnected style of language is appropriate, and will mark the difference between the oration and the peroration. 'I have done. You have heard me. The facts are before you. I ask for your judgement.'

On the Heavens

Table of Contents

Book I

1

THE science which has to do with nature clearly concerns itself for the most part with bodies and magnitudes and their properties and movements, but also with the principles of this sort of substance, as many as they may be. For of things constituted by nature some are bodies and magnitudes, some possess body and magnitude, and some are principles of things which possess these. Now a continuum is that which is divisible into parts always capable of subdivision, and a body is that which is every way divisible. A magnitude if divisible one way is a line, if two ways a surface, and if three a body. Beyond these there is no other magnitude, because the three dimensions are all that there are, and that which is divisible in three directions is divisible in all. For, as the Pythagoreans say, the world and all that is in it is determined by the number three, since beginning and middle and end give the number of an 'all', and the number they give is the triad. And so, having taken these three from nature as (so to speak) laws of it, we make further use of the number three in the worship of the Gods. Further, we use the terms in practice in this way. Of two things, or men, we say 'both', but not 'all': three is the first number to which the term 'all' has been appropriated. And in this, as we have said, we do but follow the lead which nature gives. Therefore, since 'every' and 'all' and 'complete' do not differ from one another in respect of form, but only, if at all, in their matter and in that to which they are applied, body alone among magnitudes can be complete. For it alone is determined by the three dimensions, that is, is an 'all'. But if it is divisible in three dimensions it is every way divisible, while the other magnitudes are divisible in one dimension or in two alone: for the divisibility and continuity of magnitudes depend upon the number of the dimensions, one sort being continuous in one direction, another in two, another in all. All magnitudes, then, which are divisible are also continuous. Whether we can also say that whatever is continuous is divisible does not yet, on our present grounds, appear. One thing, however, is clear. We cannot pass beyond body to a further kind, as we passed from length to surface, and from surface to body. For if we could, it would cease to be true that body is complete magnitude. We could pass beyond it only in virtue of a defect in it; and that which is complete cannot be defective, since it has being in every respect. Now bodies which are classed as parts of the whole are each complete according to our formula, since each possesses every dimension. But each is determined relatively to that part which is next to it by contact, for which reason each of them is in a sense many bodies. But the whole of which they are parts must necessarily be complete, and thus, in accordance with the meaning of the word, have being, not in some respect only, but in every respect.

2

The question as to the nature of the whole, whether it is infinite in size or limited in its total mass, is a matter for subsequent inquiry. We will now speak of those parts of the whole which are specifically distinct. Let us take this as our starting-point. All natural bodies and magnitudes we hold to be, as such, capable of locomotion; for nature, we say, is their principle of movement. But all movement that is in place, all locomotion, as we term it, is either straight or circular or a combination of these two, which are the only simple movements. And the reason of this is that these two, the straight and the circular line, are the only simple magnitudes. Now revolution about the centre is circular motion, while the upward and downward movements are in a straight line, 'upward' meaning motion away from the centre, and 'downward' motion towards it. All simple motion, then, must be motion either away from or towards or about the centre. This seems to be in exact accord with what we said above: as body found its completion in three dimensions, so its movement completes itself in three forms.

Bodies are either simple or compounded of such; and by simple bodies I mean those which possess a principle of movement in their own nature, such as fire and earth with their kinds, and whatever is akin to them. Necessarily, then, movements also will be either simple or in some sort compound-simple in the case of the simple bodies, compound in that of the composite-and in the latter case the motion will be that of the simple body which prevails in the composition. Supposing, then, that there is such a thing as simple movement, and that circular movement is an instance of it, and that both movement of a simple body is simple and simple movement is of a simple body (for if it is movement of a compound it will be in virtue of a prevailing simple element), then there must necessarily be some simple body which revolves naturally and in virtue of its own nature with a circular movement. By constraint, of course, it may be brought to move with the motion of something else different from itself, but it cannot so move naturally, since there is one sort of movement natural to each of the simple bodies. Again, if the unnatural movement is the contrary of the natural and a thing can have no more than one contrary, it will follow that circular movement, being a simple motion, must be unnatural, if it is not natural, to the body moved. If then (1) the body, whose movement is circular, is fire or some other element, its natural motion must be the contrary of the circular motion. But a single thing has a single contrary; and upward and downward motion are the contraries of one another. If, on the other hand, (2) the body moving with this circular motion which is unnatural to it is something different from the elements, there will be some other motion which is natural to it. But this cannot be. For if the natural motion is upward, it will be fire or air, and if downward, water or earth. Further, this circular motion is necessarily primary. For the perfect is naturally prior to the imperfect, and the circle is a perfect thing. This cannot be said of any straight line:-not of an infinite line; for, if it were perfect, it would have a limit and an end: nor of any finite line; for in every case there is something beyond it, since any finite line can be extended. And so, since the prior movement belongs to the body which naturally prior, and circular movement is prior to straight, and movement in a straight line belongs to simple bodies-fire moving straight upward and earthy bodies straight downward towards the centre-since this is so, it follows that circular movement also must be the movement of some simple body. For the movement of composite bodies is, as we said, determined by that simple body which preponderates in the composition. These premises clearly give the conclusion that there is in nature some bodily substance other than the formations we know, prior to them all and more divine than they. But it may also be proved as follows. We may take it that all movement is either natural or unnatural, and that the movement which is unnatural to one body is natural to another-as, for instance, is the case with the upward and downward movements, which are natural and unnatural to fire and earth respectively. It necessarily follows that circular movement, being unnatural to these bodies, is the natural movement of some other. Further, if, on the one hand, circular movement is natural to something, it must surely be some simple and primary body which is ordained to move with a natural circular motion, as fire is ordained to fly up and earth down. If, on the other hand, the movement of the rotating bodies about the centre is unnatural, it would be remarkable and indeed quite inconceivable that this movement alone should be continuous and eternal, being nevertheless contrary to nature. At any rate the evidence of all other cases goes to show that it is the unnatural which quickest passes away. And so, if, as some say, the body so moved is fire, this movement is just as unnatural to it as downward movement; for any one can see that fire moves in a straight line away from the centre. On all these grounds, therefore, we may infer with confidence that there is something beyond the bodies that are about us on this earth, different and separate from them; and that the superior glory of its nature is proportionate to its distance from this world of ours.

3

In consequence of what has been said, in part by way of assumption and in part by way of proof, it is clear that not every body either possesses lightness or heaviness. As a preliminary we must explain in what sense we are using the words 'heavy' and 'light', sufficiently, at least, for our present purpose: we can examine the terms more closely later, when we come to consider their essential nature. Let us then apply the term 'heavy' to that which naturally moves towards the centre, and 'light' to that which moves naturally away from the centre. The heaviest thing will be that which sinks to the bottom of all things that move downward, and the lightest that which rises to the surface of everything that moves upward. Now, necessarily, everything which moves either up or down possesses lightness or heaviness or both-but not both relatively to the same thing: for things are heavy and light relatively to one another; air, for instance, is light relatively to water, and water light relatively to earth. The body, then, which moves in a circle cannot possibly possess either heaviness or lightness. For neither naturally nor unnaturally can it move either towards or away from the centre. Movement in a straight line certainly does not belong to it naturally, since one sort of movement is, as we saw, appropriate to each simple body, and so we should be compelled to identify it with one of the bodies which move in this way. Suppose, then, that the movement is unnatural. In that case, if it is the downward movement which is unnatural, the upward movement will be natural; and if it is the upward which is unnatural, the downward will be natural. For we decided that of contrary movements, if the one is unnatural to anything, the other will be natural to it. But since the natural movement of the whole and of its part of earth, for instance, as a whole and of a small clod-have one and the same direction, it results, in the first place, that this body can possess no lightness or heaviness at all (for that would mean that it could move by its own nature either from or towards the centre, which, as we know, is impossible); and, secondly, that it cannot possibly move in the way of locomotion by being forced violently aside in an upward or downward direction. For neither naturally nor unnaturally can it move with any other motion but its own, either itself or any part of it, since the reasoning which applies to the whole applies also to the part.

It is equally reasonable to assume that this body will be ungenerated and indestructible and exempt from increase and alteration, since everything that comes to be comes into being from its contrary and in some substrate, and passes away likewise in a substrate by the action of the contrary into the contrary, as we explained in our opening discussions. Now the motions of contraries are contrary. If then this body can have no contrary, because there can be no contrary motion to the circular, nature seems justly to have exempted from contraries the body which was to be ungenerated and indestructible. For it is in contraries that generation and decay subsist. Again, that which is subject to increase increases upon contact with a kindred body, which is resolved into its matter. But there is nothing out of which this body can have been generated. And if it is exempt from increase and diminution, the same reasoning leads us to suppose that it is also unalterable. For alteration is movement in respect of quality; and qualitative states and dispositions, such as health and disease, do not come into being without changes of properties. But all natural bodies which change their properties we see to be subject without exception to increase and diminution. This is the case, for instance, with the bodies of animals and their parts and with vegetable bodies, and similarly also with those of the elements. And so, if the body which moves with a circular motion cannot admit of increase or diminution, it is reasonable to suppose that it is also unalterable.

The reasons why the primary body is eternal and not subject to increase or diminution, but unaging and unalterable and unmodified, will be clear from what has been said to any one who believes in our assumptions. Our theory seems to confirm experience and to be confirmed by it. For all men have some conception of the nature of the gods, and all who believe in the existence of gods at all, whether barbarian or Greek, agree in allotting the highest place to the deity, surely because they suppose that immortal is linked with immortal and regard any other supposition as inconceivable. If then there is, as there certainly is, anything divine, what we have just said about the primary bodily substance was well said. The mere evidence of the senses is enough to convince us of this, at least with human certainty. For in the whole range of time past, so far as our inherited records reach, no change appears to have taken place either in the whole scheme of the outermost heaven or in any of its proper parts. The common name, too, which has been handed down from our distant ancestors even to our own day, seems to show that they conceived of it in the fashion which we have been expressing. The same ideas, one must believe, recur in men's minds not once or twice but again and again. And so, implying that the primary body is something else beyond earth, fire, air, and water, they gave the highest place a name of its own, aither, derived from the fact that it 'runs always' for an eternity of time. Anaxagoras, however, scandalously misuses this name, taking aither as equivalent to fire.

It is also clear from what has been said why the number of what we call simple bodies cannot be greater than it is. The motion of a simple body must itself be simple, and we assert that there are only these two simple motions, the circular and the straight, the latter being subdivided into motion away from and motion towards the centre.

4

That there is no other form of motion opposed as contrary to the circular may be proved in various ways. In the first place, there is an obvious tendency to oppose the straight line to the circular. For concave and convex are a not only regarded as opposed to one another, but they are also coupled together and treated as a unity in opposition to the straight. And so, if there is a contrary to circular motion, motion in a straight line must be recognized as having the best claim to that name. But the two forms of rectilinear motion are opposed to one another by reason of their places; for up and down is a difference and a contrary opposition in place. Secondly, it may be thought that the same reasoning which holds good of the rectilinear path applies also the circular, movement from A to B being opposed as contrary to movement from B to A. But what is meant is still rectilinear motion. For that is limited to a single path, while the circular paths which pass through the same two points are infinite in number. Even if we are confined to the single semicircle and the opposition is between movement from C to D and from D to C along that semicircle, the case is no better. For the motion is the same as that along the diameter, since we invariably regard the distance between two points as the length of the straight line which joins them. It is no more satisfactory to construct a circle and treat motion 'along one semicircle as contrary to motion along the other. For example, taking a complete circle, motion from E to F on the semicircle G may be opposed to motion from F to E on the semicircle H. But even supposing these are contraries, it in no way follows that the reverse motions on the complete circumference contraries. Nor again can motion along the circle from A to B be regarded as the contrary of motion from A to C: for the motion goes from the same point towards the same point, and contrary motion was distinguished as motion from a contrary to its contrary. And even if the motion round a circle is the contrary of the reverse motion, one of the two would be ineffective: for both move to the same point, because that which moves in a circle, at whatever point it begins, must necessarily pass through all the contrary places alike. (By contrarieties of place I mean up and down, back and front, and right and left; and the contrary oppositions of movements are determined by those of places.) One of the motions, then, would be ineffective, for if the two motions were of equal strength, there would be no movement either way, and if one of the two were preponderant, the other would be inoperative. So that if both bodies were there, one of them, inasmuch as it would not be moving with its own movement, would be useless, in the sense in which a shoe is useless when it is not worn. But God and nature create nothing that has not its use.

5

This being clear, we must go on to consider the questions which remain. First, is there an infinite body, as the majority of the ancient philosophers thought, or is this an impossibility? The decision of this question, either way, is not unimportant, but rather all-important, to our search for the truth. It is this problem which has practically always been the source of the differences of those who have written about nature as a whole. So it has been and so it must be; since the least initial deviation from the truth is multiplied later a thousandfold. Admit, for instance, the existence of a minimum magnitude, and you will find that the minimum which you have introduced, small as it is, causes the greatest truths of mathematics to totter. The reason is that a principle is great rather in power than in extent; hence that which was small at the start turns out a giant at the end. Now the conception of the infinite possesses this power of principles, and indeed in the sphere of quantity possesses it in a higher degree than any other conception; so that it is in no way absurd or unreasonable that the assumption that an infinite body exists should be of peculiar moment to our inquiry. The infinite, then, we must now discuss, opening the whole matter from the beginning.

Every body is necessarily to be classed either as simple or as composite; the infinite body, therefore, will be either simple or composite.

But it is clear, further, that if the simple bodies are finite, the composite must also be finite, since that which is composed of bodies finite both in number and in magnitude is itself finite in respect of number and magnitude: its quantity is in fact the same as that of the bodies which compose it. What remains for us to consider, then, is whether any of the simple bodies can be infinite in magnitude, or whether this is impossible. Let us try the primary body first, and then go on to consider the others.

The body which moves in a circle must necessarily be finite in every respect, for the following reasons. (1) If the body so moving is infinite, the radii drawn from the centre will be infinite. But the space between infinite radii is infinite: and by the space between the radii I mean the area outside which no magnitude which is in contact with the two lines can be conceived as falling. This, I say, will be infinite: first, because in the case of finite radii it is always finite; and secondly, because in it one can always go on to a width greater than any given width; thus the reasoning which forces us to believe in infinite number, because there is no maximum, applies also to the space between the radii. Now the infinite cannot be traversed, and if the body is infinite the interval between the radii is necessarily infinite: circular motion therefore is an impossibility. Yet our eyes tell us that the heavens revolve in a circle, and by argument also we have determined that there is something to which circular movement belongs.

(2) Again, if from a finite time a finite time be subtracted, what remains must be finite and have a beginning. And if the time of a journey has a beginning, there must be a beginning also of the movement, and consequently also of the distance traversed. This applies universally. Take a line, ACE, infinite in one direction, E, and another line, BB, infinite in both directions. Let ACE describe a circle, revolving upon C as centre. In its movement it will cut BB continuously for a certain time. This will be a finite time, since the total time is finite in which the heavens complete their circular orbit, and consequently the time subtracted from it, during which the one line in its motion cuts the other, is also finite. Therefore there will be a point at which ACE began for the first time to cut BB. This, however, is impossible. The infinite, then, cannot revolve in a circle; nor could the world, if it were infinite.

(3) That the infinite cannot move may also be shown as follows. Let A be a finite line moving past the finite line, B. Of necessity A will pass clear of B and B of A at the same moment; for each overlaps the other to precisely the same extent. Now if the two were both moving, and moving in contrary directions, they would pass clear of one another more rapidly; if one were still and the other moving past it, less rapidly; provided that the speed of the latter were the same in both cases. This, however, is clear: that it is impossible to traverse an infinite line in a finite time. Infinite time, then, would be required. (This we demonstrated above in the discussion of movement.) And it makes no difference whether a finite is passing by

an infinite or an infinite by a finite. For when A is passing B, then B overlaps A and it makes no difference whether B is moved or unmoved, except that, if both move, they pass clear of one another more quickly. It is, however, quite possible that a moving line should in certain cases pass one which is stationary quicker than it passes one moving in an opposite direction. One has only to imagine the movement to be slow where both move and much faster where one is stationary. To suppose one line stationary, then, makes no difficulty for our argument, since it is quite possible for A to pass B at a slower rate when both are moving than when only one is. If, therefore, the time which the finite moving line takes to pass the other is infinite, then necessarily the time occupied by the motion of the infinite past the finite is also infinite. For the infinite to move at all is thus absolutely impossible; since the very smallest movement conceivable must take an infinity of time. Moreover the heavens certainly revolve, and they complete their circular orbit in a finite time; so that they pass round the whole extent of any line within their orbit, such as the finite line AB. The revolving body, therefore, cannot be infinite.

(4) Again, as a line which has a limit cannot be infinite, or, if it is infinite, is so only in length, so a surface cannot be infinite in that respect in which it has a limit; or, indeed, if it is completely determinate, in any respect whatever. Whether it be a square or a circle or a sphere, it cannot be infinite, any more than a foot-rule can. There is then no such thing as an infinite sphere or square or circle, and where there is no circle there can be no circular movement, and similarly where there is no infinite at all there can be no infinite movement; and from this it follows that, an infinite circle being itself an impossibility, there can be no circular motion of an infinite body.

(5) Again, take a centre C, an infinite line, AB, another infinite line at right angles to it, E, and a moving radius, CD. CD will never cease contact with E, but the position will always be something like CE, CD cutting E at F. The infinite line, therefore, refuses to complete the circle.

(6) Again, if the heaven is infinite and moves in a circle, we shall have to admit that in a finite time it has traversed the infinite. For suppose the fixed heaven infinite, and that which moves within it equal to it. It results that when the infinite body has completed its revolution, it has traversed an infinite equal to itself in a finite time. But that we know to be impossible.

(7) It can also be shown, conversely, that if the time of revolution is finite, the area traversed must also be finite; but the area traversed was equal to itself; therefore, it is itself finite.

We have now shown that the body which moves in a circle is not endless or infinite, but has its limit.

6

Further, neither that which moves towards nor that which moves away from the centre can be infinite. For the upward and downward motions are contraries and are therefore motions towards contrary places. But if one of a pair of contraries is determinate, the other must be determinate also. Now the centre is determined; for, from whatever point the body which sinks to the bottom starts its downward motion, it cannot go farther than the centre. The centre, therefore, being determinate, the upper place must also be determinate. But if these two places are determined and finite, the corresponding bodies must also be finite. Further, if up and down are determinate, the intermediate place is also necessarily determinate. For, if it is indeterminate, the movement within it will be infinite; and that we have already shown to be an impossibility. The middle region then is determinate, and consequently any body which either is in it, or might be in it, is determinate. But the bodies which move up and down may be in it, since the one moves naturally away from the centre and the other towards it.

From this alone it is clear that an infinite body is an impossibility; but there is a further point. If there is no such thing as infinite weight, then it follows that none of these bodies can be infinite. For the supposed infinite body would have to be infinite in weight. (The same argument applies to lightness: for as the one supposition involves infinite weight, so the infinity of the body which rises to the surface involves infinite lightness.) This is proved as follows. Assume the weight to be finite, and take an infinite body, AB, of the weight C. Subtract from the infinite body a finite mass, BD, the weight of which shall be E. E then is less than C, since it is the weight of a lesser mass. Suppose then that the smaller goes into the greater a certain number of times, and take BF bearing the same proportion to BD which the greater weight bears to the smaller. For you may subtract as much as you please from an infinite. If now the masses are proportionate to the weights, and the lesser weight is that of the lesser mass, the greater must be that of the greater. The weights, therefore, of the finite and of the infinite body are equal. Again, if the weight of a greater body is greater than that of a less, the weight of GB will be greater than that of FB; and thus the weight of the finite body is greater than that of the infinite. And, further, the weight of unequal masses will be the same, since the infinite and the finite cannot be equal. It does not matter whether the weights are commensurable or not. If (a) they are incommensurable the same reasoning holds. For instance, suppose E multiplied by three is rather more than C: the weight of three masses of the full size of BD will be greater than C. We thus arrive at the same impossibility as before. Again (b) we may assume weights which are commensurate; for it makes no difference whether we begin with the weight or with the mass. For example, assume the weight E to be commensurate with C, and take from the infinite mass a part BD of weight E. Then let a mass BF be taken having the same proportion to BD which the two weights have to one another. (For the mass being infinite you may subtract from it as much as you please.) These assumed bodies will be commensurate in mass and in weight alike. Nor again does it make any difference to our demonstration whether the total mass has its weight equally or unequally distributed. For it must always be Possible to take from the infinite mass a body of equal weight to BD by diminishing or increasing the size of the section to the necessary extent.

From what we have said, then, it is clear that the weight of the infinite body cannot be finite. It must then be infinite. We have therefore only to show this to be impossible in order to prove an infinite body impossible. But the impossibility of infinite weight can be shown in the following way. A given weight moves a given distance in a given time; a weight which is as great and more moves the same distance in a less time, the times being in inverse proportion to the weights. For instance, if one weight is twice another, it will take half as long over a given movement. Further, a finite weight traverses any finite distance in a finite time. It necessarily follows from this that infinite weight, if there is such a thing, being, on the one hand, as great and more than as great as the finite, will move accordingly, but being, on the other hand, compelled to move in a time inversely proportionate to its greatness, cannot move at all. The time should be less in proportion as the weight is greater. But there is no proportion between the infinite and the finite: proportion can only hold between a less and a greater finite time. And though you may say that the time of the movement can be continually diminished, yet there is no minimum. Nor, if there were, would it help us. For some finite body could have been found greater than the given finite in the same proportion which is supposed to hold between the infinite and the given finite; so that an infinite and a finite weight must have traversed an equal distance in equal time. But that is impossible. Again, whatever the time, so long as it is finite, in which the infinite performs the motion, a finite weight must necessarily move a certain finite distance in that same time. Infinite weight is therefore impossible, and the same reasoning applies also to infinite lightness. Bodies then of infinite weight and of infinite lightness are equally impossible.

That there is no infinite body may be shown, as we have shown it, by a detailed consideration of the various cases. But it may also be shown universally, not only by such reasoning as we advanced in our discussion of principles (though in that passage we have already determined universally the sense in which the existence of an infinite is to be asserted or denied), but also suitably to our present purpose in the following way. That will lead us to a further question. Even if the total mass is not infinite, it may yet be great enough to admit a plurality of universes. The question might possibly be raised whether there is any obstacle to our believing that there are other universes composed on the pattern of our own, more than one, though stopping short of infinity. First, however, let us treat of the infinite universally.

7

Every body must necessarily be either finite or infinite, and if infinite, either of similar or of dissimilar parts. If its parts are dissimilar, they must represent either a finite or an infinite number of kinds. That the kinds cannot be infinite is evident, if our original presuppositions remain unchallenged. For the primary movements being finite in number, the kinds of simple body are necessarily also finite, since the movement of a simple body is simple, and the simple movements are finite, and every natural body must always have its proper motion. Now if the infinite body is to be composed of a finite number of kinds, then each of its parts must necessarily be infinite in quantity, that is to say, the water, fire, &c., which compose it. But this is impossible, because, as we have already shown, infinite weight and lightness do not exist. Moreover it would be necessary also that their places should be infinite in extent, so that the movements too of all these bodies would be infinite. But this is not possible, if we are to hold to the truth of our original presuppositions and to the view that neither that which moves downward, nor, by the same reasoning, that which moves upward, can prolong its movement to infinity. For it is true in regard to quality, quantity, and place alike that any process of change is impossible which can have no end. I mean that if it is impossible for a thing to have come to be white, or a cubit long, or in Egypt, it is also impossible for it to be in process of coming to be any of these. It is thus impossible for a thing to be moving to a place at which in its motion it can never by any possibility arrive. Again, suppose the body to exist in dispersion, it may be maintained none the less that the total of all these scattered particles, say, of fire, is infinite. But body we saw to be that which has extension every way. How can there be several dissimilar elements, each infinite? Each would have to be infinitely extended every way.

It is no more conceivable, again, that the infinite should exist as a whole of similar parts. For, in the first place, there is no other (straight) movement beyond those mentioned: we must therefore give it one of them. And if so, we shall have to admit either infinite weight or infinite lightness. Nor, secondly, could the body whose movement is circular be infinite, since it is impossible for the infinite to move in a circle. This, indeed, would be as good as saying that the heavens are infinite, which we have shown to be impossible.

Moreover, in general, it is impossible that the infinite should move at all. If it did, it would move either naturally or by constraint: and if by constraint, it possesses also a natural motion, that is to say, there is another place, infinite like itself, to which it will move. But that is impossible.

That in general it is impossible for the infinite to be acted upon by the finite or to act upon it may be shown as follows.

(1. The infinite cannot be acted upon by the finite.) Let A be an infinite, B a finite, C the time of a given movement produced by one in the other. Suppose, then, that A was heated, or impelled, or modified in any way, or caused to undergo any sort of movement whatever, by in the time C. Let D be less than B; and, assuming that a lesser agent moves a lesser patient in an equal time, call the quantity thus modified by D, E. Then, as D is to B, so is E to some finite quantum. We assume that the alteration of equal by equal takes equal time, and the alteration of less by less or of greater by greater takes the same time, if the quantity of the patient is such as to keep the proportion which obtains between the agents,

greater and less. If so, no movement can be caused in the infinite by any finite agent in any time whatever. For a less agent will produce that movement in a less patient in an equal time, and the proportionate equivalent of that patient will be a finite quantity, since no proportion holds between finite and infinite.

(2. The infinite cannot act upon the finite.) Nor, again, can the infinite produce a movement in the finite in any time whatever. Let A be an infinite, B a finite, C the time of action. In the time C, D will produce that motion in a patient less than B, say F. Then take E, bearing the same proportion to D as the whole BF bears to F. E will produce the motion in BF in the time C. Thus the finite and infinite effect the same alteration in equal times. But this is impossible; for the assumption is that the greater effects it in a shorter time. It will be the same with any time that can be taken, so that there will no time in which the infinite can effect this movement. And, as to infinite time, in that nothing can move another or be moved by it. For such time has no limit, while the action and reaction have.

(3. There is no interaction between infinites.) Nor can infinite be acted upon in any way by infinite. Let A and B be infinites, CD being the time of the action A of upon B. Now the whole B was modified in a certain time, and the part of this infinite, E, cannot be so modified in the same time, since we assume that a less quantity makes the movement in a less time. Let E then, when acted upon by A, complete the movement in the time D. Then, as D is to CD, so is E to some finite part of B. This part will necessarily be moved by A in the time CD. For we suppose that the same agent produces a given effect on a greater and a smaller mass in longer and shorter times, the times and masses varying proportionately. There is thus no finite time in which infinites can move one another. Is their time then infinite? No, for infinite time has no end, but the movement communicated has.

If therefore every perceptible body possesses the power of acting or of being acted upon, or both of these, it is impossible that an infinite body should be perceptible. All bodies, however, that occupy place are perceptible. There is therefore no infinite body beyond the heaven. Nor again is there anything of limited extent beyond it. And so beyond the heaven there is no body at all. For if you suppose it an object of intelligence, it will be in a place-since place is what 'within' and 'beyond' denote-and therefore an object of perception. But nothing that is not in a place is perceptible.

The question may also be examined in the light of more general considerations as follows. The infinite, considered as a whole of similar parts, cannot, on the one hand, move in a circle. For there is no centre of the infinite, and that which moves in a circle moves about the centre. Nor again can the infinite move in a straight line. For there would have to be another place infinite like itself to be the goal of its natural movement and another, equally great, for the goal of its unnatural movement. Moreover, whether its rectilinear movement is natural or constrained, in either case the force which causes its motion will have to be infinite. For infinite force is force of an infinite body, and of an infinite body the force is infinite. So the motive body also will be infinite. (The proof of this is given in our discussion of movement, where it is shown that no finite thing possesses infinite power, and no infinite thing finite power.) If then that which moves naturally can also move unnaturally, there will be two infinites, one which causes, and another which exhibits the latter motion. Again, what is it that moves the infinite? If it moves itself, it must be animate. But how can it possibly be conceived as an infinite animal? And if there is something else that moves it, there will be two infinites, that which moves and that which is moved, differing in their form and power.

If the whole is not continuous, but exists, as Democritus and Leucippus think, in the form of parts separated by void, there must necessarily be one movement of all the multitude. They are distinguished, we are told, from one another by their figures; but their nature is one, like many pieces of gold separated from one another. But each piece must, as we assert, have the same motion. For a single clod moves to the same place as the whole mass of earth, and a spark to the same place as the whole mass of fire. So that if it be weight that all possess, no body is, strictly speaking, light: and if lightness be universal, none is heavy. Moreover, whatever possesses weight or lightness will have its place either at one of the extremes or in the middle region. But this is impossible while the world is conceived as infinite. And, generally, that which has no centre or extreme limit, no up or down, gives the bodies no place for their motion; and without that movement is impossible. A thing must move either naturally or unnaturally, and the two movements are determined by the proper and alien places. Again, a place in which a thing rests or to which it moves unnaturally, must be the natural place for some other body, as experience shows. Necessarily, therefore, not everything possesses weight or lightness, but some things do and some do not. From these arguments then it is clear that the body of the universe is not infinite.

8

We must now proceed to explain why there cannot be more than one heaven-the further question mentioned above. For it may be thought that we have not proved universal of bodies that none whatever can exist outside our universe, and that our argument applied only to those of indeterminate extent.

Now all things rest and move naturally and by constraint. A thing moves naturally to a place in which it rests without constraint, and rests naturally in a place to which it moves without constraint. On the other hand, a thing moves by constraint to a place in which it rests by constraint, and rests by constraint in a place to which it moves by constraint. Further, if a given movement is due to constraint, its contrary is natural. If, then, it is by constraint that earth moves from a certain place to the centre here, its movement from here to there will be natural, and if earth from there rests here without constraint, its movement hither will be natural. And the natural movement in each case is one. Further, these worlds, being similar in nature to ours, must all be composed of the same bodies as it.

Moreover each of the bodies, fire, I mean, and earth and their intermediates, must have the same power as in our world. For if these names are used equivocally, if the identity of name does not rest upon an identity of form in these elements and ours, then the whole to which they belong can only be called a world by equivocation. Clearly, then, one of the bodies will move naturally away from the centre and another towards the centre, since fire must be identical with fire, earth with earth, and so on, as the fragments of each are identical in this world. That this must be the case is evident from the principles laid down in our discussion of the movements, for these are limited in number, and the distinction of the elements depends upon the distinction of the movements. Therefore, since the movements are the same, the elements must also be the same everywhere. The particles of earth, then, in another world move naturally also to our centre and its fire to our circumference. This, however, is impossible, since, if it were true, earth must, in its own world, move upwards, and fire to the centre; in the same way the earth of our world must move naturally away from the centre when it moves towards the centre of another universe. This follows from the supposed juxtaposition of the worlds. For either we must refuse to admit the identical nature of the simple bodies in the various universes, or, admitting this, we must make the centre and the extremity one as suggested. This being so, it follows that there cannot be more worlds than one.

To postulate a difference of nature in the simple bodies according as they are more or less distant from their proper places is unreasonable. For what difference can it make whether we say that a thing is this distance away or that? One would have to suppose a difference proportionate to the distance and increasing with it, but the form is in fact the same. Moreover, the bodies must have some movement, since the fact that they move is quite evident. Are we to say then that all their movements, even those which are mutually contrary, are due to constraint? No, for a body which has no natural movement at all cannot be moved by constraint. If then the bodies have a natural movement, the movement of the particular instances of each form must necessarily have for goal a place numerically one, i.e. a particular centre or a particular extremity. If it be suggested that the goal in each case is one in form but numerically more than one, on the analogy of particulars which are many though each undifferentiated in form, we reply that the variety of goal cannot be limited to this portion or that but must extend to all alike. For all are equally undifferentiated in form, but any one is different numerically from any other. What I mean is this: if the portions in this world behave similarly both to one another and to those in another world, then the portion which is taken hence will not behave differently either from the portions in another world or from those in the same world, but similarly to them, since in form no portion differs from another. The result is that we must either abandon our present assumption or assert that the centre and the extremity are each numerically one. But this being so, the heaven, by the same evidence and the same necessary inferences, must be one only and no more.

A consideration of the other kinds of movement also makes it plain that there is some point to which earth and fire move naturally. For in general that which is moved changes from something into something, the starting-point and the goal being different in form, and always it is a finite change. For instance, to recover health is to change from disease to health, to increase is to change from smallness to greatness. Locomotion must be similar: for it also has its goal and starting-point—and therefore the starting-point and the goal of the natural movement must differ in form-just as the movement of coming to health does not take any direction which chance or the wishes of the mover may select. Thus, too, fire and earth move not to infinity but to opposite points; and since the opposition in place is between above and below, these will be the limits of their movement. (Even in circular movement there is a sort of opposition between the ends of the diameter, though the movement as a whole has no contrary: so that here too the movement has in a sense an opposed and finite goal.) There must therefore be some end to locomotion: it cannot continue to infinity.

This conclusion that local movement is not continued to infinity is corroborated by the fact that earth moves more quickly the nearer it is to the centre, and fire the nearer it is to the upper place. But if movement were infinite speed would be infinite also; and if speed then weight and lightness. For as superior speed in downward movement implies superior weight, so infinite increase of weight necessitates infinite increase of speed.

Further, it is not the action of another body that makes one of these bodies move up and the other down; nor is it constraint, like the 'extrusion' of some writers. For in that case the larger the mass of fire or earth the slower would be the upward or downward movement; but the fact is the reverse: the greater the mass of fire or earth the quicker always is its movement towards its own place. Again, the speed of the movement would not increase towards the end if it were due to constraint or extrusion; for a constrained movement always diminishes in speed as the source of constraint becomes more distant, and a body moves without constraint to the place whence it was moved by constraint.

A consideration of these points, then, gives adequate assurance of the truth of our contentions. The same could also be shown with the aid of the discussions which fall under First Philosophy, as well as from the nature of the circular movement, which must be eternal both here and in the other worlds. It is plain, too, from the following considerations that the universe must be one.

The bodily elements are three, and therefore the places of the elements will be three also; the place, first, of the body which sinks to the bottom, namely the region about the centre; the place, secondly, of the revolving body, namely the outermost place, and thirdly, the intermediate place, belonging to the intermediate body. Here in this third place will be the body which rises to the surface; since, if not here, it will be elsewhere, and it cannot be elsewhere: for we have two bodies, one weightless, one endowed with weight, and below is place of the body endowed with weight, since the region about the centre has been given to the heavy body. And its position cannot be unnatural to it, for it would have to be natural to something else, and there is nothing else. It must then occupy the intermediate place. What distinctions there are within the intermediate itself we will explain later on.

We have now said enough to make plain the character and number of the bodily elements, the place of each, and further, in general, how many in number the various places are.

9

We must show not only that the heaven is one, but also that more than one heaven is and, further, that, as exempt from decay and generation, the heaven is eternal. We may begin by raising a difficulty. From one point of view it might seem impossible that the heaven should be one and unique, since in all formations and products whether of nature or of art we can distinguish the shape in itself and the shape in combination with matter. For instance the form of the sphere is one thing and the gold or bronze sphere another; the shape of the circle again is one thing, the bronze or wooden circle another. For when we state the essential nature of the sphere or circle we do not include in the formula gold or bronze, because they do not belong to the essence, but if we are speaking of the copper or gold sphere we do include them. We still make the distinction even if we cannot conceive or apprehend any other example beside the particular thing. This may, of course, sometimes be the case: it might be, for instance, that only one circle could be found; yet none the less the difference will remain between the being of circle and of this particular circle, the one being form, the other form in matter, i.e. a particular thing. Now since the universe is perceptible it must be regarded as a particular; for everything that is perceptible subsists, as we know, in matter. But if it is a particular, there will be a distinction between the being of 'this universe' and of 'universe' unqualified. There is a difference, then, between 'this universe' and simple 'universe'; the second is form and shape, the first form in combination with matter; and any shape or form has, or may have, more than one particular instance.

On the supposition of Forms such as some assert, this must be the case, and equally on the view that no such entity has a separate existence. For in every case in which the essence is in matter it is a fact of observation that the particulars of like form are several or infinite in number. Hence there either are, or may be, more heavens than one. On these grounds, then, it might be inferred either that there are or that there might be several heavens. We must, however, return and ask how much of this argument is correct and how much not.

Now it is quite right to say that the formula of the shape apart from the matter must be different from that of the shape in the matter, and we may allow this to be true. We are not, however, therefore compelled to assert a plurality of worlds. Such a plurality is in fact impossible if this world contains the entirety of matter, as in fact it does. But perhaps our contention can be made clearer in this way. Suppose 'aquilinity' to be curvature in the nose or flesh, and flesh to be the matter of aquilinity. Suppose further, that all flesh came together into a single whole of flesh endowed with this aquiline quality. Then neither would there be, nor could there arise, any other thing that was aquiline. Similarly, suppose flesh and bones to be the matter of man, and suppose a man to be created of all flesh and all bones in indissoluble union. The possibility of another man would be removed. Whatever case you took it would be the same. The general rule is this: a thing whose essence resides in a substratum of matter can never come into being in the absence of all matter. Now the universe is certainly a particular and a material thing: if however, it is composed not of a part but of the whole of matter, then though the being of 'universe' and of 'this universe' are still distinct, yet there is no other universe, and no possibility of others being made, because all the matter is already included in this. It remains, then, only to prove that it is composed of all natural perceptible body.

First, however, we must explain what we mean by 'heaven' and in how many senses we use the word, in order to make clearer the object of our inquiry. (a) In one sense, then, we call 'heaven' the substance of the extreme circumference of the whole, or that natural body whose place is at the extreme circumference. We recognize habitually a special right to the name 'heaven' in the extremity or upper region, which we take to be the seat of all that is divine. (b) In another sense, we use this name for the body continuous with the extreme circumference which contains the moon, the sun, and some of the stars; these we say are 'in the heaven'. (c) In yet another sense we give the name to all body included within extreme circumference, since we habitually call the whole or totality 'the heaven'. The word, then, is used in three senses.

Now the whole included within the extreme circumference must be composed of all physical and sensible body, because there neither is, nor can come into being, any body outside the heaven. For if there is a natural body outside the extreme circumference it must be either a simple or a composite body, and its position must be either natural or unnatural. But it cannot be any of the simple bodies. For, first, it has been shown that that which moves in a circle cannot change its place. And, secondly, it cannot be that which moves from the centre or that which lies lowest. Naturally they could not be there, since their proper places are elsewhere; and if these are there unnaturally, the exterior place will be natural to some other body, since a place which is unnatural to one body must be natural to another: but we saw that there is no other body besides these. Then it is not possible that any simple body should be outside the heaven. But, if no simple body, neither can any mixed body be there: for the presence of the simple body is involved in the presence of the mixture. Further neither can any body come into that place: for it will do so either naturally or unnaturally, and will be either simple or composite; so that the same argument will apply, since it makes no difference whether the question is 'does A exist?' or 'could A come to exist?' From our arguments then it is evident not only that there is not, but also that there could never come to be, any bodily mass whatever outside the circumference. The world as a whole, therefore, includes all its appropriate matter, which is, as we saw, natural perceptible body. So that neither are there now, nor have there ever been, nor can there ever be formed more heavens than one, but this heaven of ours is one and unique and complete.

It is therefore evident that there is also no place or void or time outside the heaven. For in every place body can be present; and void is said to be that in which the presence of body, though not actual, is possible; and time is the number of movement. But in the absence of natural body there is no movement, and outside the heaven, as we have shown, body neither exists nor can come to exist. It is clear then that there is neither place, nor void, nor time, outside the heaven. Hence whatever is there, is of such a nature as not to occupy any place, nor does time age it; nor is there any change in any of the things which lie beyond the outermost motion; they continue through their entire duration unalterable and unmodified, living the best and most selfsufficient of lives. As a matter of fact, this word 'duration' possessed a divine significance for the ancients, for the fulfilment which includes the period of life of any creature, outside of which no natural development can fall, has been called its duration. On the same principle the fulfilment of the whole heaven, the fulfilment which includes all time and infinity, is 'duration'-a name based upon the fact that it is always-duration immortal and divine. From it derive the being and life which other things, some more or less articulately but others feebly, enjoy. So, too, in its discussions concerning the divine, popular philosophy often propounds the view that whatever is divine, whatever is primary and supreme, is necessarily unchangeable. This fact confirms what we have said. For there is nothing else stronger than it to move it-since that would mean more divine-and it has no defect and lacks none of its proper excellences. Its unceasing movement, then, is also reasonable, since everything ceases to move when it comes to its proper place, but the body whose path is the circle has one and the same place for starting-point and goal.

10

Having established these distinctions, we may now proceed to the question whether the heaven is ungenerated or generated, indestructible or destructible. Let us start with a review of the theories of other thinkers; for the proofs of a theory are difficulties for the contrary theory. Besides, those who have first heard the pleas of our adversaries will be more likely to credit the assertions which we are going to make. We shall be less open to the charge of procuring judgement by default. To give a satisfactory decision as to the truth it is necessary to be rather an arbitrator than a party to the dispute.

That the world was generated all are agreed, but, generation over, some say that it is eternal, others say that it is destructible like any other natural formation. Others again, with Empedliocles of Acragas and Heraclitus of Ephesus, believe that there is alternation in the destructive process, which takes now this direction, now that, and continues without end.

Now to assert that it was generated and yet is eternal is to assert the impossible; for we cannot reasonably attribute to anything any characteristics but those which observation detects in many or all instances. But in this case the facts point the other way: generated things are seen always to be destroyed. Further, a thing whose present state had no beginning and which could not have been other than it was at any previous moment throughout its entire duration, cannot possibly be changed. For there will have to be some cause of change, and if this had been present earlier it would have made possible another condition of that to which any other condition was impossible. Suppose that the world was formed out of elements which were formerly otherwise conditioned than as they are now. Then (1) if their condition was always so and could not have been otherwise, the world could never have come into being. And (2) if the world did come into being, then, clearly, their condition must have been capable of change and not eternal: after combination therefore they will be dispersed, just as in the past after dispersion they came into combination, and this process either has been, or could have been, indefinitely repeated. But if this is so, the world cannot be indestructible, and it does not matter whether the change of condition has actually occurred or remains a possibility.

Some of those who hold that the world, though indestructible, was yet generated, try to support their case by a parallel which is illusory. They say that in their statements about its generation they are doing what geometricians do when they construct their figures, not implying that the universe really had a beginning, but for didactic reasons facilitating understanding by exhibiting the object, like the figure, as in course of formation. The two cases, as we said, are not parallel; for, in the construction of the figure, when the various steps are completed the required figure forthwith results; but in these other demonstrations what results is not that which was required. Indeed it cannot be so; for antecedent and consequent, as assumed, are in contradiction. The ordered, it is said, arose out of the unordered; and the same thing cannot be at the same time both ordered and unordered; there must be a process and a lapse of time separating the two states. In the figure, on the other hand, there is no temporal separation. It is clear then that the universe cannot be at once eternal and generated.

To say that the universe alternately combines and dissolves is no more paradoxical than to make it eternal but varying in shape. It is as if one were to think that there was now destruction and now existence when from a child a man is generated, and from a man a child. For it is clear that when the elements come together the result is not a chance system and combination, but the very same as before-especially on the view of those who hold this theory, since they say that the contrary is the cause of each state. So that if the totality of body, which is a continuum, is now in this order or disposition and now in that, and if the combination of the whole is a world or heaven, then it will not be the world that comes into being and is destroyed, but only its dispositions.

If the world is believed to be one, it is impossible to suppose that it should be, as a whole, first generated and then destroyed, never to reappear; since before it came into being there was always present the combination prior to it, and that, we hold, could never change if it was never generated. If, on the other hand, the worlds are infinite in number the view is more plausible. But whether this is, or is not, impossible will be clear from what follows. For there are some who think it possible both for the ungenerated to be destroyed and for the generated to persist undestroyed. (This is held in the Timaeus, where Plato says that the heaven, though it was generated, will none the less exist to eternity.) So far as the heaven is concerned we have answered this view with arguments appropriate to the nature of the heaven: on the general question we shall attain clearness when we examine the matter universally.

11

We must first distinguish the senses in which we use the words 'ungenerated' and 'generated', 'destructible' and 'indestructible'. These have many meanings, and though it may make no difference to the argument, yet some confusion of mind must result from treating as uniform in its use a word which has several distinct applications. The character which is the ground of the predication will always remain obscure.

The word 'ungenerated' then is used (a) in one sense whenever something now is which formerly was not, no process of becoming or change being involved. Such is the case, according to some, with contact and motion, since there is no process of coming to be in contact or in motion. (b) It is used in another sense, when something which is capable of coming to be, with or without process, does not exist; such a thing is ungenerated in the sense that its generation is not a fact but a possibility. (c) It is also applied where there is general impossibility of any generation such that the thing now is which then was not. And 'impossibility' has two uses: first, where it is untrue to say that the thing can ever come into being, and secondly, where it cannot do so easily, quickly, or well. In the same way the word 'generated' is used, (a) first, where what formerly was not afterwards is, whether a process of becoming was or was not involved, so long as that which then was not, now is; (b) secondly, of anything capable of existing, 'capable' being defined with reference either to truth or to facility; (c) thirdly, of anything to which the passage from not being to being belongs, whether already actual, if its existence is due to a past process of becoming, or not yet actual but only possible. The uses of the words 'destructible' and 'indestructible' are similar. 'Destructible' is applied (a) to that which formerly was and afterwards either is not or might not be, whether a period of being destroyed and changed intervenes or not; and (b) sometimes we apply the word to that which a process of destruction may cause not to be; and also (c) in a third sense, to that which is easily destructible, to the 'easily destroyed', so to speak. Of the indestructible the same account holds good. It is either (a) that which now is and now is not, without any process of destruction, like contact, which without being destroyed afterwards is not, though formerly it was; or (b) that which is but might not be, or which will at some time not be, though it now is. For you exist now and so does the contact; yet both are destructible, because a time will come when it will not be true of you that you exist, nor of these things that they are in contact. Thirdly (c) in its most proper use, it is that which is, but is incapable of any destruction such that the thing which now is later ceases to be or might cease to be; or again, that which has not yet been destroyed, but in the future may cease to be. For indestructible is also used of that which is destroyed with difficulty.

This being so, we must ask what we mean by 'possible' and 'impossible'. For in its most proper use the predicate 'indestructible' is given because it is impossible that the thing should be destroyed, i.e. exist at one time and not at another. And 'ungenerated' also involves impossibility when used for that which cannot be generated, in such fashion that, while formerly it was not, later it is. An instance is a commensurable diagonal. Now when we speak of a power to move or to lift weights, we refer always to the maximum. We speak, for instance, of a power to lift a hundred talents or walk a hundred stades--though a power to effect the maximum is also a power to effect any part of the maximum--since we feel obliged in defining the power to give the limit or maximum. A thing, then, which is within it. If, for example, a man can lift a hundred talents, he can also lift two, and if he can walk a hundred stades, he can also walk two. But the power is of the maximum, and a thing said, with reference to its maximum, to be incapable of so much is also incapable of any greater amount. It is, for instance, clear that a person who cannot walk a thousand stades will also be unable to walk a thousand and one. This point need not trouble us, for we may take it as settled that what is, in the strict sense, possible is determined by a limiting maximum. Now perhaps the objection might be raised that there is no necessity in this, since he who sees a stade need not see the smaller measures contained in it, while, on the contrary, he who can see a dot or hear a small sound will perceive what is greater. This, however, does not touch our argument. The maximum may be determined either in the power or in its object. The application of this is plain. Superior sight is sight of the smaller body, but superior speed is that of the greater body.

12

Having established these distinctions we can now proceed to the sequel. If there are things capable both of being and of not being, there must be some definite maximum time of their being and not being; a time, I mean, during which continued existence is possible to them and a time during which continued nonexistence is possible. And this is true in every category, whether the thing is, for example, 'man', or 'white', or 'three cubits long', or whatever it may be. For if the time is not definite in quantity, but longer than any that can be suggested and shorter than none, then it will be possible for one and the same thing to exist for infinite time and not to exist for another infinity. This, however, is impossible.

Let us take our start from this point. The impossible and the false have not the same significance. One use of 'impossible' and 'possible', and 'false' and 'true', is hypothetical. It is impossible, for instance, on a certain hypothesis that the triangle should have its angles equal to two right angles, and on another the diagonal is commensurable. But there are also things possible and impossible, false and true, absolutely. Now it is one thing to be absolutely false, and another thing to be absolutely impossible. To say that you are standing when you are not standing is to assert a falsehood, but not an impossibility. Similarly to say that a man who is playing the harp, but not singing, is singing, is to say what is false but not impossible. To say, however, that you are at once standing and sitting, or that the diagonal is commensurable, is to say what is not only false but also impossible. Thus it is not the same thing to make a false and to make an impossible hypothesis, and from the impossible hypothesis impossible results follow. A man has, it is true, the capacity at once of sitting and of standing, because when he possesses the one he also possesses the other; but it does not follow that he can at once sit and stand, only that at another time he can do the other also. But if a thing has for infinite time more than one capacity, another time is impossible and the times must coincide. Thus if a thing which exists for infinite time is destructible, it will have the capacity of not being. Now if it exists for infinite time let this capacity be actualized; and it will be in actuality at once existent and non-existent. Thus a false conclusion would follow because a false assumption was made, but if what was assumed had not been impossible its consequence would not have been impossible.

Anything then which always exists is absolutely imperishable. It is also ungenerated, since if it was generated it will have the power for some time of not being. For as that which formerly was, but now is not, or is capable at some future time of not being, is destructible, so that which is capable of formerly not having been is generated. But in the case of that which always is, there is no time for such a capacity of not being, whether the supposed time is finite or infinite; for its capacity of being must include the finite time since it covers infinite time.

It is therefore impossible that one and the same thing should be capable of always existing and of always not-existing. And 'not always existing', the contradictory, is also excluded. Thus it is impossible for a thing always to exist and yet to be destructible. Nor, similarly, can it be generated. For of two attributes if B cannot be present without A, the impossibility A of proves the impossibility of B. What always is, then, since it is incapable of ever not being, cannot possibly be generated. But since the contradictory of 'that which is always capable of being' 'that which is not always capable of being'; while 'that which is always capable of not being' is the contrary, whose contradictory in turn is 'that which is not always capable of not being', it is necessary that the contradictories of both terms should be predicable of one and the same thing, and thus that, intermediate between what always is and what always is not, there should be that to which being and not-being are both possible; for the contradictory of each will at times be true of it unless it always exists. Hence that which not always is not will sometimes be and sometimes not be; and it is clear that this is true also of that which cannot always be but sometimes is and therefore sometimes is not. One thing, then, will have the power of being, and will thus be intermediate between the other two.

Expresed universally our argument is as follows. Let there be two attributes, A and B, not capable of being present in any one thing together, while either A or C and either B or D are capable of being present in everything. Then C and D must be predicated of everything of which neither A nor B is predicated. Let E lie between A and B; for that which is neither of two contraries is a mean between them. In E both C and D must be present, for either A or C is present everywhere and therefore in E. Since then A is impossible, C must be present, and the same argument holds of D.

Neither that which always is, therefore, nor that which always is not is either generated or destructible. And clearly whatever is generated or destructible is not eternal. If it were, it would be at once capable of always being and capable of not always being, but it has already been shown that this is impossible. Surely then whatever is ungenerated and in being must be eternal, and whatever is indestructible and in being must equally be so. (I use the words 'ungenerated' and 'indestructible' in their proper sense, 'ungenerated' for that which now is and could not at any previous time have been truly said not to be; 'indestructible' for that which now is and cannot at any future time be truly said not to be.) If, again, the two terms are coincident, if the ungenerated is indestructible, and the indestructible ungenearted, then each of them is coincident with 'eternal'; anything ungenerated is eternal and anything indestructible is eternal. This is clear too from the definition of the terms, Whatever is destructible must be generated; for it is either ungenerated, or generated, but, if ungenerated, it is by hypothesis indestructible. Whatever, further, is generated must be destructible. For it is either destructible or indestructible, but, if indestructible, it is by hypothesis ungenerated.

If, however, 'indestructible' and 'ungenerated' are not coincident, there is no necessity that either the ungenerated or the indestructible should be eternal. But they must be coincident, for the following reasons. The terms 'generated' and 'destructible' are coincident; this is obvious from our former remarks, since between what always is and what always is not there is an intermediate which is neither, and that intermediate is the generated and destructible. For whatever is either of these is capable both of being and of not being for a definite time: in either case, I mean, there is a certain period of time during which the thing is and another during which it is not. Anything therefore which is generated or destructible must be intermediate. Now let A be that which always is and B that which always is not, C the generated, and D the destructible. Then C must be intermediate between A and B. For in their case there is no time in the direction of either limit, in which either A is not or B is. But for the generated there must be such a time either actually or potentially, though not for A and B in either way. C then will be, and also not be, for a limited length of time, and this is true also of D, the destructible. Therefore each is both generated and destructible. Therefore 'generated' and 'destructible' are coincident. Now let E stand for the ungenerated, F for the generated, G for the indestructible, and H for the destructible. As for F and H, it has been shown that they are coincident. But when terms stand to one another as these do, F and H coincident, E and F never predicated of the same thing but one or other of everything, and G and H likewise, then E and G must needs be coincident. For suppose that E is not coincident with G, then F will be, since either E or F is predictable of everything. But of that of which F is predicated H will be predicable also. H will then be coincident with G, but this we saw to be impossible. And the same argument shows that G is coincident with E.

Now the relation of the ungenerated (E) to the generated (F) is the same as that of the indestructible (G) to the destructible (H). To say then that there is no reason why anything should not be generated and yet indestructible or ungenerated and yet destroyed, to imagine that in the one case generation and in the other case destruction occurs once for all, is to destroy part of the data. For (1) everything is capable of acting or being acted upon, of being or not being, either for an infinite, or for a definitely limited space of time; and the infinite time is only a possible alternative because it is after a fashion defined, as a length of time which cannot be exceeded. But infinity in one direction is neither infinite or finite. (2) Further, why, after always existing, was the thing destroyed, why, after an infinity of not being, was it generated, at one moment rather than another? If every moment is alike and the moments are infinite in number, it is clear that a generated or destructible thing existed for an infinite time. It has therefore for an infinite time the capacity of not being (since the capacity of being and the capacity of not being will be present together), if destructible, in the time before destruction, if generated, in the time after generation. If then we assume the two capacities to be actualized, opposites will be present together. (3) Further, this second capacity will be present like the first at every moment, so that the thing will have for an infinite time the capacity both of being and of not being; but this has been shown to be impossible. (4) Again, if the capacity is present prior to the activity, it will be present for all time, even while the thing was as yet ungenerated and non-existent, throughout the infinite time in which it was capable of being generated. At that time, then, when it was not, at that same time it had the capacity of being, both of being then and of being thereafter, and therefore for an infinity of time.

It is clear also on other grounds that it is impossible that the destructible should not at some time be destroyed. For otherwise it will always be at once destructible and in actuality indestructible, so that it will be at the same time capable of always existing and of not always existing. Thus the destructible is at some time actually destroyed. The generable, similarly, has been generated, for it is capable of having been generated and thus also of not always existing.

We may also see in the following way how impossible it is either for a thing which is generated to be thenceforward indestructible, or for a thing which is ungenerated and has always hitherto existed to be destroyed. Nothing that is by chance can be indestructible or ungenerated, since the products of chance and fortune are opposed to what is, or comes to be, always or usually, while anything which exists for a time infinite either absolutely or in one direction, is in existence either always or usually. That which is by chance, then, is by nature such as to exist at one time and not at another. But in things of that character the contradictory states proceed from one and the same capacity, the matter of the thing being the cause equally of its existence and of its non-existence. Hence contradictories would be present together in actuality.

Further, it cannot truly be said of a thing now that it exists last year, nor could it be said last year that it exists now. It is therefore impossible for what once did not exist later to be eternal. For in its later state it will possess the capacity of not existing, only not of not existing at a time when it exists-since then it exists in actuality-but of not existing last year or in the past. Now suppose it to be in actuality what it is capable of being. It will then be true to say now that it does not exist last year. But this is impossible. No capacity relates to being in the past, but always to being in the present or future. It is the same with the notion of an eternity of existence followed later by non-existence. In the later state the capacity will be present for that which is not there in actuality. Actualize, then, the capacity. It will be true to say now that this exists last year or in the past generally.

Considerations also not general like these but proper to the subject show it to be impossible that what was formerly eternal should later be destroyed or that what formerly was not should later be eternal. Whatever is destructible or generated is always alterable. Now alteration is due to contraries, and the things which compose the natural body are the very same that destroy it.

Book II
1

THAT the heaven as a whole neither came into being nor admits of destruction, as some assert, but is one and eternal, with no end or beginning of its total duration, containing and embracing in itself the infinity of time, we may convince ourselves not only by the arguments already set forth but also by a consideration of the views of those who differ from us in providing for its generation. If our view is a possible one, and the manner of generation which they assert is impossible, this fact will have great weight in convincing us of the immortality and eternity of the world. Hence it is well to persuade oneself of the truth of the ancient and truly traditional theories, that there is some immortal and divine thing which possesses movement, but movement such as has no limit and is rather itself the limit of all other movement. A limit is a thing which contains; and this motion, being perfect, contains those imperfect motions which have a limit and a goal, having itself no beginning or end, but unceasing through the infinity of time, and of other movements, to some the cause of their beginning, to others offering the goal. The ancients gave to the Gods the heaven or upper place, as being alone immortal; and our present argument testifies that it is indestructible and ungenerated. Further, it is unaffected by any mortal discomfort, and, in addition, effortless; for it needs no constraining necessity to keep it to its path, and prevent it from moving with some other movement more natural to itself. Such a constrained movement would necessarily involve effort the more so, the more eternal it were-and would be inconsistent with perfection. Hence we must not believe the old tale which says that the world needs some Atlas to keep it safe-a tale composed, it would seem, by men who, like later thinkers, conceived of all the upper bodies as earthy and endowed with weight, and therefore supported it in their fabulous way upon animate necessity. We must no more believe that than follow Empedocles when he says that the world, by being whirled round, received a movement quick enough to overpower its own downward tendency, and thus has been kept from destruction all this time. Nor, again, is it conceivable that it should persist eternally by the necessitation of a soul. For a soul could not live in such conditions painlessly or happily, since the movement involves constraint, being imposed on the first body, whose natural motion is different, and imposed continuously. It must therefore be uneasy and devoid of all rational satisfaction; for it could not even, like the soul of mortal animals, take recreation in the bodily relaxation of sleep. An Ixion's lot must needs possess it, without end or respite. If then, as we said, the view already stated of the first motion is a possible one, it is not only more appropriate so to conceive of its eternity, but also on this hypothesis alone are we able to advance a theory consistent with popular divinations of the divine nature. But of this enough for the present.

2

Since there are some who say that there is a right and a left in the heaven, with those who are known as Pythagoreans--to whom indeed the view really belongs--we must consider whether, if we are to apply these principles to the body of the universe, we should follow their statement of the matter or find a better way. At the start we may say that, if right and left are applicable, there are prior principles which must first be applied. These principles have been analysed in the discussion of the movements of animals, for the reason that they are proper to animal nature. For in some animals we find all such distinctions of parts as this of right and left clearly present, and in others some; but in plants we find only above and below. Now if we are to apply to the heaven such a distinction of parts, we must exept, as we have said, to find in it also the distinction which in animals is found first of them all. The distinctions are three, namely, above and below, front and its opposite, right and left-all these three oppositions we expect to find in the perfect body-and each may be called a principle. Above is the principle of length, right of breadth, front of depth. Or again we may connect them with the various movements, taking principle to mean that part, in a thing capable of movement, from which movement first begins. Growth starts from above, locomotion from the right, sense-movement from in front (for front is simply the part to which the senses are directed). Hence we must not look for above and below, right and left, front and back, in every kind of body, but only in those which, being animate, have a principle of movement within themselves. For in no inanimate thing do we observe a part from which movement originates. Some do not move at all, some move, but not indifferently in any direction; fire, for example, only upward, and earth only to the centre. It is true that we speak of above and below, right and left, in these bodies relatively to ourselves. The reference may be to our own right hands, as with the diviner, or to some similarity to our own members, such as the parts of a statue possess; or we may take the contrary spatial order, calling right that which is to our left, and left that which is to our right. We observe, however, in the things themselves none of these distinctions; indeed if they are turned round we proceed to speak of the opposite parts as right and left, a boy land below, front and back. Hence it is remarkable that the Pythagoreans should have spoken of these two principles, right and left, only, to the exclusion of the other four, which have as good a title as they. There is no less difference between above and below or front and back in animals generally than between right and left. The difference is sometimes only one of function, sometimes also one of shape; and while the distinction of above and below is characteristic of all animate things, whether plants or animals, that of right and left is not found in plants. Further, inasmuch as length is prior to breadth, if above is the principle of length, right of breadth, and if the principle of that which is prior is itself prior, then above will be prior to right, or let us say, since 'prior' is ambiguous, prior in order of generation. If, in addition, above is the region from which movement originates, right the region in which it starts, front the region to which it is directed, then on this ground too above has a certain original character as compared with the other forms of position. On these two grounds, then, they may fairly be criticized, first, for omitting the more fundamental principles, and secondly, for thinking that the two they mentioned were attributable equally to everything.

Since we have already determined that functions of this kind belong to things which possess, a principle of movement, and that the heaven is animate and possesses a principle of movement, clearly the heaven must also exhibit above and below, right and left. We need not be troubled by the question, arising from the spherical shape of the world, how there can be a distinction of right and left within it, all parts being alike and all for ever in motion. We must think of the world as of something in which right differs from left in shape as well as in other respects, which subsequently is included in a sphere. The difference of function will persist, but will appear not to by reason of the regularity of shape. In the same fashion must we conceive of the beginning of its movement. For even if it never began to move, yet it must possess a principle from which it would have begun to move if it had begun, and from which it would begin again if it came to a stand. Now by its length I mean the interval between its poles, one pole being above and the other below; for two hemispheres are specially distinguished from all others by the immobility of the poles. Further, by 'transverse' in the universe we commonly mean, not above and below, but a direction crossing the line of the poles, which, by implication, is length: for transverse motion is motion crossing motion up and down. Of the poles, that which we see above us is the lower region, and that which we do not see is the upper. For right in anything is, as we say, the region in which locomotion originates, and the rotation of the heaven originates in the region from which the stars rise. So this will be the right, and the region where they set the left. If then they begin from the right and move round to the right, the upper must be the unseen pole. For if it is the pole we see, the movement will be leftward, which we deny to be the fact. Clearly then the invisible pole is above. And those who live in the other hemisphere are above and to the right, while we are below and to the left. This is just the opposite of the view of the Pythagoreans, who make us above and on the right side and those in the other hemisphere below and on the left side; the fact being the exact opposite. Relatively, however, to the secondary revolution, I mean that of the planets, we are above and on the right and they are below and on the left. For the principle of their movement has the reverse position, since the movement itself is the contrary of the other: hence it follows that we are at its beginning and they at its end. Here we may end our discussion of the distinctions of parts created by the three dimensions and of the consequent differences of position.

3

Since circular motion is not the contrary of the reverse circular motion, we must consider why there is more than one motion, though we have to pursue our inquiries at a distance-a distance created not so much by our spatial position as by the fact that our senses enable us to perceive very few of the attributes of the heavenly bodies. But let not that deter us. The reason must be sought in the following facts. Everything which has a function exists for its function. The activity of God is immortality, i.e. eternal life. Therefore the movement of that which is divine must be eternal. But such is the heaven, viz. a divine body, and for that reason to it is given the circular body whose nature it is to move always in a circle. Why, then, is not the whole body of the heaven of the same character as that part? Because there must be something at rest at the centre of the revolving body; and of that body no part can be at rest, either elsewhere or at the centre. It could do so only if the body's natural movement were towards the centre. But the circular movement is natural, since otherwise it could not be eternal: for nothing unnatural is eternal. The unnatural is subsequent to the natural, being a derangement of the natural which occurs in the course of its generation. Earth then has to exist; for it is earth which is at rest at the centre. (At present we may take this for granted: it shall be explained later.) But if earth must exist, so must fire. For, if one of a pair of contraries naturally exists, the other, if it is really contrary, exists also naturally. In some form it must be present, since the matter of contraries is the same. Also, the positive is prior to its privation (warm, for instance, to cold), and rest and heaviness stand for the privation of lightness and movement. But further, if fire and earth exist, the intermediate bodies must exist also: each element stands in a contrary relation to every other. (This, again, we will here take for granted and try later to explain.) these four elements generation clearly is involved, since none of them can be eternal: for contraries interact with one another and destroy one another. Further, it is inconceivable that a movable body should be eternal, if its movement cannot be regarded as naturally eternal: and these bodies we know to possess movement. Thus we see that generation is necessarily involved. But if so, there must be at least one other circular motion: for a single movement of the whole heaven would necessitate an identical relation of the elements of bodies to one another. This matter also shall be cleared up in what follows: but for the present so much is clear, that the reason why there is more than one circular body is the necessity of generation, which follows on the presence of fire, which, with that of the other bodies, follows on that of earth; and earth is required because eternal movement in one body necessitates eternal rest in another.

4

The shape of the heaven is of necessity spherical; for that is the shape most appropriate to its substance and also by nature primary.

First, let us consider generally which shape is primary among planes and solids alike. Every plane figure must be either rectilinear or curvilinear. Now the rectilinear is bounded by more than one line, the curvilinear by one only. But since in any kind the one is naturally prior to the many and the simple to the complex, the circle will be the first of plane figures. Again, if by complete, as previously defined, we mean a thing outside which no part of itself can be found, and if addition is always possible to the straight line but never to the circular, clearly the line which embraces the circle is complete. If then the complete is prior to the incomplete, it follows on this ground also that the circle is primary among figures. And the sphere holds the same position among solids. For it alone is embraced by a single surface, while rectilinear solids have several. The sphere is among solids what the circle is among plane figures. Further, those who divide bodies into planes and generate them out of planes seem to bear witness to the truth of this. Alone among solids they leave the sphere undivided, as not possessing more than one surface: for the division into surfaces is not just dividing a whole by cutting it into its parts, but division of another fashion into parts different in form. It is clear, then, that the sphere is first of solid figures.

If, again, one orders figures according to their numbers, it is most natural to arrange them in this way. The circle corresponds to the number one, the triangle, being the sum of two right angles, to the number two. But if one is assigned to the triangle, the circle will not be a figure at all.

Now the first figure belongs to the first body, and the first body is that at the farthest circumference. It follows that the body which revolves with a circular movement must be spherical. The same then will be true of the body continuous with it: for that which is continuous with the spherical is spherical. The same again holds of the bodies between these and the centre. Bodies which are bounded by the spherical and in contact with it must be, as wholes, spherical; and the bodies below the sphere of the planets are contiguous with the sphere above them. The sphere then will be spherical throughout; for every body within it is contiguous and continuous with spheres.

Again, since the whole revolves, palpably and by assumption, in a circle, and since it has been shown that outside the farthest circumference there is neither void nor place, from these grounds also it will follow necessarily that the heaven is spherical. For if it is to be rectilinear in shape, it will follow that there is place and body and void without it. For a rectilinear figure as it revolves never continues in the same room, but where formerly was body, is now none, and where now is none, body will be in a moment because of the projection at the corners. Similarly, if the world had some other figure with unequal radii, if, for instance, it were lentiform, or oviform, in every case we should have to admit space and void outside the moving body, because the whole body would not always occupy the same room.

Again, if the motion of the heaven is the measure of all movements whatever in virtue of being alone continuous and regular and eternal, and if, in each kind, the measure is the minimum, and the minimum movement is the swiftest, then, clearly, the movement of the heaven must be the swiftest of all movements. Now of lines which return upon themselves the line which bounds the circle is the shortest; and that movement is the swiftest which follows the shortest line. Therefore, if the heaven moves in a circle and moves more swiftly than anything else, it must necessarily be spherical.

Corroborative evidence may be drawn from the bodies whose position is about the centre. If earth is enclosed by water, water by air, air by fire, and these similarly by the upper bodies-which while not continuous are yet contiguous with them-and if the surface of water is spherical, and that which is continuous with or embraces the spherical must itself be spherical, then on these grounds also it is clear that the heavens are spherical. But the surface of water is seen to be spherical if we take as our starting-point the fact that water naturally tends to collect in a hollow place-'hollow' meaning 'nearer the centre'. Draw from the centre the lines AB, AC, and let their extremities be joined by the straight line BC. The line AD, drawn to the base of the triangle, will be shorter than either of the radii. Therefore the place in which it terminates will be a hollow place. The water then will collect there until equality is established, that is until the line AE is equal to the two radii. Thus water forces its way to the ends of the radii, and there only will it rest: but the line which connects the extremities of the radii is circular: therefore the surface of the water

BEC is spherical.

It is plain from the foregoing that the universe is spherical. It is plain, further, that it is turned (so to speak) with a finish which no manufactured thing nor anything else within the range of our observation can even approach. For the matter of which these are composed does not admit of anything like the same regularity and finish as the substance of the enveloping body; since with each step away from earth the matter manifestly becomes finer in the same proportion as water is finer than earth.

5

Now there are two ways of moving along a circle, from A to B or from A to C, and we have already explained that these movements are not contrary to one another. But nothing which concerns the eternal can be a matter of chance or spontaneity, and the heaven and its circular motion are eternal. We must therefore ask why this motion takes one direction and not the other. Either this is itself an ultimate fact or there is an ultimate fact behind it. It may seem evidence of excessive folly or excessive zeal to try to provide an explanation of some things, or of everything, admitting no exception. The criticism, however, is not always just: one should first consider what reason there is for speaking, and also what kind of certainty is looked for, whether human merely or of a more cogent kind. When any one shall succeed in finding proofs of greater precision, gratitude will be due to him for the discovery, but at present we must be content with a probable solution. If nature always follows the best course possible, and, just as upward movement is the superior form of rectilinear movement, since the upper region is more divine than the lower, so forward movement is superior to backward, then front and back exhibits, like right and left, as we said before and as the difficulty just stated itself suggests, the distinction of prior and posterior, which provides a reason and so solves our difficulty. Supposing that nature is ordered in the best way possible, this may stand as the reason of the fact mentioned. For it is best to move with a movement simple and unceasing, and, further, in the superior of two possible directions.

6

We have next to show that the movement of the heaven is regular and not irregular. This applies only to the first heaven and the first movement; for the lower spheres exhibit a composition of several movements into one. If the movement is uneven, clearly there will be acceleration, maximum speed, and retardation, since these appear in all irregular motions. The maximum may occur either at the starting-point or at the goal or between the two; and we expect natural motion to reach its maximum at the goal, unnatural motion at the starting-point, and missiles midway between the two. But circular movement, having no beginning or limit or middle in the direct sense of the words, has neither whence nor whither nor middle: for in time it is eternal, and in length it returns upon itself without a break. If then its movement has no maximum, it can have no irregularity, since irregularity is produced by retardation and acceleration. Further, since everything that is moved is moved by something, the cause of the irregularity of movement must lie either in the mover or in the moved or both. For if the mover moved not always with the same force, or if the moved were altered and did not remain the same, or if both were to change, the result might well be an irregular movement in the moved. But none of these possibilities can be conceived as actual in the case of the heavens. As to that which is moved, we have shown that it is primary and simple and ungenerated and indestructible and generally unchanging; and the mover has an even better right to these attributes. It is the primary that moves the primary, the simple the simple, the indestructible and ungenerated that which is indestructible and ungenerated. Since then that which is moved, being a body, is nevertheless unchanging, how should the mover, which is incorporeal, be changed?

It follows then, further, that the motion cannot be irregular. For if irregularity occurs, there must be change either in the movement as a whole, from fast to slow and slow to fast, or in its parts. That there is no irregularity in the parts is obvious, since, if there were, some divergence of the stars would have taken place before now in the infinity of time, as one moved slower and another faster: but no alteration of their intervals is ever observed. Nor again is a change in the movement as a whole admissible. Retardation is always due to incapacity, and incapacity is unnatural. The incapacities of animals, age, decay, and the like, are all unnatural, due, it seems, to the fact that the whole animal complex is made up of materials which differ in respect of their proper places, and no single part occupies its own place. If therefore that which is primary contains nothing unnatural, being simple and unmixed and in its proper place and having no contrary, then it has no place for incapacity, nor, consequently, for retardation or (since acceleration involves retardation) for acceleration. Again, it is inconceivable that the mover should first show incapacity for an infinite time, and capacity afterwards for another infinity. For clearly nothing which, like incapacity, unnatural ever continues for an infinity of time; nor does the unnatural endure as long as the natural, or any form of incapacity as long as the capacity. But if the movement is retarded it must necessarily be retarded for an infinite time. Equally impossible is perpetual acceleration or perpetual retardation. For such movement would be infinite and indefinite, but every movement, in our view, proceeds from one point to another and is definite in character. Again, suppose one assumes a minimum time in less than which the heaven could not complete its movement. For, as a given walk or a given exercise on the harp cannot take any and every time, but every performance has its definite minimum time which is unsurpassable, so, one might suppose, the movement of the heaven could not be completed in any and every time. But in that case perpetual acceleration is impossible (and, equally, perpetual retardation: for the argument holds of both and each), if we may take acceleration to proceed by identical or increasing additions of speed and for an infinite time. The remaining alternative is to say that the movement exhibits an alternation of slower and faster: but this is a mere fiction and quite inconceivable. Further, irregularity of this kind would be particularly unlikely to pass unobserved, since contrast makes observation easy.

That there is one heaven, then, only, and that it is ungenerated and eternal, and further that its movement is regular, has now been sufficiently explained.

7

We have next to speak of the stars, as they are called, of their composition, shape, and movements. It would be most natural and consequent upon what has been said that each of the stars should be composed of that substance in which their path lies, since, as we said, there is an element whose natural movement is circular. In so saying we are only following the same line of thought as those who say that the stars are fiery because they believe the upper body to be fire, the presumption being that a thing is composed of the same stuff as that in which it is situated. The warmth and light which proceed from them are caused by the friction set up in the air by their motion. Movement tends to create fire in wood, stone, and iron; and with even more reason should it have that effect on air, a substance which is closer to fire than these. An example is that of missiles, which as they move are themselves fired so strongly that leaden balls are melted; and if they are fired the surrounding air must be similarly affected. Now while the missiles are heated by reason of their motion in air, which is turned into fire by the agitation produced by their movement, the upper bodies are carried on a moving sphere, so that, though they are not themselves fired, yet the air underneath the sphere of the revolving body is necessarily heated by its motion, and particularly in that part where the sun is attached to it. Hence warmth increases as the sun gets nearer or higher or overhead. Of the fact, then, that the stars are neither fiery nor move in fire, enough has been said.

8

Since changes evidently occur not only in the position of the stars but also in that of the whole heaven, there are three possibilities. Either (1) both are at rest, or (2) both are in motion, or (3) the one is at rest and the other in motion.

(1) That both should be at rest is impossible; for, if the earth is at rest, the hypothesis does not account for the observations; and we take it as granted that the earth is at rest. It remains either that both are moved, or that the one is moved and the other at rest.

(2) On the view, first, that both are in motion, we have the absurdity that the stars and the circles move with the same speed, i.e. that the ace of every star is that of the circle in it moves. For star and circle are seen to come back to the same place at the same moment; from which it follows that the star has traversed the circle

and the circle has completed its own movement, i.e. traversed its own circumference, at one and the same moment. But it is difficult to conceive that the pace of each star should be exactly proportioned to the size of its circle. That the pace of each circle should be proportionate to its size is not absurd but inevitable: but that the same should be true of the movement of the stars contained in the circles is quite incredible. For if, on the one and, we suppose that the star which moves on the greater circle is necessarily swifter, clearly we also admit that if stars shifted their position so as to exchange circles, the slower would become swifter and the swifter slower. But this would show that their movement was not their own, but due to the circles. If, on the other hand, the arrangement was a chance combination, the coincidence in every case of a greater circle with a swifter movement of the star contained in it is too much to believe. In one or two cases it might not inconceivably fall out so, but to imagine it in every case alike is a mere fiction. Besides, chance has no place in that which is natural, and what happens everywhere and in every case is no matter of chance.

(3) The same absurdity is equally plain if it is supposed that the circles stand still and that it is the stars themselves which move. For it will follow that the outer stars are the swifter, and that the pace of the stars corresponds to the size of their circles.

Since, then, we cannot reasonably suppose either that both are in motion or that the star alone moves, the remaining alternative is that the circles should move, while the stars are at rest and move with the circles to which they are attached. Only on this supposition are we involved in no absurd consequence. For, in the first place, the quicker movement of the larger circle is natural when all the circles are attached to the same centre. Whenever bodies are moving with their proper motion, the larger moves quicker. It is the same here with the revolving bodies: for the are intercepted by two radii will be larger in the larger circle, and hence it is not surprising that the revolution of the larger circle should take the same time as that of the smaller. And secondly, the fact that the heavens do not break in pieces follows not only from this but also from the proof already given of the continuity of the whole.

Again, since the stars are spherical, as our opponents assert and we may consistently admit, inasmuch as we construct them out of the spherical body, and since the spherical body has two movements proper to itself, namely rolling and spinning, it follows that if the stars have a movement of their own, it will be one of these. But neither is observed. (1) Suppose them to spin. They would then stay where they were, and not change their place, as, by observation and general consent, they do. Further, one would expect them all to exhibit the same movement: but the only star which appears to possess this movement is the sun, at sunrise or sunset, and this appearance is due not to the sun itself but to the distance from which we observe it. The visual ray being excessively prolonged becomes weak and wavering. The same reason probably accounts for the apparent twinkling of the fixed stars and the absence of twinkling in the planets. The planets are near, so that the visual ray reaches them in its full vigour, but when it comes to the fixed stars it is quivering because of the distance and its excessive extension; and its tremor produces an appearance of movement in the star: for it makes no difference whether movement is set up in the ray or in the object of vision.

(2) On the other hand, it is also clear that the stars do not roll. For rolling involves rotation: but the 'face', as it is called, of the moon is always seen. Therefore, since any movement of their own which the stars possessed would presumably be one proper to themselves, and no such movement is observed in them, clearly they have no movement of their own.

There is, further, the absurdity that nature has bestowed upon them no organ appropriate to such movement. For nature leaves nothing to chance, and would not, while caring for animals, overlook things so precious. Indeed, nature seems deliberately to have stripped them of everything which makes self-originated progression possible, and to have removed them as far as possible from things which have organs of movement. This is just why it seems proper that the whole heaven and every star should be spherical. For while of all shapes the sphere is the most convenient for movement in one place, making possible, as it does, the swiftest and most self-contained motion, for forward movement it is the most unsuitable, least of all resembling shapes which are self-moved, in that it has no dependent or projecting part, as a rectilinear figure has, and is in fact as far as possible removed in shape from ambulatory bodies. Since, therefore, the heavens have to move in one lace, and the stars are not required to move themselves forward, it is natural that both should be spherical-a shape which best suits the movement of the one and the immobility of the other.

9

From all this it is clear that the theory that the movement of the stars produces a harmony, i.e. that the sounds they make are concordant, in spite of the grace and originality with which it has been stated, is nevertheless untrue. Some thinkers suppose that the motion of bodies of that size must produce a noise, since on our earth the motion of bodies far inferior in size and in speed of movement has that effect. Also, when the sun and the moon, they say, and all the stars, so great in number and in size, are moving with so rapid a motion, how should they not produce a sound immensely great? Starting from this argument and from the observation that their speeds, as measured by their distances, are in the same ratios as musical concordances, they assert that the sound given forth by the circular movement of the stars is a harmony. Since, however, it appears unaccountable that we should not hear this music, they explain this by saying that the sound is in our ears from the very moment of birth and is thus indistinguishable from its contrary silence, since sound and silence are discriminated by mutual contrast. What happens to men, then, is just what happens to coppersmiths, who are so accustomed to the noise of the smithy that it makes no difference to them. But, as we said before, melodious and poetical as the theory is, it cannot be a true account of the facts. There is not only the absurdity of our hearing nothing, the ground of which they try to remove, but also the fact that no effect other than sensitive is produced upon us. Excessive noises, we know, shatter the solid bodies even of inanimate things: the noise of thunder, for instance, splits rocks and the strongest of bodies. But if the moving bodies are so great, and the sound which penetrates to us is proportionate to their size, that sound must needs reach us in an intensity many times that of thunder, and the force of its action must be immense. Indeed the reason why we do not hear, and show in our bodies none of the effects of violent force, is easily given: it is that there is no noise. But not only is the explanation evident; it is also a corroboration of the truth of the views we have advanced. For the very difficulty which made the Pythagoreans say that the motion of the stars produces a concord corroborates our view. Bodies which are themselves in motion, produce noise and friction: but those which are attached or fixed to a moving body, as the parts to a ship, can no more create noise, than a ship on a river moving with the stream. Yet by the same argument one might say it was absurd that on a large vessel the motion of mast and poop should not make a great noise, and the like might be said of the movement of the vessel itself. But sound is caused when a moving body is enclosed in an unmoved body, and cannot be caused by one enclosed in, and continuous with, a moving body which creates no friction. We may say, then, in this matter that if the heavenly bodies moved in a generally diffused mass of air or fire, as every one supposes, their motion would necessarily cause a noise of tremendous strength and such a noise would necessarily reach and shatter us. Since, therefore, this effect is evidently not produced, it follows that none of them can move with the motion either of animate nature or of constraint. It is as though nature had foreseen the result, that if their movement were other than it is, nothing on this earth could maintain its character.

That the stars are spherical and are not self-moved, has now been explained.

10

With their order--I mean the position of each, as involving the priority of some and the posteriority of others, and their respective distances from the extremity--with this astronomy may be left to deal, since the astronomical discussion is adequate. This discussion shows that the movements of the several stars depend, as regards the varieties of speed which they exhibit, on the distance of each from the extremity. It is established that the outermost revolution of the heavens is a simple movement and the swiftest of all, and that the movement of all other bodies is composite and relatively slow, for the reason that each is moving on its own circle with the reverse motion to that of the heavens. This at once leads us to expect that the body which is nearest to that first simple revolution should take the longest time to complete its circle, and that which is farthest from it the shortest, the others taking a longer time the nearer they are and a shorter time the farther away they are. For it is the nearest body which is most strongly influenced, and the most remote, by reason of its distance, which is least affected, the influence on the intermediate bodies varying, as the mathematicians show, with their distance.

11

With regard to the shape of each star, the most reasonable view is that they are spherical. It has been shown that it is not in their nature to move themselves, and, since nature is no wanton or random creator, clearly she will have given things which possess no movement a shape particularly unadapted to movement. Such a shape is the sphere, since it possesses no instrument of movement. Clearly then their mass will have the form of a sphere. Again, what holds of one holds of all, and the evidence of our eyes shows us that the moon is spherical. For how else should the moon as it waxes and wanes show for the most part a crescent-shaped or gibbous figure, and only at one moment a half-moon? And astronomical arguments give further confirmation; for no other hypothesis accounts for the crescent shape of the sun's eclipses. One, then, of the heavenly bodies being spherical, clearly the rest will be spherical also.

12

There are two difficulties, which may very reasonably here be raised, of which we must now attempt to state the probable solution: for we regard the zeal of one whose thirst after philosophy leads him to accept even slight indications where it is very difficult to see one's way, as a proof rather of modesty than of overconfidence.

Of many such problems one of the strangest is the problem why we find the greatest number of movements in the intermediate bodies, and not, rather, in each successive body a variety of movement proportionate to its distance from the primary motion. For we should expect, since the primary body shows one motion only, that the body which is nearest to it should move with the fewest movements, say two, and the one next after that with three, or some similar arrangement. But the opposite is the case. The movements of the sun and moon are fewer than those of some of the planets. Yet these planets are farther from the centre and thus nearer to the primary body than they, as observation has itself revealed. For we have seen the moon, half-full, pass beneath the planet Mars, which vanished on its shadow side and came forth by the bright and shining part. Similar accounts of other stars are given by the Egyptians and Babylonians, whose observations have been kept for very many years past, and from whom much of our evidence about particular stars is derived. A second difficulty which may with equal justice be raised is this. Why is it that the primary motion includes such a multitude of stars that their whole array seems to defy counting, while of the other stars each one is separated off, and in no case do we find two or more attached to the same motion?

On these questions, I say, it is well that we should seek to increase our understanding, though we have but little to go upon, and are placed at so great a distance from the facts in question. Nevertheless there are certain principles on which if we base our consideration we shall not find this difficulty by any means insoluble. We may object that we have been thinking of the stars as mere bodies, and as units with a serial order indeed but entirely inanimate; but should rather conceive them as enjoying life and action. On this view the facts cease to appear surprising. For it is natural that the best-conditioned of all things should have its good without action, that which is nearest to it should achieve it by little and simple action, and that which is farther removed by a complexity of actions, just as with men's bodies one is in good condition without exercise at all, another after a short walk, while another requires running and wrestling and hard training, and there are yet others who however hard they worked themselves could never secure this good, but only some substitute for it. To succeed often or in many things is difficult. For instance, to throw ten thousand Coan throws with the dice would be impossible, but to throw one or two is comparatively easy. In action, again, when A has to be done to get B, B to get C, and C to get D, one step or two present little difficulty, but as the series extends the difficulty grows. We must, then, think of the action of the lower stars as similar to that of animals and plants. For on our earth it is man that has the greatest variety of actions-for there are many goods that man can secure; hence his actions are various and directed to ends beyond them-while the perfectly conditioned has no need of action, since it is itself the end, and action always requires two terms, end and means. The lower animals have less variety of action than man; and plants perhaps have little action and of one kind only. For either they have but one attainable good (as indeed man has), or, if several, each contributes directly to their ultimate good. One thing then has and enjoys the ultimate good, other things attain to it, one immediately by few steps, another by many, while yet another does not even attempt to secure it but is satisfied to reach a point not far removed from that consummation. Thus, taking health as the end, there will be one thing that always possesses health, others that attain it, one by reducing flesh, another by running and thus reducing flesh, another by taking steps to enable himself to run, thus further increasing the number of movements, while another cannot attain health itself, but only running or reduction of flesh, so that one or other of these is for such a being the end. For while it is clearly best for any being to attain the real end, yet, if that cannot be, the nearer it is to the best the better will be its state. It is for this reason that the earth moves not at all and the bodies near to it with few movements. For they do not attain the final end, but only come as near to it as their share in the divine principle permits. But the first heaven finds it immediately with a single movement, and the bodies intermediate between the first and last heavens attain it indeed, but at the cost of a multiplicity of movement.

As to the difficulty that into the one primary motion is crowded a vast multitude of stars, while of the other stars each has been separately given special movements of its own, there is in the first place this reason for regarding the arrangement as a natural one. In thinking of the life and moving principle of the several heavens one must regard the first as far superior to the others. Such a superiority would be reasonable. For this single first motion has to move many of the divine bodies, while the numerous other motions move only one each, since each single planet moves with a variety of motions. Thus, then, nature makes matters equal and establishes a certain order, giving to the single motion many bodies and to the single body many motions. And there is a second reason why the other motions have each only one body, in that each of them except the last, i.e. that which contains the one star, is really moving many bodies. For this last sphere moves with many others, to which it is fixed, each sphere being actually a body; so that its movement will be a joint product. Each sphere, in fact, has its particular natural motion, to which the general movement is, as it were, added. But the force of any limited body is only adequate to moving a limited body.

The characteristics of the stars which move with a circular motion, in respect of substance and shape, movement and order, have now been sufficiently explained.

13

It remains to speak of the earth, of its position, of the question whether it is at rest or in motion, and of its shape.

I. As to its position there is some difference of opinion. Most people-all, in fact, who regard the whole heaven as finite-say it lies at the centre. But the Italian philosophers known as Pythagoreans take the contrary view. At the centre, they say, is fire, and the earth is one of the stars, creating night and day by its circular motion about the centre. They further construct another earth in opposition to ours to which they give the name counterearth. In all this they are not seeking for theories and causes to account for observed facts, but rather forcing their observations and trying to accommodate them to certain theories and opinions of their own. But there are many others who would agree that it is wrong to give the earth the central position, looking for confirmation rather to theory than to the facts of observation. Their view is that the most precious place befits the most precious thing: but fire, they say, is more precious than earth, and the limit than the intermediate, and the circumference and the centre are limits. Reasoning on this basis they take the view that it is not earth that lies at the centre of the sphere, but rather fire. The Pythagoreans have a further reason. They hold that the most important part of the world, which is the centre, should be most strictly guarded, and name it, or rather the fire which occupies that place, the 'Guardhouse of Zeus', as if the word 'centre' were quite unequivocal, and the centre of the mathematical figure were always the same with that of the thing or the natural centre. But it is better to conceive of the case of the whole heaven as analogous to that of animals, in which the centre of the animal and that of the body are different. For this reason they have no need to be so disturbed about the world, or to call in a guard for its centre: rather let them look for the centre in the other sense and tell us what it is like and where nature has set it. That centre will be something primary and precious; but to the mere position we should give the last place rather than the first. For the middle is what is defined, and what defines it is the limit, and that which contains or limits is more precious than that which is limited, see ing that the latter is the matter and the former the essence of the system.

II. As to the position of the earth, then, this is the view which some advance, and the views advanced concerning its rest or motion are similar. For here too there is no general agreement. All who deny that the earth lies at the centre think that it revolves about the centre, and not the earth only but, as we said before, the counter-earth as well. Some of them even consider it possible that there are several bodies so moving, which are invisible to us owing to the interposition of the earth. This, they say, accounts for the fact that eclipses of the moon are more frequent than eclipses of the sun: for in addition to the earth each of these moving bodies can obstruct it. Indeed, as in any case the surface of the earth is not actually a centre but distant from it a full hemisphere, there is no more difficulty, they think, in accounting for the observed facts on their view that we do not dwell at the centre, than on the common view that the earth is in the middle. Even as it is, there is nothing in the observations to suggest that we are removed from the centre by half the diameter of the earth. Others, again, say that the earth, which lies at the centre, is 'rolled', and thus in motion, about the axis of the whole heaven, So it stands written in the Timaeus.

III. There are similar disputes about the shape of the earth. Some think it is spherical, others that it is flat and drum-shaped. For evidence they bring the fact that, as the sun rises and sets, the part concealed by the earth shows a straight and not a curved edge, whereas if the earth were spherical the line of section would have to be circular. In this they leave out of account the great distance of the sun from the earth and the great size of the circumference, which, seen from a distance on these apparently small circles appears straight. Such an appearance ought not to make them doubt the circular shape of the earth. But they have another argument. They say that because it is at rest, the earth must necessarily have this shape. For there are many different ways in which the movement or rest of the earth has been conceived.

The difficulty must have occurred to every one. It would indeed be a complacent mind that felt no surprise that, while a little bit of earth, let loose in mid-air moves and will not stay still, and more there is of it the faster it moves, the whole earth, free in midair, should show no movement at all. Yet here is this great weight of earth, and it is at rest. And again, from beneath one of these moving fragments of earth, before it falls, take away the earth, and it will continue its downward movement with nothing to stop it. The difficulty then, has naturally passed into a common place of philosophy; and one may well wonder that the solutions offered are not seen to involve greater absurdities than the problem itself.

By these considerations some have been led to assert that the earth below us is infinite, saying, with Xenophanes of Colophon, that it has 'pushed its roots to infinity',-in order to save the trouble of seeking for the cause. Hence the sharp rebuke of Empedocles, in the words 'if the deeps of the earth are endless and endless the ample ether-such is the vain tale told by many a tongue, poured from the mouths of those who have seen but little of the whole. Others say the earth rests upon water. This, indeed, is the oldest theory that has been preserved, and is attributed to Thales of Miletus. It was supposed to stay still because it floated like wood and other similar substances, which are so constituted as to rest upon but not upon air. As if the same account had not to be given of the water which carries the earth as of the earth itself! It is not the nature of water, any more than of earth, to stay in mid-air: it must have something to rest upon. Again, as air is lighter than water, so is water than earth: how then can they think that the naturally lighter substance lies below the heavier? Again, if the earth as a whole is capable of floating upon water, that must obviously be the case with any part of it. But observation shows that this is not the case. Any piece of earth goes to the bottom, the quicker the larger it is. These thinkers seem to push their inquiries some way into the problem, but not so far as they might. It is what we are all inclined to do, to direct our inquiry not by the matter itself, but by the views of our opponents: and even when interrogating oneself one pushes the inquiry only to the point at which one can no longer offer any opposition. Hence a good inquirer will be one who is ready in bringing forward the objections proper to the genus, and that he will be when he has gained an understanding of all the differences.

Anaximenes and Anaxagoras and Democritus give the flatness of the earth as the cause of its staying still. Thus, they say, it does not cut, but covers like a lid, the air beneath it. This seems to be the way of flat-shaped bodies: for even the wind can scarcely move them because of their power of resistance. The same immobility, they say, is produced by the flatness of the surface which the earth presents to the air which underlies it; while the air, not having room enough to change its place because it is underneath the earth, stays there in a mass, like the water in the case of the water-clock. And they adduce an amount of evidence to prove that air, when cut off and at rest, can bear a considerable weight.

Now, first, if the shape of the earth is not flat, its flatness cannot be the cause of its immobility. But in their own account it is rather the size of the earth than its flatness that causes it to remain at rest. For the reason why the air is so closely confined that it cannot find a passage, and therefore stays where it is, is its great amount: and this amount great because the body which isolates it, the earth, is very large. This result, then, will follow, even if the earth is spherical, so long as it retains its size. So far as their arguments go, the earth will still be at rest.

In general, our quarrel with those who speak of movement in this way cannot be confined to the parts; it concerns the whole universe. One must decide at the outset whether bodies have a natural movement or not, whether there is no natural but only constrained movement. Seeing, however, that we have already decided this matter to the best of our ability, we are entitled to treat our results as representing fact. Bodies, we say, which have no natural movement, have no constrained movement; and where there is no natural and no constrained movement there will be no movement at all. This is a conclusion, the necessity of which we have already decided, and we have seen further that rest also will be inconceivable, since rest, like movement, is either natural or constrained. But if there is any natural movement, constraint will not be the sole principle of motion or of rest. If, then, it is by constraint that the earth now keeps its place, the so-called 'whirling' movement by which its parts came together at the centre was also constrained. (The form of causation supposed they all borrow from observations of liquids and of air, in which the larger and heavier bodies always move to the centre of the whirl. This is thought by all those who try to generate the heavens to explain why the earth came together at the centre. They then seek a reason for its staying there; and some say, in the manner explained, that the reason is its size and flatness, others, with Empedocles, that the motion of the heavens, moving about it at a higher speed, prevents movement of the earth, as the water in a cup, when the cup is given a circular motion, though it is often underneath the bronze, is for this same reason prevented from moving with the downward movement which is natural to it.) But suppose both the 'whirl' and its flatness (the air beneath being withdrawn) cease to prevent the earth's motion, where will the earth move to then? Its movement to the centre was constrained, and its rest at the centre is due to constraint; but there must be some motion which is natural to it. Will this be upward motion or downward or what? It must have some motion; and if upward and downward motion are alike to it, and the air above the earth does not prevent upward movement, then no more could air below it prevent downward movement. For the same cause must necessarily have the same effect on the same thing.

Further, against Empedocles there is another point which might be made. When the elements were separated off by Hate, what caused the earth to keep its place? Surely the 'whirl' cannot have been then also the cause. It is absurd too not to perceive that, while the whirling movement may have been responsible for the original coming together of the art of earth at the centre, the question remains, why now do all heavy bodies move to the earth. For the whirl surely does not come near us. Why, again, does fire move upward? Not, surely, because of the whirl. But if fire is naturally such as to move in a certain direction, clearly the same may be supposed to hold of earth. Again, it cannot be the whirl which determines the heavy and the light. Rather that movement caused the pre-existent heavy and light things to go to the middle and stay on the surface respectively. Thus, before ever the whirl began, heavy and light existed; and what can have been the ground of their distinction, or the manner and direction of their natural movements? In the infinite chaos there can have been neither above nor below, and it is by these that heavy and light are determined.

It is to these causes that most writers pay attention: but there are some, Anaximander, for instance, among the ancients, who say that the earth keeps its place because of its indifference. Motion upward and downward and sideways were all, they thought, equally inappropriate to that which is set at the centre and indifferently related to every extreme point; and to move in contrary directions at the same time was impossible: so it must needs remain still. This view is ingenious but not true. The argument would prove that everything, whatever it be, which is put at the centre, must stay there. Fire, then, will rest at the centre: for the proof turns on no peculiar property of earth. But this does not follow. The observed facts about earth are not only that it remains at the centre, but also that it moves to the centre. The place to which any fragment of earth moves must necessarily be the place to which the whole moves; and in the place to which a thing naturally moves, it will naturally rest. The reason then is not in the fact that the earth is indifferently related to every extreme point: for this would apply to any body, whereas movement to the centre is peculiar to earth. Again it is absurd to look for a reason why the earth remains at the centre and not for a reason why fire remains at the extremity. If the extremity is the natural place of fire, clearly earth must also have a natural place. But suppose that the centre is not its place, and that the reason of its remaining there is this necessity of indifference-on the analogy of the hair which, it is said, however great the tension, will not break under it, if it be evenly distributed, or of the men who, though exceedingly hungry and thirsty, and both equally, yet being equidistant from food and drink, is therefore bound to stay where he is-even so, it still remains to explain why fire stays at the extremities. It is strange, too, to ask about things staying still but not about their motion,-why, I mean, one thing, if nothing stops it, moves up, and another thing to the centre. Again, their statements are not true. It happens, indeed, to be the case that a thing to which movement this way and that is equally inappropriate is obliged to remain at the centre. But so far as their argument goes, instead of remaining there, it will move, only not as a mass but in fragments. For the argument applies equally to fire. Fire, if set at the centre, should stay there, like earth, since it will be indifferently related to every point on the extremity. Nevertheless it will move, as in fact it always does move when nothing stops it, away from the centre to the extremity. It will not, however, move in a mass to a single point on the circumference-the only possible result on the lines of the indifference theory-but rather each corresponding portion of fire to the corresponding part of the extremity, each fourth part, for instance, to a fourth part of the circumference. For since no body is a point, it will have parts. The expansion, when the body increased the place occupied, would be on the same principle as the contraction, in which the place was diminished. Thus, for all the indifference theory shows to the contrary, earth also would have moved in this manner away from the centre, unless the centre had been its natural place.

We have now outlined the views held as to the shape, position, and rest or movement of the earth.

14

Let us first decide the question whether the earth moves or is at rest. For, as we said, there are some who make it one of the stars, and others who, setting it at the centre, suppose it to be 'rolled' and in motion about the pole as axis. That both views are untenable will be clear if we take as our starting-point the fact that the earth's motion, whether the earth be at the centre or away from it, must needs be a constrained motion. It cannot be the movement of the earth itself. If it were, any portion of it would have this movement; but in fact every part moves in a straight line to the centre. Being, then, constrained and unnatural, the movement could not be eternal. But the order of the universe is eternal. Again, everything that moves with the circular movement, except the first sphere, is observed to be passed, and to move with more than one motion. The earth, then, also, whether it move about the centre or as stationary at it, must necessarily move with two motions. But if this were so, there would have to be passings and turnings of the fixed stars. Yet no such thing is observed. The same stars always rise and set in the same parts of the earth.

Further, the natural movement of the earth, part and whole alike, is the centre of the whole-whence the fact that it is now actually situated at the centre-but it might be questioned since both centres are the same, which centre it is that portions of earth and other heavy things move to. Is this their goal because it is the centre of the earth or because it is the centre of the whole? The goal, surely, must be the centre of the whole. For fire and other light things move to the extremity of the area which contains the centre. It happens, however, that the centre of the earth and of the whole is the same. Thus they do move to the centre of the earth, but accidentally, in virtue of the fact that the earth's centre lies at the centre of the whole. That the centre of the earth is the goal of their movement is indicated by the fact that heavy bodies moving towards the earth do not parallel but so as to make equal angles, and thus to a single centre, that of the earth. It is clear, then, that the earth must be at the centre and immovable, not only for the reasons already given, but also because heavy bodies forcibly thrown quite straight upward return to the point from which they started, even if they are thrown to an infinite distance. From these considerations then it is clear that the earth does not move and does not lie elsewhere than at the centre.

From what we have said the explanation of the earth's immobility is also apparent. If it is the nature of earth, as observation shows, to move from any point to the centre, as of fire contrariwise to move from the centre to the extremity, it is impossible that any portion of earth should move away from the centre except by constraint. For a single thing has a single movement, and a simple thing a simple: contrary movements cannot belong to the same thing, and movement away from the centre is the contrary of movement to it. If then no portion of earth can move away from the centre, obviously still less can the earth as a whole so move. For it is the nature of the whole to move to the point to which the part naturally moves.

Since, then, it would require a force greater than itself to move it, it must needs stay at the centre. This view is further supported by the contributions of mathematicians to astronomy, since the observations made as the shapes change by which the order of the stars is determined, are fully accounted for on the hypothesis that the earth lies at the centre. Of the position of the earth and of the manner of its rest or movement, our discussion may here end.

Its shape must necessarily be spherical. For every portion of earth has weight until it reaches the centre, and the jostling of parts greater and smaller would bring about not a waved surface, but rather compression and convergence of part and part until the centre is reached. The process should be conceived by supposing the earth to come into being in the way that some of the natural philosophers describe. Only they attribute the downward movement to constraint, and it is better to keep to the truth and say that the reason of this motion is that a thing which possesses weight is naturally endowed with a centripetal movement. When the mixture, then, was merely potential, the things that were separated off moved similarly from every side towards the centre. Whether the parts which came together at the centre were distributed at the extremities evenly, or in some other way, makes no difference. If, on the one hand, there were a similar movement from each quarter of the extremity to the single centre, it is obvious that the resulting mass would be similar on every side. For if an equal amount is added on every side the extremity of the mass will be everywhere equidistant from its centre, i.e. the figure will be spherical. But neither will it in any way affect the argument if there is not a similar accession of concurrent fragments from every side. For the greater quantity, finding a lesser in front of it, must necessarily drive it on, both having an impulse whose goal is the centre, and the greater weight driving the lesser forward till this goal is reached. In this we have also the solution of a possible difficulty. The earth, it might be argued, is at the centre and spherical in shape: if, then, a weight many times that of the earth were added to one hemisphere, the centre of the earth and of the whole will no longer be coincident. So that either the earth will not stay still at the centre, or if it does, it will be at rest without having its centre at the place to which it is still its nature to move. Such is the difficulty. A short consideration will give us an easy answer, if we first give precision to our postulate that any body endowed with weight, of whatever size, moves towards the centre. Clearly it will not stop when its edge touches the centre. The greater quantity must prevail until the body's centre occupies the centre. For that is the goal of its impulse. Now it makes no difference whether we apply this to a clod or common fragment of earth or to the earth as a whole. The fact indicated does not depend upon degrees of size but applies universally to everything that has the centripetal impulse. Therefore earth in motion, whether in a mass or in fragments, necessarily continues to move until it occupies the centre equally every way, the less being forced to equalize itself by the greater owing to the forward drive of the impulse.

If the earth was generated, then, it must have been formed in this way, and so clearly its generation was spherical; and if it is ungenerated and has remained so always, its character must be that which the initial generation, if it had occurred, would have given it. But the spherical shape, necessitated by this argument, follows also from the fact that the motions of heavy bodies always make equal angles, and are not parallel. This would be the natural form of movement towards what is naturally spherical. Either then the earth is spherical or it is at least naturally spherical. And it is right to call anything that which nature intends it to be, and which belongs to it, rather than that which it is by constraint and contrary to nature. The evidence of the senses further corroborates this. How else would eclipses of the moon show segments shaped as we see them? As it is, the shapes which the moon itself each month shows are of every kind straight, gibbous, and concave-but in eclipses the outline is always curved: and, since it is the interposition of the earth that makes the eclipse, the form of this line will be caused by the form of the earth's surface, which is therefore spherical. Again, our observations of the stars make it evident, not only that the earth is circular, but also that it is a circle of no great size. For quite a small change of position to south or north causes a manifest alteration of the horizon. There is much change, I mean, in the stars which are overhead, and the stars seen are different, as one moves northward or southward. Indeed there are some stars seen in Egypt and in the neighbourhood of Cyprus which are not seen in the northerly regions; and stars, which in the north are never beyond the range of observation, in those regions rise and set. All of which goes to show not only that the earth is circular in shape, but also that it is a sphere of no great size: for otherwise the effect of so slight a change of place would not be quickly apparent. Hence one should not be too sure of the incredibility of the view of those who conceive that there is continuity between the parts about the pillars of Hercules and the parts about India, and that in this way the ocean is one. As further evidence in favour of this they quote the case of elephants, a species occurring in each of these extreme regions, suggesting that the common characteristic of these extremes is explained by their continuity. Also, those mathematicians who try to calculate the size of the earth's circumference arrive at the figure 400,000 stades. This indicates not only that the earth's mass is spherical in shape, but also that as compared with the stars it is not of great size.

Book III

1

WE have already discussed the first heaven and its parts, the moving stars within it, the matter of which these are composed and their bodily constitution, and we have also shown that they are ungenerated and indestructible. Now things that we call natural are either substances or functions and attributes of substances. As substances I class the simple bodies-fire, earth, and the other terms of the series-and all things composed of them; for example, the heaven as a whole and its parts, animals, again, and plants and their parts. By attributes and functions I mean the movements of these and of all other things in which they have power in themselves to cause movement, and also their alterations and reciprocal transformations. It is obvious, then, that the greater part of the inquiry into nature concerns bodies: for a natural substance is either a body or a thing which cannot come into existence without body and magnitude. This appears plainly from an analysis of the character of natural things, and equally from an inspection of the instances of inquiry into nature. Since, then, we have spoken of the primary element, of its bodily constitution, and of its freedom from destruction and generation, it remains to speak of the other two. In speaking of them we shall be obliged also to inquire into generation and destruction. For if there is generation anywhere, it must be in these elements and things composed of them.

This is indeed the first question we have to ask: is generation a fact or not? Earlier speculation was at variance both with itself and with the views here put forward as to the true answer to this question. Some removed generation and destruction from the world altogether. Nothing that is, they said, is generated or destroyed, and our conviction to the contrary is an illusion. So maintained the school of Melissus and Parmenides. But however excellent their theories may otherwise be, anyhow they cannot be held to speak as students of nature. There may be things not subject to generation or any kind of movement, but if so they belong to another and a higher inquiry than the study of nature. They, however, had no idea of any form of being other than the substance of things perceived; and when they saw, what no one previously had seen, that there could be no knowledge or wisdom without some such unchanging entities, they naturally transferred what was true of them to things perceived. Others, perhaps intentionally, maintain precisely the contrary opinion to this. It has been asserted that everything in the world was subject to generation and nothing was ungenerated, but that after being generated some things remained indestructible while the rest were again destroyed. This had been asserted in the first instance by Hesiod and his followers, but afterwards outside his circle by the earliest natural philosophers. But what these thinkers maintained was that all else has been generated and, as they said, 'is flowing away, nothing having any solidity, except one single thing which persists as the basis of all these transformations. So we may interpret the statements of Heraclitus of Ephesus and many others. And some subject all bodies whatever to generation, by means of the composition and separation of planes.

Discussion of the other views may be postponed. But this last theory which composes every body of planes is, as the most superficial observation shows, in many respects in plain contradiction with mathematics. It is, however, wrong to remove the foundations of a science unless you can replace them with others more convincing. And, secondly, the same theory which composes solids of planes clearly composes planes of lines and lines of points, so that a part of a line need not be a line. This matter has been already considered in our discussion of movement, where we have shown that an indivisible length is impossible. But with respect to natural bodies there are impossibilities involved in the view which asserts indivisible lines, which we may briefly consider at this point. For the impossible consequences which result from this view in the mathematical sphere will reproduce themselves when it is applied to physical bodies, but there will be difficulties in physics which are not present in mathematics; for mathematics deals with an abstract and physics with a more concrete object. There are many attributes necessarily present in physical bodies which are necessarily excluded by indivisibility; all attributes, in fact, which are divisible. There can be nothing divisible in an indivisible thing, but the attributes of bodies are all divisible in one of two ways. They are divisible into kinds, as colour is divided into white and black, and they are divisible per accidens when that which has them is divisible. In this latter sense attributes which are simple are nevertheless divisible. Attributes of this kind will serve, therefore, to illustrate the impossibility of the view. It is impossible, if two parts of a thing have no weight, that the two together should have weight. But either all perceptible bodies or some, such as earth and water, have weight, as these thinkers would themselves admit. Now if the point has no weight, clearly the lines have not either, and, if they have not, neither have the planes. Therefore no body has weight. It is, further, manifest that their point cannot have weight. For while a heavy thing may always be heavier than something and a light thing lighter than something, a thing which is heavier or lighter than something need not be itself heavy or light, just as a large thing is larger than others, but what is larger is not always large. A thing which, judged absolutely, is small may none the less be larger than other things. Whatever, then, is heavy and also heavier than something else, must exceed this by something which is heavy. A heavy thing therefore is always divisible. But it is common ground that a point is indivisible. Again, suppose that what is heavy or weight is a dense body, and what is light rare. Dense differs from rare in containing more matter in the same cubic area. A point, then, if it may be heavy or light, may be dense or rare. But the dense is divisible while a point is indivisible. And if what is heavy must be either hard or soft, an impossible consequence is easy to draw. For a thing is soft if its surface can be pressed in, hard if it cannot; and if it can be pressed in it is divisible.

Moreover, no weight can consist of parts not possessing weight. For how, except by the merest fiction, can they specify the number and character of the parts which will produce weight? And, further, when one weight is greater than another, the difference is a third weight; from which it will follow that every indivisible part possesses weight. For suppose that a body of four points possesses weight. A body composed of more than four points will superior in weight to it, a thing which has weight. But the difference between weight and weight must be a weight, as the difference between white and whiter is white. Here the difference which makes the superior weight heavier is the single point which remains when the common number, four, is subtracted. A single point, therefore, has weight.

Further, to assume, on the one hand, that the planes can only be put in linear contact would be ridiculous. For just as there are two ways of putting lines together, namely, end to and side by side, so there must be two ways of putting planes together. Lines can be put together so that contact is linear by laying one along the other, though not by putting them end to end. But if, similarly, in putting the lanes together, superficial contact is allowed as an alternative to linear, that method will give them bodies which are not any element nor composed of elements. Again, if it is the number of planes in a body that makes one heavier than another, as the Timaeus explains, clearly the line and the point will have weight. For the three cases are, as we said before, analogous. But if the reason of differences of weight is not this, but rather the heaviness of earth and the lightness of fire, then some of the planes will be light and others heavy (which involves a similar distinction in the lines and the points); the earthplane, I mean, will be heavier than the fire-plane. In general, the result is either that there is no magnitude at all, or that all magnitude could be done away with. For a point is to a line as a line is to a plane and as a plane is to a body. Now the various forms in passing into one another will each be resolved into its ultimate constituents. It might happen therefore that nothing existed except points, and that there was no body at all. A further consideration is that if time is similarly constituted, there would be, or might be, a time at which it was done away with. For the indivisible now is like a point in a line. The same consequences follow from composing the heaven of numbers, as some of the Pythagoreans do who make all nature out of numbers. For natural bodies are manifestly endowed with weight and lightness, but an assemblage of units can neither be composed to form a body nor possess weight.

2

The necessity that each of the simple bodies should have a natural movement may be shown as follows. They manifestly move, and if they have no proper movement they must move by constraint: and the constrained is the same as the unnatural. Now an unnatural movement presupposes a natural movement which it contravenes, and which, however many the unnatural movements, is always one. For naturally a thing moves in one way, while its unnatural movements are manifold. The same may be shown, from the fact of rest. Rest, also, must either be constrained or natural, constrained in a place to which movement was constrained, natural in a place movement to which was natural. Now manifestly there is a body which is at rest at the centre. If then this rest is natural to it, clearly motion to this place is natural to it. If, on the other hand, its rest is constrained, what is hindering its motion? Something, which is at rest: but if so, we shall simply repeat the same argument; and either we shall come to an ultimate something to which rest where it is or we shall have an infinite process, which is impossible. The hindrance to its movement, then, we will suppose, is a moving thing-as Empedocles says that it is the vortex which keeps the earth still-: but in that case we ask, where would it have moved to but for the vortex? It could not move infinitely; for to traverse an infinite is impossible, and impossibilities do not happen. So the moving thing must stop somewhere, and there rest not by constraint but naturally. But a natural rest proves a natural movement to the place of rest. Hence Leucippus and Democritus, who say that the primary bodies are in perpetual movement in the void or infinite, may be asked to explain the manner of their motion and the kind of movement which is natural to them. For if the various elements are constrained by one another to move as they do, each must still have a natural movement which the constrained contravenes, and the prime mover must cause motion not by constraint but naturally. If there is no ultimate natural cause of movement and each preceding term in the series is always moved by constraint, we shall have an infinite process. The same difficulty is involved even if it is supposed, as we read in the Timaeus, that before the ordered world was made the elements moved without order. Their movement must have been due either to constraint or to their nature. And if their movement was natural, a moment's consideration shows that there was already an ordered world. For the prime mover must cause motion in virtue of its own natural movement, and the other bodies, moving without constraint, as they came to rest in their proper places, would fall into the order in which they now stand, the heavy bodies moving towards the centre and the light bodies away from it. But that is the order of their distribution in our world. There is a further question, too, which might be asked. Is it possible or impossible that bodies in unordered movement should combine in some cases into combinations like those of which bodies of nature's composing are composed, such, I mean, as bones and flesh? Yet this is what Empedocles asserts to have occurred under Love. 'Many a head', says he, 'came to birth without a neck.' The answer to the view that there are infinite bodies moving in an infinite is that, if the cause of movement is single, they must move with a single motion, and therefore not without order; and if, on the other hand, the causes are of infinite variety, their motions too must be infinitely varied. For a finite number of causes would produce a kind of order, since absence of order is not proved by diversity of direction in motions: indeed, in the world we know, not all bodies, but only bodies of the same kind, have a common goal of movement. Again, disorderly movement means in reality unnatural movement, since the order proper to perceptible things is their nature. And there is also absurdity and impossibility in the notion that the disorderly movement is infinitely continued. For the nature of things is the nature which most of them possess for most of the time. Thus their view brings them into the contrary position that disorder is natural, and order or system unnatural. But no natural fact can originate in chance. This is a point which Anaxagoras seems to have thoroughly grasped; for he starts his cosmogony from unmoved things. The others, it is true, make things collect together somehow before they try to produce motion and separation. But there is no sense in starting generation from an original state in which bodies are separated and in movement. Hence Empedocles begins after the process ruled by Love: for he could not have constructed the heaven by building it up out of bodies in separation, making them to combine by the power of Love, since our world has its constituent elements in separation, and therefore presupposes a previous state of unity and combination.

These arguments make it plain that every body has its natural movement, which is not constrained or contrary to its nature. We go on to show that there are certain bodies whose necessary impetus is that of weight and lightness. Of necessity, we assert, they must move, and a moved thing which has no natural impetus cannot move either towards or away from the centre. Suppose a body A without weight, and a body B endowed with weight. Suppose the weightless body to move the distance CD, while B in the same time moves the distance CE, which will be greater since the heavy thing must move further. Let the heavy body then be divided in the proportion CE: CD (for there is no reason why a part of B should not stand in this relation to the whole). Now if the whole moves the whole distance CE, the part must in the same time move the distance CD. A weightless body, therefore, and one which has weight will move the same distance, which is impossible. And the same argument would fit the case of lightness. Again, a body which is in motion but has neither weight nor lightness, must be moved by constraint, and must continue its constrained movement infinitely. For there will be a force which moves it, and the smaller and lighter a body is the further will a given force move it. Now let A, the weightless body, be moved the distance CE, and B, which has weight, be moved in the same time the distance CD. Dividing the heavy body in the proportion CE:CD, we subtract from the heavy body a part which will in the same time move the distance CE, since the whole moved CD: for the relative speeds of the two bodies will be in inverse ratio to their respective sizes. Thus the weightless body will move the same distance as the heavy in the same time. But this is impossible. Hence, since the motion of the weightless body will cover a greater distance than any that is suggested, it will continue infinitely. It is therefore obvious that every body must have a definite weight or lightness. But since 'nature' means a source of movement within the thing itself, while a force is a source of movement in something other than it or in itself qua other, and since movement is always due either to nature or to constraint, movement which is natural, as downward movement is to a stone, will be merely accelerated by an external force, while an unnatural movement will be due to the force alone. In either case the air is as it were instrumental to the force. For air is both light and heavy, and thus qua light produces upward motion, being propelled and set in motion by the force, and qua heavy produces a downward motion. In either case the force transmits the movement to the body by first, as it were, impregnating the air. That is why a body moved by constraint continues to move when that which gave the impulse ceases to accompany it. Otherwise, i.e. if the air were not endowed with this function, constrained movement would be impossible. And the natural movement of a body may be helped on in the same way. This discussion suffices to show (1) that all bodies are either light or heavy, and (2) how unnatural movement takes place.

From what has been said earlier it is plain that there cannot be generation either of everything or in an absolute sense of anything. It is impossible that everything should be generated, unless an extra-corporeal void is possible. For, assuming generation, the place which is to be occupied by that which is coming to be, must have been previously occupied by void in which no body was. Now it is quite possible for one body to be generated out of another, air for instance out of fire, but in the absence of any pre-existing mass generation is impossible. That which is potentially a certain kind of body may, it is true, become such in actuality, But if the potential body was not already in actuality some other kind of body, the existence of an extra-corporeal void must be admitted.

3

It remains to say what bodies are subject to generation, and why. Since in every case knowledge depends on what is primary, and the elements are the primary constituents of bodies, we must ask which of such bodies are elements, and why; and after that what is their number and character. The answer will be plain if we first explain what kind of substance an element is. An element, we take it, is a body into which other bodies may be analysed, present in them potentially or in actuality (which of these, is still disputable), and not itself divisible into bodies different in form. That, or something like it, is what all men in every case mean by element. Now if what we have described is an element, clearly there must be such bodies. For flesh and wood and all other similar bodies contain potentially fire and earth, since one sees these elements exuded from them; and, on the other hand, neither in potentiality nor in actuality does fire contain flesh or wood, or it would exude them. Similarly, even if there were only one elementary body, it would not contain them. For though it will be either flesh or bone or something else, that does not at once show that it contained these in potentiality: the further question remains, in what manner it becomes them. Now Anaxagoras opposes Empedocles' view of the elements. Empedocles says that fire and earth and the related bodies are elementary bodies of which all things are composed; but this Anaxagoras denies. His elements are the homoeomerous things, viz. flesh, bone, and the like. Earth and fire are mixtures, composed of them and all the other seeds, each consisting of a collection of all the homoeomerous bodies, separately invisible; and that explains why from these two bodies all others are generated. (To him fire and aither are the same thing.) But since every natural body has it proper movement, and movements are either simple or mixed, mixed in mixed bodies and simple in simple, there must obviously be simple bodies; for there are simple movements. It is plain, then, that there are elements, and why.

4

The next question to consider is whether the elements are finite or infinite in number, and, if finite, what their number is. Let us first show reason or denying that their number is infinite, as some suppose. We begin with the view of Anaxagoras that all the homoeomerous bodies are elements. Any one who adopts this view misapprehends the meaning of element. Observation shows that even mixed bodies are often divisible into homoeomerous parts; examples are flesh, bone, wood, and stone. Since then the composite cannot be an element, not every homoeomerous body can be an element; only, as we said before, that which is not divisible into bodies different in form. But even taking 'element' as they do, they need not assert an infinity of elements, since the hypothesis of a finite number will give identical results. Indeed even two or three such bodies serve the purpose as well, as Empedocles' attempt shows. Again, even on their view it turns out that all things are not composed of homocomerous bodies. They do not pretend that a face is composed of faces, or that any other natural conformation is composed of parts like itself. Obviously then it would be better to assume a finite number of principles. They should, in fact, be as few as possible, consistently with proving what has to be proved. This is the common demand of mathematicians, who always assume as principles things finite either in kind or in number. Again, if body is distinguished from body by the appropriate qualitative difference, and there is a limit to the number of differences (for the difference lies in qualities apprehended by sense, which are in fact finite in number, though this requires proof), then manifestly there is necessarily a limit to the number of elements.

There is, further, another view-that of Leucippus and Democritus of Abdera--the implications of which are also unacceptable. The primary masses, according to them, are infinite in number and indivisible in mass: one cannot turn into many nor many into one; and all things are generated by their combination and involution. Now this view in a sense makes things out to be numbers or composed of numbers. The exposition is not clear, but this is its real meaning. And further, they say that since the atomic bodies differ in shape, and there is an infinity of shapes, there is an infinity of simple bodies. But they have never explained in detail the shapes of the various elements, except so far to allot the sphere to fire. Air, water, and the rest they distinguished by the relative size of the atom, assuming that the atomic substance was a sort of master-seed for each and every element. Now, in the first place, they make the mistake already noticed. The principles which they assume are not limited in number, though such limitation would necessitate no other alteration in their theory. Further, if the differences of bodies are not infinite, plainly the elements will not be an infinity. Besides, a view which asserts atomic bodies must needs come into conflict with the mathematical sciences, in addition to invalidating many common opinions and apparent data of sense perception. But of these things we have already spoken in our discussion of time and movement. They are also bound to contradict themselves. For if the elements are atomic, air, earth, and water cannot be differentiated by the relative sizes of their atoms, since then they could not be generated out of one another. The extrusion of the largest atoms is a process that will in time exhaust the supply; and it is by such a process that they account for the generation of water, air, and earth from one another. Again, even on their own presuppositions it does not seem as if the clements would be infinite in number. The atoms differ in figure, and all figures are composed of pyramids, rectilinear the case of rectilinear figures, while the sphere has eight pyramidal parts. The figures must have their principles, and, whether these are one or two or more, the simple bodies must be the same in number as they. Again, if every element has its proper movement, and a simple body has a simple movement, and the number of simple movements is not infinite, because the simple motions are only two and the number of places is not infinite, on these grounds also we should have to deny that the number of elements is infinite.

5

Since the number of the elements must be limited, it remains to inquire whether there is more than one element. Some assume one only, which is according to some water, to others air, to others fire, to others again something finer than water and denser than air, an infinite body--so they say--bracing all the heavens.

Now those who decide for a single element, which is either water or air or a body finer than water and denser than air, and proceed to generate other things out of it by use of the attributes density and rarity, all alike fail to observe the fact that they are depriving the element of its priority. Generation out of the elements is, as they say, synthesis, and generation into the elements is analysis, so that the body with the finer parts must have priority in the order of nature. But they say that fire is of all bodies the finest. Hence fire will be first in the natural order. And whether the finest body is fire or not makes no difference; anyhow it must be one of the other bodies that is primary and not that which is intermediate. Again, density and rarity, as instruments of generation, are equivalent to fineness and coarseness, since the fine is rare, and coarse in their use means dense. But fineness and coarseness, again, are equivalent to greatness and smallness, since a thing with small parts is fine and a thing with large parts coarse. For that which spreads itself out widely is fine, and a thing composed of small parts is so spread out. In the end, then, they distinguish the various other substances from the element by the greatness and smallness of their parts. This method of distinction makes all judgement relative. There will be no absolute distinction between fire, water, and air, but one and the same body will be relatively to this fire, relatively to something else air. The same difficulty is involved equally in the view elements and distinguishes them by their greatness and smallness. The principle of distinction between bodies being quantity, the various sizes will be in a definite ratio, and whatever bodies are in this ratio to one another must be air, fire, earth, and water respectively. For the ratios of smaller bodies may be repeated among greater bodies.

Those who start from fire as the single element, while avoiding this difficulty, involve themselves in many others. Some of them give fire a particular shape, like those who make it a pyramid, and this on one of two grounds. The reason given may be--more crudely--that the pyramid is the most piercing of figures as fire is of bodies, or--more ingeniously--the position may be supported by the following argument. As all bodies are composed of that which has the finest parts, so all solid figures are composed of pryamids: but the finest body is fire, while among figures the pyramid is primary and has the smallest parts; and the primary body must have the primary figure: therefore fire will be a pyramid. Others, again, express no opinion on the subject of its figure, but simply regard it as the of the finest parts, which in combination will form other bodies, as the fusing of gold-dust produces solid gold. Both of these views involve the same difficulties. For (1) if, on the one hand, they make the primary body an atom, the view will be open to the objections already advanced against the atomic theory. And further the theory is inconsistent with a regard for the facts of nature. For if all bodies are quantitatively commensurable, and the relative size of the various homoeomerous masses and of their several elements are in the same ratio, so that the total mass of water, for instance, is related to the total mass of air as the elements of each are to one another, and so on, and if there is more air than water and, generally, more of the finer body than of the coarser, obviously the element of water will be smaller than that of air. But the lesser quantity is contained in the greater. Therefore the air element is divisible. And the same could be shown of fire and of all bodies whose parts are relatively fine. (2) If, on the other hand, the primary body is divisible, then (a) those who give fire a special shape will have to say that a part of fire is not fire, because a pyramid is not composed of pyramids, and also that not every body is either an element or composed of elements, since a part of fire will be neither fire nor any other element. And (b) those whose ground of distinction is size will have to recognize an element prior to the element, a regress which continues infinitely, since every body is divisible and that which has the smallest parts is the element. Further, they too will have to say that the same body is relatively to this fire and relatively to that air, to others again water and earth.

The common error of all views which assume a single element is that they allow only one natural movement, which is the same for every body. For it is a matter of observation that a natural body possesses a principle of movement. If then all bodies are one, all will have one movement. With this motion the greater their quantity the more they will move, just as fire, in proportion as its quantity is greater, moves faster with the upward motion which belongs to it. But the fact is that increase of quantity makes many things move the faster downward. For these reasons, then, as well as from the distinction already established of a plurality of natural movements, it is impossible that there should be only one element. But if the elements are not an infinity and not reducible to one, they must be several and finite in number.

6

First we must inquire whether the elements are eternal or subject to generation and destruction; for when this question has been answered their number and character will be manifest. In the first place, they cannot be eternal.

It is a matter of observation that fire, water, and every simple body undergo a process of analysis, which must either continue infinitely or stop somewhere. (1) Suppose it infinite. Then the time occupied by the process will be infinite, and also that occupied by the reverse process of synthesis. For the processes of analysis and synthesis succeed one another in the various parts. It will follow that there are two infinite times which are mutually exclusive, the time occupied by the synthesis, which is infinite, being preceded by the period of analysis. There are thus two mutually exclusive infinites, which is impossible. (2) Suppose, on the other hand, that the analysis stops somewhere. Then the body at which it stops will be either atomic or, as Empedocles seems to have intended, a divisible body which will yet never be divided. The foregoing arguments show that it cannot be an atom; but neither can it be a divisible body which analysis will never reach. For a smaller body is more easily destroyed than a larger; and a destructive process which succeeds in destroying, that is, in resolving into smaller bodies, a body of some size, cannot reasonably be expected to fail with the smaller body. Now in fire we observe a destruction of two kinds: it is destroyed by its contrary when it is quenched, and by itself when it dies out. But the effect is produced by a greater quantity upon a lesser, and the more quickly the smaller it is. The elements of bodies must therefore be subject to destruction and generation.

Since they are generated, they must be generated either from something incorporeal or from a body, and if from a body, either from one another or from something else. The theory which generates them from something incorporeal requires an extra-corporeal void. For everything that comes to be comes to be in something, and that in which the generation takes place must either be incorporeal or possess body; and if it has body, there will be two bodies in the same place at the same time, viz. that which is coming to be and that which was previously there, while if it is incorporeal, there must be an extra-corporeal void. But we have already shown that this is impossible. But, on the other hand, it is equally impossible that the elements should be generated from some kind of body. That would involve a body distinct from the elements and prior to them. But if this body possesses weight or lightness, it will be one of the elements; and if it has no tendency to movement, it will be an immovable or mathematical entity, and therefore not in a place at all. A place in which a thing is at rest is a place in which it might move, either by constraint, i.e. unnaturally, or in the absence of constraint, i.e. naturally. If, then, it is in a place and somewhere, it will be one of the elements; and if it is not in a place, nothing can come from it, since that which comes into being and that out of which it comes must needs be together. The elements therefore cannot be generated from something incorporeal nor from a body which is not an element, and the only remaining alternative is that they are generated from one another.

7

We must, therefore, turn to the question, what is the manner of their generation from one another? Is it as Empedocles and Democritus say, or as those who resolve bodies into planes say, or is there yet another possibility? (1) What the followers of Empedocles do, though without observing it themselves, is to reduce the generation of elements out of one another to an illusion. They make it a process of excretion from a body of what was in it all the time-as though generation required a vessel rather than a material-so that it involves no change of anything. And even if this were accepted, there are other implications equally unsatisfactory. We do not expect a mass of matter to be made heavier by compression. But they will be bound to maintain this, if they say that water is a body present in air and excreted from air, since air becomes heavier when it turns into water. Again, when the mixed body is divided, they can show no reason why one of the constituents must by itself take up more room than the body did: but when water turns into air, the room occupied is increased. The fact is that the finer body takes up more room, as is obvious in any case of transformation. As the liquid is converted into vapour or air the vessel which contains it is often burst because it does not contain room enough. Now, if there is no void at all, and if, as those who take this view say, there is no expansion of bodies, the impossibility of this is manifest: and if there is void and expansion, there is no accounting for the fact that the body which results from division cfpies of necessity a greater space. It is inevitable, too, that generation of one out of another should come to a stop, since a finite quantum cannot contain an infinity of finite quanta. When earth produces water something is taken away from the earth, for the process is one of excretion. The same thing happens again when the residue produces water. But this can only go on for ever, if the finite body contains an infinity, which is impossible. Therefore the generation of elements out of one another will not always continue.

(2) We have now explained that the mutual transformations of the elements cannot take place by means of excretion. The remaining alternative is that they should be generated by changing into one another. And this in one of two ways, either by change of shape, as the same wax takes the shape both of a sphere and of a cube, or, as some assert, by resolution into planes. (a) Generation by change of shape would necessarily involve the assertion of atomic bodies. For if the particles were divisible there would be a part of fire which was not fire and a part of earth which was not earth, for the reason that not every part of a pyramid is a pyramid nor of a cube a cube. But if (b) the process is resolution into planes, the first difficulty is that the elements cannot all be generated out of one another. This they are obliged to assert, and do assert. It is absurd, because it is unreasonable that one element alone should have no part in the transformations, and also contrary to the observed data of sense, according to which all alike change into one another. In fact their explanation of the observations is not consistent with the observations. And the reason is that their ultimate principles are wrongly assumed: they had certain predetermined views, and were resolved to bring everything into line with them. It seems that perceptible things require perceptible principles, eternal things

eternal principles, corruptible things corruptible principles; and, in general, every subject matter principles homogeneous with itself. But they, owing to their love for their principles, fall into the attitude of men who undertake the defence of a position in argument. In the confidence that the principles are true they are ready to accept any consequence of their application. As though some principles did not require to be judged from their results, and particularly from their final issue! And that issue, which in the case of productive knowledge is the product, in the knowledge of nature is the unimpeachable evidence of the senses as to each fact.

The result of their view is that earth has the best right to the name element, and is alone indestructible; for that which is indissoluble is indestructible and elementary, and earth alone cannot be dissolved into any body but itself. Again, in the case of those elements which do suffer dissolution, the 'suspension' of the triangles is unsatisfactory. But this takes place whenever one is dissolved into another, because of the numerical inequality of the triangles which compose them. Further, those who hold these views must needs suppose that generation does not start from a body. For what is generated out of planes cannot be said to have been generated from a body. And they must also assert that not all bodies are divisible, coming thus into conflict with our most accurate sciences, namely the mathematical, which assume that even the intelligible is divisible, while they, in their anxiety to save their hypothesis, cannot even admit this of every perceptible thing. For any one who gives each element a shape of its own, and makes this the ground of distinction between the substances, has to attribute to them indivisibility; since division of a pyramid or a sphere must leave somewhere at least a residue which is not sphere or a pyramid. Either, then, a part of fire is not fire, so that there is a body prior to the element-for every body is either an element or composed of elements-or not every body is divisible.

8

In general, the attempt to give a shape to each of the simple bodies is unsound, for the reason, first, that they will not succeed in filling the whole. It is agreed that there are only three plane figures which can fill a space, the triangle, the square, and the hexagon, and only two solids, the pyramid and the cube. But the theory needs more than these because the elements which it recognizes are more in number. Secondly, it is manifest that the simple bodies are often given a shape by the place in which they are included, particularly water and air. In such a case the shape of the element cannot persist; for, if it did, the contained mass would not be in continuous contact with the containing body; while, if its shape is changed, it will cease to be water, since the distinctive quality is shape. Clearly, then, their shapes are not fixed. Indeed, nature itself seems to offer corroboration of this theoretical conclusion. Just as in other cases the substratum must be formless and unshapen-for thus the 'all-receptive', as we read in the Timaeus, will be best for modelling-so the elements should be conceived as a material for composite things; and that is why they can put off their qualitative distinctions and pass into one another. Further, how can they account for the generation of flesh and bone or any other continuous body? The elements alone cannot produce them because their collocation cannot produce a continuum. Nor can the composition of planes; for this produces the elements themselves, not bodies made up of them. Any one then who insists upon an exact statement of this kind of theory, instead of assenting after a passing glance at it, will see that it removes generation from the world.

Further, the very properties, powers, and motions, to which they paid particular attention in allotting shapes, show the shapes not to be in accord with the bodies. Because fire is mobile and productive of heat and combustion, some made it a sphere, others a pyramid. These shapes, they thought, were the most mobile because they offer the fewest points of contact and are the least stable of any; they were also the most apt to produce warmth and combustion, because the one is angular throughout while the other has the most acute angles, and the angles, they say, produce warmth and combustion. Now, in the first place, with regard to movement both are in error. These may be the figures best adapted to movement; they are not, however, well adapted to the movement of fire, which is an upward and rectilinear movement, but rather to that form of circular movement which we call rolling. Earth, again, they call a cube because it is stable and at rest. But it rests only in its own place, not anywhere; from any other it moves if nothing hinders, and fire and the other bodies do the same. The obvious inference, therefore, is that fire and each several element is in a foreign place a sphere or a pyramid, but in its own a cube. Again, if the possession of angles makes a body produce heat and combustion, every element produces heat, though one may do so more than another. For they all possess angles, the octahedron and dodecahedron as well as the pyramid; and Democritus makes even the sphere a kind of angle, which cuts things because of its mobility. The difference, then, will be one of degree: and this is plainly false. They must also accept the inference that the mathematical produce heat and combustion, since they too possess angles and contain atomic spheres and pyramids, especially if there are, as they allege, atomic figures. Anyhow if these functions belong to some of these things and not to others, they should explain the difference, instead of speaking in quite general terms as they do. Again, combustion of a body produces fire, and fire is a sphere or a pyramid. The body, then, is turned into spheres or pyramids. Let us grant that these figures may reasonably be supposed to cut and break up bodies as fire does; still it remains quite inexplicable that a pyramid must needs produce pyramids or a sphere spheres. One might as well postulate that a knife or a saw divides things into knives or saws. It is also ridiculous to think only of division when allotting fire its shape. Fire is generally thought of as combining and connecting rather than as separating. For though it separates bodies different in kind, it combines those which are the same; and the combining is essential to it, the functions of connecting and uniting being a mark of fire, while the separating is incidental. For the expulsion of the foreign body is an incident in the compacting of the homogeneous. In choosing the shape, then, they should have thought either of both functions or preferably of the combining function. In addition, since hot and cold are contrary powers, it is impossible to allot any shape to the cold. For the shape given must be the contrary of that given to the hot, but there is no contrariety between figures. That is why they have all left the cold out, though properly either all or none should have their distinguishing figures. Some of them, however, do attempt to explain this power, and they contradict themselves. A body of large particles, they say, is cold because instead of penetrating through the passages it crushes. Clearly, then, that which is hot is that which penetrates these passages, or in other words that which has fine particles. It results that hot and cold are distinguished not by the figure but by the size of the particles. Again, if the pyramids are unequal in size, the large ones will not be fire, and that figure will produce not combustion but its contrary.

From what has been said it is clear that the difference of the elements does not depend upon their shape. Now their most important differences are those of property, function, and power; for every natural body has, we maintain, its own functions, properties, and powers. Our first business, then, will be to speak of these, and that inquiry will enable us to explain the differences of each from each.

Book IV

1

WE have now to consider the terms 'heavy' and 'light'. We must ask what the bodies so called are, how they are constituted, and what is the reason of their possessing these powers. The consideration of these questions is a proper part of the theory of movement, since we call things heavy and light because they have the power of being moved naturally in a certain way. The activities corresponding to these powers have not been given any name, unless it is thought that 'impetus' is such a name. But because the inquiry into nature is concerned with movement, and these things have in themselves some spark (as it were) of movement, all inquirers avail themselves of these powers, though in all but a few cases without exact discrimination. We must then first look at whatever others have said, and formulate the questions which require settlement in the interests of this inquiry, before we go on to state our own view of the matter.

Language recognizes (a) an absolute, (b) a relative heavy and light. Of two heavy things, such as wood and bronze, we say that the one is relatively light, the other relatively heavy. Our predecessors have not dealt at all with the absolute use, of the terms, but only with the relative. I mean, they do not explain what the heavy is or what the light is, but only the relative heaviness and lightness of things possessing weight. This can be made clearer as follows. There are things whose constant nature it is to move away from the centre, while others move constantly towards the centre; and of these movements that which is away from the centre I call upward movement and that which is towards it I call downward movement. (The view, urged by some, that there is no up and no down in the heaven, is absurd. There can be, they say, no up and no down, since the universe is similar every way, and from any point on the earth's surface a man by advancing far enough will come to stand foot to foot with himself. But the extremity of the whole, which we call 'above', is in position above and in nature primary. And since the universe has an extremity and a centre, it must clearly have an up and down. Common usage is thus correct, though inadequate. And the reason of its inadequacy is that men think that the universe is not similar every way. They recognize only the hemisphere which is over us. But if they went on to think of the world as formed on this pattern all round, with a centre identically related to each point on the extremity, they would have to admit that the extremity was above and the centre below.) By absolutely light, then, we mean that which moves upward or to the extremity, and by absolutely heavy that which moves downward or to the centre. By lighter or relatively light we mean that one, of two bodies endowed with weight and equal in bulk, which is exceeded by the other in the speed of its natural downward movement.

<h1 style="text-align:center">2</h1>

Those of our predecessors who have entered upon this inquiry have for the most part spoken of light and heavy things only in the sense in which one of two things both endowed with weight is said to be the lighter. And this treatment they consider a sufficient analysis also of the notions of absolute heaviness, to which their account does not apply. This, however, will become clearer as we advance. One use of the terms 'lighter' and 'heavier' is that which is set forth in writing in the Timaeus, that the body which is composed of the greater number of identical parts is relatively heavy, while that which is composed of a smaller number is relatively light. As a larger quantity of lead or of bronze is heavier than a smaller-and this holds good of all homogeneous masses, the superior weight always depending upon a numerical superiority of equal parts-in precisely the same way, they assert, lead is heavier than wood. For all bodies, in spite of the general opinion to the contrary, are composed of identical parts and of a single material. But this analysis says nothing of the absolutely heavy and light. The facts are that fire is always light and moves upward, while earth and all earthy things move downwards or towards the centre. It cannot then be the fewness of the triangles (of which, in their view, all these bodies are composed) which disposes fire to move upward. If it were, the greater the quantity of fire the slower it would move, owing to the increase of weight due to the increased number of triangles. But the palpable fact, on the contrary, is that the greater the quantity, the lighter the mass is and the quicker its upward movement: and, similarly, in the reverse movement from above downward, the small mass will move quicker and the large slower. Further, since to be lighter is to have fewer of these homogeneous parts and to be heavier is to have more, and air, water, and fire are composed of the same triangles, the only difference being in the number of such parts, which must therefore explain any distinction of relatively light and heavy between these bodies, it follows that there must be a certain quantum of air which is heavier than water. But the facts are directly opposed to this. The larger the quantity of air the more readily it moves upward, and any portion of air without exception will rise up out of the water.

So much for one view of the distinction between light and heavy. To others the analysis seems insufficient; and their views on the subject, though they belong to an older generation than ours, have an air of novelty. It is apparent that there are bodies which, when smaller in bulk than others, yet exceed them in weight. It is therefore obviously insufficient to say that bodies of equal weight are composed of an equal number of primary parts: for that would give equality of bulk. Those who maintain that the primary or atomic parts, of which bodies endowed with weight are composed, are planes, cannot so speak without absurdity; but those who regard them as solids are in a better position to assert that of such bodies the larger is the heavier. But since in composite bodies the weight obviously does not correspond in this way to the bulk, the lesser bulk being often superior in weight (as, for instance, if one be wool and the other bronze), there are some who think and say that the cause is to be found elsewhere. The void, they say, which is imprisoned in bodies, lightens them and sometimes makes the larger body the lighter. The reason is that there is more void. And this would also account for the fact that a body composed of a number of solid parts equal to, or even smaller than, that of another is sometimes larger in bulk than it. In short, generally and in every case a body is relatively light when it contains a relatively large amount of void. This is the way they put it themselves, but their account requires an addition. Relative lightness must depend not only on an excess of void, but also an a defect of solid: for if the ratio of solid to void exceeds a certain proportion, the relative lightness will disappear. Thus fire, they say, is the lightest of things just for this reason that it has the most void. But it would follow that a large mass of gold, as containing more void than a small mass of fire, is lighter than it, unless it also contains many times as much solid. The addition is therefore necessary.

Of those who deny the existence of a void some, like Anaxagoras and Empedocles, have not tried to analyse the notions of light and heavy at all; and those who, while still denying the existence of a void, have attempted this, have failed to explain why there are bodies which are absolutely heavy and light, or in other words why some move upward and others downward. The fact, again, that the body of greater bulk is sometimes lighter than smaller bodies is one which they have passed over in silence, and what they have said gives no obvious suggestion for reconciling their views with the observed facts.

But those who attribute the lightness of fire to its containing so much void are necessarily involved in practically the same difficulties. For though fire be supposed to contain less solid than any other body, as well as more void, yet there will be a certain quantum of fire in which the amount of solid or plenum is in excess of the solids contained in some small quantity of earth. They may reply that there is an excess of void also. But the question is, how will they discriminate the absolutely heavy? Presumably, either by its excess of solid or by its defect of void. On the former view there could be an amount of earth so small as to contain less solid than a large mass of fire. And similarly, if the distinction rests on the amount of void, there will be a body, lighter than the absolutely light, which nevertheless moves downward as constantly as the other moves upward. But that cannot be so, since the absolutely light is always lighter than bodies which have weight and move downward, while, on the other hand, that which is lighter need not be light, because in common speech we distinguish a lighter and a heavier (viz. water and earth) among bodies endowed with weight. Again, the suggestion of a certain ratio between the void and the solid in a body is no more equal to solving the problem before us. The manner of speaking will issue in a similar impossibility. For any two portions of fire, small or great, will exhibit the same ratio of solid to void, but the upward movement of the greater is quicker than that of the less, just as the downward movement of a mass of gold or lead, or of any other body endowed with weight, is quicker in proportion to its size. This, however, should not be the case if the ratio is the ground of distinction between heavy things and light. There is also an absurdity in attributing the upward movement of bodies to a void which does not itself move. If, however, it is the nature of a void to move upward and of a plenum to move downward, and therefore each causes a like movement in other things, there was no need to raise the question why composite bodies are some light and some heavy; they had only to explain why these two things are themselves light and heavy respectively, and to give, further, the reason why the plenum and the void are not eternally separated. It is also unreasonable to imagine a place for the void, as if the void were not itself a kind of place. But if the void is to move, it must have a place out of which and into which the change carries it. Also what is the cause of its movement? Not, surely, its voidness: for it is not the void only which is moved, but also the solid.

Similar difficulties are involved in all other methods of distinction, whether they account for the relative lightness and heaviness of bodies by distinctions of size, or proceed on any other principle, so long as they attribute to each the same matter, or even if they recognize more than one matter, so long as that means only a pair of contraries. If there is a single matter, as with those who compose things of triangles, nothing can be absolutely heavy or light: and if there is one matter and its contrary-the void, for instance, and the plenum-no reason can be given for the relative lightness and heaviness of the bodies intermediate between the absolutely light and heavy when compared either with one another or with these themselves. The view which bases the distinction upon differences of size is more like a mere fiction than those previously mentioned, but, in that it is able to make distinctions between the four elements, it is in a stronger position for meeting the foregoing difficulties. Since, however, it imagines that these bodies which differ in size are all made of one substance, it implies, equally with the view that there is but one matter, that there is nothing absolutely light and nothing which moves upward (except as being passed by other things or forced up by them); and since a multitude of small atoms are heavier than a few large ones, it will follow that much air or fire is heavier than a little water or earth, which is impossible.

3

These, then, are the views which have been advanced by others and the terms in which they state them. We may begin our own statement by settling a question which to some has been the main difficulty-the question why some bodies move always and naturally upward and others downward, while others again move both upward and downward. After that we will inquire into light and heavy and of the various phenomena connected with them. The local movement of each body into its own place must be regarded as similar to what happens in connexion with other forms of generation and change. There are, in fact, three kinds of movement, affecting respectively the size, the form, and the place of a thing, and in each it is observable that change proceeds from a contrary to a contrary or to something intermediate: it is never the change of any chance subject in any chance direction, nor, similarly, is the relation of the mover to its object fortuitous: the thing altered is different from the thing increased, and precisely the same difference holds between that which produces alteration and that which produces increase. In the same manner it must be thought that produces local motion and that which is so moved are not fortuitously related. Now, that which produces upward and downward movement is that which produces weight and lightness, and that which is moved is that which is potentially heavy or light, and the movement of each body to its own place is motion towards its own form. (It is best to interpret in this sense the common statement of the older writers that 'like moves to like'. For the words are not in every sense true to fact. If one were to remove the earth to where the moon now is, the various fragments of earth would each move not towards it but to the place in which it now is. In general, when a number of similar and undifferentiated bodies are moved with the same motion this result is necessarily produced, viz. that the place which is the natural goal of the movement of each single part is also that of the whole. But since the place of a thing is the boundary of that which contains it, and the continent of all things that move upward or downward is the extremity and the centre, and this boundary comes to be, in a sense, the form of that which is contained, it is to its like that a body moves when it moves to its own place. For the successive members of the series are like one another: water, I mean, is like air and air like fire, and between intermediates the relation may be converted, though not between them and the extremes; thus air is like water, but water is like earth: for the relation of each outer body to that which is next within it is that of form to matter.) Thus to ask why fire moves upward and earth downward is the same as to ask why the healable, when moved and changed qua healable, attains health and not whiteness; and similar questions might be asked concerning any other subject of aletion. Of course the subject of increase, when changed qua increasable, attains not health but a superior size. The same applies in the other cases. One thing changes in quality, another in quantity: and so in place, a light thing goes upward, a heavy thing downward. The only difference is that in the last case, viz. that of the heavy and the light, the bodies are thought to have a spring of change within themselves, while the subjects of healing and increase are thought to be moved purely from without. Sometimes, however, even they change of themselves, i.e. in response to a slight external movement reach health or increase, as the case may be. And since the same thing which is healable is also receptive of disease, it depends on whether it is moved qua healable or qua liable to disease whether the motion is towards health or towards disease. But the reason why the heavy and the light appear more than these things to contain within themselves the source of their movements is that their matter is nearest to being. This is indicated by the fact that locomotion belongs to bodies only when isolated from other bodies, and is generated last of the several kinds of movement; in order of being then it will be first. Now whenever air comes into being out of water, light out of heavy, it goes to the upper place. It is forthwith light: becoming is at an end, and in that place it has being. Obviously, then, it is a potentiality, which, in its passage to actuality, comes into that place and quantity and quality which belong to its actuality. And the same fact explains why what is already actually fire or earth moves, when nothing obstructs it, towards its own place. For motion is equally immediate in the case of nutriment, when nothing hinders, and in the case of the thing healed, when nothing stays the healing. But the movement is also due to the original creative force and to that which removes the hindrance or off which the moving thing rebounded, as was explained in our opening discussions, where we tried to show how none of these things moves itself. The reason of the various motions of the various bodies, and the meaning of the motion of a body to its own place, have now been explained.

4

We have now to speak of the distinctive properties of these bodies and of the various phenomena connected with them. In accordance with general conviction we may distinguish the absolutely heavy, as that which sinks to the bottom of all things, from the absolutely light, which is that which rises to the surface of all things. I use the term 'absolutely', in view of the generic character of 'light' and 'heavy', in order to confine the application to bodies which do not combine lightness and heaviness. It is apparent, I mean, that fire, in whatever quantity, so long as there is no external obstacle moves upward, and earth downward; and, if the quantity is increased, the movement is the same, though swifter. But the heaviness and lightness of bodies which combine these qualities is different from this, since while they rise to the surface of some bodies they sink to the bottom of others. Such are air and water. Neither of them is absolutely either light or heavy. Both are lighter than earth-for any portion of either rises to the surface of it-but heavier than fire, since a portion of either, whatever its quantity, sinks to the bottom of fire; compared together, however, the one has absolute weight, the other absolute lightness, since air in any quantity rises to the surface of water, while water in any quantity sinks to the bottom of air. Now other bodies are severally light and heavy, and evidently in them the attributes are due to the difference of their uncompounded parts: that is to say, according as the one or the other happens to preponderate the bodies will be heavy and light respectively. Therefore we need only speak of these parts, since they are primary and all else consequential: and in so doing we shall be following the advice which we gave to those whose attribute heaviness to the presence of plenum and lightness to that of void. It is due to the properties of the elementary bodies that a body which is regarded as light in one place is regarded as heavy in another, and vice versa. In air, for instance, a talent's weight of wood is heavier than a mina of lead, but in water the wood is the lighter. The reason is that all the elements except fire have weight and all but earth lightness. Earth, then, and bodies in which earth preponderates, must needs have weight everywhere, while water is heavy anywhere but in earth, and air is heavy when not in water or earth. In its own place each of these bodies has weight except fire, even air. Of this we have evidence in the fact that a bladder when inflated weighs more than when empty. A body, then, in which air preponderates over earth and water, may well be lighter than something in water and yet heavier than it in air, since such a body does not rise in air but rises to the surface in water.

The following account will make it plain that there is an absolutely light and an absolutely heavy body. And by absolutely light I mean one which of its own nature always moves upward, by absolutely heavy one which of its own nature always moves downward, if no obstacle is in the way. There are, I say, these two kinds of body, and it is not the case, as some maintain, that all bodies have weight. Different views are in fact agreed that there is a heavy body, which moves uniformly towards the centre. But is also similarly a light body. For we see with our eyes, as we said before, that earthy things sink to the bottom of all things and move towards the centre. But the centre is a fixed point. If therefore there is some body which rises to the surface of all things-and we observe fire to move upward even in air itself, while the air remains at rest-clearly this body is moving towards the extremity. It cannot then have any weight. If it had, there would be another body in which it sank: and if that had weight, there would be yet another which moved to the extremity and thus rose to the surface of all moving things. In fact, however, we have no evidence of such a body. Fire, then, has no weight. Neither has earth any lightness, since it sinks to the bottom of all things, and that which sinks moves to the centre. That there is a centre towards which the motion of heavy things, and away from which that of light things is directed, is manifest in many ways. First, because no movement can continue to infinity. For what cannot be can no more come-to-be than be, and movement is a coming to-be in one place from another. Secondly, like the upward movement of fire, the downward movement of earth and all heavy things makes equal angles on every side with the earth's surface: it must therefore be directed towards the centre. Whether it is really the centre of the earth and not rather that of the whole to which it moves, may be left to another inquiry, since these are coincident. But since that which sinks to the bottom of all things moves to the centre, necessarily that which rises to the surface moves to the extremity of the region in which the movement of these bodies takes place. For the centre is opposed as contrary to the extremity, as that which sinks is opposed to that which rises to the surface. This also gives a reasonable ground for the duality of heavy and light in the spatial duality centre and extremity. Now there is also the intermediate region to which each name is given in opposition to the other extreme. For that which is intermediate between the two is in a sense both extremity and centre. For this reason there is another heavy and light; namely, water and air. But in our view the continent pertains to form and the contained to matter: and this distinction is present in every genus. Alike in the sphere of quality and in that of quantity there is that which corresponds rather to form and that which corresponds to matter. In the same way, among spatial distinctions, the above belongs to the determinate, the below to matter. The same holds, consequently, also of the matter itself of that which is heavy and light: as potentially possessing the one character, it is matter for the heavy, and as potentially possessing the other, for the light. It is the same matter, but its being is different, as that which is receptive of disease is the same as that which is receptive of health, though in being different from it, and therefore diseasedness is different from healthiness.

5

A thing then which has the one kind of matter is light and always moves upward, while a thing which has the opposite matter is heavy and always moves downward. Bodies composed of kinds of matter different from these but having relatively to each other the character which these have absolutely, possess both the upward and the downward motion. Hence air and water each have both lightness and weight, and water sinks to the bottom of all things except earth, while air rises to the surface of all things except fire. But since there is one body only which rises to the surface of all things and one only which sinks to the bottom of all things, there must needs be two other bodies which sink in some bodies and rise to the surface of others. The kinds of matter, then, must be as numerous as these bodies, i.e. four, but though they are four there must be a common matter of all-particularly if they pass into one another-which in each is in being different. There is no reason why there should not be one or more intermediates between the contraries, as in the case of colour; for 'intermediate' and 'mean' are capable of more than one application.

Now in its own place every body endowed with both weight and lightness has weight--whereas earth has weight everywhere-but they only have lightness among bodies to whose surface they rise. Hence when a support is withdrawn such a body moves downward until it reaches the body next below it, air to the place of water and water to that of earth. But if the fire above air is removed, it will not move upward to the place of fire, except by constraint; and in that way water also may be drawn up, when the upward movement of air which has had a common surface with it is swift enough to overpower the downward impulse of the water. Nor does water move upward to the place of air, except in the manner just described. Earth is not so affected at all, because a common surface is not possible to it. Hence water is drawn up into the vessel to which fire is applied, but not earth. As earth fails to move upward, so fire fails to move downward when air is withdrawn from beneath it: for fire has no weight even in its own place, as earth has no lightness. The other two move downward when the body beneath is withdrawn because, while the absolutely heavy is that which sinks to the bottom of all things, the relatively heavy sinks to its own place or to the surface of the body in which it rises, since it is similar in matter to it.

It is plain that one must suppose as many distinct species of matter as there are bodies. For if, first, there is a single matter of all things, as, for instance, the void or the plenum or extension or the triangles, either all things will move upward or all things will move downward, and the second motion will be abolished. And so, either there will be no absolutely light body, if superiority of weight is due to superior size or number of the constituent bodies or to the fullness of the body: but the contrary is a matter of observation, and it has been shown that the downward and upward movements are equally constant and universal: or, if the matter in question is the void or something similar, which moves uniformly upward, there will be nothing to move uniformly downward. Further, it will follow that the intermediate bodies move downward in some cases quicker than earth: for air in sufficiently large quantity will contain a larger number of triangles or solids or particles. It is, however, manifest that no portion of air whatever moves downward. And the same reasoning applies to lightness, if that is supposed to depend on superiority of quantity of matter. But if, secondly, the kinds of matter are two, it will be difficult to make the intermediate bodies behave as air and water behave. Suppose, for example, that the two asserted are void and plenum. Fire, then, as moving upward, will be void, earth, as moving downward, plenum; and in air, it will be said, fire preponderates, in water, earth. There will then be a quantity of water containing more fire than a little air, and a large amount of air will contain more earth than a little water: consequently we shall have to say that air in a certain quantity moves downward more quickly than a little water. But such a thing has never been observed anywhere. Necessarily, then, as fire goes up because it has something, e.g. void, which other things do not have, and earth goes downward because it has plenum, so air goes to its own place above water because it has something else, and water goes downward because of some special kind of body. But if the two bodies are one matter, or two matters both present in each, there will be a certain quantity of each at which water will excel a little air in the upward movement and air excel water in the downward movement, as we have already often said.

6

The shape of bodies will not account for their moving upward or downward in general, though it will account for their moving faster or slower. The reasons for this are not difficult to see. For the problem thus raised is why a flat piece of iron or lead floats upon water, while smaller and less heavy things, so long as they are round or long-a needle, for instance-sink down; and sometimes a thing floats because it is small, as with gold dust and the various earthy and dusty materials which throng the air. With regard to these questions, it is wrong to accept the explanation offered by Democritus. He says that the warm bodies moving up out of the water hold up heavy bodies which are broad, while the narrow ones fall through, because the bodies which offer this resistance are not numerous. But this would be even more likely to happen in air-an objection which he himself raises. His reply to the objection is feeble. In the air, he says, the 'drive' (meaning by drive the movement of the upward moving bodies) is not uniform in direction. But since some continua are easily divided and others less easily, and things which produce division differ similarly in the case with which they produce it, the explanation must be found in this fact. It is the easily bounded, in proportion as it is easily bounded, which is easily divided; and air is more so than water, water than earth. Further, the smaller the quantity in each kind, the more easily it is divided and disrupted. Thus the reason why broad things keep their place is because they cover so wide a surface and the greater quantity is less easily disrupted. Bodies of the opposite shape sink down because they occupy so little of the surface, which is therefore easily parted. And these considerations apply with far greater force to air, since it is so much more easily divided than water. But since there are two factors, the force responsible for the downward motion of the heavy body and the disruption-resisting force of the continuous surface, there must be some ratio between the two. For in proportion as the force applied by the heavy thing towards disruption and division exceeds that which resides in the continuum, the quicker will it force its way down; only if the force of the heavy thing is the weaker, will it ride upon the surface.

We have now finished our examination of the heavy and the light and of the phenomena connected with them.

Politics

Table of Contents

BOOK I

I

EVERY STATE is a community of some kind, and every community is established with a view to some good; for mankind always act in order to obtain that which they think good. But, if all communities aim at some good, the state or political community, which is the highest of all, and which embraces all the rest, aims at good in a greater degree than any other, and at the highest good.

Some people think that the qualifications of a statesman, king, householder, and master are the same, and that they differ, not in kind, but only in the number of their subjects. For example, the ruler over a few is called a master; over more, the manager of a household; over a still larger number, a statesman or king, as if there were no difference between a great household and a small state. The distinction which is made between the king and the statesman is as follows: When the government is personal, the ruler is a king; when, according to the rules of the political science, the citizens rule and are ruled in turn, then he is called a statesman.

But all this is a mistake; for governments differ in kind, as will be evident to any one who considers the matter according to the method which has hitherto guided us. As in other departments of science, so in politics, the compound should always be resolved into the simple elements or least parts of the whole. We must therefore look at the elements of which the state is composed, in order that we may see in what the different kinds of rule differ from one another, and whether any scientific result can be attained about each one of them.

II

He who thus considers things in their first growth and origin, whether a state or anything else, will obtain the clearest view of them. In the first place there must be a union of those who cannot exist without each other; namely, of male and female, that the race may continue (and this is a union which is formed, not of deliberate purpose, but because, in common with other animals and with plants, mankind have a natural desire to leave behind them an image of themselves), and of natural ruler and subject, that both may be preserved. For that which can foresee by the exercise of mind is by nature intended to be lord and master, and that which can with its body give effect to such foresight is a subject, and by nature a slave; hence master and slave have the same interest. Now nature has distinguished between the female and the slave. For she is not niggardly, like the smith who fashions the Delphian knife for many uses; she makes each thing for a single use, and every instrument is best made when intended for one and not for many uses. But among barbarians no distinction is made between women and slaves, because there is no natural ruler among them: they are a community of slaves, male and female. Wherefore the poets say,

It is meet that Hellenes should rule over barbarians;

as if they thought that the barbarian and the slave were by nature one.

Out of these two relationships between man and woman, master and slave, the first thing to arise is the family, and Hesiod is right when he says,

First house and wife and an ox for the plough,

for the ox is the poor man's slave. The family is the association established by nature for the supply of men's everyday wants, and the members of it are called by Charondas 'companions of the cupboard,' and by Epimenides the Cretan, 'companions of the manger.' But when several families are united, and the association aims at something more than the supply of daily needs, the first society to be formed is the village. And the most natural form of the village appears to be that of a colony from the family, composed of the children and grandchildren, who are said to be suckled 'with the same milk.' And this is the reason why Hellenic states were originally governed by kings; because the Hellenes were under royal rule before they came together, as the barbarians still are. Every family is ruled by the eldest, and therefore in the colonies of the family the kingly form of government prevailed because they were of the same blood. As Homer says:

Each one gives law to his children and to his wives.

For they lived dispersedly, as was the manner in ancient times. Wherefore men say that the Gods have a king, because they themselves either are or were in ancient times under the rule of a king. For they imagine, not only the forms of the Gods, but their ways of life to be like their own.

When several villages are united in a single complete community, large enough to be nearly or quite self-sufficing, the state comes into existence, originating in the bare needs of life, and continuing in existence for the sake of a good life. And therefore, if the earlier forms of society are natural, so is the state, for it is the end of them, and the nature of a thing is its end. For what each thing is when fully developed, we call its nature, whether we are speaking of a man, a horse, or a family. Besides, the final cause and end of a thing is the best, and to be self-sufficing is the end and the best.

Hence it is evident that the state is a creation of nature, and that man is by nature a political animal. And he who by nature and not by mere accident is without a state, is either a bad man or above humanity; he is like the

Tribeless, lawless, heartless one,

whom Homer denounces — the natural outcast is forthwith a lover of war; he may be compared to an isolated piece at draughts.

Now, that man is more of a political animal than bees or any other gregarious animals is evident. Nature, as we often say, makes nothing in vain, and man is the only animal whom she has endowed with the gift of speech. And whereas mere voice is but an indication of pleasure or pain, and is therefore found in other animals (for their nature attains to the perception of pleasure and pain and the intimation of them to one another, and no further), the power of speech is intended to set forth the expedient and inexpedient, and therefore likewise the just and the unjust. And it is a characteristic of man that he alone has any sense of good and evil, of just and unjust, and the like, and the association of living beings who have this sense makes a family and a state.

Further, the state is by nature clearly prior to the family and to the individual, since the whole is of necessity prior to the part; for example, if the whole body be destroyed, there will be no foot or hand, except in an equivocal sense, as we might speak of a stone hand; for when destroyed the hand will be no better than that. But things are defined by their working and power; and we ought not to say that they are the same when they no longer have their proper quality, but only that they have the same name. The proof that the state is a creation of nature and prior to the individual is that the individual, when isolated, is not self-sufficing; and therefore he is like a part in relation to the whole. But he who is unable to live in society, or who has no need because he is sufficient for himself, must be either a beast or a god: he is no part of a state. A social instinct is implanted in all men by nature, and yet he who first founded the state was the greatest of benefactors. For man, when perfected, is the best of animals, but, when separated from law and justice, he is the worst of all; since armed injustice is the more dangerous, and he is equipped at birth with arms, meant to be used by intelligence and virtue, which he may use for the worst ends. Wherefore, if he have not virtue, he is the most unholy and the most savage of animals, and the most full of lust and gluttony. But justice is the bond of men in states, for the administration of justice, which is the determination of what is just, is the principle of order in political society.

III

Seeing then that the state is made up of households, before speaking of the state we must speak of the management of the household. The parts of household management correspond to the persons who compose the household, and a complete household consists of slaves and freemen. Now we should begin by examining everything in its fewest possible elements; and the first and fewest possible parts of a family are master and slave, husband and wife, father and children. We have therefore to consider what each of these three relations is and ought to be: I mean the relation of master and servant, the marriage relation (the conjunction of man and wife has no name of its own), and thirdly, the procreative relation (this also has no proper name). And there is another element of a household, the so-called art of getting wealth, which, according to some, is identical with household management, according to others, a principal part of it; the nature of this art will also have to be considered by us.

Let us first speak of master and slave, looking to the needs of practical life and also seeking to attain some better theory of their relation than exists at present. For some are of opinion that the rule of a master is a science, and that the management of a household, and the mastership of slaves, and the political and royal rule, as I was saying at the outset, are all the same. Others affirm that the rule of a master over slaves is contrary to nature, and that the distinction between slave and freeman exists by law only, and not by nature; and being an interference with nature is therefore unjust.

IV

Property is a part of the household, and the art of acquiring property is a part of the art of managing the household; for no man can live well, or indeed live at all, unless he be provided with necessaries. And as in the arts which have a definite sphere the workers must have their own proper instruments for the accomplishment of their work, so it is in the management of a household. Now instruments are of various sorts; some are living, others lifeless; in the rudder, the pilot of a ship has a lifeless, in the look-out man, a living instrument; for in the arts the servant is a kind of instrument. Thus, too, a possession is an instrument for maintaining life. And so, in the arrangement of the family, a slave is a living possession, and property a number of such instruments; and the servant is himself an instrument which takes precedence of all other instruments. For if every instrument could accomplish its own work, obeying or anticipating the will of others, like the statues of Daedalus, or the tripods of Hephaestus, which, says the poet,

of their own accord entered the assembly of the Gods;

if, in like manner, the shuttle would weave and the plectrum touch the lyre without a hand to guide them, chief workmen would not want servants, nor masters slaves. Here, however, another distinction must be drawn; the instruments commonly so called are instruments of production, whilst a possession is an instrument of action. The shuttle, for example, is not only of use; but something else is made by it, whereas of a garment or of a bed there is only the use. Further, as production and action are different in kind, and both require instruments, the instruments which they employ must likewise differ in kind. But life is action and not production, and therefore the slave is the minister of action. Again, a possession is spoken of as a part is spoken of; for the part is not only a part of something else, but wholly belongs to it; and this is also true of a possession. The master is only the master of the slave; he does not belong to him, whereas the slave is not only the slave of his master, but wholly belongs to him. Hence we see what is the nature and office of a slave; he who is by nature not his own but another's man, is by nature a slave; and he may be said to be another's man who, being a human being, is also a possession. And a possession may be defined as an instrument of action, separable from the possessor.

V

But is there any one thus intended by nature to be a slave, and for whom such a condition is expedient and right, or rather is not all slavery a violation of nature?

There is no difficulty in answering this question, on grounds both of reason and of fact. For that some should rule and others be ruled is a thing not only necessary, but expedient; from the hour of their birth, some are marked out for subjection, others for rule.

And there are many kinds both of rulers and subjects (and that rule is the better which is exercised over better subjects — for example, to rule over men is better than to rule over wild beasts; for the work is better which is executed by better workmen, and where one man rules and another is ruled, they may be said to have a work); for in all things which form a composite whole and which are made up of parts, whether continuous or discrete, a distinction between the ruling and the subject element comes to light. Such a duality exists in living creatures, but not in them only; it originates in the constitution of the universe; even in things which have no life there is a ruling principle, as in a musical mode. But we are wandering from the subject. We will therefore restrict ourselves to the living creature, which, in the first place, consists of soul and body: and of these two, the one is by nature the ruler, and the other the subject. But then we must look for the intentions of nature in things which retain their nature, and not in things which are corrupted. And therefore we must study the man who is in the most perfect state both of body and soul, for in him we shall see the true relation of the two; although in bad or corrupted natures the body will often appear to rule over the soul, because they are in an evil and unnatural condition. At all events we may firstly observe in living creatures both a despotical and a constitutional rule; for the soul rules the body with a despotical rule, whereas the intellect rules the appetites with a constitutional and royal rule. And it is clear that the rule of the soul over the body, and of the mind and the rational element over the passionate, is natural and expedient; whereas the equality of the two or the rule of the inferior is always hurtful. The same holds good of animals in relation to men; for tame animals have a better nature than wild, and all tame animals are better off when they are ruled by man; for then they are preserved. Again, the male is by nature superior, and the female inferior; and the one rules, and the other is ruled; this principle, of necessity, extends to all mankind.

Where then there is such a difference as that between soul and body, or between men and animals (as in the case of those whose business is to use their body, and who can do nothing better), the lower sort are by nature slaves, and it is better for them as for all inferiors that they should be under the rule of a master. For he who can be, and therefore is, another's and he who participates in rational principle enough to apprehend, but not to have, such a principle, is a slave by nature. Whereas the lower animals cannot even apprehend a principle; they obey their instincts. And indeed the use made of slaves and of tame animals is not very different; for both with their bodies minister to the needs of life. Nature would like to distinguish between the bodies of freemen and slaves, making the one strong for servile labor, the other upright, and although useless for such services, useful for political life in the arts both of war and peace. But the opposite often happens — that some have the souls and others have the bodies of freemen. And doubtless if men differed from one another in the mere forms of their bodies as much as the statues of the Gods do from men, all would acknowledge that the inferior class should be slaves of the superior. And if this is true of the body, how much more just that a similar distinction should exist in the soul? but the beauty of the body is seen, whereas the beauty of the soul is not seen. It is clear, then, that some men are by nature free, and others slaves, and that for these latter slavery is both expedient and right.

VI

But that those who take the opposite view have in a certain way right on their side, may be easily seen. For the words slavery and slave are used in two senses. There is a slave or slavery by law as well as by nature. The law of which I speak is a sort of convention — the law by which whatever is taken in war is supposed to belong to the victors. But this right many jurists impeach, as they would an orator who brought forward an unconstitutional measure: they detest the notion that, because one man has the power of doing violence and is superior in brute strength, another shall be his slave and subject. Even among philosophers there is a difference of opinion. The origin of the dispute, and what makes the views invade each other's territory, is as follows: in some sense virtue, when furnished with means, has actually the greatest power of exercising force; and as superior power is only found where there is superior excellence of some kind, power seems to imply virtue, and the dispute to be simply one about justice (for it is due to one party identifying justice with goodwill while the other identifies it with the mere rule of the stronger). If these views are thus set out separately, the other views have no force or plausibility against the view that the superior in virtue ought to rule, or be master. Others, clinging, as they think, simply to a principle of justice (for law and custom are a sort of justice), assume that slavery in accordance with the custom of war is justified by law, but at the same moment they deny this. For what if the cause of the war be unjust? And again, no one would ever say he is a slave who is unworthy to be a slave. Were this the case, men of the highest rank would be slaves and the children of slaves if they or their parents chance to have been taken captive and sold. Wherefore Hellenes do not like to call Hellenes slaves, but confine the term to barbarians. Yet, in using this language, they really mean the natural slave of whom we spoke at first; for it must be admitted that some are slaves everywhere, others nowhere. The same principle applies to nobility. Hellenes regard themselves as noble everywhere, and not only in their own country, but they deem the barbarians noble only when at home, thereby implying that there are two sorts of nobility and freedom, the one absolute, the other relative. The Helen of Theodectes says:

Who would presume to call me servant who am on both sides sprung from the stem of the Gods?

What does this mean but that they distinguish freedom and slavery, noble and humble birth, by the two principles of good and evil? They think that as men and animals beget men and animals, so from good men a good man springs. But this is what nature, though she may intend it, cannot always accomplish.

We see then that there is some foundation for this difference of opinion, and that all are not either slaves by nature or freemen by nature, and also that there is in some cases a marked distinction between the two classes, rendering it expedient and right for the one to be slaves and the others to be masters: the one practicing obedience, the others exercising the authority and lordship which nature intended them to have. The abuse of this authority is injurious to both; for the interests of part and whole, of body and soul, are the same, and the slave is a part of the master, a living but separated part of his bodily frame. Hence, where the relation of master and slave between them is natural they are friends and have a common interest, but where it rests merely on law and force the reverse is true.

VII

The previous remarks are quite enough to show that the rule of a master is not a constitutional rule, and that all the different kinds of rule are not, as some affirm, the same with each other. For there is one rule exercised over subjects who are by nature free, another over subjects who are by nature slaves. The rule of a household is a monarchy, for every house is under one head: whereas constitutional rule is a government of freemen and equals. The master is not called a master because he has science, but because he is of a certain character, and the same remark applies to the slave and the freeman. Still there may be a science for the master and science for the slave. The science of the slave would be such as the man of Syracuse taught, who made money by instructing slaves in their ordinary duties. And such a knowledge may be carried further, so as to include cookery and similar menial arts. For some duties are of the more necessary, others of the more honorable sort; as the proverb says, 'slave before slave, master before master.' But all such branches of knowledge are servile. There is likewise a science of the master, which teaches the use of slaves; for the master as such is concerned, not with the acquisition, but with the use of them. Yet this so-called science is not anything great or wonderful; for the master need only know how to order that which the slave must know how to execute. Hence those who are in a position which places them above toil have stewards who attend to their households while they occupy themselves with philosophy or with politics. But the art of acquiring slaves, I mean of justly acquiring them, differs both from the art of the master and the art of the slave, being a species of hunting or war. Enough of the distinction between master and slave.

VIII

Let us now inquire into property generally, and into the art of getting wealth, in accordance with our usual method, for a slave has been shown to be a part of property. The first question is whether the art of getting wealth is the same with the art of managing a household or a part of it, or instrumental to it; and if the last, whether in the way that the art of making shuttles is instrumental to the art of weaving, or in the way that the casting of bronze is instrumental to the art of the statuary, for they are not instrumental in the same way, but the one provides tools and the other material; and by material I mean the substratum out of which any work is made; thus wool is the material of the weaver, bronze of the statuary. Now it is easy to see that the art of household management is not identical with the art of getting wealth, for the one uses the material which the other provides. For the art which uses household stores can be no other than the art of household management. There is, however, a doubt whether the art of getting wealth is a part of household management or a distinct art. If the getter of wealth has to consider whence wealth and property can be procured, but there are many sorts of property and riches, then are husbandry, and the care and provision of food in general, parts of the wealth-getting art or distinct arts? Again, there are many sorts of food, and therefore there are many kinds of lives both of animals and men; they must all have food, and the differences in their food have made differences in their ways of life. For of beasts, some are gregarious, others are solitary; they live in the way which is best adapted to sustain them, accordingly as they are carnivorous or herbivorous or omnivorous: and their habits are determined for them by nature in such a manner that they may obtain with greater facility the food of their choice. But, as different species have different tastes, the same things are not naturally pleasant to all of them; and therefore the lives of carnivorous or herbivorous animals further differ among themselves. In the lives of men too there is a great difference. The laziest are shepherds, who lead an idle life, and get their subsistence without trouble from tame animals; their flocks having to wander from place to place in search of pasture, they are compelled to follow them, cultivating a sort of living farm. Others support themselves by hunting, which is of different kinds. Some, for example, are brigands, others, who dwell near lakes or marshes or rivers or a sea in which there are fish, are fishermen, and others live by the pursuit of birds or wild beasts. The greater number obtain a living from the cultivated fruits of the soil. Such are the modes of subsistence which prevail among those whose industry springs up of itself, and whose food is not acquired by exchange and retail trade — there is the shepherd, the husbandman, the brigand, the fisherman, the hunter. Some gain a comfortable maintenance out of two employments, eking out the deficiencies of one of them by another: thus the life of a shepherd may be combined with that of a brigand, the life of a farmer with that of a hunter. Other modes of life are similarly combined in any way which the needs of men may require. Property, in the sense of a bare livelihood, seems to be given by nature herself to all, both when they are first born, and when they are grown up. For some animals bring forth, together with their offspring, so much food as will last until they are able to supply themselves; of this the vermiparous or oviparous animals are an instance; and the viviparous animals have up to a certain time a supply of food for their young in themselves, which is called milk. In like manner we may infer that, after the birth of animals, plants exist for their sake, and that the other animals exist for the sake of man, the tame for use and food, the wild, if not all at least the greater part of them, for food, and for the provision of clothing and various instruments. Now if nature makes nothing incomplete, and nothing in vain, the inference must be that she has made all animals for the sake of man. And so, in one point of view, the art of war is a natural art of acquisition, for the art of acquisition includes hunting, an art which we ought to practice against wild beasts, and against men who, though intended by nature to be governed, will not submit; for war of such a kind is naturally just.

Of the art of acquisition then there is one kind which by nature is a part of the management of a household, in so far as the art of household management must either find ready to hand, or itself provide, such things necessary to life, and useful for the community of the family or state, as can be stored. They are the elements of true riches; for the amount of property which is needed for a good life is not unlimited, although Solon in one of his poems says that

No bound to riches has been fixed for man.

But there is a boundary fixed, just as there is in the other arts; for the instruments of any art are never unlimited, either in number or size, and riches may be defined as a number of instruments to be used in a household or in a state. And so we see that there is a natural art of acquisition which is practiced by managers of households and by statesmen, and what is the reason of this.

IX

There is another variety of the art of acquisition which is commonly and rightly called an art of wealth-getting, and has in fact suggested the notion that riches and property have no limit. Being nearly connected with the preceding, it is often identified with it. But though they are not very different, neither are they the same. The kind already described is given by nature, the other is gained by experience and art.

Let us begin our discussion of the question with the following considerations: Of everything which we possess there are two uses: both belong to the thing as such, but not in the same manner, for one is the proper, and the other the improper or secondary use of it. For example, a shoe is used for wear, and is used for exchange; both are uses of the shoe. He who gives a shoe in exchange for money or food to him who wants one, does indeed use the shoe as a shoe, but this is not its proper or primary purpose, for a shoe is not made to be an object of barter. The same may be said of all possessions, for the art of exchange extends to all of them, and it arises at first from what is natural, from the circumstance that some have too little, others too much. Hence we may infer that retail trade is not a natural part of the art of getting wealth; had it been so, men would have ceased to exchange when they had enough. In the first community, indeed, which is the family, this art is obviously of no use, but it begins to be useful when the society increases. For the members of the family originally had all things in common; later, when the family divided into parts, the parts shared in many things, and different parts in different things, which they had to give in exchange for what they wanted, a kind of barter which is still practiced among barbarous nations who exchange with one another the necessaries of life and nothing more; giving and receiving wine, for example, in exchange for coin, and the like. This sort of barter is not part of the wealth-getting art and is not contrary to nature, but is needed for the satisfaction of men's natural wants. The other or more complex form of exchange grew, as might have been inferred, out of the simpler. When the inhabitants of one country became more dependent on those of another, and they imported what they needed, and exported what they had too much of, money necessarily came into use. For the various necessaries of life are not easily carried about, and hence men agreed to employ in their dealings with each other something which was intrinsically useful and easily applicable to the purposes of life, for example, iron, silver, and the like. Of this the value was at first measured simply by size and weight, but in process of time they put a stamp upon it, to save the trouble of weighing and to mark the value.

When the use of coin had once been discovered, out of the barter of necessary articles arose the other art of wealth getting, namely, retail trade; which was at first probably a simple matter, but became more complicated as soon as men learned by experience whence and by what exchanges the greatest profit might be made. Originating in the use of coin, the art of getting wealth is generally thought to be chiefly concerned with it, and to be the art which produces riches and wealth; having to consider how they may be accumulated. Indeed, riches is assumed by many to be only a quantity of coin, because the arts of getting wealth and retail trade are concerned with coin. Others maintain that coined money is a mere sham, a thing not natural, but conventional only, because, if the users substitute another commodity for it, it is worthless, and because it is not useful as a means to any of the necessities of life, and, indeed, he who is rich in coin may often be in want of necessary food. But how can that be wealth of which a man may have a great abundance and yet perish with hunger, like Midas in the fable, whose insatiable prayer turned everything that was set before him into gold?

Hence men seek after a better notion of riches and of the art of getting wealth than the mere acquisition of coin, and they are right. For natural riches and the natural art of wealth-getting are a different thing; in their true form they are part of the management of a household; whereas retail trade is the art of producing wealth, not in every way, but by exchange. And it is thought to be concerned with coin; for coin is the unit of exchange and the measure or limit of it. And there is no bound to the riches which spring from this art of wealth getting. As in the art of medicine there is no limit to the pursuit of health, and as in the other arts there is no limit to the pursuit of their several ends, for they aim at accomplishing their ends to the uttermost (but of the means there is a limit, for the end is always the limit), so, too, in this art of wealth-getting there is no limit of the end, which is riches of the spurious kind, and the acquisition of wealth. But the art of wealth-getting which consists in household management, on the other hand, has a limit; the unlimited acquisition of wealth is not its business. And, therefore, in one point of view, all riches must have a limit; nevertheless, as a matter of fact, we find the opposite to be the case; for all getters of wealth increase their hoard of coin without limit. The source of the confusion is the near connection between the two kinds of wealth-getting; in either, the instrument is the same, although the use is different, and so they pass into one another; for each is a use of the same property, but with a difference: accumulation is the end in the one case, but there is a further end in the other. Hence some persons are led to believe that getting wealth is the object of household management, and the whole idea of their lives is that they ought either to increase their money without limit, or at any rate not to lose it. The origin of this disposition in men is that they are intent upon living only, and not upon living well; and, as their desires are unlimited they also desire that the means of gratifying them should be without limit. Those who do aim at a good life seek the means of obtaining bodily pleasures; and, since the enjoyment of these appears to depend on property, they are absorbed in getting wealth: and so there arises the second species of wealth-getting. For, as their enjoyment is in excess, they seek an art which produces the excess of enjoyment; and, if they are not able to supply their pleasures by the art of getting wealth, they try other arts, using in turn every faculty in a manner contrary to nature. The quality of courage, for example, is not intended to make wealth, but to inspire confidence; neither is this the aim of the general's or of the physician's art; but the one aims at victory and the other at health. Nevertheless, some men turn every quality or art into a means of getting wealth; this they conceive to be the end, and to the promotion of the end they think all things must contribute.

Thus, then, we have considered the art of wealth-getting which is unnecessary, and why men want it; and also the necessary art of wealth-getting, which we have seen to be different from the other, and to be a natural part of the art of managing a household, concerned with the provision of food, not, however, like the former kind, unlimited, but having a limit.

X

And we have found the answer to our original question, Whether the art of getting wealth is the business of the manager of a household and of the statesman or not their business? viz., that wealth is presupposed by them. For as political science does not make men, but takes them from nature and uses them, so too nature provides them with earth or sea or the like as a source of food. At this stage begins the duty of the manager of a household, who has to order the things which nature supplies; he may be compared to the weaver who has not to make but to use wool, and to know, too, what sort of wool is good and serviceable or bad and unserviceable. Were this otherwise, it would be difficult to see why the art of getting wealth is a part of the management of a household and the art of medicine not; for surely the members of a household must have health just as they must have life or any other necessary. The answer is that as from one point of view the master of the house and the ruler of the state have to consider about health, from another point of view not they but the physician; so in one way the art of household management, in another way the subordinate art, has to consider about wealth. But, strictly speaking, as I have already said, the means of life must be provided beforehand by nature; for the business of nature is to furnish food to that which is born, and the food of the offspring is always what remains over of that from which it is produced. Wherefore the art of getting wealth out of fruits and animals is always natural.

There are two sorts of wealth-getting, as I have said; one is a part of household management, the other is retail trade: the former necessary and honorable, while that which consists in exchange is justly censured; for it is unnatural, and a mode by which men gain from one another. The most hated sort, and with the greatest reason, is usury, which makes a gain out of money itself, and not from the natural object of it. For money was intended to be used in exchange, but not to increase at interest. And this term interest, which means the birth of money from money, is applied to the breeding of money because the offspring resembles the parent. Wherefore of all modes of getting wealth this is the most unnatural.

XI

Enough has been said about the theory of wealth-getting; we will now proceed to the practical part. The discussion of such matters is not unworthy of philosophy, but to be engaged in them practically is illiberal and irksome. The useful parts of wealth-getting are, first, the knowledge of livestock — which are most profitable, and where, and how — as, for example, what sort of horses or sheep or oxen or any other animals are most likely to give a return. A man ought to know which of these pay better than others, and which pay best in particular places, for some do better in one place and some in another. Secondly, husbandry, which may be either tillage or planting, and the keeping of bees and of fish, or fowl, or of any animals which may be useful to man. These are the divisions of the true or proper art of wealth-getting and come first. Of the other, which consists in exchange, the first and most important division is commerce (of which there are three kinds — the provision of a ship, the conveyance of goods, exposure for sale — these again differing as they are safer or more profitable), the second is usury, the third, service for hire — of this, one kind is employed in the mechanical arts, the other in unskilled and bodily labor. There is still a third sort of wealth getting intermediate between this and the first or natural mode which is partly natural, but is also concerned with exchange, viz., the industries that make their profit from the earth, and from things growing from the earth which, although they bear no fruit, are nevertheless profitable; for example, the cutting of timber and all mining. The art of mining, by which minerals are obtained, itself has many branches, for there are various kinds of things dug out of the earth. Of the several divisions of wealth-getting I now speak generally; a minute consideration of them might be useful in practice, but it would be tiresome to dwell upon them at greater length now.

Those occupations are most truly arts in which there is the least element of chance; they are the meanest in which the body is most deteriorated, the most servile in which there is the greatest use of the body, and the most illiberal in which there is the least need of excellence.

Works have been written upon these subjects by various persons; for example, by Chares the Parian, and Apollodorus the Lemnian, who have treated of Tillage and Planting, while others have treated of other branches; any one who cares for such matters may refer to their writings. It would be well also to collect the scattered stories of the ways in which individuals have succeeded in amassing a fortune; for all this is useful to persons who value the art of getting wealth. There is the anecdote of Thales the Milesian and his financial device, which involves a principle of universal application, but is attributed to him on account of his reputation for wisdom. He was reproached for his poverty, which was supposed to show that philosophy was of no use. According to the story, he knew by his skill in the stars while it was yet winter that there would be a great harvest of olives in the coming year; so, having a little money, he gave deposits for the use of all the olive-presses in Chios and Miletus, which he hired at a low price because no one bid against him. When the harvest-time came, and many were wanted all at once and of a sudden, he let them out at any rate which he pleased, and made a quantity of money. Thus he showed the world that philosophers can easily be rich if they like, but that their ambition is of another sort. He is supposed to have given a striking proof of his wisdom, but, as I was saying, his device for getting wealth is of universal application, and is nothing but the creation of a monopoly. It is an art often practiced by cities when they are want of money; they make a monopoly of provisions.

There was a man of Sicily, who, having money deposited with him, bought up an the iron from the iron mines; afterwards, when the merchants from their various markets came to buy, he was the only seller, and without much increasing the price he gained 200 per cent. Which when Dionysius heard, he told him that he might take away his money, but that he must not remain at Syracuse, for he thought that the man had discovered a way of making money which was injurious to his own interests. He made the same discovery as Thales; they both contrived to create a monopoly for themselves. And statesmen as well ought to know these things; for a state is often as much in want of money and of such devices for obtaining it as a household, or even more so; hence some public men devote themselves entirely to finance.

XII

Of household management we have seen that there are three parts — one is the rule of a master over slaves, which has been discussed already, another of a father, and the third of a husband. A husband and father, we saw, rules over wife and children, both free, but the rule differs, the rule over his children being a royal, over his wife a constitutional rule. For although there may be exceptions to the order of nature, the male is by nature fitter for command than the female, just as the elder and full-grown is superior to the younger and more immature. But in most constitutional states the citizens rule and are ruled by turns, for the idea of a constitutional state implies that the natures of the citizens are equal, and do not differ at all. Nevertheless, when one rules and the other is ruled we endeavor to create a difference of outward forms and names and titles of respect, which may be illustrated by the saying of Amasis about his foot-pan. The relation of the male to the female is of this kind, but there the inequality is permanent. The rule of a father over his children is royal, for he rules by virtue both of love and of the respect due to age, exercising a kind of royal power. And therefore Homer has appropriately called Zeus 'father of Gods and men,' because he is the king of them all. For a king is the natural superior of his subjects, but he should be of the same kin or kind with them, and such is the relation of elder and younger, of father and son.

XIII

Thus it is clear that household management attends more to men than to the acquisition of inanimate things, and to human excellence more than to the excellence of property which we call wealth, and to the virtue of freemen more than to the virtue of slaves. A question may indeed be raised, whether there is any excellence at all in a slave beyond and higher than merely instrumental and ministerial qualities — whether he can have the virtues of temperance, courage, justice, and the like; or whether slaves possess only bodily and ministerial qualities. And, whichever way we answer the question, a difficulty arises; for, if they have virtue, in what will they differ from freemen? On the other hand, since they are men and share in rational principle, it seems absurd to say that they have no virtue. A similar question may be raised about women and children, whether they too have virtues: ought a woman to be temperate and brave and just, and is a child to be called temperate, and intemperate, or note So in general we may ask about the natural ruler, and the natural subject, whether they have the same or different virtues. For if a noble nature is equally required in both, why should one of them always rule, and the other always be ruled? Nor can we say that this is a question of degree, for the difference between ruler and subject is a difference of kind, which the difference of more and less never is. Yet how strange is the supposition that the one ought, and that the other ought not, to have virtue! For if the ruler is intemperate and unjust, how can he rule well? If the subject, how can he obey well? If he be licentious and cowardly, he will certainly not do his duty. It is evident, therefore, that both of them must have a share of virtue, but varying as natural subjects also vary among themselves. Here the very constitution of the soul has shown us the way; in it one part naturally rules, and the other is subject, and the virtue of the ruler we in maintain to be different from that of the subject; the one being the virtue of the rational, and the other of the irrational part. Now, it is obvious that the same principle applies generally, and therefore almost all things rule and are ruled according to nature. But the kind of rule differs; the freeman rules over the slave after another manner from that in which the male rules over the female, or the man over the child; although the parts of the soul are present in an of them, they are present in different degrees. For the slave has no deliberative faculty at all; the woman has, but it is without authority, and the child has, but it is immature. So it must necessarily be supposed to be with the moral virtues also; all should partake of them, but only in such manner and degree as is required by each for the fulfillment of his duty. Hence the ruler ought to have moral virtue in perfection, for his function, taken absolutely, demands a master artificer, and rational principle is such an artificer; the subjects, oil the other hand, require only that measure of virtue which is proper to each of them. Clearly, then, moral virtue belongs to all of them; but the temperance of a man and of a woman, or the courage and justice of a man and of a woman, are not, as Socrates maintained, the same; the courage of a man is shown in commanding, of a woman in obeying. And this holds of all other virtues, as will be more clearly seen if we look at them in detail, for those who say generally that virtue consists in a good disposition of the soul, or in doing rightly, or the like, only deceive themselves. Far better than such definitions is their mode of speaking, who, like Gorgias, enumerate the virtues. All classes must be deemed to have their special attributes; as the poet says of women,

Silence is a woman's glory,

but this is not equally the glory of man. The child is imperfect, and therefore obviously his virtue is not relative to himself alone, but to the perfect man and to his teacher, and in like manner the virtue of the slave is relative to a master. Now we determined that a slave is useful for the wants of life, and therefore he will obviously require only so much virtue as will prevent him from failing in his duty through cowardice or lack of self-control. Some one will ask whether, if what we are saying is true, virtue will not be required also in the artisans, for they often fail in their work through the lack of self control? But is there not a great difference in the two cases? For the slave shares in his master's life; the artisan is less closely connected with him, and only attains excellence in proportion as he becomes a slave. The meaner sort of mechanic has a special and separate slavery; and whereas the slave exists by nature, not so the shoemaker or other artisan. It is manifest, then, that the master ought to be the source of such excellence in the slave, and not a mere possessor of the art of mastership which trains the slave in his duties. Wherefore they are mistaken who forbid us to converse with slaves and say that we should employ command only, for slaves stand even more in need of admonition than children.

So much for this subject; the relations of husband and wife, parent and child, their several virtues, what in their intercourse with one another is good, and what is evil, and how we may pursue the good and good and escape the evil, will have to be discussed when we speak of the different forms of government. For, inasmuch as every family is a part of a state, and these relationships are the parts of a family, and the virtue of the part must have regard to the virtue of the whole, women and children must be trained by education with an eye to the constitution, if the virtues of either of them are supposed to make any difference in the virtues of the state. And they must make a difference: for the children grow up to be citizens, and half the free persons in a state are women.

Of these matters, enough has been said; of what remains, let us speak at another time. Regarding, then, our present inquiry as complete, we will make a new beginning. And, first, let us examine the various theories of a perfect state.

BOOK II
I

OUR PURPOSE is to consider what form of political community is best of all for those who are most able to realize their ideal of life. We must therefore examine not only this but other constitutions, both such as actually exist in well-governed states, and any theoretical forms which are held in esteem; that what is good and useful may be brought to light. And let no one suppose that in seeking for something beyond them we are anxious to make a sophistical display at any cost; we only undertake this inquiry because all the constitutions with which we are acquainted are faulty.

We will begin with the natural beginning of the subject. Three alternatives are conceivable: The members of a state must either have (1) all things or (2) nothing in common, or (3) some things in common and some not. That they should have nothing in common is clearly impossible, for the constitution is a community, and must at any rate have a common place — one city will be in one place, and the citizens are those who share in that one city. But should a well ordered state have all things, as far as may be, in common, or some only and not others? For the citizens might conceivably have wives and children and property in common, as Socrates proposes in the Republic of Plato. Which is better, our present condition, or the proposed new order of society.

II

There are many difficulties in the community of women. And the principle on which Socrates rests the necessity of such an institution evidently is not established by his arguments. Further, as a means to the end which he ascribes to the state, the scheme, taken literally is impracticable, and how we are to interpret it is nowhere precisely stated. I am speaking of the premise from which the argument of Socrates proceeds, 'that the greater the unity of the state the better.' Is it not obvious that a state may at length attain such a degree of unity as to be no longer a state? since the nature of a state is to be a plurality, and in tending to greater unity, from being a state, it becomes a family, and from being a family, an individual; for the family may be said to be more than the state, and the individual than the family. So that we ought not to attain this greatest unity even if we could, for it would be the destruction of the state. Again, a state is not made up only of so many men, but of different kinds of men; for similars do not constitute a state. It is not like a military alliance The usefulness of the latter depends upon its quantity even where there is no difference in quality (for mutual protection is the end aimed at), just as a greater weight of anything is more useful than a less (in like manner, a state differs from a nation, when the nation has not its population organized in villages, but lives an Arcadian sort of life); but the elements out of which a unity is to be formed differ in kind. Wherefore the principle of compensation, as I have already remarked in the Ethics, is the salvation of states. Even among freemen and equals this is a principle which must be maintained, for they cannot an rule together, but must change at the end of a year or some other period of time or in some order of succession. The result is that upon this plan they all govern; just as if shoemakers and carpenters were to exchange their occupations, and the same persons did not always continue shoemakers and carpenters. And since it is better that this should be so in politics as well, it is clear that while there should be continuance of the same persons in power where this is possible, yet where this is not possible by reason of the natural equality of the citizens, and at the same time it is just that an should share in the government (whether to govern be a good thing or a bad), an approximation to this is that equals should in turn retire from office and should, apart from official position, be treated alike. Thus the one party rule and the others are ruled in turn, as if they were no longer the same persons. In like manner when they hold office there is a variety in the offices held. Hence it is evident that a city is not by nature one in that sense which some persons affirm; and that what is said to be the greatest good of cities is in reality their destruction; but surely the good of things must be that which preserves them. Again, in another point of view, this extreme unification of the state is clearly not good; for a family is more self-sufficing than an individual, and a city than a family, and a city only comes into being when the community is large enough to be self-sufficing. If then self-sufficiency is to be desired, the lesser degree of unity is more desirable than the greater.

III

But, even supposing that it were best for the community to have the greatest degree of unity, this unity is by no means proved to follow from the fact 'of all men saying "mine" and "not mine" at the same instant of time,' which, according to Socrates, is the sign of perfect unity in a state. For the word 'all' is ambiguous. If the meaning be that every individual says 'mine' and 'not mine' at the same time, then perhaps the result at which Socrates aims may be in some degree accomplished; each man will call the same person his own son and the same person his wife, and so of his property and of all that falls to his lot. This, however, is not the way in which people would speak who had their had their wives and children in common; they would say 'all' but not 'each.' In like manner their property would be described as belonging to them, not severally but collectively. There is an obvious fallacy in the term 'all': like some other words, 'both,' 'odd,' 'even,' it is ambiguous, and even in abstract argument becomes a source of logical puzzles. That all persons call the same thing mine in the sense in which each does so may be a fine thing, but it is impracticable; or if the words are taken in the other sense, such a unity in no way conduces to harmony. And there is another objection to the proposal. For that which is common to the greatest number has the least care bestowed upon it. Every one thinks chiefly of his own, hardly at all of the common interest; and only when he is himself concerned as an individual. For besides other considerations, everybody is more inclined to neglect the duty which he expects another to fulfill; as in families many attendants are often less useful than a few. Each citizen will have a thousand sons who will not be his sons individually but anybody will be equally the son of anybody, and will therefore be neglected by all alike. Further, upon this principle, every one will use the word 'mine' of one who is prospering or the reverse, however small a fraction he may himself be of the whole number; the same boy will be 'so and so's son,' the son of each of the thousand, or whatever be the number of the citizens; and even about this he will not be positive; for it is impossible to know who chanced to have a child, or whether, if one came into existence, it has survived. But which is better — for each to say 'mine' in this way, making a man the same relation to two thousand or ten thousand citizens, or to use the word 'mine' in the ordinary and more restricted sense? For usually the same person is called by one man his own son whom another calls his own brother or cousin or kinsman — blood relation or connection by marriage either of himself or of some relation of his, and yet another his clansman or tribesman; and how much better is it to be the real cousin of somebody than to be a son after Plato's fashion! Nor is there any way of preventing brothers and children and fathers and mothers from sometimes recognizing one another; for children are born like their parents, and they will necessarily be finding indications of their relationship to one another. Geographers declare such to be the fact; they say that in part of Upper Libya, where the women are common, nevertheless the children who are born are assigned to their respective fathers on the ground of their likeness. And some women, like the females of other animals — for example, mares and cows — have a strong tendency to produce offspring resembling their parents, as was the case with the Pharsalian mare called Honest.

IV

Other evils, against which it is not easy for the authors of such a community to guard, will be assaults and homicides, voluntary as well as involuntary, quarrels and slanders, all which are most unholy acts when committed against fathers and mothers and near relations, but not equally unholy when there is no relationship. Moreover, they are much more likely to occur if the relationship is unknown, and, when they have occurred, the customary expiations of them cannot be made. Again, how strange it is that Socrates, after having made the children common, should hinder lovers from carnal intercourse only, but should permit love and familiarities between father and son or between brother and brother, than which nothing can be more unseemly, since even without them love of this sort is improper. How strange, too, to forbid intercourse for no other reason than the violence of the pleasure, as though the relationship of father and son or of brothers with one another made no difference.

This community of wives and children seems better suited to the husbandmen than to the guardians, for if they have wives and children in common, they will be bound to one another by weaker ties, as a subject class should be, and they will remain obedient and not rebel. In a word, the result of such a law would be just the opposite of which good laws ought to have, and the intention of Socrates in making these regulations about women and children would defeat itself. For friendship we believe to be the greatest good of states and the preservative of them against revolutions; neither is there anything which Socrates so greatly lauds as the unity of the state which he and all the world declare to be created by friendship. But the unity which he commends would be like that of the lovers in the Symposium, who, as Aristophanes says, desire to grow together in the excess of their affection, and from being two to become one, in which case one or both would certainly perish. Whereas in a state having women and children common, love will be watery; and the father will certainly not say 'my son,' or the son 'my father.' As a little sweet wine mingled with a great deal of water is imperceptible in the mixture, so, in this sort of community, the idea of relationship which is based upon these names will be lost; there is no reason why the so-called father should care about the son, or the son about the father, or brothers about one another. Of the two qualities which chiefly inspire regard and affection — that a thing is your own and that it is your only one — neither can exist in such a state as this.

Again, the transfer of children as soon as they are born from the rank of husbandmen or of artisans to that of guardians, and from the rank of guardians into a lower rank, will be very difficult to arrange; the givers or transferrers cannot but know whom they are giving and transferring, and to whom. And the previously mentioned evils, such as assaults, unlawful loves, homicides, will happen more often amongst those who are transferred to the lower classes, or who have a place assigned to them among the guardians; for they will no longer call the members of the class they have left brothers, and children, and fathers, and mothers, and will not, therefore, be afraid of committing any crimes by reason of consanguinity. Touching the community of wives and children, let this be our conclusion.

V

Next let us consider what should be our arrangements about property: should the citizens of the perfect state have their possessions in common or not? This question may be discussed separately from the enactments about women and children. Even supposing that the women and children belong to individuals, according to the custom which is at present universal, may there not be an advantage in having and using possessions in common? Three cases are possible: (1) the soil may be appropriated, but the produce may be thrown for consumption into the common stock; and this is the practice of some nations. Or (2), the soil may be common, and may be cultivated in common, but the produce divided among individuals for their private use; this is a form of common property which is said to exist among certain barbarians. Or (3), the soil and the produce may be alike common.

When the husbandmen are not the owners, the case will be different and easier to deal with; but when they till the ground for themselves the question of ownership will give a world of trouble. If they do not share equally enjoyments and toils, those who labor much and get little will necessarily complain of those who labor little and receive or consume much. But indeed there is always a difficulty in men living together and having all human relations in common, but especially in their having common property. The partnerships of fellow-travelers are an example to the point; for they generally fall out over everyday matters and quarrel about any trifle which turns up. So with servants: we are most able to take offense at those with whom we most we most frequently come into contact in daily life.

These are only some of the disadvantages which attend the community of property; the present arrangement, if improved as it might be by good customs and laws, would be far better, and would have the advantages of both systems. Property should be in a certain sense common, but, as a general rule, private; for, when everyone has a distinct interest, men will not complain of one another, and they will make more progress, because every one will be attending to his own business. And yet by reason of goodness, and in respect of use, 'Friends,' as the proverb says, 'will have all things common.' Even now there are traces of such a principle, showing that it is not impracticable, but, in well-ordered states, exists already to a certain extent and may be carried further. For, although every man has his own property, some things he will place at the disposal of his friends, while of others he shares the use with them. The Lacedaemonians, for example, use one another's slaves, and horses, and dogs, as if they were their own; and when they lack provisions on a journey, they appropriate what they find in the fields throughout the country. It is clearly better that property should be private, but the use of it common; and the special business of the legislator is to create in men this benevolent disposition. Again, how immeasurably greater is the pleasure, when a man feels a thing to be his own; for surely the love of self is a feeling implanted by nature and not given in vain, although selfishness is rightly censured; this, however, is not the mere love of self, but the love of self in excess, like the miser's love of money; for all, or almost all, men love money and other such objects in a measure. And further, there is the greatest pleasure in doing a kindness or service to friends or guests or companions, which can only be rendered when a man has private property. These advantages are lost by excessive unification of the state. The exhibition of two virtues, besides, is visibly annihilated in such a state: first, temperance towards women (for it is an honorable action to abstain from another's wife for temperance' sake); secondly, liberality in the matter of property. No one, when men have all things in common, will any longer set an example of liberality or do any liberal action; for liberality consists in the use which is made of property.

Such legislation may have a specious appearance of benevolence; men readily listen to it, and are easily induced to believe that in some wonderful manner everybody will become everybody's friend, especially when some one is heard denouncing the evils now existing in states, suits about contracts, convictions for perjury, flatteries of rich men and the like, which are said to arise out of the possession of private property. These evils, however, are due to a very different cause — the wickedness of human nature. Indeed, we see that there is much more quarrelling among those who have all things in common, though there are not many of them when compared with the vast numbers who have private property.

Again, we ought to reckon, not only the evils from which the citizens will be saved, but also the advantages which they will lose. The life which they are to lead appears to be quite impracticable. The error of Socrates must be attributed to the false notion of unity from which he starts. Unity there should be, both of the family and of the state, but in some respects only. For there is a point at which a state may attain such a degree of unity as to be no longer a state, or at which, without actually ceasing to exist, it will become an inferior state, like harmony passing into unison, or rhythm which has been reduced to a single foot. The state, as I was saying, is a plurality which should be united and made into a community by education; and it is strange that the author of a system of education which he thinks will make the state virtuous, should expect to improve his citizens by regulations of this sort, and not by philosophy or by customs and laws, like those which prevail at Sparta and Crete respecting common meals, whereby the legislator has made property common. Let us remember that we should not disregard the experience of ages; in the multitude of years these things, if they were good, would certainly not have been unknown; for almost everything has been found out, although sometimes they are not put together; in other cases men do not use the knowledge which they have. Great light would be thrown on this subject if we could see such a form of government in the actual process of construction; for the legislator could not form a state at all without distributing and dividing its constituents into associations for common meals, and into phratries and tribes. But all this legislation ends only in forbidding agriculture to the guardians, a prohibition which the Lacedaemonians try to enforce already.

But, indeed, Socrates has not said, nor is it easy to decide, what in such a community will be the general form of the state. The citizens who are not guardians are the majority, and about them nothing has been determined: are the husbandmen, too, to have their property in common? Or is each individual to have his own? And are the wives and children to be individual or common. If, like the guardians, they are to have all things in common, what do they differ from them, or what will they gain by submitting to their government? Or, upon what principle would they submit, unless indeed the governing class adopt the ingenious policy of the Cretans, who give their slaves the same institutions as their own, but forbid them gymnastic exercises and the possession of arms. If, on the other hand, the inferior classes are to be like other cities in respect of marriage and property, what will be the form of the community? Must it not contain two states in one, each hostile to the other He makes the guardians into a mere occupying garrison, while the husbandmen and artisans and the rest are the real citizens. But if so the suits and quarrels, and all the evils which Socrates affirms to exist in other states, will exist equally among them. He says indeed that, having so good an education, the citizens will not need many laws, for example laws about the city or about the markets; but then he confines his education to the guardians. Again, he makes the husbandmen owners of the property upon condition of their paying a tribute. But in that case they are likely to be much more unmanageable and conceited than the Helots, or Penestae, or slaves in general. And whether community of wives and property be necessary for the lower equally with the higher class or not, and the questions akin to this, what will be the education, form of government, laws of the lower class, Socrates has nowhere determined: neither is it easy to discover this, nor is their character of small importance if the common life of the guardians is to be maintained.

Again, if Socrates makes the women common, and retains private property, the men will see to the fields, but who will see to the house? And who will do so if the agricultural class have both their property and their wives in common? Once more: it is absurd to argue, from the analogy of the animals, that men and women should follow the same pursuits, for animals have not to manage a household. The government, too, as constituted by Socrates, contains elements of danger; for he makes the same persons always rule. And if this is often a cause of disturbance among the meaner sort, how much more among high-spirited warriors? But that the persons whom he makes rulers must be the same is evident; for the gold which the God mingles in the souls of men is not at one time given to one, at another time to another, but always to the same: as he says, 'God mingles gold in some, and silver in others, from their very birth; but brass and iron in those who are meant to be artisans and husbandmen.' Again, he deprives the guardians even of happiness, and says that the legislator ought to make the whole state happy. But the whole cannot be happy unless most, or all, or some of its parts enjoy happiness. In this respect happiness is not like the even principle in numbers, which may exist only in the whole, but in neither of the parts; not so happiness. And if the guardians are not happy, who are? Surely not the artisans, or the common people. The Republic of which Socrates discourses has all these difficulties, and others quite as great.

VI

The same, or nearly the same, objections apply to Plato's later work, the Laws, and therefore we had better examine briefly the constitution which is therein described. In the Republic, Socrates has definitely settled in all a few questions only; such as the community of women and children, the community of property, and the constitution of the state. The population is divided into two classes — one of husbandmen, and the other of warriors; from this latter is taken a third class of counselors and rulers of the state. But Socrates has not determined whether the husbandmen and artisans are to have a share in the government, and whether they, too, are to carry arms and share in military service, or not. He certainly thinks that the women ought to share in the education of the guardians, and to fight by their side. The remainder of the work is filled up with digressions foreign to the main subject, and with discussions about the education of the guardians. In the Laws there is hardly anything but laws; not much is said about the constitution. This, which he had intended to make more of the ordinary type, he gradually brings round to the other or ideal form. For with the exception of the community of women and property, he supposes everything to be the same in both states; there is to be the same education; the citizens of both are to live free from servile occupations, and there are to be common meals in both. The only difference is that in the Laws, the common meals are extended to women, and the warriors number 5000, but in the Republic only 1000.

The discourses of Socrates are never commonplace; they always exhibit grace and originality and thought; but perfection in everything can hardly be expected. We must not overlook the fact that the number of 5000 citizens, just now mentioned, will require a territory as large as Babylon, or some other huge site, if so many persons are to be supported in idleness, together with their women and attendants, who will be a multitude many times as great. In framing an ideal we may assume what we wish, but should avoid impossibilities.

It is said that the legislator ought to have his eye directed to two points — the people and the country. But neighboring countries also must not be forgotten by him, firstly because the state for which he legislates is to have a political and not an isolated life. For a state must have such a military force as will be serviceable against her neighbors, and not merely useful at home. Even if the life of action is not admitted to be the best, either for individuals or states, still a city should be formidable to enemies, whether invading or retreating.

There is another point: Should not the amount of property be defined in some way which differs from this by being clearer? For Socrates says that a man should have so much property as will enable him to live temperately, which is only a way of saying 'to live well'; this is too general a conception. Further, a man may live temperately and yet miserably. A better definition would be that a man must have so much property as will enable him to live not only temperately but liberally; if the two are parted, liberally will combine with luxury; temperance will be associated with toil. For liberality and temperance are the only eligible qualities which have to do with the use of property. A man cannot use property with mildness or courage, but temperately and liberally he may; and therefore the practice of these virtues is inseparable from property. There is an inconsistency, too, in too, in equalizing the property and not regulating the number of the citizens; the population is to remain unlimited, and he thinks that it will be sufficiently equalized by a certain number of marriages being unfruitful, however many are born to others, because he finds this to be the case in existing states. But greater care will be required than now; for among ourselves, whatever may be the number of citizens, the property is always distributed among them, and therefore no one is in want; but, if the property were incapable of division as in the Laws, the supernumeraries, whether few or many, would get nothing. One would have thought that it was even more necessary to limit population than property; and that the limit should be fixed by calculating the chances of mortality in the children, and of sterility in married persons. The neglect of this subject, which in existing states is so common, is a never-failing cause of poverty among the citizens; and poverty is the parent of revolution and crime. Pheidon the Corinthian, who was one of the most ardent legislators, thought that the families and the number of citizens ought to remain the same, although originally all the lots may have been of different sizes: but in the Laws the opposite principle is maintained. What in our opinion is the right arrangement will have to be explained hereafter.

There is another omission in the Laws: Socrates does not tell us how the rulers differ from their subjects; he only says that they should be related as the warp and the woof, which are made out of different wools. He allows that a man's whole property may be increased fivefold, but why should not his land also increase to a certain extent? Again, will the good management of a household be promoted by his arrangement of homesteads? For he assigns to each individual two homesteads in separate places, and it is difficult to live in two houses.

The whole system of government tends to be neither democracy nor oligarchy, but something in a mean between them, which is usually called a polity, and is composed of the heavy-armed soldiers. Now, if he intended to frame a constitution which would suit the greatest number of states, he was very likely right, but not if he meant to say that this constitutional form came nearest to his first or ideal state; for many would prefer the Lacedaemonian, or, possibly, some other more aristocratic government. Some, indeed, say that the best constitution is a combination of all existing forms, and they praise the Lacedaemonian because it is made up of oligarchy, monarchy, and democracy, the king forming the monarchy, and the council of elders the oligarchy while the democratic element is represented by the Ephors; for the Ephors are selected from the people. Others, however, declare the Ephoralty to be a tyranny, and find the element of democracy in the common meals and in the habits of daily life. In the Laws it is maintained that the best constitution is made up of democracy and tyranny, which are either not constitutions at all, or are the worst of all. But they are nearer the truth who combine many forms; for the constitution is better which is made up of more numerous elements. The constitution proposed in the Laws has no element of monarchy at all; it is nothing but oligarchy and democracy, leaning rather to oligarchy. This is seen in the mode of appointing magistrates; for although the appointment of them by lot from among those who have been already selected combines both elements, the way in which the rich are compelled by law to attend the assembly and vote for magistrates or discharge other political duties, while the rest may do as they like, and the endeavor to have the greater number of the magistrates appointed out of the richer classes and the highest officers selected from those who have the greatest incomes, both these are oligarchical features. The oligarchical principle prevails also in the choice of the council, for all are compelled to choose, but the compulsion extends only to the choice out of the first class, and of an equal number out of the second class and out of the third class, but not in this latter case to all the voters but to those of the first three classes; and the selection of candidates out of the fourth class is only compulsory on the first and second. Then, from the persons so chosen, he says that there ought to be an equal number of each class selected. Thus a preponderance will be given to the better sort of people, who have the larger incomes, because many of the lower classes, not being compelled will not vote. These considerations, and others which will be adduced when the time comes for examining similar polities, tend to show that states like Plato's should not be composed of democracy and monarchy. There is also a danger in electing the magistrates out of a body who are themselves elected; for, if but a small number choose to combine, the elections will always go as they desire. Such is the constitution which is described in the Laws.

VII

Other constitutions have been proposed; some by private persons, others by philosophers and statesmen, which all come nearer to established or existing ones than either of Plato's. No one else has introduced such novelties as the community of women and children, or public tables for women: other legislators begin with what is necessary. In the opinion of some, the regulation of property is the chief point of all, that being the question upon which all revolutions turn. This danger was recognized by Phaleas of Chalcedon, who was the first to affirm that the citizens of a state ought to have equal possessions. He thought that in a new colony the equalization might be accomplished without difficulty, not so easily when a state was already established; and that then the shortest way of compassing the desired end would be for the rich to give and not to receive marriage portions, and for the poor not to give but to receive them.

Plato in the Laws was of opinion that, to a certain extent, accumulation should be allowed, forbidding, as I have already observed, any citizen to possess more than five times the minimum qualification But those who make such laws should remember what they are apt to forget — that the legislator who fixes the amount of property should also fix the number of children; for, if the children are too many for the property, the law must be broken. And, besides the violation of the law, it is a bad thing that many from being rich should become poor; for men of ruined fortunes are sure to stir up revolutions. That the equalization of property exercises an influence on political society was clearly understood even by some of the old legislators. Laws were made by Solon and others prohibiting an individual from possessing as much land as he pleased; and there are other laws in states which forbid the sale of property: among the Locrians, for example, there is a law that a man is not to sell his property unless he can prove unmistakably that some misfortune has befallen him. Again, there have been laws which enjoin the preservation of the original lots. Such a law existed in the island of Leucas, and the abrogation of it made the constitution too democratic, for the rulers no longer had the prescribed qualification. Again, where there is equality of property, the amount may be either too large or too small, and the possessor may be living either in luxury or penury. Clearly, then, the legislator ought not only to aim at the equalization of properties, but at moderation in their amount. Further, if he prescribe this moderate amount equally to all, he will be no nearer the mark; for it is not the possessions but the desires of mankind which require to be equalized, and this is impossible, unless a sufficient education is provided by the laws. But Phaleas will probably reply that this is precisely what he means; and that, in his opinion, there ought to be in states, not only equal property, but equal education. Still he should tell precisely what he means; and that, in his opinion, there ought to be in be in having one and the same for all, if it is of a sort that predisposes men to avarice, or ambition, or both. Moreover, civil troubles arise, not only out of the inequality of property, but out of the inequality of honor, though in opposite ways. For the common people quarrel about the inequality of property, the higher class about the equality of honor; as the poet says,

"The bad and good alike in honor share."

There are crimes of which the motive is want; and for these Phaleas expects to find a cure in the equalization of property, which will take away from a man the temptation to be a highwayman, because he is hungry or cold. But want is not the sole incentive to crime; men also wish to enjoy themselves and not to be in a state of desire — they wish to cure some desire, going beyond the necessities of life, which preys upon them; nay, this is not the only reason — they may desire superfluities in order to enjoy pleasures unaccompanied with pain, and therefore they commit crimes.

Now what is the cure of these three disorders? Of the first, moderate possessions and occupation; of the second, habits of temperance; as to the third, if any desire pleasures which depend on themselves, they will find the satisfaction of their desires nowhere but in philosophy; for all other pleasures we are dependent on others. The fact is that the greatest crimes are caused by excess and not by necessity. Men do not become tyrants in order that they may not suffer cold; and hence great is the honor bestowed, not on him who kills a thief, but on him who kills a tyrant. Thus we see that the institutions of Phaleas avail only against petty crimes.

There is another objection to them. They are chiefly designed to promote the internal welfare of the state. But the legislator should consider also its relation to neighboring nations, and to all who are outside of it. The government must be organized with a view to military strength; and of this he has said not a word. And so with respect to property: there should not only be enough to supply the internal wants of the state, but also to meet dangers coming from without. The property of the state should not be so large that more powerful neighbors may be tempted by it, while the owners are unable to repel the invaders; nor yet so small that the state is unable to maintain a war even against states of equal power, and of the same character. Phaleas has not laid down any rule; but we should bear in mind that abundance of wealth is an advantage. The best limit will probably be, that a more powerful neighbor must have no inducement to go to war with you by reason of the excess of your wealth, but only such as he would have had if you had possessed less. There is a story that Eubulus, when Autophradates was going to besiege Atarneus, told him to consider how long the operation would take, and then reckon up the cost which would be incurred in the time. 'For,' said he, 'I am willing for a smaller sum than that to leave Atarneus at once.' These words of Eubulus made an impression on Autophradates, and he desisted from the siege.

The equalization of property is one of the things that tend to prevent the citizens from quarrelling. Not that the gain in this direction is very great. For the nobles will be dissatisfied because they think themselves worthy of more than an equal share of honors; and this is often found to be a cause of sedition and revolution. And the avarice of mankind is insatiable; at one time two obols was pay enough; but now, when this sum has become customary, men always want more and more without end; for it is of the nature of desire not to be satisfied, and most men live only for the gratification of it. The beginning of reform is not so much to equalize property as to train the nobler sort of natures not to desire more, and to prevent the lower from getting more; that is to say, they must be kept down, but not ill-treated. Besides, the equalization proposed by Phaleas is imperfect; for he only equalizes land, whereas a man may be rich also in slaves, and cattle, and money, and in the abundance of what are called his movables. Now either all these things must be equalized, or some limit must be imposed on them, or they must an be let alone. It would appear that Phaleas is legislating for a small city only, if, as he supposes, all the artisans are to be public slaves and not to form a supplementary part of the body of citizens. But if there is a law that artisans are to be public slaves, it should only apply to those engaged on public works, as at Epidamnus, or at Athens on the plan which Diophantus once introduced.

From these observations any one may judge how far Phaleas was wrong or right in his ideas.

VIII

Hippodamus, the son of Euryphon, a native of Miletus, the same who invented the art of planning cities, and who also laid out the Piraeus — a strange man, whose fondness for distinction led him into a general eccentricity of life, which made some think him affected (for he would wear flowing hair and expensive ornaments; but these were worn on a cheap but warm garment both in winter and summer); he, besides aspiring to be an adept in the knowledge of nature, was the first person not a statesman who made inquiries about the best form of government.

The city of Hippodamus was composed of 10,000 citizens divided into three parts — one of artisans, one of husbandmen, and a third of armed defenders of the state. He also divided the land into three parts, one sacred, one public, the third private: the first was set apart to maintain the customary worship of the Gods, the second was to support the warriors, the third was the property of the husbandmen. He also divided laws into three classes, and no more, for he maintained that there are three subjects of lawsuits — insult, injury, and homicide. He likewise instituted a single final court of appeal, to which all causes seeming to have been improperly decided might be referred; this court he formed of elders chosen for the purpose. He was further of opinion that the decisions of the courts ought not to be given by the use of a voting pebble, but that every one should have a tablet on which he might not only write a simple condemnation, or leave the tablet blank for a simple acquittal; but, if he partly acquitted and partly condemned, he was to distinguish accordingly. To the existing law he objected that it obliged the judges to be guilty of perjury, whichever way they voted. He also enacted that those who discovered anything for the good of the state should be honored; and he provided that the children of citizens who died in battle should be maintained at the public expense, as if such an enactment had never been heard of before, yet it actually exists at Athens and in other places. As to the magistrates, he would have them all elected by the people, that is, by the three classes already mentioned, and those who were elected were to watch over the interests of the public, of strangers, and of orphans. These are the most striking points in the constitution of Hippodamus. There is not much else.

The first of these proposals to which objection may be taken is the threefold division of the citizens. The artisans, and the husbandmen, and the warriors, all have a share in the government. But the husbandmen have no arms, and the artisans neither arms nor land, and therefore they become all but slaves of the warrior class. That they should share in all the offices is an impossibility; for generals and guardians of the citizens, and nearly all the principal magistrates, must be taken from the class of those who carry arms. Yet, if the two other classes have no share in the government, how can they be loyal citizens? It may be said that those who have arms must necessarily be masters of both the other classes, but this is not so easily accomplished unless they are numerous; and if they are, why should the other classes share in the government at all, or have power to appoint magistrates? Further, what use are farmers to the city? Artisans there must be, for these are wanted in every city, and they can live by their craft, as elsewhere; and the husbandmen too, if they really provided the warriors with food, might fairly have a share in the government. But in the republic of Hippodamus they are supposed to have land of their own, which they cultivate for their private benefit. Again, as to this common land out of which the soldiers are maintained, if they are themselves to be the cultivators of it, the warrior class will be identical with the husbandmen, although the legislator intended to make a distinction between them. If, again, there are to be other cultivators distinct both from the husbandmen, who have land of their own, and from the warriors, they will make a fourth class, which has no place in the state and no share in anything. Or, if the same persons are to cultivate their own lands, and those of the public as well, they will have difficulty in supplying the quantity of produce which will maintain two households: and why, in this case, should there be any division, for they might find food themselves and give to the warriors from the same land and the same lots? There is surely a great confusion in all this.

Neither is the law to commended which says that the judges, when a simple issue is laid before them, should distinguish in their judgement; for the judge is thus converted into an arbitrator. Now, in an arbitration, although the arbitrators are many, they confer with one another about the decision, and therefore they can distinguish; but in courts of law this is impossible, and, indeed, most legislators take pains to prevent the judges from holding any communication with one another. Again, will there not be confusion if the judge thinks that damages should be given, but not so much as the suitor demands? He asks, say, for twenty minae, and the judge allows him ten minae (or in general the suitor asks for more and the judge allows less), while another judge allows five, another four minae. In this way they will go on splitting up the damages, and some will grant the whole and others nothing: how is the final reckoning to be taken? Again, no one contends that he who votes for a simple acquittal or condemnation perjures himself, if the indictment has been laid in an unqualified form; and this is just, for the judge who acquits does not decide that the defendant owes nothing, but that he does not owe the twenty minae. He only is guilty of perjury who thinks that the defendant ought not to pay twenty minae, and yet condemns him.

To honor those who discover anything which is useful to the state is a proposal which has a specious sound, but cannot safely be enacted by law, for it may encourage informers, and perhaps even lead to political commotions. This question involves another. It has been doubted whether it is or is not expedient to make any changes in the laws of a country, even if another law be better. Now, if an changes are inexpedient, we can hardly assent to the proposal of Hippodamus; for, under pretense of doing a public service, a man may introduce measures which are really destructive to the laws or to the constitution. But, since we have touched upon this subject, perhaps we had better go a little into detail, for, as I was saying, there is a difference of opinion, and it may sometimes seem desirable to make changes. Such changes in the other arts and sciences have certainly been beneficial; medicine, for example, and gymnastic, and every other art and craft have departed from traditional usage. And, if politics be an art, change must be necessary in this as in any other art. That improvement has occurred is shown by the fact that old customs are exceedingly simple and barbarous. For the ancient Hellenes went about armed and bought their brides of each other. The remains of ancient laws which have come down to us are quite absurd; for example, at Cumae there is a law about murder, to the effect that if the accuser produce a certain number of witnesses from among his own kinsmen, the accused shall be held guilty. Again, men in general desire the good, and not merely what their fathers had. But the primeval inhabitants, whether they were born of the earth or were the survivors of some destruction, may be supposed to have been no better than ordinary or even foolish people among ourselves (such is certainly the tradition concerning the earth-born men); and it would be ridiculous to rest contented with their notions. Even when laws have been written down, they ought not always to remain unaltered. As in other sciences, so in politics, it is impossible that all things should be precisely set down in writing; for enactments must be universal, but actions are concerned with particulars. Hence we infer that sometimes and in certain cases laws may be changed; but when we look at the matter from another point of view, great caution would seem to be required. For the habit of lightly changing the laws is an evil, and, when the advantage is small, some errors both of lawgivers and rulers had better be left; the citizen will not gain so much by making the change as he will lose by the habit of disobedience. The analogy of the arts is false; a change in a law is a very different thing from a change in an art. For the law has no power to command obedience except that of habit, which can only be given by time, so that a readiness to change from old to new laws enfeebles the power of the law. Even if we admit that the laws are to be changed, are they all to be changed, and in every state? And are they to be changed by anybody who likes, or only by certain persons? These are very important questions; and therefore we had better reserve the discussion of them to a more suitable occasion.

IX

In the governments of Lacedaemon and Crete, and indeed in all governments, two points have to be considered: first, whether any particular law is good or bad, when compared with the perfect state; secondly, whether it is or is not consistent with the idea and character which the lawgiver has set before his citizens. That in a well-ordered state the citizens should have leisure and not have to provide for their daily wants is generally acknowledged, but there is a difficulty in seeing how this leisure is to be attained. The Thessalian Penestae have often risen against their masters, and the Helots in like manner against the Lacedaemonians, for whose misfortunes they are always lying in wait. Nothing, however, of this kind has as yet happened to the Cretans; the reason probably is that the neighboring cities, even when at war with one another, never form an alliance with rebellious serfs, rebellions not being for their interest, since they themselves have a dependent population. Whereas all the neighbors of the Lacedaemonians, whether Argives, Messenians, or Arcadians, were their enemies. In Thessaly, again, the original revolt of the slaves occurred because the Thessalians were still at war with the neighboring Achaeans, Perrhaebians, and Magnesians. Besides, if there were no other difficulty, the treatment or management of slaves is a troublesome affair; for, if not kept in hand, they are insolent, and think that they are as good as their masters, and, if harshly treated, they hate and conspire against them. Now it is clear that when these are the results the citizens of a state have not found out the secret of managing their subject population.

Again, the license of the Lacedaemonian women defeats the intention of the Spartan constitution, and is adverse to the happiness of the state. For, a husband and wife being each a part of every family, the state may be considered as about equally divided into men and women; and, therefore, in those states in which the condition of the women is bad, half the city may be regarded as having no laws. And this is what has actually happened at Sparta; the legislator wanted to make the whole state hardy and temperate, and he has carried out his intention in the case of the men, but he has neglected the women, who live in every sort of intemperance and luxury. The consequence is that in such a state wealth is too highly valued, especially if the citizen fall under the dominion of their wives, after the manner of most warlike races, except the Celts and a few others who openly approve of male loves. The old mythologer would seem to have been right in uniting Ares and Aphrodite, for all warlike races are prone to the love either of men or of women. This was exemplified among the Spartans in the days of their greatness; many things were managed by their women. But what difference does it make whether women rule, or the rulers are ruled by women? The result is the same. Even in regard to courage, which is of no use in daily life, and is needed only in war, the influence of the Lacedaemonian women has been most mischievous. The evil showed itself in the Theban invasion, when, unlike the women other cities, they were utterly useless and caused more confusion than the enemy. This license of the Lacedaemonian women existed from the earliest times, and was only what might be expected. For, during the wars of the Lacedaemonians, first against the Argives, and afterwards against the Arcadians and Messenians, the men were long away from home, and, on the return of peace, they gave themselves into the legislator's hand, already prepared by the discipline of a soldier's life (in which there are many elements of virtue), to receive his enactments. But, when Lycurgus, as tradition says, wanted to bring the women under his laws, they resisted, and he gave up the attempt. These then are the causes of what then happened, and this defect in the constitution is clearly to be attributed to them. We are not, however, considering what is or is not to be excused, but what is right or wrong, and the disorder of the women, as I have already said, not only gives an air of indecorum to the constitution considered in itself, but tends in a measure to foster avarice.

The mention of avarice naturally suggests a criticism on the inequality of property. While some of the Spartan citizen have quite small properties, others have very large ones; hence the land has passed into the hands of a few. And this is due also to faulty laws; for, although the legislator rightly holds up to shame the sale or purchase of an inheritance, he allows anybody who likes to give or bequeath it. Yet both practices lead to the same result. And nearly two-fifths of the whole country are held by women; this is owing to the number of heiresses and to the large dowries which are customary. It would surely have been better to have given no dowries at all, or, if any, but small or moderate ones. As the law now stands, a man may bestow his heiress on any one whom he pleases, and, if he die intestate, the privilege of giving her away descends to his heir. Hence, although the country is able to maintain 1500 cavalry and 30,000 hoplites, the whole number of Spartan citizens fell below 1000. The result proves the faulty nature of their laws respecting property; for the city sank under a single defeat; the want of men was their ruin. There is a tradition that, in the days of their ancient kings, they were in the habit of giving the rights of citizenship to strangers, and therefore, in spite of their long wars, no lack of population was experienced by them; indeed, at one time Sparta is said to have numbered not less than 10,000 citizens Whether this statement is true or not, it would certainly have been better to have maintained their numbers by the equalization of property. Again, the law which relates to the procreation of children is adverse to the correction of this inequality. For the legislator, wanting to have as many Spartans as he could, encouraged the citizens to have large families; and there is a law at Sparta that the father of three sons shall be exempt from military service, and he who has four from all the burdens of the state. Yet it is obvious that, if there were many children, the land being distributed as it is, many of them must necessarily fall into poverty.

The Lacedaemonian constitution is defective in another point; I mean the Ephoralty. This magistracy has authority in the highest matters, but the Ephors are chosen from the whole people, and so the office is apt to fall into the hands of very poor men, who, being badly off, are open to bribes. There have been many examples at Sparta of this evil in former times; and quite recently, in the matter of the Andrians, certain of the Ephors who were bribed did their best to ruin the state. And so great and tyrannical is their power, that even the kings have been compelled to court them, so that, in this way as well together with the royal office, the whole constitution has deteriorated, and from being an aristocracy has turned into a democracy. The Ephoralty certainly does keep the state together; for the people are contented when they have a share in the highest office, and the result, whether due to the legislator or to chance, has been advantageous. For if a constitution is to be permanent, all the parts of the state must wish that it should exist and the same arrangements be maintained. This is the case at Sparta, where the kings desire its permanence because they have due honor in their own persons; the nobles because they are represented in the council of elders (for the office of elder is a reward of virtue); and the people, because all are eligible to the Ephoralty. The election of Ephors out of the whole people is perfectly right, but ought not to be carried on in the present fashion, which is too childish. Again, they have the decision of great causes, although they are quite ordinary men, and therefore they should not determine them merely on their own judgment, but according to written rules, and to the laws. Their way of life, too, is not in accordance with the spirit of the constitution — they have a deal too much license; whereas, in the case of the other citizens, the excess of strictness is so intolerable that they run away from the law into the secret indulgence of sensual pleasures.

Again, the council of elders is not free from defects. It may be said that the elders are good men and well trained in manly virtue; and that, therefore, there is an advantage to the state in having them. But that judges of important causes should hold office for life is a disputable thing, for the mind grows old as well as the body. And when men have been educated in such a manner that even the legislator himself cannot trust them, there is real danger. Many of the elders are well known to have taken bribes and to have been guilty of partiality in public affairs. And therefore they ought not to be irresponsible; yet at Sparta they are so. But (it may be replied), 'All magistracies are accountable to the Ephors.' Yes, but this prerogative is too great for them, and we maintain that the control should be exercised in some other manner. Further, the mode in which the Spartans elect their elders is childish; and it is improper that the person to be elected should canvass for the office; the worthiest should be appointed, whether he chooses or not. And here the legislator clearly indicates the same intention which appears in other parts of his constitution; he would have his citizens ambitious, and he has reckoned upon this quality in the election of the elders; for no one would ask to be elected if he were not. Yet ambition and avarice, almost more than any other passions, are the motives of crime.

Whether kings are or are not an advantage to states, I will consider at another time; they should at any rate be chosen, not as they are now, but with regard to their personal life and conduct. The legislator himself obviously did not suppose that he could make them really good men; at least he shows a great distrust of their virtue. For this reason the Spartans used to join enemies with them in the same embassy, and the quarrels between the kings were held to be conservative of the state.

Neither did the first introducer of the common meals, called 'phiditia,' regulate them well. The entertainment ought to have been provided at the public cost, as in Crete; but among the Lacedaemonians every one is expected to contribute, and some of them are too poor to afford the expense; thus the intention of the legislator is frustrated. The common meals were meant to be a popular institution, but the existing manner of regulating them is the reverse of popular. For the very poor can scarcely take part in them; and, according to ancient custom, those who cannot contribute are not allowed to retain their rights of citizenship.

The law about the Spartan admirals has often been censured, and with justice; it is a source of dissension, for the kings are perpetual generals, and this office of admiral is but the setting up of another king.

The charge which Plato brings, in the Laws, against the intention of the legislator, is likewise justified; the whole constitution has regard to one part of virtue only — the virtue of the soldier, which gives victory in war. So long as they were at war, therefore, their power was preserved, but when they had attained empire they fell for of the arts of peace they knew nothing, and had never engaged in any employment higher than war. There is another error, equally great, into which they have fallen. Although they truly think that the goods for which men contend are to be acquired by virtue rather than by vice, they err in supposing that these goods are to be preferred to the virtue which gains them.

Once more: the revenues of the state are ill-managed; there is no money in the treasury, although they are obliged to carry on great wars, and they are unwilling to pay taxes. The greater part of the land being in the hands of the Spartans, they do not look closely into one another's contributions. The result which the legislator has produced is the reverse of beneficial; for he has made his city poor, and his citizens greedy.

Enough respecting the Spartan constitution, of which these are the principal defects.

X

The Cretan constitution nearly resembles the Spartan, and in some few points is quite as good; but for the most part less perfect in form. The older constitutions are generally less elaborate than the later, and the Lacedaemonian is said to be, and probably is, in a very great measure, a copy of the Cretan. According to tradition, Lycurgus, when he ceased to be the guardian of King Charillus, went abroad and spent most of his time in Crete. For the two countries are nearly connected; the Lyctians are a colony of the Lacedaemonians, and the colonists, when they came to Crete, adopted the constitution which they found existing among the inhabitants. Even to this day the Perioeci, or subject population of Crete, are governed by the original laws which Minos is supposed to have enacted. The island seems to be intended by nature for dominion in Hellas, and to be well situated; it extends right across the sea, around which nearly all the Hellenes are settled; and while one end is not far from the Peloponnese, the other almost reaches to the region of Asia about Triopium and Rhodes. Hence Minos acquired the empire of the sea, subduing some of the islands and colonizing others; at last he invaded Sicily, where he died near Camicus.

The Cretan institutions resemble the Lacedaemonian. The Helots are the husbandmen of the one, the Perioeci of the other, and both Cretans and Lacedaemonians have common meals, which were anciently called by the Lacedaemonians not 'phiditia' but 'andria'; and the Cretans have the same word, the use of which proves that the common meals originally came from Crete. Further, the two constitutions are similar; for the office of the Ephors is the same as that of the Cretan Cosmi, the only difference being that whereas the Ephors are five, the Cosmi are ten in number. The elders, too, answer to the elders in Crete, who are termed by the Cretans the council. And the kingly office once existed in Crete, but was abolished, and the Cosmi have now the duty of leading them in war. All classes share in the ecclesia, but it can only ratify the decrees of the elders and the Cosmi.

The common meals of Crete are certainly better managed than the Lacedaemonian; for in Lacedaemon every one pays so much per head, or, if he fails, the law, as I have already explained, forbids him to exercise the rights of citizenship. But in Crete they are of a more popular character. There, of all the fruits of the earth and cattle raised on the public lands, and of the tribute which is paid by the Perioeci, one portion is assigned to the Gods and to the service of the state, and another to the common meals, so that men, women, and children are all supported out of a common stock. The legislator has many ingenious ways of securing moderation in eating, which he conceives to be a gain; he likewise encourages the separation of men from women, lest they should have too many children, and the companionship of men with one another — whether this is a good or bad thing I shall have an opportunity of considering at another time. But that the Cretan common meals are better ordered than the Lacedaemonian there can be no doubt.

On the other hand, the Cosmi are even a worse institution than the Ephors, of which they have all the evils without the good. Like the Ephors, they are any chance persons, but in Crete this is not counterbalanced by a corresponding political advantage. At Sparta every one is eligible, and the body of the people, having a share in the highest office, want the constitution to be permanent. But in Crete the Cosmi are elected out of certain families, and not out of the whole people, and the elders out of those who have been Cosmi.

The same criticism may be made about the Cretan, which has been already made about the Lacedaemonian elders. Their irresponsibility and life tenure is too great a privilege, and their arbitrary power of acting upon their own judgment, and dispensing with written law, is dangerous. It is no proof of the goodness of the institution that the people are not discontented at being excluded from it. For there is no profit to be made out of the office as out of the Ephoralty, since, unlike the Ephors, the Cosmi, being in an island, are removed from temptation.

The remedy by which they correct the evil of this institution is an extraordinary one, suited rather to a close oligarchy than to a constitutional state. For the Cosmi are often expelled by a conspiracy of their own colleagues, or of private individuals; and they are allowed also to resign before their term of office has expired. Surely all matters of this kind are better regulated by law than by the will of man, which is a very unsafe rule. Worst of all is the suspension of the office of Cosmi, a device to which the nobles often have recourse when they will not submit to justice. This shows that the Cretan government, although possessing some of the characteristics of a constitutional state, is really a close oligarchy.

The nobles have a habit, too, of setting up a chief; they get together a party among the common people and their own friends and then quarrel and fight with one another. What is this but the temporary destruction of the state and dissolution of society? A city is in a dangerous condition when those who are willing are also able to attack her. But, as I have already said, the island of Crete is saved by her situation; distance has the same effect as the Lacedaemonian prohibition of strangers; and the Cretans have no foreign dominions. This is the reason why the Perioeci are contented in Crete, whereas the Helots are perpetually revolting. But when lately foreign invaders found their way into the island, the weakness of the Cretan constitution was revealed. Enough of the government of Crete.

XI

The Carthaginians are also considered to have an excellent form of government, which differs from that of any other state in several respects, though it is in some very like the Lacedaemonian. Indeed, all three states — the Lacedaemonian, the Cretan, and the Carthaginian — nearly resemble one another, and are very different from any others. Many of the Carthaginian institutions are excellent The superiority of their constitution is proved by the fact that the common people remain loyal to the constitution the Carthaginians have never had any rebellion worth speaking of, and have never been under the rule of a tyrant.

Among the points in which the Carthaginian constitution resembles the Lacedaemonian are the following: The common tables of the clubs answer to the Spartan phiditia, and their magistracy of the 104 to the Ephors; but, whereas the Ephors are any chance persons, the magistrates of the Carthaginians are elected according to merit — this is an improvement. They have also their kings and their gerusia, or council of elders, who correspond to the kings and elders of Sparta. Their kings, unlike the Spartan, are not always of the same family, nor that an ordinary one, but if there is some distinguished family they are selected out of it and not appointed by seniority — this is far better. Such officers have great power, and therefore, if they are persons of little worth, do a great deal of harm, and they have already done harm at Lacedaemon.

Most of the defects or deviations from the perfect state, for which the Carthaginian constitution would be censured, apply equally to all the forms of government which we have mentioned. But of the deflections from aristocracy and constitutional government, some incline more to democracy and some to oligarchy. The kings and elders, if unanimous, may determine whether they will or will not bring a matter before the people, but when they are not unanimous, the people decide on such matters as well. And whatever the kings and elders bring before the people is not only heard but also determined by them, and any one who likes may oppose it; now this is not permitted in Sparta and Crete. That the magistrates of five who have under them many important matters should be co-opted, that they should choose the supreme council of 100, and should hold office longer than other magistrates (for they are virtually rulers both before and after they hold office) — these are oligarchical features; their being without salary and not elected by lot, and any similar points, such as the practice of having all suits tried by the magistrates, and not some by one class of judges or jurors and some by another, as at Lacedaemon, are characteristic of aristocracy. The Carthaginian constitution deviates from aristocracy and inclines to oligarchy, chiefly on a point where popular opinion is on their side. For men in general think that magistrates should be chosen not only for their merit, but for their wealth: a man, they say, who is poor cannot rule well — he has not the leisure. If, then, election of magistrates for their wealth be characteristic of oligarchy, and election for merit of aristocracy, there will be a third form under which the constitution of Carthage is comprehended; for the Carthaginians choose their magistrates, and particularly the highest of them — their kings and generals — with an eye both to merit and to wealth.

But we must acknowledge that, in thus deviating from aristocracy, the legislator has committed an error. Nothing is more absolutely necessary than to provide that the highest class, not only when in office, but when out of office, should have leisure and not disgrace themselves in any way; and to this his attention should be first directed. Even if you must have regard to wealth, in order to secure leisure, yet it is surely a bad thing that the greatest offices, such as those of kings and generals, should be bought. The law which allows this abuse makes wealth of more account than virtue, and the whole state becomes avaricious. For, whenever the chiefs of the state deem anything honorable, the other citizens are sure to follow their example; and, where virtue has not the first place, their aristocracy cannot be firmly established. Those who have been at the expense of purchasing their places will be in the habit of repaying themselves; and it is absurd to suppose that a poor and honest man will be wanting to make gains, and that a lower stamp of man who has incurred a great expense will not. Wherefore they should rule who are able to rule best. And even if the legislator does not care to protect the good from poverty, he should at any rate secure leisure for them when in office.

It would seem also to be a bad principle that the same person should hold many offices, which is a favorite practice among the Carthaginians, for one business is better done by one man. The legislator should see to this and should not appoint the same person to be a flute-player and a shoemaker. Hence, where the state is large, it is more in accordance both with constitutional and with democratic principles that the offices of state should be distributed among many persons. For, as I said, this arrangement is fairer to all, and any action familiarized by repetition is better and sooner performed. We have a proof in military and naval matters; the duties of command and of obedience in both these services extend to all.

The government of the Carthaginians is oligarchical, but they successfully escape the evils of oligarchy by enriching one portion of the people after another by sending them to their colonies. This is their panacea and the means by which they give stability to the state. Accident favors them, but the legislator should be able to provide against revolution without trusting to accidents. As things are, if any misfortune occurred, and the bulk of the subjects revolted, there would be no way of restoring peace by legal methods.

Such is the character of the Lacedaemonian, Cretan, and Carthaginian constitutions, which are justly celebrated.

XII

Of those who have treated of governments, some have never taken any part at all in public affairs, but have passed their lives in a private station; about most of them, what was worth telling has been already told. Others have been lawgivers, either in their own or in foreign cities, whose affairs they have administered; and of these some have only made laws, others have framed constitutions; for example, Lycurgus and Solon did both. Of the Lacedaemonian constitution I have already spoken. As to Solon, he is thought by some to have been a good legislator, who put an end to the exclusiveness of the oligarchy, emancipated the people, established the ancient Athenian democracy, and harmonized the different elements of the state. According to their view, the council of Areopagus was an oligarchical element, the elected magistracy, aristocratical, and the courts of law, democratical. The truth seems to be that the council and the elected magistracy existed before the time of Solon, and were retained by him, but that he formed the courts of law out of an the citizens, thus creating the democracy, which is the very reason why he is sometimes blamed. For in giving the supreme power to the law courts, which are elected by lot, he is thought to have destroyed the non-democratic element. When the law courts grew powerful, to please the people who were now playing the tyrant the old constitution was changed into the existing democracy. Ephialtes and Pericles curtailed the power of the Areopagus; Pericles also instituted the payment of the juries, and thus every demagogue in turn increased the power of the democracy until it became what we now see. All this is true; it seems, however, to be the result of circumstances, and not to have been intended by Solon. For the people, having been instrumental in gaining the empire of the sea in the Persian War, began to get a notion of itself, and followed worthless demagogues, whom the better class opposed. Solon, himself, appears to have given the Athenians only that power of electing to offices and calling to account the magistrates which was absolutely necessary; for without it they would have been in a state of slavery and enmity to the government. All the magistrates he appointed from the notables and the men of wealth, that is to say, from the pentacosio-medimni, or from the class called zeugitae, or from a third class of so-called knights or cavalry. The fourth class were laborers who had no share in any magistracy.

Mere legislators were Zaleucus, who gave laws to the Epizephyrian Locrians, and Charondas, who legislated for his own city of Catana, and for the other Chalcidian cities in Italy and Sicily. Some people attempt to make out that Onomacritus was the first person who had any special skill in legislation, and that he, although a Locrian by birth, was trained in Crete, where he lived in the exercise of his prophetic art; that Thales was his companion, and that Lycurgus and Zaleucus were disciples of Thales, as Charondas was of Zaleucus. But their account is quite inconsistent with chronology.

There was also Philolaus, the Corinthian, who gave laws to the Thebans. This Philolaus was one of the family of the Bacchiadae, and a lover of Diocles, the Olympic victor, who left Corinth in horror of the incestuous passion which his mother Halcyone had conceived for him, and retired to Thebes, where the two friends together ended their days. The inhabitants still point out their tombs, which are in full view of one another, but one is visible from the Corinthian territory, the other not. Tradition says the two friends arranged them thus, Diocles out of horror at his misfortunes, so that the land of Corinth might not be visible from his tomb; Philolaus that it might. This is the reason why they settled at Thebes, and so Philolaus legislated for the Thebans, and, besides some other enactments, gave them laws about the procreation of children, which they call the 'Laws of Adoption.' These laws were peculiar to him, and were intended to preserve the number of the lots.

In the legislation of Charondas there is nothing remarkable, except the suits against false witnesses. He is the first who instituted denunciation for perjury. His laws are more exact and more precisely expressed than even those of our modern legislators.

(Characteristic of Phaleas is the equalization of property; of Plato, the community of women, children, and property, the common meals of women, and the law about drinking, that the sober shall be masters of the feast; also the training of soldiers to acquire by practice equal skill with both hands, so that one should be as useful as the other.)

Draco has left laws, but he adapted them to a constitution which already existed, and there is no peculiarity in them which is worth mentioning, except the greatness and severity of the punishments.

Pittacus, too, was only a lawgiver, and not the author of a constitution; he has a law which is peculiar to him, that, if a drunken man do something wrong, he shall be more heavily punished than if he were sober; he looked not to the excuse which might be offered for the drunkard, but only to expediency, for drunken more often than sober people commit acts of violence.

Androdamas of Rhegium gave laws to the Chalcidians of Thrace. Some of them relate to homicide, and to heiresses; but there is nothing remarkable in them.

And here let us conclude our inquiry into the various constitutions which either actually exist, or have been devised by theorists.

BOOK III

I

HE who would inquire into the essence and attributes of various kinds of governments must first of all determine 'What is a state?' At present this is a disputed question. Some say that the state has done a certain act; others, no, not the state, but the oligarchy or the tyrant. And the legislator or statesman is concerned entirely with the state; a constitution or government being an arrangement of the inhabitants of a state. But a state is composite, like any other whole made up of many parts; these are the citizens, who compose it. It is evident, therefore, that we must begin by asking, Who is the citizen, and what is the meaning of the term? For here again there may be a difference of opinion. He who is a citizen in a democracy will often not be a citizen in an oligarchy. Leaving out of consideration those who have been made citizens, or who have obtained the name of citizen any other accidental manner, we may say, first, that a citizen is not a citizen because he lives in a certain place, for resident aliens and slaves share in the place; nor is he a citizen who has no legal right except that of suing and being sued; for this right may be enjoyed under the provisions of a treaty. Nay, resident aliens in many places do not possess even such rights completely, for they are obliged to have a patron, so that they do but imperfectly participate in citizenship, and we call them citizens only in a qualified sense, as we might apply the term to children who are too young to be on the register, or to old men who have been relieved from state duties. Of these we do not say quite simply that they are citizens, but add in the one case that they are not of age, and in the other, that they are past the age, or something of that sort; the precise expression is immaterial, for our meaning is clear. Similar difficulties to those which I have mentioned may be raised and answered about deprived citizens and about exiles. But the citizen whom we are seeking to define is a citizen in the strictest sense, against whom no such exception can be taken, and his special characteristic is that he shares in the administration of justice, and in offices. Now of offices some are discontinuous, and the same persons are not allowed to hold them twice, or can only hold them after a fixed interval; others have no limit of time — for example, the office of a dicast or ecclesiast. It may, indeed, be argued that these are not magistrates at all, and that their functions give them no share in the government. But surely it is ridiculous to say that those who have the power do not govern. Let us not dwell further upon this, which is a purely verbal question; what we want is a common term including both dicast and ecclesiast. Let us, for the sake of distinction, call it 'indefinite office,' and we will assume that those who share in such office are citizens. This is the most comprehensive definition of a citizen, and best suits all those who are generally so called.

But we must not forget that things of which the underlying principles differ in kind, one of them being first, another second, another third, have, when regarded in this relation, nothing, or hardly anything, worth mentioning in common. Now we see that governments differ in kind, and that some of them are prior and that others are posterior; those which are faulty or perverted are necessarily posterior to those which are perfect. (What we mean by perversion will be hereafter explained.) The citizen then of necessity differs under each form of government; and our definition is best adapted to the citizen of a democracy; but not necessarily to other states. For in some states the people are not acknowledged, nor have they any regular assembly, but only extraordinary ones; and suits are distributed by sections among the magistrates. At Lacedaemon, for instance, the Ephors determine suits about contracts, which they distribute among themselves, while the elders are judges of homicide, and other causes are decided by other magistrates. A similar principle prevails at Carthage; there certain magistrates decide all causes. We may, indeed, modify our definition of the citizen so as to include these states. In them it is the holder of a definite, not of an indefinite office, who legislates and judges, and to some or all such holders of definite offices is reserved the right of deliberating or judging about some things or about all things. The conception of the citizen now begins to clear up.

He who has the power to take part in the deliberative or judicial administration of any state is said by us to be a citizens of that state; and, speaking generally, a state is a body of citizens sufficing for the purposes of life.

II

But in practice a citizen is defined to be one of whom both the parents are citizens; others insist on going further back; say to two or three or more ancestors. This is a short and practical definition but there are some who raise the further question: How this third or fourth ancestor came to be a citizen? Gorgias of Leontini, partly because he was in a difficulty, partly in irony, said — 'Mortars are what is made by the mortar-makers, and the citizens of Larissa are those who are made by the magistrates; for it is their trade to make Larissaeans.' Yet the question is really simple, for, if according to the definition just given they shared in the government, they were citizens. This is a better definition than the other. For the words, 'born of a father or mother who is a citizen,' cannot possibly apply to the first inhabitants or founders of a state.

There is a greater difficulty in the case of those who have been made citizens after a revolution, as by Cleisthenes at Athens after the expulsion of the tyrants, for he enrolled in tribes many metics, both strangers and slaves. The doubt in these cases is, not who is, but whether he who is ought to be a citizen; and there will still be a furthering the state, whether a certain act is or is not an act of the state; for what ought not to be is what is false. Now, there are some who hold office, and yet ought not to hold office, whom we describe as ruling, but ruling unjustly. And the citizen was defined by the fact of his holding some kind of rule or office — he who holds a judicial or legislative office fulfills our definition of a citizen. It is evident, therefore, that the citizens about whom the doubt has arisen must be called citizens.

III

Whether they ought to be so or not is a question which is bound up with the previous inquiry. For a parallel question is raised respecting the state, whether a certain act is or is not an act of the state; for example, in the transition from an oligarchy or a tyranny to a democracy. In such cases persons refuse to fulfill their contracts or any other obligations, on the ground that the tyrant, and not the state, contracted them; they argue that some constitutions are established by force, and not for the sake of the common good. But this would apply equally to democracies, for they too may be founded on violence, and then the acts of the democracy will be neither more nor less acts of the state in question than those of an oligarchy or of a tyranny. This question runs up into another: on what principle shall we ever say that the state is the same, or different? It would be a very superficial view which considered only the place and the inhabitants (for the soil and the population may be separated, and some of the inhabitants may live in one place and some in another). This, however, is not a very serious difficulty; we need only remark that the word 'state' is ambiguous.

It is further asked: When are men, living in the same place, to be regarded as a single city — what is the limit? Certainly not the wall of the city, for you might surround all Peloponnesus with a wall. Like this, we may say, is Babylon, and every city that has the compass of a nation rather than a city; Babylon, they say, had been taken for three days before some part of the inhabitants became aware of the fact. This difficulty may, however, with advantage be deferred to another occasion; the statesman has to consider the size of the state, and whether it should consist of more than one nation or not.

Again, shall we say that while the race of inhabitants, as well as their place of abode, remain the same, the city is also the same, although the citizens are always dying and being born, as we call rivers and fountains the same, although the water is always flowing away and coming again Or shall we say that the generations of men, like the rivers, are the same, but that the state changes? For, since the state is a partnership, and is a partnership of citizens in a constitution, when the form of government changes, and becomes different, then it may be supposed that the state is no longer the same, just as a tragic differs from a comic chorus, although the members of both may be identical. And in this manner we speak of every union or composition of elements as different when the form of their composition alters; for example, a scale containing the same sounds is said to be different, accordingly as the Dorian or the Phrygian mode is employed. And if this is true it is evident that the sameness of the state consists chiefly in the sameness of the constitution, and it may be called or not called by the same name, whether the inhabitants are the same or entirely different. It is quite another question, whether a state ought or ought not to fulfill engagements when the form of government changes.

IV

There is a point nearly allied to the preceding: Whether the virtue of a good man and a good citizen is the same or not. But, before entering on this discussion, we must certainly first obtain some general notion of the virtue of the citizen. Like the sailor, the citizen is a member of a community. Now, sailors have different functions, for one of them is a rower, another a pilot, and a third a look-out man, a fourth is described by some similar term; and while the precise definition of each individual's virtue applies exclusively to him, there is, at the same time, a common definition applicable to them all. For they have all of them a common object, which is safety in navigation. Similarly, one citizen differs from another, but the salvation of the community is the common business of them all. This community is the constitution; the virtue of the citizen must therefore be relative to the constitution of which he is a member. If, then, there are many forms of government, it is evident that there is not one single virtue of the good citizen which is perfect virtue. But we say that the good man is he who has one single virtue which is perfect virtue. Hence it is evident that the good citizen need not of necessity possess the virtue which makes a good man.

The same question may also be approached by another road, from a consideration of the best constitution. If the state cannot be entirely composed of good men, and yet each citizen is expected to do his own business well, and must therefore have virtue, still inasmuch as all the citizens cannot be alike, the virtue of the citizen and of the good man cannot coincide. All must have the virtue of the good citizen — thus, and thus only, can the state be perfect; but they will not have the virtue of a good man, unless we assume that in the good state all the citizens must be good.

Again, the state, as composed of unlikes, may be compared to the living being: as the first elements into which a living being is resolved are soul and body, as soul is made up of rational principle and appetite, the family of husband and wife, property of master and slave, so of all these, as well as other dissimilar elements, the state is composed; and, therefore, the virtue of all the citizens cannot possibly be the same, any more than the excellence of the leader of a chorus is the same as that of the performer who stands by his side. I have said enough to show why the two kinds of virtue cannot be absolutely and always the same.

But will there then be no case in which the virtue of the good citizen and the virtue of the good man coincide? To this we answer that the good ruler is a good and wise man, and that he who would be a statesman must be a wise man. And some persons say that even the education of the ruler should be of a special kind; for are not the children of kings instructed in riding and military exercises? As Euripides says:

"No subtle arts for me, but what the state requires."

As though there were a special education needed by a ruler. If then the virtue of a good ruler is the same as that of a good man, and we assume further that the subject is a citizen as well as the ruler, the virtue of the good citizen and the virtue of the good man cannot be absolutely the same, although in some cases they may; for the virtue of a ruler differs from that of a citizen. It was the sense of this difference which made Jason say that 'he felt hungry when he was not a tyrant,' meaning that he could not endure to live in a private station. But, on the other hand, it may be argued that men are praised for knowing both how to rule and how to obey, and he is said to be a citizen of approved virtue who is able to do both. Now if we suppose the virtue of a good man to be that which rules, and the virtue of the citizen to include ruling and obeying, it cannot be said that they are equally worthy of praise. Since, then, it is sometimes thought that the ruler and the ruled must learn different things and not the same, but that the citizen must know and share in them both, the inference is obvious. There is, indeed, the rule of a master, which is concerned with menial offices — the master need not know how to perform these, but may employ others in the execution of them: the other would be degrading; and by the other I mean the power actually to do menial duties, which vary much in character and are executed by various classes of slaves, such, for example, as handicraftsmen, who, as their name signifies, live by the labor of their hands: under these the mechanic is included. Hence in ancient times, and among some nations, the working classes had no share in the government — a privilege which they only acquired under the extreme democracy. Certainly the good man and the statesman and the good citizen ought not to learn the crafts of inferiors except for their own occasional use; if they habitually practice them, there will cease to be a distinction between master and slave.

This is not the rule of which we are speaking; but there is a rule of another kind, which is exercised over freemen and equals by birth — a constitutional rule, which the ruler must learn by obeying, as he would learn the duties of a general of cavalry by being under the orders of a general of cavalry, or the duties of a general of infantry by being under the orders of a general of infantry, and by having had the command of a regiment and of a company. It has been well said that 'he who has never learned to obey cannot be a good commander.' The two are not the same, but the good citizen ought to be capable of both; he should know how to govern like a freeman, and how to obey like a freeman — these are the virtues of a citizen. And, although the temperance and justice of a ruler are distinct from those of a subject, the virtue of a good man will include both; for the virtue of the good man who is free and also a subject, e.g., his justice, will not be one but will comprise distinct kinds, the one qualifying him to rule, the other to obey, and differing as the temperance and courage of men and women differ. For a man would be thought a coward if he had no more courage than a courageous woman, and a woman would be thought loquacious if she imposed no more restraint on her conversation than the good man; and indeed their part in the management of the household is different, for the duty of the one is to acquire, and of the other to preserve. Practical wisdom only is characteristic of the ruler: it would seem that all other virtues must equally belong to ruler and subject. The virtue of the subject is certainly not wisdom, but only true opinion; he may be compared to the maker of the flute, while his master is like the flute-player or user of the flute.

From these considerations may be gathered the answer to the question, whether the virtue of the good man is the same as that of the good citizen, or different, and how far the same, and how far different.

V

There still remains one more question about the citizen: Is he only a true citizen who has a share of office, or is the mechanic to be included? If they who hold no office are to be deemed citizens, not every citizen can have this virtue of ruling and obeying; for this man is a citizen And if none of the lower class are citizens, in which part of the state are they to be placed? For they are not resident aliens, and they are not foreigners. May we not reply, that as far as this objection goes there is no more absurdity in excluding them than in excluding slaves and freedmen from any of the above-mentioned classes? It must be admitted that we cannot consider all those to be citizens who are necessary to the existence of the state; for example, children are not citizen equally with grown-up men, who are citizens absolutely, but children, not being grown up, are only citizens on a certain assumption. Nay, in ancient times, and among some nations the artisan class were slaves or foreigners, and therefore the majority of them are so now. The best form of state will not admit them to citizenship; but if they are admitted, then our definition of the virtue of a citizen will not apply to every citizen nor to every free man as such, but only to those who are freed from necessary services. The necessary people are either slaves who minister to the wants of individuals, or mechanics and laborers who are the servants of the community. These reflections carried a little further will explain their position; and indeed what has been said already is of itself, when understood, explanation enough.

Since there are many forms of government there must be many varieties of citizen and especially of citizens who are subjects; so that under some governments the mechanic and the laborer will be citizens, but not in others, as, for example, in aristocracy or the so-called government of the best (if there be such an one), in which honors are given according to virtue and merit; for no man can practice virtue who is living the life of a mechanic or laborer. In oligarchies the qualification for office is high, and therefore no laborer can ever be a citizen; but a mechanic may, for an actual majority of them are rich. At Thebes there was a law that no man could hold office who had not retired from business for ten years. But in many states the law goes to the length of admitting aliens; for in some democracies a man is a citizen though his mother only be a citizen; and a similar principle is applied to illegitimate children; the law is relaxed when there is a dearth of population. But when the number of citizens increases, first the children of a male or a female slave are excluded; then those whose mothers only are citizens; and at last the right of citizenship is confined to those whose fathers and mothers are both citizens.

Hence, as is evident, there are different kinds of citizens; and he is a citizen in the highest sense who shares in the honors of the state. Compare Homer's words, 'like some dishonored stranger'; he who is excluded from the honors of the state is no better than an alien. But when his exclusion is concealed, then the object is that the privileged class may deceive their fellow inhabitants.

As to the question whether the virtue of the good man is the same as that of the good citizen, the considerations already adduced prove that in some states the good man and the good citizen are the same, and in others different. When they are the same it is not every citizen who is a good man, but only the statesman and those who have or may have, alone or in conjunction with others, the conduct of public affairs.

VI

Having determined these questions, we have next to consider whether there is only one form of government or many, and if many, what they are, and how many, and what are the differences between them.

A constitution is the arrangement of magistracies in a state, especially of the highest of all. The government is everywhere sovereign in the state, and the constitution is in fact the government. For example, in democracies the people are supreme, but in oligarchies, the few; and, therefore, we say that these two forms of government also are different: and so in other cases.

First, let us consider what is the purpose of a state, and how many forms of government there are by which human society is regulated. We have already said, in the first part of this treatise, when discussing household management and the rule of a master, that man is by nature a political animal. And therefore, men, even when they do not require one another's help, desire to live together; not but that they are also brought together by their common interests in proportion as they severally attain to any measure of well-being. This is certainly the chief end, both of individuals and of states. And also for the sake of mere life (in which there is possibly some noble element so long as the evils of existence do not greatly overbalance the good) mankind meet together and maintain the political community. And we all see that men cling to life even at the cost of enduring great misfortune, seeming to find in life a natural sweetness and happiness.

There is no difficulty in distinguishing the various kinds of authority; they have been often defined already in discussions outside the school. The rule of a master, although the slave by nature and the master by nature have in reality the same interests, is nevertheless exercised primarily with a view to the interest of the master, but accidentally considers the slave, since, if the slave perish, the rule of the master perishes with him. On the other hand, the government of a wife and children and of a household, which we have called household management, is exercised in the first instance for the good of the governed or for the common good of both parties, but essentially for the good of the governed, as we see to be the case in medicine, gymnastic, and the arts in general, which are only accidentally concerned with the good of the artists themselves. For there is no reason why the trainer may not sometimes practice gymnastics, and the helmsman is always one of the crew. The trainer or the helmsman considers the good of those committed to his care. But, when he is one of the persons taken care of, he accidentally participates in the advantage, for the helmsman is also a sailor, and the trainer becomes one of those in training. And so in politics: when the state is framed upon the principle of equality and likeness, the citizens think that they ought to hold office by turns. Formerly, as is natural, every one would take his turn of service; and then again, somebody else would look after his interest, just as he, while in office, had looked after theirs. But nowadays, for the sake of the advantage which is to be gained from the public revenues and from office, men want to be always in office. One might imagine that the rulers, being sickly, were only kept in health while they continued in office; in that case we may be sure that they would be hunting after places. The conclusion is evident: that governments which have a regard to the common interest are constituted in accordance with strict principles of justice, and are therefore true forms; but those which regard only the interest of the rulers are all defective and perverted forms, for they are despotic, whereas a state is a community of freemen.

VII

Having determined these points, we have next to consider how many forms of government there are, and what they are; and in the first place what are the true forms, for when they are determined the perversions of them will at once be apparent. The words constitution and government have the same meaning, and the government, which is the supreme authority in states, must be in the hands of one, or of a few, or of the many. The true forms of government, therefore, are those in which the one, or the few, or the many, govern with a view to the common interest; but governments which rule with a view to the private interest, whether of the one or of the few, or of the many, are perversions. For the members of a state, if they are truly citizens, ought to participate in its advantages. Of forms of government in which one rules, we call that which regards the common interests, kingship or royalty; that in which more than one, but not many, rule, aristocracy; and it is so called, either because the rulers are the best men, or because they have at heart the best interests of the state and of the citizens. But when the citizens at large administer the state for the common interest, the government is called by the generic name — a constitution. And there is a reason for this use of language. One man or a few may excel in virtue; but as the number increases it becomes more difficult for them to attain perfection in every kind of virtue, though they may in military virtue, for this is found in the masses. Hence in a constitutional government the fighting-men have the supreme power, and those who possess arms are the citizens.

Of the above-mentioned forms, the perversions are as follows: of royalty, tyranny; of aristocracy, oligarchy; of constitutional government, democracy. For tyranny is a kind of monarchy which has in view the interest of the monarch only; oligarchy has in view the interest of the wealthy; democracy, of the needy: none of them the common good of all.

VIII

But there are difficulties about these forms of government, and it will therefore be necessary to state a little more at length the nature of each of them. For he who would make a philosophical study of the various sciences, and does not regard practice only, ought not to overlook or omit anything, but to set forth the truth in every particular. Tyranny, as I was saying, is monarchy exercising the rule of a master over the political society; oligarchy is when men of property have the government in their hands; democracy, the opposite, when the indigent, and not the men of property, are the rulers. And here arises the first of our difficulties, and it relates to the distinction drawn. For democracy is said to be the government of the many. But what if the many are men of property and have the power in their hands? In like manner oligarchy is said to be the government of the few; but what if the poor are fewer than the rich, and have the power in their hands because they are stronger? In these cases the distinction which we have drawn between these different forms of government would no longer hold good.

Suppose, once more, that we add wealth to the few and poverty to the many, and name the governments accordingly — an oligarchy is said to be that in which the few and the wealthy, and a democracy that in which the many and the poor are the rulers — there will still be a difficulty. For, if the only forms of government are the ones already mentioned, how shall we describe those other governments also just mentioned by us, in which the rich are the more numerous and the poor are the fewer, and both govern in their respective states?

The argument seems to show that, whether in oligarchies or in democracies, the number of the governing body, whether the greater number, as in a democracy, or the smaller number, as in an oligarchy, is an accident due to the fact that the rich everywhere are few, and the poor numerous. But if so, there is a misapprehension of the causes of the difference between them. For the real difference between democracy and oligarchy is poverty and wealth. Wherever men rule by reason of their wealth, whether they be few or many, that is an oligarchy, and where the poor rule, that is a democracy. But as a fact the rich are few and the poor many; for few are well-to-do, whereas freedom is enjoyed by an, and wealth and freedom are the grounds on which the oligarchical and democratical parties respectively claim power in the state.

IX

Let us begin by considering the common definitions of oligarchy and democracy, and what is justice oligarchical and democratical. For all men cling to justice of some kind, but their conceptions are imperfect and they do not express the whole idea. For example, justice is thought by them to be, and is, equality, not.

however, for however, for but only for equals. And inequality is thought to be, and is, justice; neither is this for all, but only for unequals. When the persons are omitted, then men judge erroneously. The reason is that they are passing judgment on themselves, and most people are bad judges in their own case. And whereas justice implies a relation to persons as well as to things, and a just distribution, as I have already said in the Ethics, implies the same ratio between the persons and between the things, they agree about the equality of the things, but dispute about the equality of the persons, chiefly for the reason which I have just given — because they are bad judges in their own affairs; and secondly, because both the parties to the argument are speaking of a limited and partial justice, but imagine themselves to be speaking of absolute justice. For the one party, if they are unequal in one respect, for example wealth, consider themselves to be unequal in all; and the other party, if they are equal in one respect, for example free birth, consider themselves to be equal in all. But they leave out the capital point. For if men met and associated out of regard to wealth only, their share in the state would be proportioned to their property, and the oligarchical doctrine would then seem to carry the day. It would not be just that he who paid one mina should have the same share of a hundred minae, whether of the principal or of the profits, as he who paid the remaining ninety-nine. But a state exists for the sake of a good life, and not for the sake of life only: if life only were the object, slaves and brute animals might form a state, but they cannot, for they have no share in happiness or in a life of free choice. Nor does a state exist for the sake of alliance and security from injustice, nor yet for the sake of exchange and mutual intercourse; for then the Tyrrhenians and the Carthaginians, and all who have commercial treaties with one another, would be the citizens of one state. True, they have agreements about imports, and engagements that they will do no wrong to one another, and written articles of alliance. But there are no magistrates common to the contracting parties who will enforce their engagements; different states have each their own magistracies. Nor does one state take care that the citizens of the other are such as they ought to be, nor see that those who come under the terms of the treaty do no wrong or wickedness at an, but only that they do no injustice to one another. Whereas, those who care for good government take into consideration virtue and vice in states. Whence it may be further inferred that virtue must be the care of a state which is truly so called, and not merely enjoys the name: for without this end the community becomes a mere alliance which differs only in place from alliances of which the members live apart; and law is only a convention, 'a surety to one another of justice,' as the sophist Lycophron says, and has no real power to make the citizens

This is obvious; for suppose distinct places, such as Corinth and Megara, to be brought together so that their walls touched, still they would not be one city, not even if the citizens had the right to intermarry, which is one of the rights peculiarly characteristic of states. Again, if men dwelt at a distance from one another, but not so far off as to have no intercourse, and there were laws among them that they should not wrong each other in their exchanges, neither would this be a state. Let us suppose that one man is a carpenter, another a husbandman, another a shoemaker, and so on, and that their number is ten thousand: nevertheless, if they have nothing in common but exchange, alliance, and the like, that would not constitute a state. Why is this? Surely not because they are at a distance from one another: for even supposing that such a community were to meet in one place, but that each man had a house of his own, which was in a manner his state, and that they made alliance with one another, but only against evil-doers; still an accurate thinker would not deem this to be a state, if their intercourse with one another was of the same character after as before their union. It is clear then that a state is not a mere society, having a common place, established for the prevention of mutual crime and for the sake of exchange. These are conditions without which a state cannot exist; but all of them together do not constitute a state, which is a community of families and aggregations of families in well-being, for the sake of a perfect and self-sufficing life. Such a community can only be established among those who live in the same place and intermarry. Hence arise in cities family connections, brotherhoods, common sacrifices, amusements which draw men together. But these are created by friendship, for the will to live together is friendship. The end of the state is the good life, and these are the means towards it. And the state is the union of families and villages in a perfect and self-sufficing life, by which we mean a happy and honorable life.

Our conclusion, then, is that political society exists for the sake of noble actions, and not of mere companionship. Hence they who contribute most to such a society have a greater share in it than those who have the same or a greater freedom or nobility of birth but are inferior to them in political virtue; or than those who exceed them in wealth but are surpassed by them in virtue.

From what has been said it will be clearly seen that all the partisans of different forms of government speak of a part of justice only.

X

There is also a doubt as to what is to be the supreme power in the state: Is it the multitude? Or the wealthy? Or the good? Or the one best man? Or a tyrant? Any of these alternatives seems to involve disagreeable consequences. If the poor, for example, because they are more in number, divide among themselves the property of the rich — is not this unjust? No, by heaven (will be the reply), for the supreme authority justly willed it. But if this is not injustice, pray what is? Again, when in the first division all has been taken, and the majority divide anew the property of the minority, is it not evident, if this goes on, that they will ruin the state? Yet surely, virtue is not the ruin of those who possess her, nor is justice destructive of a state; and therefore this law of confiscation clearly cannot be just. If it were, all the acts of a tyrant must of necessity be just; for he only coerces other men by superior power, just as the multitude coerce the rich. But is it just then that the few and the wealthy should be the rulers? And what if they, in like manner, rob and plunder the people — is this just? if so, the other case will likewise be just. But there can be no doubt that all these things are wrong and unjust.

Then ought the good to rule and have supreme power? But in that case everybody else, being excluded from power, will be dishonored. For the offices of a state are posts of honor; and if one set of men always holds them, the rest must be deprived of them. Then will it be well that the one best man should rule? Nay, that is still more oligarchical, for the number of those who are dishonored is thereby increased. Some one may say that it is bad in any case for a man, subject as he is to all the accidents of human passion, to have the supreme power, rather than the law. But what if the law itself be democratical or oligarchical, how will that help us out of our difficulties? Not at all; the same consequences will follow.

XI

Most of these questions may be reserved for another occasion. The principle that the multitude ought to be supreme rather than the few best is one that is maintained, and, though not free from difficulty, yet seems to contain an element of truth. For the many, of whom each individual is but an ordinary person, when they meet together may very likely be better than the few good, if regarded not individually but collectively, just as a feast to which many contribute is better than a dinner provided out of a single purse. For each individual among the many has a share of virtue and prudence, and when they meet together, they become in a manner one man, who has many feet, and hands, and senses; that is a figure of their mind and disposition. Hence the many are better judges than a single man of music and poetry; for some understand one part, and some another, and among them they understand the whole. There is a similar combination of qualities in good men, who differ from any individual of the many, as the beautiful are said to differ from those who are not beautiful, and works of art from realities, because in them the scattered elements are combined, although, if taken separately, the eye of one person or some other feature in another person would be fairer than in the picture. Whether this principle can apply to every democracy, and to all bodies of men, is not clear. Or rather, by heaven, in some cases it is impossible of application; for the argument would equally hold about brutes; and wherein, it will be asked, do some men differ from brutes? But there may be bodies of men about whom our statement is nevertheless true. And if so, the difficulty which has been already raised, and also another which is akin to it — viz., what power should be assigned to the mass of freemen and citizens, who are not rich and have no personal merit — are both solved. There is still a danger in allowing them to share the great offices of state, for their folly will lead them into error, and their dishonesty into crime. But there is a danger also in not letting them share, for a state in which many poor men are excluded from office will necessarily be full of enemies. The only way of escape is to assign to them some deliberative and judicial functions. For this reason Solon and certain other legislators give them the power of electing to offices, and of calling the magistrates to account, but they do not allow them to hold office singly.

When they meet together their perceptions are quite good enough, and combined with the better class they are useful to the state (just as impure food when mixed with what is pure sometimes makes the entire mass more wholesome than a small quantity of the pure would be), but each individual, left to himself, forms an imperfect judgment. On the other hand, the popular form of government involves certain difficulties. In the first place, it might be objected that he who can judge of the healing of a sick man would be one who could himself heal his disease, and make him whole — that is, in other words, the physician; and so in all professions and arts. As, then, the physician ought to be called to account by physicians, so ought men in general to be called to account by their peers. But physicians are of three kinds: there is the ordinary practitioner, and there is the physician of the higher class, and thirdly the intelligent man who has studied the art: in all arts there is such a class; and we attribute the power of judging to them quite as much as to professors of the art. Secondly, does not the same principle apply to elections? For a right election can only be made by those who have knowledge; those who know geometry, for example, will choose a geometrician rightly, and those who know how to steer, a pilot; and, even if there be some occupations and arts in which private persons share in the ability to choose, they certainly cannot choose better than those who know. So that, according to this argument, neither the election of magistrates, nor the calling of them to account, should be entrusted to the many. Yet possibly these objections are to a great extent met by our old answer, that if the people are not utterly degraded, although individually they may be worse judges than those who have special knowledge — as a body they are as good or better. Moreover, there are some arts whose products are not judged of solely, or best, by the artists themselves, namely those arts whose products are recognized even by those who do not possess the art; for example, the knowledge of the house is not limited to the builder only; the user, or, in other words, the master, of the house will be even a better judge than the builder, just as the pilot will judge better of a rudder than the carpenter, and the guest will judge better of a feast than the cook.

This difficulty seems now to be sufficiently answered, but there is another akin to it. That inferior persons should have authority in greater matters than the good would appear to be a strange thing, yet the election and calling to account of the magistrates is the greatest of all. And these, as I was saying, are functions which in some states are assigned to the people, for the assembly is supreme in all such matters. Yet persons of any age, and having but a small property qualification, sit in the assembly and deliberate and judge, although for the great officers of state, such as treasurers and generals, a high qualification is required. This difficulty may be solved in the same manner as the preceding, and the present practice of democracies may be really defensible. For the power does not reside in the dicast, or senator, or ecclesiast, but in the court, and the senate, and the assembly, of which individual senators, or ecclesiasts, or dicasts, are only parts or members. And for this reason the many may claim to have a higher authority than the few; for the people, and the senate, and the courts consist of many persons, and their property collectively is greater than the property of one or of a few individuals holding great offices. But enough of this.

The discussion of the first question shows nothing so clearly as that laws, when good, should be supreme; and that the magistrate or magistrates should regulate those matters only on which the laws are unable to speak with precision owing to the difficulty of any general principle embracing all particulars. But what are good laws has not yet been clearly explained; the old difficulty remains. The goodness or badness, justice or injustice, of laws varies of necessity with the constitutions of states. This, however, is clear, that the laws must be adapted to the constitutions. But if so, true forms of government will of necessity have just laws, and perverted forms of government will have unjust laws.

XII

In all sciences and arts the end is a good, and the greatest good and in the highest degree a good in the most authoritative of all — this is the political science of which the good is justice, in other words, the common interest. All men think justice to be a sort of equality; and to a certain extent they agree in the philosophical distinctions which have been laid down by us about Ethics. For they admit that justice is a thing and has a relation to persons, and that equals ought to have equality. But there still remains a question: equality or inequality of what? Here is a difficulty which calls for political speculation. For very likely some persons will say that offices of state ought to be unequally distributed according to superior excellence, in whatever respect, of the citizen, although there is no other difference between him and the rest of the community; for that those who differ in any one respect have different rights and claims. But, surely, if this is true, the complexion or height of a man, or any other advantage, will be a reason for his obtaining a greater share of political rights. The error here lies upon the surface, and may be illustrated from the other arts and sciences. When a number of flute players are equal in their art, there is no reason why those of them who are better born should have better flutes given to them; for they will not play any better on the flute, and the superior instrument should be reserved for him who is the superior artist. If what I am saying is still obscure, it will be made clearer as we proceed. For if there were a superior flute-player who was far inferior in birth and beauty, although either of these may be a greater good than the art of flute-playing, and may excel flute-playing in a greater ratio than he excels the others in his art, still he ought to have the best flutes given to him, unless the advantages of wealth and birth contribute to excellence in flute-playing, which they do not. Moreover, upon this principle any good may be compared with any other. For if a given height may be measured wealth and against freedom, height in general may be so measured. Thus if A excels in height more than B in virtue, even if virtue in general excels height still more, all goods will be commensurable; for if a certain amount is better than some other, it is clear that some other will be equal. But since no such comparison can be made, it is evident that there is good reason why in politics men do not ground their claim to office on every sort of inequality any more than in the arts. For if some be slow, and others swift, that is no reason why the one should have little and the others much; it is in gymnastics contests that such excellence is rewarded. Whereas the rival claims of candidates for office can only be based on the possession of elements which enter into the composition of a state. And therefore the noble, or free-born, or rich, may with good reason claim office; for holders of offices must be freemen and taxpayers: a state can be no more composed entirely of poor men than entirely of slaves. But if wealth and freedom are necessary elements, justice and valor are equally so; for without the former qualities a state cannot exist at all, without the latter not well.

XIII

If the existence of the state is alone to be considered, then it would seem that all, or some at least, of these claims are just; but, if we take into account a good life, then, as I have already said, education and virtue have superior claims. As, however, those who are equal in one thing ought not to have an equal share in all, nor those who are unequal in one thing to have an unequal share in all, it is certain that all forms of government which rest on either of these principles are perversions. All men have a claim in a certain sense, as I have already admitted, but all have not an absolute claim. The rich claim because they have a greater share in the land, and land is the common element of the state; also they are generally more trustworthy in contracts. The free claim under the same title as the noble; for they are nearly akin. For the noble are citizens in a truer sense than the ignoble, and good birth is always valued in a man's own home and country. Another reason is, that those who are sprung from better ancestors are likely to be better men, for nobility is excellence of race. Virtue, too, may be truly said to have a claim, for justice has been acknowledged by us to be a social virtue, and it implies all others. Again, the many may urge their claim against the few; for, when taken collectively, and compared with the few, they are stronger and richer and better. But, what if the good, the rich, the noble, and the other classes who make up a state, are all living together in the same city, Will there, or will there not, be any doubt who shall rule? No doubt at all in determining who ought to rule in each of the above-mentioned forms of government. For states are characterized by differences in their governing bodies-one of them has a government of the rich, another of the virtuous, and so on. But a difficulty arises when all these elements co-exist. How are we to decide? Suppose the virtuous to be very few in number: may we consider their numbers in relation to their duties, and ask whether they are enough to administer the state, or so many as will make up a state? Objections may be urged against all the aspirants to political power. For those who found their claims on wealth or family might be thought to have no basis of justice; on this principle, if any one person were richer than all the rest, it is clear that he ought to be ruler of them. In like manner he who is very distinguished by his birth ought to have the superiority over all those who claim on the ground that they are freeborn. In an aristocracy, or government of the best, a like difficulty occurs about virtue; for if one citizen be better than the other members of the government, however good they may be, he too, upon the same principle of justice, should rule over them. And if the people are to be supreme because they are stronger than the few, then if one man, or more than one, but not a majority, is stronger than the many, they ought to rule, and not the many.

All these considerations appear to show that none of the principles on which men claim to rule and to hold all other men in subjection to them are strictly right. To those who claim to be masters of the government on the ground of their virtue or their wealth, the many might fairly answer that they themselves are often better and richer than the few — I do not say individually, but collectively. And another ingenious objection which is sometimes put forward may be met in a similar manner. Some persons doubt whether the legislator who desires to make the justest laws ought to legislate with a view to the good of the higher classes or of the many, when the case which we have mentioned occurs. Now what is just or right is to be interpreted in the sense of 'what is equal'; and that which is right in the sense of being equal is to be considered with reference to the advantage of the state, and the common good of the citizens. And a citizen is one who shares in governing and being governed. He differs under different forms of government, but in the best state he is one who is able and willing to be governed and to govern with a view to the life of virtue.

If, however, there be some one person, or more than one, although not enough to make up the full complement of a state, whose virtue is so pre-eminent that the virtues or the political capacity of all the rest admit of no comparison with his or theirs, he or they can be no longer regarded as part of a state; for justice will not be done to the superior, if he is reckoned only as the equal of those who are so far inferior to him in virtue and in political capacity. Such an one may truly be deemed a God among men. Hence we see that legislation is necessarily concerned only with those who are equal in birth and in capacity; and that for men of pre-eminent virtue there is no law — they are themselves a law. Any would be ridiculous who attempted to make laws for them: they would probably retort what, in the fable of Antisthenes, the lions said to the hares, when in the council of the beasts the latter began haranguing and claiming equality for all. And for this reason democratic states have instituted ostracism; equality is above all things their aim, and therefore they ostracized and banished from the city for a time those who seemed to predominate too much through their wealth, or the number of their friends, or through any other political influence. Mythology tells us that the Argonauts left Heracles behind for a similar reason; the ship Argo would not take him because she feared that he would have been too much for the rest of the crew. Wherefore those who denounce tyranny and blame the counsel which Periander gave to Thrasybulus cannot be held altogether just in their censure. The story is that Periander, when the herald was sent to ask counsel of him, said nothing, but only cut off the tallest ears of corn till he had brought the field to a level. The herald did not know the meaning of the action, but came and reported what he had seen to Thrasybulus, who understood that he was to cut off the principal men in the state; and this is a policy not only expedient for tyrants or in practice confined to them, but equally necessary in oligarchies and democracies. Ostracism is a measure of the same kind, which acts by disabling and banishing the most prominent citizens. Great powers do the same to whole cities and nations, as the Athenians did to the Samians, Chians, and Lesbians; no sooner had they obtained a firm grasp of the empire, than they humbled their allies contrary to treaty; and the Persian king has repeatedly crushed the Medes, Babylonians, and other nations, when their spirit has been stirred by the recollection of their former greatness.

The problem is a universal one, and equally concerns all forms of government, true as well as false; for, although perverted forms with a view to their own interests may adopt this policy, those which seek the common interest do so likewise. The same thing may be observed in the arts and sciences; for the painter will not allow the figure to have a foot which, however beautiful, is not in proportion, nor will the shipbuilder allow the stem or any other part of the vessel to be unduly large, any more than the chorus-master will allow any one who sings louder or better than all the rest to sing in the choir. Monarchs, too, may practice compulsion and still live in harmony with their cities, if their own government is for the interest of the state. Hence where there is an acknowledged superiority the argument in favor of ostracism is based upon a kind of political justice. It would certainly be better that the legislator should from the first so order his state as to have no need of such a remedy. But if the need arises, the next best thing is that he should endeavor to correct the evil by this or some similar measure. The principle, however, has not been fairly applied in states; for, instead of looking to the good of their own constitution, they have used ostracism for factious purposes. It is true that under perverted forms of government, and from their special point of view, such a measure is just and expedient, but it is also clear that it is not absolutely just. In the perfect state there would be great doubts about the use of it, not when applied to excess in strength, wealth, popularity, or the like, but when used against some one who is pre-eminent in virtue — what is to be done with him? Mankind will not say that such an one is to be expelled and exiled; on the other hand, he ought not to be a subject — that would be as if mankind should claim to rule over Zeus, dividing his offices among them. The only alternative is that all should joyfully obey such a ruler, according to what seems to be the order of nature, and that men like him should be kings in their state for life.

XIV

The preceding discussion, by a natural transition, leads to the consideration of royalty, which we admit to be one of the true forms of government. Let us see whether in order to be well governed a state or country should be under the rule of a king or under some other form of government; and whether monarchy, although good for some, may not be bad for others. But first we must determine whether there is one species of royalty or many. It is easy to see that there are many, and that the manner of government is not the same in all of them.

Of royalties according to law, (1) the Lacedaemonian is thought to answer best to the true pattern; but there the royal power is not absolute, except when the kings go on an expedition, and then they take the command. Matters of religion are likewise committed to them. The kingly office is in truth a kind of generalship, irresponsible and perpetual. The king has not the power of life and death, except in a specified case, as for instance, in ancient times, he had it when upon a campaign, by right of force. This custom is described in Homer. For Agamemnon is patient when he is attacked in the assembly, but when the army goes out to battle he has the power even of life and death. Does he not say — 'When I find a man skulking apart from the battle, nothing shall save him from the dogs and vultures, for in my hands is death'?

This, then, is one form of royalty — a generalship for life: and of such royalties some are hereditary and others elective.

(2) There is another sort of monarchy not uncommon among the barbarians, which nearly resembles tyranny. But this is both legal and hereditary. For barbarians, being more servile in character than Hellenes, and Asiadics than Europeans, do not rebel against a despotic government. Such royalties have the nature of tyrannies because the people are by nature slaves; but there is no danger of their being overthrown, for they are hereditary and legal. Wherefore also their guards are such as a king and not such as a tyrant would employ, that is to say, they are composed of citizens, whereas the guards of tyrants are mercenaries. For kings rule according to law over voluntary subjects, but tyrants over involuntary; and the one are guarded by their fellow-citizens the others are guarded against them.

These are two forms of monarchy, and there was a third (3) which existed in ancient Hellas, called an Aesymnetia or dictatorship. This may be defined generally as an elective tyranny, which, like the barbarian monarchy, is legal, but differs from it in not being hereditary. Sometimes the office was held for life, sometimes for a term of years, or until certain duties had been performed. For example, the Mytilenaeans elected Pittacus leader against the exiles, who were headed by Antimenides and Alcaeus the poet. And Alcaeus himself shows in one of his banquet odes that they chose Pittacus tyrant, for he reproaches his fellow-citizens for 'having made the low-born Pittacus tyrant of the spiritless and ill-fated city, with one voice shouting his praises.'

These forms of government have always had the character of tyrannies, because they possess despotic power; but inasmuch as they are elective and acquiesced in by their subjects, they are kingly.

(4) There is a fourth species of kingly rule — that of the heroic times — which was hereditary and legal, and was exercised over willing subjects. For the first chiefs

were benefactors of the people in arts or arms; they either gathered them into a community, or procured land for them; and thus they became kings of voluntary subjects, and their power was inherited by their descendants. They took the command in war and presided over the sacrifices, except those which required a priest. They also decided causes either with or without an oath; and when they swore, the form of the oath was the stretching out of their sceptre. In ancient times their power extended continuously to all things whatsoever, in city and country, as well as in foreign parts; but at a later date they relinquished several of these privileges, and others the people took from them, until in some states nothing was left to them but the sacrifices; and where they retained more of the reality they had only the right of leadership in war beyond the border.

These, then, are the four kinds of royalty. First the monarchy of the heroic ages; this was exercised over voluntary subjects, but limited to certain functions; the king was a general and a judge, and had the control of religion The second is that of the barbarians, which is a hereditary despotic government in accordance with law. A third is the power of the so-called Aesynmete or Dictator; this is an elective tyranny. The fourth is the Lacedaemonian, which is in fact a generalship, hereditary and perpetual. These four forms differ from one another in the manner which I have described.

(5) There is a fifth form of kingly rule in which one has the disposal of all, just as each nation or each state has the disposal of public matters; this form corresponds to the control of a household. For as household management is the kingly rule of a house, so kingly rule is the household management of a city, or of a nation, or of many nations.

XV

Of these forms we need only consider two, the Lacedaemonian and the absolute royalty; for most of the others lie in a region between them, having less power than the last, and more than the first. Thus the inquiry is reduced to two points: first, is it advantageous to the state that there should be a perpetual general, and if so, should the office be confined to one family, or open to the citizens in turn? Secondly, is it well that a single man should have the supreme power in all things? The first question falls under the head of laws rather than of constitutions; for perpetual generalship might equally exist under any form of government, so that this matter may be dismissed for the present. The other kind of royalty is a sort of constitution; this we have now to consider, and briefly to run over the difficulties involved in it. We will begin by inquiring whether it is more advantageous to be ruled by the best man or by the best laws.

The advocates of royalty maintain that the laws speak only in general terms, and cannot provide for circumstances; and that for any science to abide by written rules is absurd. In Egypt the physician is allowed to alter his treatment after the fourth day, but if sooner, he takes the risk. Hence it is clear that a government acting according to written laws is plainly not the best. Yet surely the ruler cannot dispense with the general principle which exists in law; and this is a better ruler which is free from passion than that in which it is innate. Whereas the law is passionless, passion must ever sway the heart of man. Yes, it may be replied, but then on the other hand an individual will be better able to deliberate in particular cases.

The best man, then, must legislate, and laws must be passed, but these laws will have no authority when they miss the mark, though in all other cases retaining their authority. But when the law cannot determine a point at all, or not well, should the one best man or should all decide? According to our present practice assemblies meet, sit in judgment, deliberate, and decide, and their judgments an relate to individual cases. Now any member of the assembly, taken separately, is certainly inferior to the wise man. But the state is made up of many individuals. And as a feast to which all the guests contribute is better than a banquet furnished by a single man, so a multitude is a better judge of many things than any individual.

Again, the many are more incorruptible than the few; they are like the greater quantity of water which is less easily corrupted than a little. The individual is liable to be overcome by anger or by some other passion, and then his judgment is necessarily perverted; but it is hardly to be supposed that a great number of persons would all get into a passion and go wrong at the same moment. Let us assume that they are the freemen, and that they never act in violation of the law, but fill up the gaps which the law is obliged to leave. Or, if such virtue is scarcely attainable by the multitude, we need only suppose that the majority are good men and good citizens, and ask which will be the more incorruptible, the one good ruler, or the many who are all good? Will not the many? But, you will say, there may be parties among them, whereas the one man is not divided against himself. To which we may answer that their character is as good as his. If we call the rule of many men, who are all of them good, aristocracy, and the rule of one man royalty, then aristocracy will be better for states than royalty, whether the government is supported by force or not, provided only that a number of men equal in virtue can be found.

The first governments were kingships, probably for this reason, because of old, when cities were small, men of eminent virtue were few. Further, they were made kings because they were benefactors, and benefits can only be bestowed by good men. But when many persons equal in merit arose, no longer enduring the pre-eminence of one, they desired to have a commonwealth, and set up a constitution. The ruling class soon deteriorated and enriched themselves out of the public treasury; riches became the path to honor, and so oligarchies naturally grew up. These passed into tyrannies and tyrannies into democracies; for love of gain in the ruling classes was always tending to diminish their number, and so to strengthen the masses, who in the end set upon their masters and established democracies. Since cities have increased in size, no other form of government appears to be any longer even easy to establish.

Even supposing the principle to be maintained that kingly power is the best thing for states, how about the family of the king? Are his children to succeed him? If they are no better than anybody else, that will be mischievous. But, says the lover of royalty, the king, though he might, will not hand on his power to his children. That, however, is hardly to be expected, and is too much to ask of human nature. There is also a difficulty about the force which he is to employ; should a king have guards about him by whose aid he may be able to coerce the refractory? If not, how will he administer his kingdom? Even if he be the lawful sovereign who does nothing arbitrarily or contrary to law, still he must have some force wherewith to maintain the law. In the case of a limited monarchy there is not much difficulty in answering this question; the king must have such force as will be more than a match for one or more individuals, but not so great as that of the people. The ancients observe this principle when they have guards to any one whom they appointed dictator or tyrant. Thus, when Dionysius asked the Syracusans to allow him guards, somebody advised that they should give him only such a number.

XVI

At this place in the discussion there impends the inquiry respecting the king who acts solely according to his own will he has now to be considered. The so-called limited monarchy, or kingship according to law, as I have already remarked, is not a distinct form of government, for under all governments, as, for example, in a democracy or aristocracy, there may be a general holding office for life, and one person is often made supreme over the administration of a state. A magistracy of this kind exists at Epidamnus, and also at Opus, but in the latter city has a more limited power. Now, absolute monarchy, or the arbitrary rule of a sovereign over an the citizens, in a city which consists of equals, is thought by some to be quite contrary to nature; it is argued that those who are by nature equals must have the same natural right and worth, and that for unequals to have an equal share, or for equals to have an uneven share, in the offices of state, is as bad as for different bodily constitutions to have the same food and clothing. Wherefore it is thought to be just that among equals every one be ruled as well as rule, and therefore that an should have their turn. We thus arrive at law; for an order of succession implies law. And the rule of the law, it is argued, is preferable to that of any individual. On the same principle, even if it be better for certain individuals to govern, they should be made only guardians and ministers of the law. For magistrates there must be — this is admitted; but then men say that to give authority to any one man when all are equal is unjust. Nay, there may indeed be cases which the law seems unable to determine, but in such cases can a man? Nay, it will be replied, the law trains officers for this express purpose, and appoints them to determine matters which are left undecided by it, to the best of their judgment. Further, it permits them to make any amendment of the existing laws which experience suggests. Therefore he who bids the law rule may be deemed to bid God and Reason alone rule, but he who bids man rule adds an element of the beast; for desire is a wild beast, and passion perverts the minds of rulers, even when they are the best of men. The law is reason unaffected by desire. We are told that a patient should call in a physician; he will not get better if he is doctored out of a book. But the parallel of the arts is clearly not in point; for the physician does nothing contrary to rule from motives of friendship; he only cures a patient and takes a fee; whereas magistrates do many things from spite and partiality. And, indeed, if a man suspected the physician of being in league with his enemies to destroy him for a bribe, he would rather have recourse to the book. But certainly physicians, when they are sick, call in other physicians, and training-masters, when they are in training, other training-masters, as if they could not judge judge truly about their own case and might be influenced by their feelings. Hence it is evident that in seeking for justice men seek for the mean or neutral, for the law is the mean. Again, customary laws have more weight, and relate to more important matters, than written laws, and a man may be a safer ruler than the written law, but not safer than the customary law.

Again, it is by no means easy for one man to superintend many things; he will have to appoint a number of subordinates, and what difference does it make whether these subordinates always existed or were appointed by him because he needed theme If, as I said before, the good man has a right to rule because he is better, still two good men are better than one: this is the old saying, two going together, and the prayer of Agamemnon,

"Would that I had ten such councillors!"

And at this day there are magistrates, for example judges, who have authority to decide some matters which the law is unable to determine, since no one doubts that the law would command and decide in the best manner whatever it could. But some things can, and other things cannot, be comprehended under the law, and this is the origin of the nexted question whether the best law or the best man should rule. For matters of detail about which men deliberate cannot be included in legislation. Nor does any one deny that the decision of such matters must be left to man, but it is argued that there should be many judges, and not one only. For every ruler who has been trained by the law judges well; and it would surely seem strange that a person should see better with two eyes, or hear better with two ears, or act better with two hands or feet, than many with many; indeed, it is already the practice of kings to make to themselves many eyes and ears and hands and feet. For they make colleagues of those who are the friends of themselves and their governments. They must be friends of the monarch and of his government; if not his friends, they will not do what he wants; but friendship implies likeness and equality; and, therefore, if he thinks that his friends ought to rule, he must think that those who are equal to himself and like himself ought to rule equally with himself. These are the principal controversies relating to monarchy.

XVII

But may not all this be true in some cases and not in others? for there is by nature both a justice and an advantage appropriate to the rule of a master, another to kingly rule, another to constitutional rule; but there is none naturally appropriate to tyranny, or to any other perverted form of government; for these come into being contrary to nature. Now, to judge at least from what has been said, it is manifest that, where men are alike and equal, it is neither expedient nor just that one man should be lord of all, whether there are laws, or whether there are no laws, but he himself is in the place of law. Neither should a good man be lord over good men, nor a bad man over bad; nor, even if he excels in virtue, should he have a right to rule, unless in a particular case, at which I have already hinted, and to which I will once more recur. But first of all, I must determine what natures are suited for government by a king, and what for an aristocracy, and what for a constitutional government.

A people who are by nature capable of producing a race superior in the virtue needed for political rule are fitted for kingly government; and a people submitting to be ruled as freemen by men whose virtue renders them capable of political command are adapted for an aristocracy; while the people who are suited for constitutional freedom are those among whom there naturally exists a warlike multitude able to rule and to obey in turn by a law which gives office to the well-to-do according to their desert. But when a whole family or some individual, happens to be so pre-eminent in virtue as to surpass all others, then it is just that they should be the royal family and supreme over all, or that this one citizen should be king of the whole nation. For, as I said before, to give them authority is not only agreeable to that ground of right which the founders of all states, whether aristocratical, or oligarchical, or again democratical, are accustomed to put forward (for these all recognize the claim of excellence, although not the same excellence), but accords with the principle already laid down. For surely it would not be right to kill, or ostracize, or exile such a person, or require that he should take his turn in being governed. The whole is naturally superior to the part, and he who has this pre-eminence is in the relation of a whole to a part. But if so, the only alternative is that he should have the supreme power, and that mankind should obey him, not in turn, but always. These are the conclusions at which we arrive respecting royalty and its various forms, and this is the answer to the question, whether it is or is not advantageous to states, and to which, and how.

XVIII

We maintain that the true forms of government are three, and that the best must be that which is administered by the best, and in which there is one man, or a whole family, or many persons, excelling all the others together in virtue, and both rulers and subjects are fitted, the one to rule, the others to be ruled, in such a manner as to attain the most eligible life. We showed at the commencement of our inquiry that the virtue of the good man is necessarily the same as the virtue of the citizen of the perfect state. Clearly then in the same manner, and by the same means through which a man becomes truly good, he will frame a state that is to be ruled by an aristocracy or by a king, and the same education and the same habits will be found to make a good man and a man fit to be a statesman or a king.

Having arrived at these conclusions, we must proceed to speak of the perfect state, and describe how it comes into being and is established.

BOOK IV
I

IN all arts and sciences which embrace the whole of any subject, and do not come into being in a fragmentary way, it is the province of a single art or science to consider all that appertains to a single subject. For example, the art of gymnastic considers not only the suitableness of different modes of training to different bodies (2), but what sort is absolutely the best (1); (for the absolutely best must suit that which is by nature best and best furnished with the means of life), and also what common form of training is adapted to the great majority of men (4). And if a man does not desire the best habit of body, or the greatest skill in gymnastics, which might be attained by him, still the trainer or the teacher of gymnastic should be able to impart any lower degree of either (3). The same principle equally holds in medicine and shipbuilding, and the making of clothes, and in the arts generally.

Hence it is obvious that government too is the subject of a single science, which has to consider what government is best and of what sort it must be, to be most in accordance with our aspirations, if there were no external impediment, and also what kind of government is adapted to particular states. For the best is often unattainable, and therefore the true legislator and statesman ought to be acquainted, not only with (1) that which is best in the abstract, but also with (2) that which is best relatively to circumstances. We should be able further to say how a state may be constituted under any given conditions (3); both how it is originally formed and, when formed, how it may be longest preserved; the supposed state being so far from having the best constitution that it is unprovided even with the conditions necessary for the best; neither is it the best under the circumstances, but of an inferior type.

He ought, moreover, to know (4) the form of government which is best suited to states in general; for political writers, although they have excellent ideas, are often unpractical. We should consider, not only what form of government is best, but also what is possible and what is easily attainable by all. There are some who would have none but the most perfect; for this many natural advantages are required. Others, again, speak of a more attainable form, and, although they reject the constitution under which they are living, they extol some one in particular, for example the Lacedaemonian. Any change of government which has to be introduced should be one which men, starting from their existing constitutions, will be both willing and able to adopt, since there is quite as much trouble in the reformation of an old constitution as in the establishment of a new one, just as to unlearn is as hard as to learn. And therefore, in addition to the qualifications of the statesman already mentioned, he should be able to find remedies for the defects of existing constitutions, as has been said before. This he cannot do unless he knows how many forms of government there are. It is often supposed that there is only one kind of democracy and one of oligarchy. But this is a mistake; and, in order to avoid such mistakes, we must ascertain what differences there are in the constitutions of states, and in how many ways they are combined. The same political insight will enable a man to know which laws are the best, and which are suited to different constitutions; for the laws are, and ought to be, relative to the constitution, and not the constitution to the laws. A constitution is the organization of offices in a state, and determines what is to be the governing body, and what is the end of each community. But laws are not to be confounded with the principles of the constitution; they are the rules according to which the magistrates should administer the state, and proceed against offenders. So that we must know the varieties, and the number of varieties, of each form of government, if only with a view to making laws. For the same laws cannot be equally suited to all oligarchies or to all democracies, since there is certainly more than one form both of democracy and of oligarchy.

II

In our original discussion about governments we divided them into three true forms: kingly rule, aristocracy, and constitutional government, and three corresponding perversions — tyranny, oligarchy, and democracy. Of kingly rule and of aristocracy, we have already spoken, for the inquiry into the perfect state is the same thing with the discussion of the two forms thus named, since both imply a principle of virtue provided with external means. We have already determined in what aristocracy and kingly rule differ from one another, and when the latter should be established. In what follows we have to describe the so-called constitutional government, which bears the common name of all constitutions, and the other forms, tyranny, oligarchy, and democracy.

It is obvious which of the three perversions is the worst, and which is the next in badness. That which is the perversion of the first and most divine is necessarily the worst. And just as a royal rule, if not a mere name, must exist by virtue of some great personal superiority in the king, so tyranny, which is the worst of governments, is necessarily the farthest removed from a well-constituted form; oligarchy is little better, for it is a long way from aristocracy, and democracy is the most tolerable of the three.

A writer who preceded me has already made these distinctions, but his point of view is not the same as mine. For he lays down the principle that when all the constitutions are good (the oligarchy and the rest being virtuous), democracy is the worst, but the best when all are bad. Whereas we maintain that they are in any case defective, and that one oligarchy is not to be accounted better than another, but only less bad.

Not to pursue this question further at present, let us begin by determining (1) how many varieties of constitution there are (since of democracy and oligarchy there are several): (2) what constitution is the most generally acceptable, and what is eligible in the next degree after the perfect state; and besides this what other there is which is aristocratical and well-constituted, and at the same time adapted to states in general; (3) of the other forms of government to whom each is suited. For democracy may meet the needs of some better than oligarchy, and conversely. In the next place (4) we have to consider in what manner a man ought to proceed who desires to establish some one among these various forms, whether of democracy or of oligarchy; and lastly, (5) having briefly discussed these subjects to the best of our power, we will endeavor to ascertain the modes of ruin and preservation both of constitutions generally and of each separately, and to what causes they are to be attributed.

III

The reason why there are many forms of government is that every state contains many elements. In the first place we see that all states are made up of families, and in the multitude of citizen there must be some rich and some poor, and some in a middle condition; the rich are heavy-armed, and the poor not. Of the common people, some are husbandmen, and some traders, and some artisans. There are also among the notables differences of wealth and property — for example, in the number of horses which they keep, for they cannot afford to keep them unless they are rich. And therefore in old times the cities whose strength lay in their cavalry were oligarchies, and they used cavalry in wars against their neighbors; as was the practice of the Eretrians and Chalcidians, and also of the Magnesians on the river Maeander, and of other peoples in Asia. Besides differences of wealth there are differences of rank and merit, and there are some other elements which were mentioned by us when in treating of aristocracy we enumerated the essentials of a state. Of these elements, sometimes all, sometimes the lesser and sometimes the greater number, have a share in the government. It is evident then that there must be many forms of government, differing in kind, since the parts of which they are composed differ from each other in kind. For a constitution is an organization of offices, which all the citizens distribute among themselves, according to the power which different classes possess, for example the rich or the poor, or according to some principle of equality which includes both. There must therefore be as many forms of government as there are modes of arranging the offices, according to the superiorities and differences of the parts of the state.

There are generally thought to be two principal forms: as men say of the winds that there are but two — north and south, and that the rest of them are only variations of these, so of governments there are said to be only two forms — democracy and oligarchy. For aristocracy is considered to be a kind of oligarchy, as being the rule of a few, and the so-called constitutional government to be really a democracy, just as among the winds we make the west a variation of the north, and the east of the south wind. Similarly of musical modes there are said to be two kinds, the Dorian and the Phrygian; the other arrangements of the scale are comprehended under one or other of these two. About forms of government this is a very favorite notion. But in either case the better and more exact way is to distinguish, as I have done, the one or two which are true forms, and to regard the others as perversions, whether of the most perfectly attempered mode or of the best form of government: we may compare the severer and more overpowering modes to the oligarchical forms, and the more relaxed and gentler ones to the democratic.

IV

It must not be assumed, as some are fond of saying, that democracy is simply that form of government in which the greater number are sovereign, for in oligarchies, and indeed in every government, the majority rules; nor again is oligarchy that form of government in which a few are sovereign. Suppose the whole population of a city to be 1300, and that of these 1000 are rich, and do not allow the remaining 300 who are poor, but free, and in an other respects their equals, a share of the government — no one will say that this is a democracy. In like manner, if the poor were few and the masters of the rich who outnumber them, no one would ever call such a government, in which the rich majority have no share of office, an oligarchy. Therefore we should rather say that democracy is the form of government in which the free are rulers, and oligarchy in which the rich; it is only an accident that the free are the many and the rich are the few. Otherwise a government in which the offices were given according to stature, as is said to be the case in Ethiopia, or according to beauty, would be an oligarchy; for the number of tall or good-looking men is small. And yet oligarchy and democracy are not sufficiently distinguished merely by these two characteristics of wealth and freedom. Both of them contain many other elements, and therefore we must carry our analysis further, and say that the government is not a democracy in which the freemen, being few in number, rule over the many who are not free, as at Apollonia, on the Ionian Gulf, and at Thera; (for in each of these states the nobles, who were also the earliest settlers, were held in chief honor, although they were but a few out of many). Neither is it a democracy when the rich have the government because they exceed in number; as was the case formerly at Colophon, where the bulk of the inhabitants were possessed of large property before the Lydian War. But the form of government is a democracy when the free, who are also poor and the majority, govern, and an oligarchy when the rich and the noble govern, they being at the same time few in number.

I have said that there are many forms of government, and have explained to what causes the variety is due. Why there are more than those already mentioned, and what they are, and whence they arise, I will now proceed to consider, starting from the principle already admitted, which is that every state consists, not of one, but of many parts. If we were going to speak of the different species of animals, we should first of all determine the organs which are indispensable to every animal, as for example some organs of sense and the instruments of receiving and digesting food, such as the mouth and the stomach, besides organs of locomotion. Assuming now that there are only so many kinds of organs, but that there may be differences in them — I mean different kinds of mouths, and stomachs, and perceptive and locomotive organs — the possible combinations of these differences will necessarily furnish many variedes of animals. (For animals cannot be the same which have different kinds of mouths or of ears.) And when all the combinations are exhausted, there will be as many sorts of animals as there are combinations of the necessary organs. The same, then, is true of the forms of government which have been described; states, as I have repeatedly said, are composed, not of one, but of many elements. One element is the food-producing class, who are called husbandmen; a second, the class of mechanics who practice the arts without which a city cannot exist; of these arts some are absolutely necessary, others contribute to luxury or to the grace of life. The third class is that of traders, and by traders I mean those who are engaged in buying and selling, whether in commerce or in retail trade. A fourth class is that of the serfs or laborers. The warriors make up the fifth class, and they are as necessary as any of the others, if the country is not to be the slave of every invader. For how can a state which has any title to the name be of a slavish nature? The state is independent and self-sufficing, but a slave is the reverse of independent. Hence we see that this subject, though ingeniously, has not been satisfactorily treated in the Republic. Socrates says that a state is made up of four sorts of people who are absolutely necessary; these are a weaver, a husbandman, a shoemaker, and a builder; afterwards, finding that they are not enough, he adds a smith, and again a herdsman, to look after the necessary animals; then a merchant, and then a retail trader. All these together form the complement of the first state, as if a state were established merely to supply the necessaries of life, rather than for the sake of the good, or stood equally in need of shoemakers and of husbandmen. But he does not admit into the state a military class until the country has increased in size, and is beginning to encroach on its neighbor's land, whereupon they go to war. Yet even amongst his four original citizens, or whatever be the number of those whom he associates in the state, there must be some one who will dispense justice and determine what is just. And as the soul may be said to be more truly part of an animal than the body, so the higher parts of states, that is to say, the warrior class, the class engaged in the administration of justice, and that engaged in deliberation, which is the special business of political common sense — these are more essential to the state than the parts which minister to the necessaries of life. Whether their several functions are the functions of different citizens, or of the same — for it may often happen that the same persons are both warriors and husbandmen — is immaterial to the argument. The higher as well as the lower elements are to be equally considered parts of the state, and if so, the military element at any rate must be included. There are also the wealthy who minister to the state with their property; these form the seventh class. The eighth class is that of magistrates and of officers; for the state cannot exist without rulers. And therefore some must be able to take office and to serve the state, either always or in turn. There only remains the class of those who deliberate and who judge between disputants; we were just now distinguishing them. If presence of all these elements, and their fair and equitable organization, is necessary to states, then there must also be persons who have the ability of statesmen. Different functions appear to be often combined in the same individual; for example, the warrior may also be a husbandman, or an artisan; or, again, the councillor a judge. And all claim to possess political ability, and think that they are quite competent to fill most offices. But the same persons cannot be rich and poor at the same time. For this reason the rich and the poor are regarded in an especial sense as parts of a state. Again, because the rich are generally few in number, while the poor are many, they appear to be antagonistic, and as the one or the other prevails they form the government. Hence arises the common opinion that there are two kinds of government — democracy and oligarchy.

I have already explained that there are many forms of constitution, and to what causes the variety is due. Let me now show that there are different forms both of democracy and oligarchy, as will indeed be evident from what has preceded. For both in the common people and in the notables various classes are included; of the common people, one class are husbandmen, another artisans; another traders, who are employed in buying and selling; another are the seafaring class, whether engaged in war or in trade, as ferrymen or as fishermen. (In many places any one of these classes forms quite a large population; for example, fishermen at Tarentum and Byzantium, crews of triremes at Athens, merchant seamen at Aegina and Chios, ferrymen at Tenedos.) To the classes already mentioned may be added day-laborers, and those who, owing to their needy circumstances, have no leisure, or those who are not of free birth on both sides; and there may be other classes as well. The notables again may be divided according to their wealth, birth, virtue, education, and similar differences.

Of forms of democracy first comes that which is said to be based strictly on equality. In such a democracy the law says that it is just for the poor to have no more advantage than the rich; and that neither should be masters, but both equal. For if liberty and equality, as is thought by some, are chiefly to be found in democracy, they will be best attained when all persons alike share in the government to the utmost. And since the people are the majority, and the opinion of the majority is decisive, such a government must necessarily be a democracy. Here then is one sort of democracy. There is another, in which the magistrates are elected according to a certain property qualification, but a low one; he who has the required amount of property has a share in the government, but he who loses his property loses his rights. Another kind is that in which all the citizens who are under no disqualification share in the government, but still the law is supreme. In another, everybody, if he be only a citizen, is admitted to the government, but the law is supreme as before. A fifth form of democracy, in other respects the same, is that in which, not the law, but the multitude, have the supreme power, and supersede the law by their decrees. This is a state of affairs brought about by the demagogues. For in democracies which are subject to the law the best citizens hold the first place, and there are no demagogues; but where the laws are not supreme, there demagogues spring up. For the people becomes a monarch, and is many in one; and the many have the power in their hands, not as individuals, but collectively. Homer says that 'it is not good to have a rule of many,' but whether he means this corporate rule, or the rule of many individuals, is uncertain. At all events this sort of democracy, which is now a monarch, and no longer under the control of law, seeks to exercise monarchical sway, and grows into a despot; the flatterer is held in honor; this sort of democracy being relatively to other democracies what tyranny is to other forms of monarchy. The spirit of both is the same, and they alike exercise a despotic rule over the better citizens. The decrees of the demos correspond to the edicts of the tyrant; and the demagogue is to the one what the flatterer is to the other. Both have great power; the flatterer with the tyrant, the demagogue with democracies of the kind which we are describing. The demagogues make the decrees of the people override the laws, by referring all things to the popular assembly. And therefore they grow great, because the people have all things in their hands, and they hold in their hands the votes of the people, who are too ready to listen to them. Further, those who have any complaint to bring against the magistrates say, 'Let the people be judges'; the people are too happy to accept the invitation; and so the authority of every office is undermined. Such a democracy is fairly open to the objection that it is not a constitution at all; for where the laws have no authority, there is no constitution. The law ought to be supreme over all, and the magistracies should judge of particulars, and only this should be considered a constitution. So that if democracy be a real form of government, the sort of system in which all things are regulated by decrees is clearly not even a democracy in the true sense of the word, for decrees relate only to particulars.

These then are the different kinds of democracy.

BOOK V

Of oligarchies, too, there are different kinds: one where the property qualification for office is such that the poor, although they form the majority, have no share in the government, yet he who acquires a qualification may obtain a share. Another sort is when there is a qualification for office, but a high one, and the vacancies in the governing body are fired by co-optation. If the election is made out of all the qualified persons, a constitution of this kind inclines to an aristocracy, if out of a privileged class, to an oligarchy. Another sort of oligarchy is when the son succeeds the father. There is a fourth form, likewise hereditary, in which the magistrates are supreme and not the law. Among oligarchies this is what tyranny is among monarchies, and the last-mentioned form of democracy among democracies; and in fact this sort of oligarchy receives the name of a dynasty (or rule of powerful families).

These are the different sorts of oligarchies and democracies. It should, however, be remembered that in many states the constitution which is established by law, although not democratic, owing to the education and habits of the people may be administered democratically, and conversely in other states the established constitution may incline to democracy, but may be administered in an oligarchical spirit. This most often happens after a revolution: for governments do not change at once; at first the dominant party are content with encroaching a little upon their opponents. The laws which existed previously continue in force, but the authors of the revolution have the power in their hands.

BOOK VI

From what has been already said we may safely infer that there are so many different kinds of democracies and of oligarchies. For it is evident that either all the classes whom we mentioned must share in the government, or some only and not others. When the class of husbandmen and of those who possess moderate fortunes have the supreme power, the government is administered according to law. For the citizens being compelled to live by their labor have no leisure; and so they set up the authority of the law, and attend assemblies only when necessary. They all obtain a share in the government when they have acquired the qualification which is fixed by the law — the absolute exclusion of any class would be a step towards oligarchy; hence all who have acquired the property qualification are admitted to a share in the constitution. But leisure cannot be provided for them unless there are revenues to support them. This is one sort of democracy, and these are the causes which give birth to it. Another kind is based on the distinction which naturally comes next in order; in this, every one to whose birth there is no objection is eligible, but actually shares in the government only if he can find leisure. Hence in such a democracy the supreme power is vested in the laws, because the state has no means of paying the citizens. A third kind is when all freemen have a right to share in the government, but do not actually share, for the reason which has been already given; so that in this form again the law must rule. A fourth kind of democracy is that which comes latest in the history of states. In our own day, when cities have far outgrown their original size, and their revenues have increased, all the citizens have a place in the government, through the great preponderance of the multitude; and they all, including the poor who receive pay, and therefore have leisure to exercise their rights, share in the administration. Indeed, when they are paid, the common people have the most leisure, for they are not hindered by the care of their property, which often fetters the rich, who are thereby prevented from taking part in the assembly or in the courts, and so the state is governed by the poor, who are a majority, and not by the laws.

So many kinds of democracies there are, and they grow out of these necessary causes.

Of oligarchies, one form is that in which the majority of the citizens have some property, but not very much; and this is the first form, which allows to any one who obtains the required amount the right of sharing in the government. The sharers in the government being a numerous body, it follows that the law must govern, and not individuals. For in proportion as they are further removed from a monarchical form of government, and in respect of property have neither so much as to be able to live without attending to business, nor so little as to need state support, they must admit the rule of law and not claim to rule themselves. But if the men of property in the state are fewer than in the former case, and own more property, there arises a second form of oligarchy. For the stronger they are, the more power they claim, and having this object in view, they themselves select those of the other classes who are to be admitted to the government; but, not being as yet strong enough to rule without the law, they make the law represent their wishes. When this power is intensified by a further diminution of their numbers and increase of their property, there arises a third and further stage of oligarchy, in which the governing class keep the offices in their own hands, and the law ordains that the son shall succeed the father. When, again, the rulers have great wealth and numerous friends, this sort of family despotism approaches a monarchy; individuals rule and not the law. This is the fourth sort of oligarchy, and is analogous to the last sort of democracy.

BOOK VII

There are still two forms besides democracy and oligarchy; one of them is universally recognized and included among the four principal forms of government, which are said to be (1) monarchy, (2) oligarchy, (3) democracy, and (4) the so-called aristocracy or government of the best. But there is also a fifth, which retains the generic name of polity or constitutional government; this is not common, and therefore has not been noticed by writers who attempt to enumerate the different kinds of government; like Plato, in their books about the state, they recognize four only. The term 'aristocracy' is rightly applied to the form of government which is described in the first part of our treatise; for that only can be rightly called aristocracy which is a government formed of the best men absolutely, and not merely of men who are good when tried by any given standard. In the perfect state the good man is absolutely the same as the good citizen; whereas in other states the good citizen is only good relatively to his own form of government. But there are some states differing from oligarchies and also differing from the so-called polity or constitutional government; these are termed aristocracies, and in them the magistrates are certainly chosen, both according to their wealth and according to their merit. Such a form of government differs from each of the two just now mentioned, and is termed an aristocracy. For indeed in states which do not make virtue the aim of the community, men of merit and reputation for virtue may be found. And so where a government has regard to wealth, virtue, and numbers, as at Carthage, that is aristocracy; and also where it has regard only to two out of the three, as at Lacedaemon, to virtue and numbers, and the two principles of democracy and virtue temper each other. There are these two forms of aristocracy in addition to the first and perfect state, and there is a third form, viz., the constitutions which incline more than the so-called polity towards oligarchy.

BOOK VIII

I have yet to speak of the so-called polity and of tyranny. I put them in this order, not because a polity or constitutional government is to be regarded as a perversion any more than the above mentioned aristocracies. The truth is, that they an fall short of the most perfect form of government, and so they are reckoned among perversions, and the really perverted forms are perversions of these, as I said in the original discussion. Last of all I will speak of tyranny, which I place last in the series because I am inquiring into the constitutions of states, and this is the very reverse of a constitution

Having explained why I have adopted this order, I will proceed to consider constitutional government; of which the nature will be clearer now that oligarchy and democracy have been defined. For polity or constitutional government may be described generally as a fusion of oligarchy and democracy; but the term is usually applied to those forms of government which incline towards democracy, and the term aristocracy to those which incline towards oligarchy, because birth and education are commonly the accompaniments of wealth. Moreover, the rich already possess the external advantages the want of which is a temptation to crime, and hence they are called noblemen and gentlemen. And inasmuch as aristocracy seeks to give predominance to the best of the citizens, people say also of oligarchies that they are composed of noblemen and gentlemen. Now it appears to be an impossible thing that the state which is governed not by the best citizens but by the worst should be well-governed, and equally impossible that the state which is ill-governed should be governed by the best. But we must remember that good laws, if they are not obeyed, do not constitute good government. Hence there are two parts of good government; one is the actual obedience of citizens to the laws, the other part is the goodness of the laws which they obey; they may obey bad laws as well as good. And there may be a further subdivision; they may obey either the best laws which are attainable to them, or the best absolutely.

The distribution of offices according to merit is a special characteristic of aristocracy, for the principle of an aristocracy is virtue, as wealth is of an oligarchy, and freedom of a democracy. In all of them there of course exists the right of the majority, and whatever seems good to the majority of those who share in the government has authority. Now in most states the form called polity exists, for the fusion goes no further than the attempt to unite the freedom of the poor and the wealth of the rich, who commonly take the place of the noble. But as there are three grounds on which men claim an equal share in the government, freedom, wealth, and virtue (for the fourth or good birth is the result of the two last, being only ancient wealth and virtue), it is clear that the admixture of the two elements, that is to say, of the rich and poor, is to be called a polity or constitutional government; and the union of the three is to be called aristocracy or the government of the best, and more than any other form of government, except the true and ideal, has a right to this name.

Thus far I have shown the existence of forms of states other than monarchy, democracy, and oligarchy, and what they are, and in what aristocracies differ from one another, and polities from aristocracies — that the two latter are not very unlike is obvious.

IX

Next we have to consider how by the side of oligarchy and democracy the so-called polity or constitutional government springs up, and how it should be organized. The nature of it will be at once understood from a comparison of oligarchy and democracy; we must ascertain their different characteristics, and taking a portion from each, put the two together, like the parts of an indenture. Now there are three modes in which fusions of government may be affected. In the first mode we must combine the laws made by both governments, say concerning the administration of justice. In oligarchies they impose a fine on the rich if they do not serve as judges, and to the poor they give no pay; but in democracies they give pay to the poor and do not fine the rich. Now (1) the union of these two modes is a common or middle term between them, and is therefore characteristic of a constitutional government, for it is a combination of both. This is one mode of uniting the two elements. Or (2) a mean may be taken between the enactments of the two: thus democracies require no property qualification, or only a small one, from members of the assembly, oligarchies a high one; here neither of these is the common term, but a mean between them. (3) There is a third mode, in which something is borrowed from the oligarchical and something from the democratical principle. For example, the appointment of magistrates by lot is thought to be democratical, and the election of them oligarchical; democratical again when there is no property qualification, oligarchical when there is. In the aristocratical or constitutional state, one element will be taken from each — from oligarchy the principle of electing to offices, from democracy the disregard of qualification. Such are the various modes of combination.

There is a true union of oligarchy and democracy when the same state may be termed either a democracy or an oligarchy; those who use both names evidently feel that the fusion is complete. Such a fusion there is also in the mean; for both extremes appear in it. The Lacedaemonian constitution, for example, is often described as a democracy, because it has many democratical features. In the first place the youth receive a democratical education. For the sons of the poor are brought up with with the sons of the rich, who are educated in such a manner as to make it possible for the sons of the poor to be educated by them. A similar equality prevails in the following period of life, and when the citizens are grown up to manhood the same rule is observed; there is no distinction between the rich and poor. In like manner they all have the same food at their public tables, and the rich wear only such clothing as any poor man can afford. Again, the people elect to one of the two greatest offices of state, and in the other they share; for they elect the Senators and share in the Ephoralty. By others the Spartan constitution is said to be an oligarchy, because it has many oligarchical elements. That all offices are filled by election and none by lot, is one of these oligarchical characteristics; that the power of inflicting death or banishment rests with a few persons is another; and there are others. In a well attempted polity there should appear to be both elements and yet neither; also the government should rely on itself, and not on foreign aid, and on itself not through the good will of a majority — they might be equally well-disposed when there is a vicious form of government — but through the general willingness of all classes in the state to maintain the constitution.

Enough of the manner in which a constitutional government, and in which the so-called aristocracies ought to be framed.

X

Of the nature of tyranny I have still to speak, in order that it may have its place in our inquiry (since even tyranny is reckoned by us to be a form of government), although there is not much to be said about it. I have already in the former part of this treatise discussed royalty or kingship according to the most usual meaning of the term, and considered whether it is or is not advantageous to states, and what kind of royalty should be established, and from what source, and how.

When speaking of royalty we also spoke of two forms of tyranny, which are both according to law, and therefore easily pass into royalty. Among barbarians there are elected monarchs who exercise a despotic power; despotic rulers were also elected in ancient Hellas, called Aesymnetes or Dictators. These monarchies, when compared with one another, exhibit certain differences. And they are, as I said before, royal, in so far as the monarch rules according to law over willing subjects; but they are tyrannical in so far as he is despotic and rules according to his own fancy. There is also a third kind of tyranny, which is the most typical form, and is the counterpart of the perfect monarchy. This tyranny is just that arbitrary power of an individual which is responsible to no one, and governs all alike, whether equals or better, with a view to its own advantage, not to that of its subjects, and therefore against their will. No freeman, if he can escape from it, will endure such a government.

The kinds of tyranny are such and so many, and for the reasons which I have given.

XI

We have now to inquire what is the best constitution for most states, and the best life for most men, neither assuming a standard of virtue which is above ordinary persons, nor an education which is exceptionally favored by nature and circumstances, nor yet an ideal state which is an aspiration only, but having regard to the life in which the majority are able to share, and to the form of government which states in general can attain. As to those aristocracies, as they are called, of which we were just now speaking, they either lie beyond the possibilities of the greater number of states, or they approximate to the so-called constitutional government, and therefore need no separate discussion. And in fact the conclusion at which we arrive respecting all these forms rests upon the same grounds. For if what was said in the Ethics is true, that the happy life is the life according to virtue lived without impediment, and that virtue is a mean, then the life which is in a mean, and in a mean attainable by every one, must be the best. And the same the same principles of virtue and vice are characteristic of cities and of constitutions; for the constitution is in a figure the life of the city.

Now in all states there are three elements: one class is very rich, another very poor, and a third in a mean. It is admitted that moderation and the mean are best, and therefore it will clearly be best to possess the gifts of fortune in moderation; for in that condition of life men are most ready to follow rational principle. But he who greatly excels in beauty, strength, birth, or wealth, or on the other hand who is very poor, or very weak, or very much disgraced, finds it difficult to follow rational principle. Of these two the one sort grow into violent and great criminals, the others into rogues and petty rascals. And two sorts of offenses correspond to them, the one committed from violence, the other from roguery. Again, the middle class is least likely to shrink from rule, or to be over-ambitious for it; both of which are injuries to the state. Again, those who have too much of the goods of fortune, strength, wealth, friends, and the like, are neither willing nor able to submit to authority. The evil begins at home; for when they are boys, by reason of the luxury in which they are brought up, they never learn, even at school, the habit of obedience. On the other hand, the very poor, who are in the opposite extreme, are too degraded. So that the one class cannot obey, and can only rule despotically; the other knows not how to command and must be ruled like slaves. Thus arises a city, not of freemen, but of masters and slaves, the one despising, the other envying; and nothing can be more fatal to friendship and good fellowship in states than this: for good fellowship springs from friendship; when men are at enmity with one another, they would rather not even share the same path. But a city ought to be composed, as far as possible, of equals and similars; and these are generally the middle classes. Wherefore the city which is composed of middle-class citizens is necessarily best constituted in respect of the elements of which we say the fabric of the state naturally consists. And this is the class of citizens which is most secure in a state, for they do not, like the poor, covet their neighbors' goods; nor do others covet theirs, as the poor covet the goods of the rich; and as they neither plot against others, nor are themselves plotted against, they pass through life safely. Wisely then did Phocylides pray — 'Many things are best in the mean; I desire to be of a middle condition in my city.'

Thus it is manifest that the best political community is formed by citizens of the middle class, and that those states are likely to be well-administered in which the middle class is large, and stronger if possible than both the other classes, or at any rate than either singly; for the addition of the middle class turns the scale, and prevents either of the extremes from being dominant. Great then is the good fortune of a state in which the citizens have a moderate and sufficient property; for where some possess much, and the others nothing, there may arise an extreme democracy, or a pure oligarchy; or a tyranny may grow out of either extreme — either out of the most rampant democracy, or out of an oligarchy; but it is not so likely to arise out of the middle constitutions and those akin to them. I will explain the reason of this hereafter, when I speak of the revolutions of states. The mean condition of states is clearly best, for no other is free from faction; and where the middle class is large, there are least likely to be factions and dissensions. For a similar reason large states are less liable to faction than small ones, because in them the middle class is large; whereas in small states it is easy to divide all the citizens into two classes who are either rich or poor, and to leave nothing in the middle. And democracies are safer and more permanent than oligarchies, because they have a middle class which is more numerous and has a greater share in the government; for when there is no middle class, and the poor greatly exceed in number, troubles arise, and the state soon comes to an end. A proof of the superiority of the middle dass is that the best legislators have been of a middle condition; for example, Solon, as his own verses testify; and Lycurgus, for he was not a king; and Charondas, and almost all legislators.

These considerations will help us to understand why most governments are either democratical or oligarchical. The reason is that the middle class is seldom numerous in them, and whichever party, whether the rich or the common people, transgresses the mean and predominates, draws the constitution its own way, and thus arises either oligarchy or democracy. There is another reason — the poor and the rich quarrel with one another, and whichever side gets the better, instead of establishing a just or popular government, regards political supremacy as the prize of victory, and the one party sets up a democracy and the other an oligarchy. Further, both the parties which had the supremacy in Hellas looked only to the interest of their own form of government, and established in states, the one, democracies, and the other, oligarchies; they thought of their own advantage, of the public not at all. For these reasons the middle form of government has rarely, if ever, existed, and among a very few only. One man alone of all who ever ruled in Hellas was induced to give this middle constitution to states. But it has now become a habit among the citizens of states, not even to care about equality; all men are seeking for dominion, or, if conquered, are willing to submit.

What then is the best form of government, and what makes it the best, is evident; and of other constitutions, since we say that there are many kinds of democracy and many of oligarchy, it is not difficult to see which has the first and which the second or any other place in the order of excellence, now that we have determined which is the best. For that which is nearest to the best must of necessity be better, and that which is furthest from it worse, if we are judging absolutely and not relatively to given conditions: I say 'relatively to given conditions,' since a particular government may be preferable, but another form may be better for some people.

XII

We have now to consider what and what kind of government is suitable to what and what kind of men. I may begin by assuming, as a general principle common to all governments, that the portion of the state which desires the permanence of the constitution ought to be stronger than that which desires the reverse. Now every city is composed of quality and quantity. By quality I mean freedom, wealth, education, good birth, and by quantity, superiority of numbers. Quality may exist in one of the classes which make up the state, and quantity in the other. For example, the meanly-born may be more in number than the well-born, or the poor than the rich, yet they may not so much exceed in quantity as they fall short in quality; and therefore there must be a comparison of quantity and quality. Where the number of the poor is more than proportioned to the wealth of the rich, there will naturally be a democracy, varying in form with the sort of people who compose it in each case. If, for example, the husbandmen exceed in number, the first form of democracy will then arise; if the artisans and laboring class, the last; and so with the intermediate forms. But where the rich and the notables exceed in quality more than they fall short in quantity, there oligarchy arises, similarly assuming various forms according to the kind of superiority possessed by the oligarchs.

The legislator should always include the middle class in his government; if he makes his laws oligarchical, to the middle class let him look; if he makes them democratical, he should equally by his laws try to attach this class to the state. There only can the government ever be stable where the middle class exceeds one or both of the others, and in that case there will be no fear that the rich will unite with the poor against the rulers. For neither of them will ever be willing to serve the other, and if they look for some form of government more suitable to both, they will find none better than this, for the rich and the poor will never consent to rule in turn, because they mistrust one another. The arbiter is always the one trusted, and he who is in the middle is an arbiter. The more perfect the admixture of the political elements, the more lasting will be the constitution. Many even of those who desire to form aristocratical governments make a mistake, not only in giving too much power to the rich, but in attempting to overreach the people. There comes a time when out of a false good there arises a true evil, since the encroachments of the rich are more destructive to the constitution than those of the people.

XIII

The devices by which oligarchies deceive the people are five in number; they relate to (1) the assembly; (2) the magistracies; (3) the courts of law; (4) the use of arms; (5) gymnastic exercises. (1) The assemblies are thrown open to all, but either the rich only are fined for non-attendance, or a much larger fine is inflicted upon them. (2) to the magistracies, those who are qualified by property cannot decline office upon oath, but the poor may. (3) In the law courts the rich, and the rich only, are fined if they do not serve, the poor are let off with impunity, or, as in the laws of Charondas, a larger fine is inflicted on the rich, and a smaller one on the poor. In some states all citizen who have registered themselves are allowed to attend the assembly and to try causes; but if after registration they do not attend either in the assembly or at the courts, heavy fines are imposed upon them. The intention is that through fear of the fines they may avoid registering themselves, and then they cannot sit in the law-courts or in the assembly. concerning (4) the possession of arms, and (5) gymnastic exercises, they legislate in a similar spirit. For the poor are not obliged to have arms, but the rich are fined for not having them; and in like manner no penalty is inflicted on the poor for non-attendance at the gymnasium, and consequently, having nothing to fear, they do not attend, whereas the rich are liable to a fine, and therefore they take care to attend.

These are the devices of oligarchical legislators, and in democracies they have counter devices. They pay the poor for attending the assemblies and the law-courts, and they inflict no penalty on the rich for non-attendance. It is obvious that he who would duly mix the two principles should combine the practice of both, and provide that the poor should be paid to attend, and the rich fined if they do not attend, for then all will take part; if there is no such combination, power will be in the hands of one party only. The government should be confined to those who carry arms. As to the property qualification, no absolute rule can be laid down, but we must see what is the highest qualification sufficiently comprehensive to secure that the number of those who have the rights of citizens exceeds the number of those excluded. Even if they have no share in office, the poor, provided only that they are not outraged or deprived of their property, will be quiet enough.

But to secure gentle treatment for the poor is not an easy thing, since a ruling class is not always humane. And in time of war the poor are apt to hesitate unless they are fed; when fed, they are willing enough to fight. In some states the government is vested, not only in those who are actually serving, but also in those who have served; among the Malians, for example, the governing body consisted of the latter, while the magistrates were chosen from those actually on service. And the earliest government which existed among the Hellenes, after the overthrow of the kingly power, grew up out of the warrior class, and was originally taken from the knights (for strength and superiority in war at that time depended on cavalry; indeed, without discipline, infantry are useless, and in ancient times there was no military knowledge or tactics, and therefore the strength of armies lay in their cavalry). But when cities increased and the heavy armed grew in strength, more had a share in the government; and this is the reason why the states which we call constitutional governments have been hitherto called democracies. Ancient constitutions, as might be expected, were oligarchical and royal; their population being small they had no considerable middle class; the people were weak in numbers and organization, and were therefore more contented to be governed.

I have explained why there are various forms of government, and why there are more than is generally supposed; for democracy, as well as other constitutions, has more than one form: also what their differences are, and whence they arise, and what is the best form of government, speaking generally and to whom the various forms of government are best suited; all this has now been explained.

XIV

Having thus gained an appropriate basis of discussion, we will proceed to speak of the points which follow next in order. We will consider the subject not only in general but with reference to particular constitutions. All constitutions have three elements, concerning which the good lawgiver has to regard what is expedient for each constitution. When they are well-ordered, the constitution is well-ordered, and as they differ from one another, constitutions differ. There is (1) one element which deliberates about public affairs; secondly (2) that concerned with the magistrates — the question being, what they should be, over what they should exercise authority, and what should be the mode of electing to them; and thirdly (3) that which has judicial power.

The deliberative element has authority in matters of war and peace, in making and unmaking alliances; it passes laws, inflicts death, exile, confiscation, elects magistrates and audits their accounts. These powers must be assigned either all to all the citizens or an to some of them (for example, to one or more magistracies, or different causes to different magistracies), or some of them to all, and others of them only to some. That all things should be decided by all is characteristic of democracy; this is the sort of equality which the people desire. But there are various ways in which all may share in the government; they may deliberate, not all in one body, but by turns, as in the constitution of Telecles the Milesian. There are other constitutions in which the boards of magistrates meet and deliberate, but come into office by turns, and are elected out of the tribes and the very smallest divisions of the state, until every one has obtained office in his turn. The citizens, on the other hand, are assembled only for the purposes of legislation, and to consult about the constitution, and to hear the edicts of the magistrates. In another variety of democracy the citizen form one assembly, but meet only to elect magistrates, to pass laws, to advise about war and peace, and to make scrutinies. Other matters are referred severally to special magistrates, who are elected by vote or by lot out of all the citizens Or again, the citizens meet about election to offices and about scrutinies, and deliberate concerning war or alliances while other matters are administered by the magistrates, who, as far as is possible, are elected by vote. I am speaking of those magistracies in which special knowledge is required. A fourth form of democracy is when all the citizens meet to deliberate about everything, and the magistrates decide nothing, but only make the preliminary inquiries; and that is the way in which the last and worst form of democracy, corresponding, as we maintain, to the close family oligarchy and to tyranny, is at present administered. All these modes are democratical.

On the other hand, that some should deliberate about all is oligarchical. This again is a mode which, like the democratical has many forms. When the deliberative class being elected out of those who have a moderate qualification are numerous and they respect and obey the prohibitions of the law without altering it, and any one who has the required qualification shares in the government, then, just because of this moderation, the oligarchy inclines towards polity. But when only selected individuals and not the whole people share in the deliberations of the state, then, although, as in the former case, they observe the law, the government is a pure oligarchy. Or, again, when those who have the power of deliberation are self-elected, and son succeeds father, and they and not the laws are supreme — the government is of necessity oligarchical. Where, again, particular persons have authority in particular matters — for example, when the whole people decide about peace and war and hold scrutinies, but the magistrates regulate everything else, and they are elected by vote — there the government is an aristocracy. And if some questions are decided by magistrates elected by vote, and others by magistrates elected by lot, either absolutely or out of select candidates, or elected partly by vote, partly by lot — these practices are partly characteristic of an aristocratical government, and party of a pure constitutional government.

These are the various forms of the deliberative body; they correspond to the various forms of government. And the government of each state is administered according to one or other of the principles which have been laid down. Now it is for the interest of democracy, according to the most prevalent notion of it (I am speaking of that extreme form of democracy in which the people are supreme even over the laws), with a view to better deliberation to adopt the custom of oligarchies respecting courts of law. For in oligarchies the rich who are wanted to be judges are compelled to attend under pain of a fine, whereas in deinocracies the poor are paid to attend. And this practice of oligarchies should be adopted by democracies in their public assemblies, for they will advise better if they all deliberate together — the people with the notables and the notables with the people. It is also a good plan that those who deliberate should be elected by vote or by lot in equal numbers out of the different classes; and that if the people greatly exceed in number those who have political training, pay should not be given to all, but only to as many as would balance the number of the notables, or that the number in excess should be eliminated by lot. But in oligarchies either certain persons should be co-opted from the mass, or a class of officers should be appointed such as exist in some states who are termed probuli and guardians of the law; and the citizens should occupy themselves exclusively with matters on which these have previously deliberated; for so the people will have a share in the deliberations of the state, but will not be able to disturb the principles of the constitution. Again, in oligarchies either the people ought to accept the measures of the government, or not to pass anything contrary to them; or, if all are allowed to share in counsel, the decision should rest with the magistrates. The opposite of what is done in constitutional governments should be the rule in oligarchies; the veto of the majority should be final, their assent not final, but the proposal should be referred back to the magistrates. Whereas in constitutional governments they take the contrary course; the few have the negative, not the affirmative power; the affirmation of everything rests with the multitude.

These, then, are our conclusions respecting the deliberative, that is, the supreme element in states.

XV

Next we will proceed to consider the distribution of offices; this too, being a part of politics concerning which many questions arise: What shall their number be? Over what shall they preside, and what shall be their duration? Sometimes they last for six months, sometimes for less; sometimes they are annual, while in other cases offices are held for still longer periods. Shall they be for life or for a long term of years; or, if for a short term only, shall the same persons hold them over and over again, or once only? Also about the appointment to them — from whom are they to be chosen, by whom, and how? We should first be in a position to say what are the possible varieties of them, and then we may proceed to determine which are suited to different forms of government. But what are to be included under the term 'offices'? That is a question not quite so easily answered. For a political community requires many officers; and not every one who is chosen by vote or by lot is to be regarded as a ruler. In the first place there are the priests, who must be distinguished from political officers; masters of choruses and heralds, even ambassadors, are elected by vote. Some duties of superintendence again are political, extending either to all the citizens in a single sphere of action, like the office of the general who superintends them when they are in the field, or to a section of them only, like the inspectorships of women or of youth. Other offices are concerned with household management, like that of the corn measurers who exist in many states and are elected officers. There are also menial offices which the rich have executed by their slaves. Speaking generally, those are to be called offices to which the duties are assigned of deliberating about certain measures and of judging and commanding, especially the last; for to command is the especial duty of a magistrate. But the question is not of any importance in practice; no one has ever brought into court the meaning of the word, although such problems have a speculative interest.

What kinds of offices, and how many, are necessary to the existence of a state, and which, if not necessary, yet conduce to its well being are much more important considerations, affecting all constitutions, but more especially small states. For in great states it is possible, and indeed necessary, that every office should have a special function; where the citizens are numerous, many may hold office. And so it happens that some offices a man holds a second time only after a long interval, and others he holds once only; and certainly every work is better done which receives the sole, and not the divided attention of the worker. But in small states it is necessary to combine many offices in a few hands, since the small number of citizens does not admit of many holding office: for who will there be to succeed them? And yet small states at times require the same offices and laws as large ones; the difference is that the one want them often, the others only after long intervals. Hence there is no reason why the care of many offices should not be imposed on the same person, for they will not interfere with each other. When the population is small, offices should be like the spits which also serve to hold a lamp. We must first ascertain how many magistrates are necessary in every state, and also how many are not exactly necessary, but are nevertheless useful, and then there will be no difficulty in seeing what offices can be combined in one. We should also know over which matters several local tribunals are to have jurisdiction, and in which authority should be centralized: for example, should one person keep order in the market and another in some other place, or should the same person be responsible everywhere? Again, should offices be divided according to the subjects with which they deal, or according to the persons with whom they deal: I mean to say, should one person see to good order in general, or one look after the boys, another after the women, and so on? Further, under different constitutions, should the magistrates be the same or different? For example, in democracy, oligarchy, aristocracy, monarchy, should there be the same magistrates, although they are elected, not out of equal or similar classes of citizen but differently under different constitutions — in aristocracies, for example, they are chosen from the educated, in oligarchies from the wealthy, and in democracies from the free — or are there certain differences in the offices answering to them as well, and may the same be suitable to some, but different offices to others? For in some states it may be convenient that the same office should have a more extensive, in other states a narrower sphere. Special offices are peculiar to certain forms of government: for example that of probuli, which is not a democratic office, although a bule or council is. There must be some body of men whose duty is to prepare measures for the people in order that they may not be diverted from their business; when these are few in number, the state inclines to an oligarchy: or rather the probuli must always be few, and are therefore an oligarchical element. But when both institutions exist in a state, the probuli are a check on the council; for the counselors is a democratic element, but the probuli are oligarchical. Even the power of the council disappears when democracy has taken that extreme form in which the people themselves are always meeting and deliberating about everything. This is the case when the members of the assembly receive abundant pay; for they have nothing to do and are always holding assemblies and deciding everything for themselves. A magistracy which controls the boys or the women, or any similar office, is suited to an aristocracy rather than to a democracy; for how can the magistrates prevent the wives of the poor from going out of doors? Neither is it an oligarchical office; for the wives of the oligarchs are too fine to be controlled.

Enough of these matters. I will now inquire into appointments to offices. The varieties depend on three terms, and the combinations of these give all possible modes: first, who appoints? secondly, from whom? and thirdly, how? Each of these three admits of three varieties: (A) All the citizens, or (B) only some, appoint. Either (1) the magistrates are chosen out of all or (2) out of some who are distinguished either by a property qualification, or by birth, or merit, or for some special reason, as at Megara only those were eligible who had returned from exile and fought together against the democracy. They may be appointed either (a) by vote or (b) by lot. Again, these several varieties may be coupled, I mean that (C) some officers may be elected by some, others by all, and (3) some again out of some, and others out of all, and (c) some by vote and others by lot. Each variety of these terms admits of four modes.

For either (A 1 a) all may appoint from all by vote, or (A 1 b) all from all by lot, or (A 2 a) all from some by vote, or (A 2 b) all from some by lot (and from all, either by sections, as, for example, by tribes, and wards, and phratries, until all the citizens have been gone through; or the citizens may be in all cases eligible indiscriminately); or again (A 1 c, A 2 c) to some offices in the one way, to some in the other. Again, if it is only some that appoint, they may do so either (B 1 a) from all by vote, or (B 1 b) from all by lot, or (B 2 a) from some by vote, or (B 2 b) from some by lot, or to some offices in the one way, to others in the other, i.e., (B 1 c) from all, to some offices by vote, to some by lot, and (B 2 C) from some, to some offices by vote, to some by lot. Thus the modes that arise, apart from two (C, 3) out of the three couplings, number twelve. Of these systems two are popular, that all should appoint from all (A 1 a) by vote or (A 1 b) by lot — or (A 1 c) by both. That all should not appoint at once, but should appoint from all or from some either by lot or by vote or by both, or appoint to some offices from all and to others from some ('by both' meaning to some offices by lot, to others by vote), is characteristic of a polity. And (B 1 c) that some should appoint from all, to some offices by vote, to others by lot, is also characteristic of a polity, but more oligarchical than the former method. And (A 3 a, b, c, B 3 a, b, c) to appoint from both, to some offices from all, to others from some, is characteristic of a polity with a leaning towards aristocracy. That (B 2) some should appoint from some is oligarchical — even (B 2 b) that some should appoint from some by lot (and if this does not actually occur, it is none the less oligarchical in character), or (B 2 C) that some should appoint from some by both. (B 1 a) that some should appoint from all, and (A 2 a) that all should appoint from some, by vote, is aristocratic.

These are the different modes of constituting magistrates, and these correspond to different forms of government: which are proper to which, or how they ought to be established, will be evident when we determine the nature of their powers. By powers I mean such powers as a magistrate exercises over the revenue or in defense of the country; for there are various kinds of power: the power of the general, for example, is not the same with that which regulates contracts in the market.

XVI

Of the three parts of government, the judicial remains to be considered, and this we shall divide on the same principle. There are three points on which the variedes of law-courts depend: The persons from whom they are appointed, the matters with which they are concerned, and the manner of their appointment. I mean, (1) are the judges taken from all, or from some only? (2) how many kinds of law-courts are there? (3) are the judges chosen by vote or by lot?

First, let me determine how many kinds of law-courts there are. There are eight in number: One is the court of audits or scrutinies; a second takes cognizance of ordinary offenses against the state; a third is concerned with treason against the constitution; the fourth determines disputes respecting penalties, whether raised by magistrates or by private persons; the fifth decides the more important civil cases; the sixth tries cases of homicide, which are of various kinds, (a) premeditated, (b) involuntary, (c) cases in which the guilt is confessed but the justice is disputed; and there may be a fourth court (d) in which murderers who have fled from justice are tried after their return; such as the Court of Phreatto is said to be at Athens. But cases of this sort rarely happen at all even in large cities. The different kinds of homicide may be tried either by the same or by different courts. (7) There are courts for strangers: of these there are two subdivisions, (a) for the settlement of their disputes with one another, (b) for the settlement of disputes between them and the citizens. And besides all these there must be (8) courts for small suits about sums of a drachma up to five drachmas, or a little more, which have to be determined, but they do not require many judges.

Nothing more need be said of these small suits, nor of the courts for homicide and for strangers: I would rather speak of political cases, which, when mismanaged, create division and disturbances in constitutions.

Now if all the citizens judge, in all the different cases which I have distinguished, they may be appointed by vote or by lot, or sometimes by lot and sometimes by vote. Or when a single class of causes are tried, the judges who decide them may be appointed, some by vote, and some by lot. These then are the four modes of appointing judges from the whole people, and there will be likewise four modes, if they are elected from a part only; for they may be appointed from some by vote and judge in all causes; or they may be appointed from some by lot and judge in all causes; or they may be elected in some cases by vote, and in some cases taken by lot, or some courts, even when judging the same causes, may be composed of members some appointed by vote and some by lot. These modes, then, as was said, answer to those previously mentioned.

Once more, the modes of appointment may be combined; I mean, that some may be chosen out of the whole people, others out of some, some out of both; for example, the same tribunal may be composed of some who were elected out of all, and of others who were elected out of some, either by vote or by lot or by both.

In how many forms law-courts can be established has now been considered. The first form, viz., that in which the judges are taken from all the citizens, and in which all causes are tried, is democratical; the second, which is composed of a few only who try all causes, oligarchical; the third, in which some courts are taken from all classes, and some from certain classes only, aristocratical and constitutional.

On Generation and Corruption

Table of Contents

Book I

1

OUR next task is to study coming-to-be and passing-away. We are to distinguish the causes, and to state the definitions, of these processes considered in general-as changes predicable uniformly of all the things that come-to-be and pass-away by nature. Further, we are to study growth and 'alteration'. We must inquire what each of them is; and whether 'alteration' is to be identified with coming-to-be, or whether to these different names there correspond two separate processes with distinct natures.

On this question, indeed, the early philosophers are divided. Some of them assert that the so-called 'unqualified coming-to-be' is 'alteration', while others maintain that 'alteration' and coming-to-be are distinct. For those who say that the universe is one something (i.e. those who generate all things out of one thing) are bound to assert that coming-to-be is 'alteration', and that whatever 'comes-to-be' in the proper sense of the term is 'being altered': but those who make the matter of things more than one must distinguish coming-to-be from 'alteration'. To this latter class belong Empedocles, Anaxagoras, and Leucippus. And yet Anaxagoras himself failed to understand his own utterance. He says, at all events, that coming-to-be and passing-away are the same as 'being altered':' yet, in common with other thinkers, he affirms that the elements are many. Thus Empedocles holds that the corporeal elements are four, while all the elements-including those which initiate movement-are six in number; whereas Anaxagoras agrees with Leucippus and Democritus that the elements are infinite.

(Anaxagoras posits as elements the 'homoeomeries', viz. bone, flesh, marrow, and everything else which is such that part and whole are the same in name and nature; while Democritus and Leucippus say that there are indivisible bodies, infinite both in number and in the varieties of their shapes, of which everything else is composed-the compounds differing one from another according to the shapes, 'positions', and 'groupings' of their constituents.)

For the views of the school of Anaxagoras seem diametrically opposed to those of the followers of Empedocles. Empedocles says that Fire, Water, Air, and Earth are four elements, and are thus 'simple' rather than flesh, bone, and bodies which, like these, are 'homoeomeries'. But the followers of Anaxagoras regard the 'homoeomeries' as 'simple' and elements, whilst they affirm that Earth, Fire, Water, and Air are composite; for each of these is (according to them) a 'common seminary' of all the 'homoeomeries'.

Those, then, who construct all things out of a single element, must maintain that coming-to-be and passing-away are 'alteration'. For they must affirm that the underlying something always remains identical and one; and change of such a substratum is what we call 'altering' Those, on the other hand, who make the ultimate kinds of things more than one, must maintain that 'alteration' is distinct from coming-to-be: for coming-to-be and passing-away result from the consilience and the dissolution of the many kinds. That is why Empedocles too uses language to this effect, when he says 'There is no coming-to-be of anything, but only a mingling and a divorce of what has been mingled'. Thus it is clear (i) that to describe coming-to-be and passing-away in these terms is in accordance with their fundamental assumption, and (ii) that they do in fact so describe them: nevertheless, they too must recognize 'alteration' as a fact distinct from coming to-be, though it is impossible for them to do so consistently with what they say.

That we are right in this criticism is easy to perceive. For 'alteration' is a fact of observation. While the substance of the thing remains unchanged, we see it 'altering' just as we see in it the changes of magnitude called 'growth' and 'diminution'. Nevertheless, the statements of those who posit more 'original reals' than one make 'alteration' impossible. For 'alteration, as we assert, takes place in respect to certain qualities: and these qualities (I mean, e.g. hot-cold, white-black, dry-moist, soft-hard, and so forth) are, all of them, differences characterizing the 'elements'. The actual words of Empedocles may be quoted in illustration

The sun everywhere bright to see, and hot,
The rain everywhere dark and cold;

and he distinctively characterizes his remaining elements in a similar manner. Since, therefore, it is not possible for Fire to become Water, or Water to become Earth, neither will it be possible for anything white to become black, or anything soft to become hard; and the same argument applies to all the other qualities. Yet this is what 'alteration' essentially is.

It follows, as an obvious corollary, that a single matter must always be assumed as underlying the contrary 'poles' of any change whether change of place, or growth and diminution, or 'alteration'; further, that the being of this matter and the being of 'alteration' stand and fall together. For if the change is 'alteration', then the substratum is a single element; i.e. all things which admit of change into one another have a single matter. And, conversely, if the substratum of the changing things is one, there is 'alteration'.

Empedocles, indeed, seems to contradict his own statements as well as the observed facts. For he denies that any one of his elements comes-to-be out of any other, insisting on the contrary that they are the things out of which everything else comes-to-be; and yet (having brought the entirety of existing things, except Strife, together into one) he maintains, simultaneously with this denial, that each thing once more comes-to-be out of the One. Hence it was clearly out of a One that this came-to-be Water, and that Fire, various portions of it being separated off by certain characteristic differences or qualities-as indeed he calls the sun 'white and hot', and the earth 'heavy and hard'. If, therefore, these characteristic differences be taken away (for they can be taken away, since they came-to-be), it will clearly be inevitable for Earth to come to-be out of Water and Water out of Earth, and for each of the other elements to undergo a similar transformation-not only then, but also now-if, and because, they change their qualities. And, to judge by what he says, the qualities are such that they can be 'attached' to things and can again be 'separated' from them, especially since Strife and Love are still fighting with one another for the mastery. It was owing to this same conflict that the elements were generated from a One at the former period. I say 'generated', for presumably Fire, Earth, and Water had no distinctive existence at all while merged in one.

There is another obscurity in the theory Empedocles. Are we to regard the One as his 'original real'? Or is it the Many-i.e. Fire and Earth, and the bodies co-ordinate with these? For the One is an 'element' in so far as it underlies the process as matter-as that out of which Earth and Fire come-to-be through a change of qualities due to 'the motion'. On the other hand, in so far as the One results from composition (by a consilience of the Many), whereas they result from disintegration the Many are more 'elementary' than the One, and prior to it in their nature.

2

We have therefore to discuss the whole subject of 'unqualified' coming-to-be and passing-away; we have to inquire whether these changes do or do not occur and, if they occur, to explain the precise conditions of their occurrence. We must also discuss the remaining forms of change, viz. growth and 'alteration'. For though, no doubt, Plato investigated the conditions under which things come-to-be and pass-away, he confined his inquiry to these changes; and he discussed not all coming-to-be, but only that of the elements. He asked no questions as to how flesh or bones, or any of the other similar compound things, come-to-be; nor again did he examine the conditions under which 'alteration' or growth are attributable to things.

A similar criticism applies to all our predecessors with the single exception of Democritus. Not one of them penetrated below the surface or made a thorough examination of a single one of the problems. Democritus, however, does seem not only to have thought carefully about all the problems, but also to be distinguished from the outset by his method. For, as we are saying, none of the other philosophers made any definite statement about growth, except such as any amateur might have made. They said that things grow 'by the accession of like to like', but they did not proceed to explain the manner of this accession. Nor did they give any account of 'combination': and they neglected almost every single one of the remaining problems, offering no explanation, e.g. of 'action' or 'passion' how in physical actions one thing acts and the other undergoes action. Democritus and Leucippus, however, postulate the 'figures', and make 'alteration' and coming-to-be result from them. They explain coming-to-be and passing-away by their 'dissociation' and 'association', but 'alteration' by their 'grouping' and 'Position'. And since they thought that the 'truth lay in the appearance, and the appearances are conflicting and infinitely many, they made the 'figures' infinite in number. Hence-owing to the changes of the compound-the same thing seems different and conflicting to different people: it is 'transposed' by a small additional ingredient, and appears utterly other by the 'transposition' of a single constituent. For Tragedy and Comedy are both composed of the same letters.

Since almost all our predecessors think (i) that coming-to-be is distinct from 'alteration', and (ii) that, whereas things 'alter' by change of their qualities, it is by 'association' and 'dissociation' that they come-to-be and pass-away, we must concentrate our attention on these theses. For they lead to many perplexing and well-grounded dilemmas. If, on the one hand, coming-to-be is 'association', many impossible consequences result: and yet there are other arguments, not easy to unravel, which force the conclusion upon us that coming-to-be cannot possibly be anything else. If, on the other hand, coming-to-be is not 'association', either there is no such thing as coming-to-be at all or it is 'alteration': or else we must endeavour to unravel this dilemma too-and a stubborn one we shall find it. The fundamental question, in dealing with all these difficulties, is this: 'Do things come-to-be and "alter" and grow, and undergo the contrary changes, because the primary "reals" are indivisible magnitudes? Or is no magnitude indivisible?' For the answer we give to this question makes the greatest difference. And again, if the primary 'reals' are indivisible magnitudes, are these bodies, as Democritus and Leucippus maintain? Or are they planes, as is asserted in the Timaeus?

To resolve bodies into planes and no further-this, as we have also remarked elsewhere, in itself a paradox. Hence there is more to be said for the view that there are indivisible bodies. Yet even these involve much of paradox. Still, as we have said, it is possible to construct 'alteration' and coming-to-be with them, if one 'transposes' the same by 'turning' and 'intercontact', and by 'the varieties of the figures', as Democritus does. (His denial of the reality of colour is a corollary from this position: for, according to him, things get coloured by 'turning' of the 'figures'.) But the possibility of such a construction no longer exists for those who divide bodies into planes. For nothing except solids results from putting planes together: they do not even attempt to generate any quality from them.

Lack of experience diminishes our power of taking a comprehensive view of the admitted facts. Hence those who dwell in intimate association with nature and its phenomena grow more and more able to formulate, as the foundations of their theories, principles such as to admit of a wide and coherent development: while those whom devotion to abstract discussions has rendered unobservant of the facts are too ready to dogmatize on the basis of a few observations. The rival treatments of the subject now before us will serve to illustrate how great is the difference between a 'scientific' and a 'dialectical' method of inquiry. For, whereas the Platonists argue that there must be atomic magnitudes 'because otherwise "The Triangle" will be more than one', Democritus would appear to have been convinced by arguments appropriate to the subject, i.e. drawn from the science of nature. Our meaning will become clear as we proceed. For to suppose that a body (i.e. a magnitude) is divisible through and through, and that this division is possible, involves a difficulty. What will there be in the body which escapes the division?

If it is divisible through and through, and if this division is possible, then it might be, at one and the same moment, divided through and through, even though the dividings had not been effected simultaneously: and the actual occurrence of this result would involve no impossibility. Hence the same principle will apply whenever a body is by nature divisible through and through, whether by bisection, or generally by any method whatever: nothing impossible will have resulted if it has actually been divided-not even if it has been divided into innumerable parts, themselves divided innumerable times. Nothing impossible will have resulted, though perhaps nobody in fact could so divide it.

Since, therefore, the body is divisible through and through, let it have been divided. What, then, will remain? A magnitude? No: that is impossible, since then there will be something not divided, whereas ex hypothesi the body was divisible through and through. But if it be admitted that neither a body nor a magnitude will remain, and yet division is to take place, the constituents of the body will either be points (i.e. without magnitude) or absolutely nothing. If its constituents are nothings, then it might both come-to-be out of nothings and exist as a composite of nothings: and thus presumably the whole body will be nothing but an appearance. But if it consists of points, a similar absurdity will result: it will not possess any magnitude. For when the points were in contact and coincided to form a single magnitude, they did not make the whole any bigger (since, when the body was divided into two or more parts, the whole was not a bit smaller or bigger than it was before the division): hence, even if all the points be put together, they will not make any magnitude.

But suppose that, as the body is being divided, a minute section-a piece of sawdust, as it were-is extracted, and that in this sense-a body 'comes away' from the magnitude, evading the division. Even then the same argument applies. For in what sense is that section divisible? But if what 'came away' was not a body but a separable form or quality, and if the magnitude is 'points or contacts thus qualified': it is paradoxical that a magnitude should consist of elements, which are not magnitudes. Moreover, where will the points be? And are they motionless or moving? And every contact is always a contact of two somethings, i.e. there is always something besides the contact or the division or the point.

These, then, are the difficulties resulting from the supposition that any and every body, whatever its size, is divisible through and through. There is, besides, this further consideration. If, having divided a piece of wood or anything else, I put it together, it is again equal to what it was, and is one. Clearly this is so, whatever the point at which I cut the wood. The wood, therefore, has been divided potentially through and through. What, then, is there in the wood besides the division? For even if we suppose there is some quality, yet how is the wood dissolved into such constituents and how does it come-to-be out of them? Or how are such constituents separated so as to exist apart from one another? Since, therefore, it is impossible for magnitudes to consist of contacts or points, there must be indivisible bodies and magnitudes. Yet, if we do postulate the latter, we are confronted with equally impossible consequences, which we have examined in other works.' But we must try to disentangle these perplexities, and must therefore formulate the whole problem over again.

On the one hand, then, it is in no way paradoxical that every perceptible body should be indivisible as well as divisible at any and every point. For the second predicate will attach to it potentially, but the first actually. On the other hand, it would seem to be impossible for a body to be, even potentially, divisible at all points simultaneously. For if it were possible, then it might actually occur, with the result, not that the body would simultaneously be actually both (indivisible and divided), but that it would be simultaneously divided at any and every point. Consequently, nothing will remain and the body will have passed-away into what is incorporeal: and so it might come-to-be again either out of points or absolutely out of nothing. And how is that possible?

But now it is obvious that a body is in fact divided into separable magnitudes which are smaller at each division-into magnitudes which fall apart from one another and are actually separated. Hence (it is urged) the process of dividing a body part by part is not a 'breaking up' which could continue ad infinitum; nor can a body be simultaneously divided at every point, for that is not possible; but there is a limit, beyond which the 'breaking up' cannot proceed. The necessary consequence-especially if coming-to-be and passing-away are to take place by 'association' and 'dissociation' respectively-is that a body must contain atomic magnitudes which are invisible. Such is the argument which is believed to establish the necessity of atomic magnitudes: we must now show that it conceals a faulty inference, and exactly where it conceals it.

For, since point is not 'immediately-next' to point, magnitudes are 'divisible through and through' in one sense, and yet not in another. When, however, it is admitted that a magnitude is 'divisible through and through', it is thought there is a point not only anywhere, but also everywhere, in it: hence it is supposed to follow, from the admission, that the magnitude must be divided away into nothing. For it is supposed—there is a point everywhere within it, so that it consists either of contacts or of points. But it is only in one sense that the magnitude is 'divisible through and through', viz. in so far as there is one point anywhere within it and all its points are everywhere within it if you take them singly one by one. But there are not more points than one anywhere within it, for the points are not 'consecutive': hence it is not simultaneously 'divisible through and through'. For if it were, then, if it be divisible at its centre, it will be divisible also at a point 'immediately-next' to its centre. But it is not so divisible: for position is not 'immediately-next' to position, nor point to point-in other words, division is not 'immediately-next' to division, nor composition to composition.

Hence there are both 'association' and 'dissociation', though neither (a) into, and out of, atomic magnitudes (for that involves many impossibilities), nor (b) so that division takes place through and through-for this would have resulted only if point had been 'immediately-next' to point: but 'dissociation' takes place into small (i.e. relatively small) parts, and 'association' takes place out of relatively small parts.

It is wrong, however, to suppose, as some assert, that coming-to-be and passing-away in the unqualified and complete sense are distinctively defined by 'association' and 'dissociation', while the change that takes place in what is continuous is 'alteration'. On the contrary, this is where the whole error lies. For unqualified coming-to-be and passing-away are not effected by 'association' and 'dissociation'. They take place when a thing changes, from this to that, as a whole. But the philosophers we are criticizing suppose that all such change is 'alteration': whereas in fact there is a difference. For in that which underlies the change there is a factor corresponding to the definition and there is a material factor. When, then, the change is in these constitutive factors, there will be coming-to-be or passing-away: but when it is in the thing's qualities, i.e. a change of the thing per accidens, there will be 'alteration'.

'Dissociation' and 'association' affect the thing's susceptibility to passing-away. For if water has first been 'dissociated' into smallish drops, air comes-to-be out of it more quickly: while, if drops of water have first been 'associated', air comes-to-be more slowly. Our doctrine will become clearer in the sequel.' Meantime, so much may be taken as established-viz. that coming-to-be cannot be 'association', at least not the kind of 'association' some philosophers assert it to be.

3

Now that we have established the preceding distinctions, we must first consider whether there is anything which comes-to-be and passes-away in the unqualified sense: or whether nothing comes-to-be in this strict sense, but everything always comes-to-be something and out of something-I mean, e.g. comes-to-be-healthy out of being-ill and ill out of being-healthy, comes-to-be-small out of being big and big out of being-small, and so on in every other instance. For if there is to be coming-to-be without qualification, 'something' must-without qualification-'come-to-be out of not-being', so that it would be true to say that 'not-being is an attribute of some things'. For qualified coming-to-be is a process out of qualified not-being (e.g. out of not-white or not-beautiful), but unqualified coming-to-be is a process out of unqualified not-being.

Now 'unqualified' means either (i) the primary predication within each Category, or (ii) the universal, i.e. the all-comprehensive, predication. Hence, if 'unqualified not-being' means the negation of 'being' in the sense of the primary term of the Category in question, we shall have, in 'unqualified coming-to-be', a coming-to-be of a substance out of not-substance. But that which is not a substance or a 'this' clearly cannot possess predicates drawn from any of the other Categories either—e.g. we cannot attribute to it any quality, quantity, or position. Otherwise, properties would admit of existence in separation from substances. If, on the other hand, 'unqualified not-being' means 'what is not in any sense at all', it will be a universal negation of all forms of being, so that what comes-to-be will have to come-to-be out of nothing.

Although we have dealt with these problems at greater length in another work, where we have set forth the difficulties and established the distinguishing definitions, the following concise restatement of our results must here be offered: In one sense things come-to-be out of that which has no 'being' without qualification: yet in another sense they come-to-be always out of what is'. For coming-to-be necessarily implies the pre-existence of something which potentially 'is', but actually 'is not'; and this something is spoken of both as 'being' and as 'not-being'.

These distinctions may be taken as established: but even then it is extraordinarily difficult to see how there can be 'unqualified coming-to-be' (whether we suppose it to occur out of what potentially 'is', or in some other way), and we must recall this problem for further examination. For the question might be raised whether substance (i.e. the 'this') comes-to-be at all. Is it not rather the 'such', the 'so great', or the 'somewhere', which comes-to-be? And the same question might be raised about 'passing-away' also. For if a substantial thing comes-to-be, it is clear that there will 'be' (not actually, but potentially) a substance, out of which its coming-to-be will proceed and into which the thing that is passing-away will necessarily change. Then will any predicate belonging to the remaining Categories attach actually to this presupposed substance? In other words, will that which is only potentially a 'this' (which only potentially is), while without the qualification 'potentially' it is not a 'this' (i.e. is not), possess, e.g. any determinate size or quality or position? For (i) if it possesses none of these determinations actually, but all of them only potentially, the result is first that a being, which is not a determinate being, is capable of separate existence; and in addition that coming-to-be proceeds out of nothing pre-existing-a thesis which, more than any other, preoccupied and alarmed the earliest philosophers. On the other hand (ii) if, although it is not a 'this somewhat' or a substance, it is to possess some of the remaining determinations quoted above, then (as we said)' properties will be separable from substances.

We must therefore concentrate all our powers on the discussion of these difficulties and on the solution of a further question—viz. What is the cause of the perpetuity of coming-to-be? Why is there always unqualified, as well as partial, coming-to-be? Cause' in this connexion has two senses. It means (i) the source from which, as we say, the process 'originates', and (ii) the matter. It is the material cause that we have here to state. For, as to the other cause, we have already explained (in our treatise on Motion that it involves (a) something immovable through all time and (b) something always being moved. And the accurate treatment of the first of these-of the immovable 'originative source'-belongs to the province of the other, or 'prior', philosophy: while as regards 'that which sets everything else in motion by being itself continuously moved', we shall have to explain later' which amongst the so-called 'specific' causes exhibits this character. But at present we are to state the material cause-the cause classed under the head of matter-to which it is due that passing-away and coming-to-be never fail to occur in Nature. For perhaps, if we succeed in clearing up this question, it will simultaneously become clear what account we ought to give of that which perplexed us just now, i.e. of unqualified passing-away and coming-to-be.

Our new question too—viz. 'what is the cause of the unbroken continuity of coming-to-be?'-is sufficiently perplexing, if in fact what passes-away vanishes into 'what is not' and 'what is not' is nothing (since 'what is not' is neither a thing, nor possessed of a quality or quantity, nor in any place). If, then, some one of the things 'which are' constantly disappearing, why has not the whole of 'what is' been used up long ago and vanished away assuming of course that the material of all the several comings-to-be was finite? For, presumably, the unfailing continuity of coming-to-be cannot be attributed to the infinity of the material. That is impossible, for nothing is actually infinite. A thing is infinite only potentially, i.e. the dividing of it can continue indefinitely: so that we should have to suppose there is only one kind of coming-to-be in the world-viz. one which never fails, because it is such that what comes-to-be is on each successive occasion smaller than before. But in fact this is not what we see occurring.

Why, then, is this form of change necessarily ceaseless? Is it because the passing-away of this is a coming-to-be of something else, and the coming-to-be of this a passing-away of something else?

The cause implied in this solution must no doubt be considered adequate to account for coming-to-be and passing-away in their general character as they occur in all existing things alike. Yet, if the same process is a coming to-be of this but a passing-away of that, and a passing-away of this but a coming-to-be of that, why are some things said to come-to-be and pass-away without qualification, but others only with a qualification?

The distinction must be investigated once more, for it demands some explanation. (It is applied in a twofold manner.) For (i) we say 'it is now passing-away' without qualification, and not merely 'this is passing-away': and we call this change 'coming-to-be', and that 'passing-away', without qualification. And (ii) so-and-so 'comes-to-be-something', but does not 'come-to-be' without qualification; for we say that the student 'comes-to-be-learned', not 'comes-to-be' without qualification.

(i) Now we often divide terms into those which signify a 'this somewhat' and those which do not. And (the first form of) the distinction, which we are investigating, results from a similar division of terms: for it makes a difference into what the changing thing changes. Perhaps, e.g. the passage into Fire is 'coming-to-be' unqualified, but 'passing-away-of-something' (e.g. Earth): whilst the coming-to-be of Earth is qualified (not unqualified) 'coming-to-be', though unqualified 'passing-away' (e.g. of Fire). This would be the case on the theory set forth in Parmenides: for he says that the things into which change takes place are two, and he asserts that these two, viz. what is and what is not, are Fire and Earth. Whether we postulate these, or other things of a similar kind, makes no difference. For we are trying to discover not what undergoes these changes, but what is their characteristic manner. The passage, then, into what 'is' not except with a qualification is unqualified passing-away, while the passage into what 'is' without qualification is unqualified coming-to-be. Hence whatever the contrasted 'poles' of the changes may be whether Fire and Earth, or some other couple-the one of them will be 'a being' and the other 'a not-being'.

We have thus stated one characteristic manner in which unqualified will be distinguished from qualified coming-to-be and passing-away: but they are also distinguished according to the special nature of the material of the changing thing. For a material, whose constitutive differences signify more a 'this somewhat', is itself more 'substantial' or 'real': while a material, whose constitutive differences signify privation, is 'not real'. (Suppose, e.g. that 'the hot' is a positive predication, i.e. a 'form', whereas 'cold' is a privation, and that Earth and Fire differ from one another by these constitutive differences.)

The opinion, however, which most people are inclined to prefer, is that the distinction depends upon the difference between 'the perceptible' and 'the imperceptible'. Thus, when there is a change into perceptible material, people say there is 'coming-to-be'; but when there is a change into invisible material, they call it 'passing-away'. For they distinguish 'what is' and 'what is not' by their perceiving and not-perceiving, just as what is knowable 'is' and what is unknowable 'is not'-perception on their view having the force of knowledge. Hence, just as they deem themselves to live and to 'be' in virtue of their perceiving or their capacity to perceive, so too they deem the things to 'be' qua perceived or perceptible-and in this they are in a sense on the track of the truth, though what they actually say is not true.

Thus unqualified coming-to-be and passingaway turn out to be different according to common opinion from what they are in truth. For Wind and Air are in truth more real more a 'this somewhat' or a 'form'-than Earth. But they are less real to perception which explains why things are commonly said to 'pass-away' without qualification when they change into Wind and Air, and to 'come-to-be' when they change into what is tangible, i.e. into Earth.

We have now explained why there is 'unqualified coming-to-be' (though it is a passing-away-of-something) and 'unqualified passing-away (though it is a coming-to-be-of-something). For this distinction of appellation depends upon a difference in the material out of which, and into which, the changes are effected. It depends either upon whether the material is or is not 'substantial', or upon whether it is more or less 'substantial', or upon whether it is more or less perceptible.

(ii) But why are some things said to 'come to-be' without qualification, and others only to 'come-to-be-so-and-so', in cases different from the one we have been considering where two things come-to-be reciprocally out of one another? For at present we have explained no more than this:-why, when two things change reciprocally into one another, we do not attribute coming-to-be and passing-away uniformly to them both, although every coming-to-be is a passing-away of something else and every passing-away some other thing's coming-to-be. But the question subsequently formulated involves a different problem-viz. why, although the learning thing is said to 'come-to-be-learned' but not to 'come-to-be' without qualification, yet the growing thing is said to 'come-to-be'.

The distinction here turns upon the difference of the Categories. For some things signify a this somewhat, others a such, and others a so-much. Those things, then, which do not signify substance, are not said to 'come-to-be' without qualification, but only to 'come-to-be-so-and-so'. Nevertheless, in all changing things alike, we speak of 'coming-to-be' when the thing comes-to-be something in one of the two Columns—e.g. in Substance, if it comes-to-be Fire but not if it comes-to-be Earth; and in Quality, if it comes-to-be learned but not when it comes-to-be ignorant.

We have explained why some things come to-be without qualification, but not others both in general, and also when the changing things are substances and nothing else; and we have stated that the substratum is the material cause of the continuous occurrence of coming to-be, because it is such as to change from contrary to contrary and because, in substances, the coming-to-be of one thing is always a passing-away of another, and the passing-away of one thing is always another's coming-to-be. But there is no need even to discuss the other question we raised-viz. why coming-to-be continues though things are constantly being destroyed. For just as people speak of 'a passing-away' without qualification when a thing has passed into what is imperceptible and what in that sense 'is not', so also they speak of 'a coming-to-be out of a not-being' when a thing emerges from an imperceptible. Whether, therefore, the substratum is or is not something, what comes-to-be emerges out of a 'not-being': so that a thing comes-to-be out of a not-being' just as much as it 'passes-away into what is not'. Hence it is reasonable enough that coming-to-be should never fail. For coming-to-be is a passing-away of 'what is not' and passing-away is a coming to-be of 'what is not'.

But what about that which 'is' not except with a qualification? Is it one of the two contrary poles of the chang-e.g. Earth (i.e. the heavy) a 'not-being', but Fire (i.e. the light) a 'being'? Or, on the contrary, does what is 'include Earth as well as Fire, whereas what is not' is matter-the matter of Earth and Fire alike? And again, is the matter of each different? Or is it the same, since otherwise they would not come-to-be reciprocally out of one another, i.e. contraries out of contraries? For these things-Fire, Earth, Water, Air-are characterized by 'the contraries'.

Perhaps the solution is that their matter is in one sense the same, but in another sense different. For that which underlies them, whatever its nature may be qua underlying them, is the same: but its actual being is not the same. So much, then, on these topics.

4

Next we must state what the difference is between coming-to-be and 'alteration'-for we maintain that these changes are distinct from one another.

Since, then, we must distinguish (a) the substratum, and (b) the property whose nature it is to be predicated of the substratum; and since change of each of these occurs; there is 'alteration' when the substratum is perceptible and persists, but changes in its own properties, the properties in question being opposed to one another either as contraries or as intermediates. The body, e.g. although persisting as the same body, is now healthy and now ill; and the bronze is now spherical and at another time angular, and yet remains the same bronze. But when nothing perceptible persists in its identity as a substratum, and the thing changes as a whole (when e.g. the seed as a whole is converted into blood, or water into air, or air as a whole into water), such an occurrence is no longer 'alteration'. It is a coming-to-be of one substance and a passing-away of the other-especially if the change proceeds from an imperceptible something to something perceptible (either to touch or to all the senses), as when water comes-to-be out of, or passes-away into, air: for air is pretty well imperceptible. If, however, in such cases, any property (being one of a pair of contraries) persists, in the thing that has come-to-be, the same as it was in the thing which has passed-away-if, e.g. when water comes-to-be out of air, both are transparent or cold-the second thing, into which the first changes, must not be a property of this persistent identical something. Otherwise the change will be 'alteration.' Suppose, e.g. that the musical man passed-away and an unmusical man came-tobe, and that the man persists as something identical. Now, if 'musicalness and unmusicalness' had not been a property essentially inhering in man, these changes would have been a coming-to-be of unmusicalness and a passing-away of musicalness: but in fact 'musicalness and unmusicalness' are a property of the persistent identity, viz. man. (Hence, as regards man, these changes are 'modifications'; though, as regards musical man and unmusical man, they are a passing-away and a coming-to-be.) Consequently such changes are 'alteration.' When the change from contrary to contrary is in quantity, it is 'growth and diminution'; when it is in place, it is 'motion'; when it is in property, i.e. in quality, it is 'alteration': but, when nothing persists, of which the resultant is a property (or an 'accident' in any sense of the term), it is 'coming-to-be', and the converse change is 'passing-away'.

'Matter', in the most proper sense of the term, is to be identified with the substratum which is receptive of coming-to-be and passing-away: but the substratum of the remaining kinds of change is also, in a certain sense, 'matter', because all these substrata are receptive of 'contrarieties' of some kind. So much, then, as an answer to the questions (i) whether coming-to-be 'is' or 'is not'—i.e. what are the precise conditions of its occurrence and (ii) what 'alteration' is: but we have still to treat of growth.

5

We must explain (i) wherein growth differs from coming-to-be and from 'alteration', and ii) what is the process of growing and the process of diminishing in each and all of the things that grow and diminish.

Hence our first question is this: Do these changes differ from one another solely because of a difference in their respective 'spheres'? In other words, do they differ because, while a change from this to that (viz. from potential to actual substance) is coming-to-be, a change in the sphere of magnitude is growth and one in the sphere of quality is 'alteration'-both growth and 'alteration' being changes from what is-potentially to what is-actually magnitude and quality respectively? Or is there also a difference in the manner of the change, since it is evident that, whereas neither what is 'altering' nor what is coming-to-be necessarily changes its place, what is growing or diminishing changes its spatial position of necessity, though in a different manner from that in which the moving thing does so? For that which is being moved changes its place as a whole: but the growing thing changes its place like a metal that is being beaten, retaining its position as a whole while its parts change their places. They change their places, but not in the same way as the parts of a revolving globe. For the parts of the globe change their places while the whole continues to occupy an equal place: but the parts of the rowing thing expand over an ever-increasing place and the parts of the diminishing thing contract within an ever-diminishing area.

It is clear, then, that these changes-the changes of that which is coming-to-be, of that which is 'altering', and of that which is growing-differ in manner as well as in sphere. But how are we to conceive the 'sphere' of the change which is growth and diminution? The sphere' of growing and diminishing is believed to be magnitude. Are we to suppose that body and magnitude come-to-be out of something which, though potentially magnitude and body, is actually incorporeal and devoid of magnitude? And since this description may be understood in two different ways, in which of these two ways are we to apply it to the process of growth? Is the matter, out of which growth takes place, (i) 'separate' and existing alone by itself, or (ii) 'separate' but contained in another body?

Perhaps it is impossible for growth to take place in either of these ways. For since the matter is 'separate', either (a) it will occupy no place (as if it were a point), or (b) it will be a 'void', i.e. a non-perceptible body. But the first of these alternatives is impossible. For since what comes-to-be out of this incorporeal and sizeless something will always be 'somewhere', it too must be 'somewhere'-either intrinsically or indirectly. And the second alternative necessarily implies that the matter is contained in some other body. But if it is to be 'in' another body and yet remains 'separate' in such a way that it is in no sense a part of that body (neither a part of its substantial being nor an 'accident' of it), many impossibilities will result. It is as if we were to suppose that when, e.g. air comes-to-be out of water the process were due not to a change of the but to the matter of the air being 'contained in' the water as in a vessel. This is impossible. For (i) there is nothing to prevent an indeterminate number of matters being thus 'contained in' the water, so that they might come-to-be actually an indeterminate quantity of air; and (ii) we do not in fact see air coming-to-be out of water in this fashion, viz. withdrawing out of it and leaving it unchanged.

It is therefore better to suppose that in all instances of coming-to-be the matter is inseparable, being numerically identical and one with the 'containing' body, though isolable from it by definition. But the same reasons also forbid us to regard the matter, out of which the body comes-to-be, as points or lines. The matter is that of which points and lines are limits, and it is something that can never exist without quality and without form.

Now it is no doubt true, as we have also established elsewhere,' that one thing 'comes-to-be' (in the unqualified sense) out of another thing: and further it is true that the efficient cause of its coming-to-be is either (i) an actual thing (which is the same as the effect either generically-or the efficient cause of the coming-to-be of a hard thing is not a hard thing or specifically, as e.g. fire is the efficient cause of the coming-to-be of fire or one man of the birth of another), or (ii) an actuality. Nevertheless, since there is also a matter out of which corporeal substance itself comes-to-be (corporeal substance, however, already characterized as such-and-such a determinate body, for there is no such thing as body in general), this same matter is also the matter of magnitude and quality-being separable from these matters by definition, but not separable in place unless Qualities are, in their turn, separable.

It is evident, from the preceding development and discussion of difficulties, that growth is not a change out of something which, though potentially a magnitude, actually possesses no magnitude. For, if it were, the 'void' would exist in separation; but we have explained in a former work' that this is impossible. Moreover, a change of that kind is not peculiarly distinctive of growth, but characterizes coming-to-be as such or in general. For growth is an increase, and diminution is a lessening, of the magnitude which is there already-that, indeed, is why the growing thing must possess some magnitude. Hence growth must not be regarded as a process from a matter without magnitude to an actuality of magnitude: for this would be a body's coming-to-be rather than its growth.

We must therefore come to closer quarters with the subject of our inquiry. We must grapple' with it (as it were) from its beginning, and determine the precise character of the growing and diminishing whose causes we are investigating.

It is evident (i) that any and every part of the growing thing has increased, and that similarly in diminution every part has become smaller: also (ii) that a thing grows by the accession, and diminishes by the departure, of something. Hence it must grow by the accession either (a) of something incorporeal or (b) of a body. Now, if (a) it grows by the accession of something incorporeal, there will exist separate a void: but (as we have stated before)' is impossible for a matter of magnitude to exist 'separate'. If, on the other hand (b) it grows by the accession of a body, there will be two bodies-that which grows and that which increases it-in the same place: and this too is impossible.

But neither is it open to us to say that growth or diminution occurs in the way in which e.g. air is generated from water. For, although the volume has then become greater, the change will not be growth, but a coming to-be of the one-viz. of that into which the change is taking place-and a passing-away of the contrasted body. It is not a growth of either. Nothing grows in the process; unless indeed there be something common to both things (to that which is coming-to-be and to that which passed-away), e.g. 'body', and this grows. The water has not grown, nor has the air: but the former has passed-away and the latter has come-to-be, and-if anything has grown-there has been a growth of 'body.' Yet this too is impossible. For our account of growth must preserve the characteristics of that which is growing and diminishing. And these characteristics are three: (i) any and every part of the growing magnitude is made bigger (e.g. if flesh grows, every particle of the flesh gets bigger), (ii) by the accession of something, and (iii) in such a way that the growing thing is preserved and persists. For whereas a thing does not persist in the processes of unqualified coming-to-be or passing-away, that which grows or 'alters' persists in its identity through the 'altering' and through the growing or diminishing, though the quality (in 'alteration') and the size (in growth) do not remain the same. Now if the generation of air from water is to be regarded as growth, a thing might grow without the accession (and without the persistence) of anything, and diminish without the departure of anything-and that which grows need not persist. But this characteristic must be preserved. for the growth we are discussing has been assumed to be thus characterized.

One might raise a further difficulty. What is 'that which grows'? Is it that to which something is added? If, e.g. a man grows in his shin, is it the shin which is greater-but not that 'whereby' he grows, viz. not the food? Then why have not both 'grown'? For when A is added to B, both A and B are greater, as when you mix wine with water; for each ingredient is alike increased in volume. Perhaps the explanation is that the substance of the one remains unchanged, but the substance of the other (viz. of the food) does not. For indeed, even in the mixture of wine and water, it is the prevailing ingredient which is said to have increased in volume. We say, e.g. that the wine has increased, because the whole mixture acts as wine but not as water. A similar principle applies also to 'alteration'. Flesh is said to have been 'altered' if, while its character and substance remain, some one of its essential properties, which was not there before, now qualifies it: on the other hand, that 'whereby' it has been 'altered' may have undergone no change, though sometimes it too has been affected. The altering agent, however, and the originative source of the process are in the growing thing and in that which is being 'altered': for the efficient cause is in these. No doubt the food, which has come in, may sometimes expand as well as the body that has consumed it (that is so, e.g. if, after having come in, a food is converted into wind), but when it has undergone this change it has passed-away: and the efficient cause is not in the food.

We have now developed the difficulties sufficiently and must therefore try to find a solution of the problem. Our solution must preserve intact the three characteristics of growth-that the growing thing persists, that it grows by the accession (and diminishes by the departure) of something, and further that every perceptible particle of it has become either larger or smaller. We must recognize also (a) that the growing body is not 'void' and that yet there are not two magnitudes in the same place, and (b) that it does not grow by the accession of something incorporeal.

Two preliminary distinctions will prepare us to grasp the cause of growth. We must note (i) that the organic parts grow by the growth of the tissues (for every organ is composed of these as its constituents); and (ii) that flesh, bone, and every such part-like every other thing which has its form immersed in matter-has a twofold nature: for the form as well as the matter is called 'flesh' or 'bone'.

Now, that any and every part of the tissue qua form should grow-and grow by the accession of something-is possible, but not that any and every part of the tissue qua matter should do so. For we must think of the tissue after the image of flowing water that is measured by one and the same measure: particle after particle comes-to-be, and each successive particle is different. And it is in this sense that the matter of the flesh grows, some flowing out and some flowing in fresh; not in the sense that fresh matter accedes to every particle of it. There is, however, an accession to every part of its figure or 'form'.

That growth has taken place proportionally, is more manifest in the organic parts—e.g. in the hand. For there the fact that the matter is distinct from the form is more manifest than in flesh, i.e. than in the tissues. That is why there is a greater tendency to suppose that a corpse still possesses flesh and bone than that it still has a hand or an arm.

Hence in one sense it is true that any and every part of the flesh has grown; but in another sense it is false. For there has been an accession to every part of the flesh in respect to its form, but not in respect to its matter. The whole, however, has become larger. And this increase is due (a) on the one hand to the accession of something, which is called 'food' and is said to be 'contrary' to flesh, but (b) on the other hand to the transformation of this food into the same form as that of flesh as if, e.g. 'moist' were to accede to 'dry' and, having acceded, were to be transformed and to become 'dry'. For in one sense 'Like grows by Like', but in another sense 'Unlike grows by Unlike'.

One might discuss what must be the character of that 'whereby' a thing grows. Clearly it must be potentially that which is growing-potentially flesh, e.g. if it is flesh that is growing. Actually, therefore, it must be 'other' than the growing thing. This 'actual other', then, has passed-away and come-to-be flesh. But it has not been transformed into flesh alone by itself (for that would have been a coming-to-be, not a growth): on the contrary, it is the growing thing which has come-to-be flesh (and grown) by the food. In what way, then, has the food been modified by the growing thing? Perhaps we should say that it has been 'mixed' with it, as if one were to pour water into wine and the wine were able to convert the new ingredient into wine. And as fire lays hold of the inflammable, so the active principle of growth, dwelling in the growing thing that which is actually flesh), lays hold of an acceding food which is potentially flesh and converts it into actual flesh. The acceding food, therefore, must be together with the growing thing: for if it were apart from it, the change would be a coming-to-be. For it is possible to produce fire by piling logs on to the already burning fire. That is 'growth'. But when the logs themselves are set on fire, that is 'coming-to-be'.

'Quantum-in-general' does not come-to-be any more than 'animal' which is neither man nor any other of the specific forms of animal: what 'animal-in-general' is in coming-to-be, that 'quantum-in-general' is in growth. But what does come-to-be in growth is flesh or bone-or a hand or arm (i.e. the tissues of these organic parts). Such things come-to-be, then, by the accession not of quantified-flesh but of a quantified-something. In so far as this acceding food is potentially the double result e.g. is potentially so-much-flesh-it produces growth: for it is bound to become actually both so-much and flesh. But in so far as it is potentially flesh only, it nourishes: for it is thus that 'nutrition' and 'growth' differ by their definition. That is why a body's' nutrition' continues so long as it is kept alive (even when it is diminishing), though not its 'growth'; and why nutrition, though 'the same' as growth, is yet different from it in its actual being. For in so far as that which accedes is potentially 'so much-flesh' it tends to increase flesh: whereas, in so far as it is potentially 'flesh' only, it is nourishment.

The form of which we have spoken is a kind of power immersed in matter-a duct, as it were. If, then, a matter accedes-a matter, which is potentially a duct and also potentially possesses determinate quantity the ducts to which it accedes will become bigger. But if it is no longer able to act-if it has been weakened by the continued influx of matter, just as water, continually mixed in greater and greater quantity with wine, in the end makes the wine watery and converts it into water-then it will cause a diminution of the quantum; though still the form persists.

(In discussing the causes of coming-to-be) we must first investigate the matter, i.e. the so-called 'elements'. We must ask whether they really are elements or not, i.e. whether each of them is eternal or whether there is a sense in which they come-to-be: and, if they do come-to-be, whether all of them come-to-be in the same manner reciprocally out of one another, or whether one amongst them is something primary. Hence we must begin by explaining certain preliminary matters, about which the statements now current are vague.

For all (the pluralist philosophers)—those who generate the 'elements' as well as those who generate the bodies that are compounded of the elements—make use of 'dissociation' and 'association', and of 'action' and 'passion'. Now 'association' is 'combination'; but the precise meaning of the process we call 'combining' has not been explained. Again, (all the monists make use of 'alteration': but) without an agent and a patient there cannot be 'altering' any more than there can be 'dissociating' and 'associating'. For not only those who postulate a plurality of elements employ their reciprocal action and passion to generate the compounds: those who derive things from a single element are equally compelled to introduce 'acting'. And in this respect Diogenes is right when he argues that 'unless all things were derived from one, reciprocal action and passion could not have occurred'. The hot thing, e.g. would not be cooled and the cold thing in turn be warmed: for heat and cold do not change reciprocally into one another, but what changes (it is clear) is the substratum. Hence, whenever there is action and passion between two things, that which underlies them must be a single something. No doubt, it is not true to say that all things are of this character: but it is true of all things between which there is reciprocal action and passion.

But if we must investigate 'action-passion' and 'combination', we must also investigate 'contact'. For action and passion (in the proper sense of the terms) can only occur between things which are such as to touch one another; nor can things enter into combination at all unless they have come into a certain kind of contact. Hence we must give a definite account of these three things—of 'contact', 'combination', and 'acting'.

Let us start as follows. All things which admit of 'combination' must be capable of reciprocal contact: and the same is true of any two things, of which one 'acts' and the other 'suffers action' in the proper sense of the terms. For this reason we must treat of 'contact' first. every term which possesses a variety of meaning includes those various meanings either owing to a mere coincidence of language, or owing to a real order of derivation in the different things to which it is applied: but, though this may be taken to hold of 'contact' as of all such terms, it is nevertheless true that contact' in the proper sense applies only to things which have 'position'. And 'position' belongs only to those things which also have a Place': for in so far as we attribute 'contact' to the mathematical things, we must also attribute 'place' to them, whether they exist in separation or in some other fashion. Assuming, therefore, that 'to touch' is—as we have defined it in a previous work—'to have the extremes together', only those things will touch one another which, being separate magnitudes and possessing position, have their extremes 'together'. And since position belongs only to those things which also have a 'place', while the primary differentiation of 'place' is the above' and 'the below' (and the similar pairs of opposites), all things which touch one another will have 'weight' or 'lightness' either both these qualities or one or the other of them. But bodies which are heavy or light are such as to 'act' and 'suffer action'. Hence it is clear that those things are by nature such as to touch one another, which (being separate magnitudes) have their extremes 'together' and are able to move, and be moved by, one another.

The manner in which the 'mover' moves the moved' not always the same: on the contrary, whereas one kind of 'mover' can only impart motion by being itself moved, another kind can do so though remaining itself unmoved. Clearly therefore we must recognize a corresponding variety in speaking of the 'acting' thing too: for the 'mover' is said to 'act' (in a sense) and the 'acting' thing to 'impart motion'. Nevertheless there is a difference and we must draw a distinction. For not every 'mover' can 'act', if (a) the term 'agent' is to be used in contrast to 'patient' and (b) 'patient' is to be applied only to those things whose motion is a 'qualitative affection'—i.e. a quality, like white' or 'hot', in respect to which they are moved' only in the sense that they are 'altered': on the contrary, to 'impart motion' is a wider term than to 'act'. Still, so much, at any rate, is clear: the things which are 'such as to impart motion', if that description be interpreted in one sense, will touch the things which are 'such as to be moved by them'-while they will not touch them, if the description be interpreted in a different sense. But the disjunctive definition of 'touching' must include and distinguish (a) 'contact in general' as the relation between two things which, having position, are such that one is able to impart motion and the other to be moved, and (b) 'reciprocal contact' as the relation between two things, one able to impart motion and the other able to be moved in such a way that 'action and passion' are predicable of them.

As a rule, no doubt, if A touches B, B touches A. For indeed practically all the 'movers' within our ordinary experience impart motion by being moved: in their case, what touches inevitably must, and also evidently does, touch something which reciprocally touches it. Yet, if A moves B, it is possible as we sometimes express it-for A 'merely to touch' B, and that which touches need not touch a something which touches it. Nevertheless it is commonly supposed that 'touching' must be reciprocal. The reason of this belief is that 'movers' which belong to the same kind as the 'moved' impart motion by being moved. Hence if anything imparts motion without itself being moved, it may touch the 'moved' and yet itself be touched by nothing-for we say sometimes that the man who grieves us 'touches' us, but not that we 'touch' him.

The account just given may serve to distinguish and define the 'contact' which occurs in the things of Nature.

7

Next in order we must discuss 'action' and 'passion'. The traditional theories on the subject are conflicting. For (i) most thinkers are unanimous in maintaining (a) that 'like' is always unaffected by 'like', because (as they argue) neither of two 'likes' is more apt than the other either to act or to suffer action, since all the properties which belong to the one belong identically and in the same degree to the other; and (b) that 'unlikes', i.e. 'differents', are by nature such as to act and suffer action reciprocally. For even when the smaller fire is destroyed by the greater, it suffers this effect (they say) owing to its 'contrariety' since the great is contrary to the small. But (ii) Democritus dissented from all the other thinkers and maintained a theory peculiar to himself. He asserts that agent and patient are identical, i.e. 'like'. It is not possible (he says) that 'others', i.e. 'differents', should suffer action from one another: on the contrary, even if two things, being 'others', do act in some way on one another, this happens to them not qua 'others' but qua possessing an identical property.

Such, then, are the traditional theories, and it looks as if the statements of their advocates were in manifest conflict. But the reason of this conflict is that each group is in fact stating a part, whereas they ought to have taken a comprehensive view of the subject as a whole. For (i) if A and B are 'like'-absolutely and in all respects without difference from one another —it is reasonable to infer that neither is in any way affected by the other. Why, indeed, should either of them tend to act any more than the other? Moreover, if 'like' can be affected by 'like', a thing can also be affected by itself: and yet if that were so-if 'like' tended in fact to act qua 'like'-there would be nothing indestructible or immovable, for everything would move itself. And (ii) the same consequence follows if A and B are absolutely 'other', i.e. in no respect identical. Whiteness could not be affected in any way by line nor line by whiteness—except perhaps 'coincidentally', viz. if the line happened to be white or black: for unless two things either are, or are composed of, 'contraries', neither drives the other out of its natural condition. But (iii) since only those things which either involve a 'contrariety' or are 'contraries'—and not any things selected at random-are such as to suffer action and to act, agent and patient must be 'like' (i.e. identical) in kind and yet 'unlike' (i.e. contrary) in species. (For it is a law of nature that body is affected by body, flavour by flavour, colour by colour, and so in general what belongs to any kind by a member of the same kind-the reason being that 'contraries' are in every case within a single identical kind, and it is 'contraries' which reciprocally act and suffer action.) Hence agent and patient must be in one sense identical, but in another sense other than (i.e. 'unlike') one another. And since (a) patient and agent are generically identical (i.e. 'like') but specifically 'unlike', while (b) it is 'contraries' that exhibit this character: it is clear that 'contraries' and their 'intermediates' are such as to suffer action and to act reciprocally-for indeed it is these that constitute the entire sphere of passing-away and coming-to-be.

We can now understand why fire heats and the cold thing cools, and in general why the active thing assimilates to itself the patient. For agent and patient are contrary to one another, and coming-to-be is a process into the contrary: hence the patient must change into the agent, since it is only thus that coming-to be will be a process into the contrary. And, again, it is intelligible that the advocates of both views, although their theories are not the same, are yet in contact with the nature of the facts. For sometimes we speak of the substratum as suffering action (e.g. of 'the man' as being healed, being warmed and chilled, and similarly in all the other cases), but at other times we say 'what is cold is 'being warmed', 'what is sick is being healed': and in both these ways of speaking we express the truth, since in one sense it is the 'matter', while in another sense it is the 'contrary', which suffers action. (We make the same distinction in speaking of the agent: for sometimes we say that 'the man', but at other times that 'what is hot', produces heat.) Now the one group of thinkers supposed that agent and patient must possess something identical, because they fastened their attention on the substratum: while the other group maintained the opposite because their attention was concentrated on the 'contraries'. We must conceive the same account to hold of action and passion as that which is true of 'being moved' and 'imparting motion'. For the 'mover', like the 'agent', has two meanings. Both (a) that which contains the originative source of the motion is thought to 'impart motion' (for the originative source is first amongst the causes), and also (b) that which is last, i.e. immediately next to the moved thing and to the coming-to-be. A similar distinction holds also of the agent: for we speak not only (a) of the doctor, but also (b) of the wine, as healing. Now, in motion, there is nothing to prevent the firs; mover being unmoved (indeed, as regards some 'first' movers' this is actually necessary) although the last mover always imparts motion by being itself moved: and, in action, there is nothing to prevent the first agent being unaffected, while the last agent only acts by suffering action itself. For agent and patient have not the same matter, agent acts without being affected: thus the art of healing produces health without itself being acted upon in any way by that which is being healed. But (b) the food, in acting, is itself in some way acted upon: for, in acting, it is simultaneously heated or cooled or otherwise affected. Now the art of healing corresponds to an 'originative source', while the food corresponds to 'the last' (i.e. 'continuous') mover.

Those active powers, then, whose forms are not embodied in matter, are unaffected: but those whose forms are in matter are such as to be affected in acting. For we maintain that one and the same 'matter' is equally, so to say, the basis of either of the two opposed things-being as it were a 'kind'; and that that which can he hot must be made hot, provided the heating agent is there, i.e. comes near. Hence (as we have said) some of the active powers are unaffected while others are such as to be affected; and what holds of motion is true also of the active powers. For as in motion 'the first mover' is unmoved, so among the active powers 'the first agent' is unaffected.

The active power is a 'cause' in the sense of that from which the process originates: but the end, for the sake of which it takes place, is not 'active'. (That is why health is not 'active', except metaphorically.) For when the agent is there, the patient he-comes something: but when 'states' are there, the patient no longer becomes but already is-and 'forms' (i.e. lends') are a kind of 'state'. As to the 'matter', it (qua matter) is passive. Now fire contains 'the hot' embodied in matter: but a 'hot' separate from matter (if such a thing existed) could not suffer any action. Perhaps, indeed, it is impossible that 'the hot' should exist in separation from matter: but if there are any entities thus separable, what we are saying would be true of them.

We have thus explained what action and passion are, what things exhibit them, why they do so, and in what manner. We must go on to discuss how it is possible for action and passion to take place.

8

Some philosophers think that the 'last' agent-the 'agent' in the strictest sense-enters in through certain pores, and so the patient suffers action. It is in this way, they assert, that we see and hear and exercise all our other senses. Moreover, according to them, things are seen through air and water and other transparent bodies, because such bodies possess pores, invisible indeed owing to their minuteness, but close-set and arranged in rows: and the more transparent the body, the more frequent and serial they suppose its pores to be. Such was the theory which some philosophers (including Empedocles) advanced in regard to the structure of certain bodies. They do not restrict it to the bodies which act and suffer action: but 'combination' too, they say, takes place 'only between bodies whose pores are in reciprocal symmetry'. The most systematic and consistent theory, however, and one that applied to all bodies, was advanced by Leucippus and Democritus: and, in maintaining it, they took as their starting-point what naturally comes first.

For some of the older philosophers thought that 'what is' must of necessity be 'one' and immovable. The void, they argue, 'is not': but unless there is a void with a separate being of its own, 'what is' cannot be moved-nor again can it be 'many', since there is nothing to keep things apart. And in this respect, they insist, the view that the universe is not 'continuous' but 'discretes-in-contact' is no better than the view that there are 'many' (and not 'one') and a void. For (suppose that the universe is discretes-in-contact. Then), if it is divisible through and through, there is no 'one', and therefore no 'many' either, but the Whole is void; while to maintain that it is divisible at some points, but not at others, looks like an arbitrary fiction. For up to what limit is it divisible? And for what reason is part of the Whole indivisible, i.e. a plenum, and part divided? Further, they maintain, it is equally necessary to deny the existence of motion.

Reasoning in this way, therefore, they were led to transcend sense-perception, and to disregard it on the ground that 'one ought to follow the argument': and so they assert that the universe is 'one' and immovable. Some of them add that it is 'infinite', since the limit (if it had one) would be a limit against the void.

There were, then, certain thinkers who, for the reasons we have stated, enunciated views of this kind as their theory of 'The Truth'.... Moreover, although these opinions appear to follow logically in a dialectical discussion, yet to believe them seems next door to madness when one considers the facts. For indeed no lunatic seems to be so far out of his senses as to suppose that fire and ice are 'one': it is only between what is right and what seems right from habit, that some people are mad enough to see no difference.

Leucippus, however, thought he had a theory which harmonized with sense-perception and would not abolish either coming-to-be and passing-away or motion and the multiplicity of things. He made these concessions to the facts of perception: on the other hand, he conceded to the Monists that there could be no motion without a void. The result is a theory which he states as follows: 'The void is a "not being", and no part of "what is" is a "not-being"; for what "is" in the strict sense of the term is an absolute plenum. This plenum, however, is not "one": on the contrary, it is a many" infinite in number and invisible owing to the minuteness of their bulk. The "many" move in the void (for there is a void): and by coming together they produce "coming to-be", while by separating they produce "passing-away". Moreover, they act and suffer action wherever they chance to be in contact (for there they are not "one"), and they generate by being put together and becoming intertwined. From the genuinely-one, on the other hand, there never could have come-to-be a multiplicity, nor from the genuinely-many a "one": that is impossible. But' (just as Empedocles and some of the other philosophers say that things suffer action through their pores, so) 'all "alteration" and all "passion" take place in the way that has been explained: breaking-up (i.e. passing-away) is effected by means of the void, and so too is growth-solids creeping in to fill the void places.' Empedocles too is practically bound to adopt the same theory as Leucippus. For he must say that there are certain solids which, however, are indivisible-unless there are continuous pores all through the body. But this last alternative is impossible: for then there will be nothing solid in the body (nothing beside the poroc) but all of it will be void. It is necessary, therefore, for his 'contiguous discretes' to be indivisible, while the intervals between them-which he calls 'pores'-must be void. But this is precisely Leucippus' theory of action and passion.

Such, approximately, are the current explanations of the manner in which some things 'act' while others 'suffer action'. And as regards the Atomists, it is not only clear what their explanation is: it is also obvious that it follows with tolerable consistency from the assumptions they employ. But there is less obvious consistency in the explanation offered by the other thinkers. It is not clear, for instance, how, on the theory of Empedocles, there is to be 'passing-away' as well as 'alteration'. For the primary bodies of the Atomists-the primary constituents of which bodies are composed, and the ultimate elements into which they are dissolved-are indivisible, differing from one another only in figure. In the philosophy of Empedocles, on the other hand, it is evident that all the other bodies down to the 'elements' have their coming-to-be and their passing-away: but it is not clear how the 'elements' themselves, severally in their aggregated masses, come-to-be and pass-away. Nor is it possible for Empedocles to explain how they do so, since he does not assert that Fire too (and similarly every one of his other 'elements') possesses 'elementary constituents' of itself.

Such an assertion would commit him to doctrines like those which Plato has set forth in the Timaeus. For although both Plato and Leucippus postulate elementary constituents that are indivisible and distinctively characterized by figures, there is this great difference between the two theories: the 'indivisibles' of Leucippus (i) are solids, while those of Plato are planes, and (ii) are characterized by an infinite variety of figures, while the characterizing figures employed by Plato are limited in number. Thus the 'comings-to-be' and the 'dissociations' result from the 'indivisibles' (a) according to Leucippus through the void and through contact (for it is at the point of contact that each of the composite bodies is divisible), but (b) according to Plato in virtue of contact alone, since he denies there is a void.

Now we have discussed 'indivisible planes' in the preceding treatise.' But with regard to the assumption of 'indivisible solids', although we must not now enter upon a detailed study of its consequences, the following criticisms fall within the compass of a short digression: i. The Atomists are committed to the view that every 'indivisible' is incapable alike of receiving a sensible property (for nothing can 'suffer action' except through the void) and of producing one-no 'indivisible' can be, e.g. either hard or cold. Yet it is surely a paradox that an exception is made of 'the hot'-'the hot' being assigned as peculiar to the spherical figure: for, that being so, its 'contrary' also ('the cold') is bound to belong to another of the figures. If, however, these properties (heat and cold) do belong to the 'indivisibles', it is a further paradox that they should not possess heaviness and lightness, and hardness and softness. And yet Democritus says 'the more any indivisible exceeds, the heavier it is'-to which we must clearly add 'and the hotter it is'. But if that is their character, it is impossible they should not be affected by one another: the 'slightly-hot indivisible', e.g. will inevitably suffer action from one which far exceeds it in heat. Again, if any 'indivisible' is 'hard', there must also be one which is 'soft': but 'the soft' derives its very name from the fact that it suffers a certain action-for 'soft' is that which yields to pressure.

II. But further, not only is it paradoxical (i) that no property except figure should belong to the 'indivisibles': it is also paradoxical (ii) that, if other properties do belong to them, one only of these additional properties should attach to each-e.g. that this 'indivisible' should be cold and that 'indivisible' hot. For, on that supposition, their substance would not even be uniform. And it is equally impossible (iii) that more than one of these additional properties should belong to the single 'indivisible'. For, being indivisible, it will possess these properties in the same point-so that, if it 'suffers action' by being chilled, it will also, qua chilled, 'act' or 'suffer action' in some other way. And the same line of argument applies to all the other properties too: for the difficulty we have just raised confronts, as a necessary consequence, all who advocate 'indivisibles' (whether solids or planes), since their 'indivisibles' cannot become either 'rarer' or 'derser' inasmuch as there is no void in them.

III. It is a further paradox that there should be small 'indivisibles', but not large ones. For it is natural enough, from the ordinary point of view, that the larger bodies should be more liable to fracture than the small ones, since they (viz. the large bodies) are easily broken up because they collide with many other bodies. But why should indivisibility as such be the property of small, rather than of large, bodies?

IV. Again, is the substance of all those solids uniform, or do they fall into sets which differ from one another-as if, e.g. some of them, in their aggregated bulk, were 'fiery', others earthy'? For (i) if all of them are uniform in substance, what is it that separated one from another? Or why, when they come into contact, do they not coalesce into one, as drops of water run together when drop touches drop (for the two cases are precisely parallel)? On the other hand (ii) if they fall into differing sets, how are these characterized? It is clear, too, that these, rather than the 'figures', ought to be postulated as 'original reals', i.e. causes from which the phenomena result. Moreover, if they differed in substance, they would both act and suffer action on coming into reciprocal contact.

V. Again, what is it which sets them moving? For if their 'mover' is other than themselves, they are such as to 'suffer action'. If, on the other hand, each of them sets itself in motion, either (a) it will be divisible ('imparting motion' qua this, 'being moved' qua that), or (b) contrary properties will attach to it in the same respect—i.e. 'matter' will be identical in-potentiality as well as numerically-identical.

As to the thinkers who explain modification of property through the movement facilitated by the pores, if this is supposed to occur notwithstanding the fact that the pores are filled, their postulate of pores is superfluous. For if the whole body suffers action under these conditions, it would suffer action in the same way even if it had no pores but were just its own continuous self. Moreover, how can their account of 'vision through a medium' be correct? It is impossible for (the visual ray) to penetrate the transparent bodies at their 'contacts'; and impossible for it to pass through their pores if every pore be full. For how will that differ from having no pores at all? The body will be uniformly 'full' throughout. But, further, even if these passages, though they must contain bodies, are 'void', the same consequence will follow once more. And if they are 'too minute to admit any body', it is absurd to suppose there is a 'minute' void and yet to deny the existence of a 'big' one (no matter how small the 'big' may be), or to imagine 'the void' means anything else than a body's place-whence it clearly follows that to every body there will correspond a void of equal cubic capacity.

As a general criticism we must urge that to postulate pores is superfluous. For if the agent produces no effect by touching the patient, neither will it produce any by passing through its pores. On the other hand, if it acts by contact, then-even without pores-some things will 'suffer action' and others will 'act', provided they are by nature adapted for reciprocal action and passion. Our arguments have shown that it is either false or futile to advocate pores in the sense in which some thinkers conceive them. But since bodies are divisible through and through, the postulate of pores is ridiculous: for, qua divisible, a body can fall into separate parts.

9

Let explain the way in which things in fact possess the power of generating, and of acting and suffering action: and let us start from the principle we have often enunciated. For, assuming the distinction between (a) that which is potentially and (b) that which is actually such-and-such, it is the nature of the first, precisely in so far as it is what it is, to suffer action through and through, not merely to be susceptible in some parts while insusceptible in others. But its susceptibility varies in degree, according as it is more or less; such-and such, and one would be more justified in speaking of 'pores' in this connexion: for instance, in the metals there are veins of 'the susceptible' stretching continuously through the substance.

So long, indeed, as any body is naturally coherent and one, it is insusceptible. So, too, bodies are insusceptible so long as they are not in contact either with one another or with other bodies which are by nature such as to act and suffer action. (To illustrate my meaning: Fire heats not only when in contact, but also from a distance. For the fire heats the air, and the air-being by nature such as both to act and suffer action-heats the body.) But the supposition that a body is 'susceptible in some parts, but insusceptible in others' (is only possible for those who hold an erroneous view concerning the divisibility of magnitudes. For us) the following account results from the distinctions we established at the beginning. For (i) if magnitudes are not divisible through and through-if, on the contrary, there are indivisible solids or planes-then indeed no body would be susceptible through and through :but neither would any be continuous. Since, however, (ii) this is false, i.e. since every body is divisible, there is no difference between 'having been divided into parts which remain in contact' and 'being divisible'. For if a body 'can be separated at the contacts' (as some thinkers express it), then, even though it has not yet been divided, it will be in a state of dividedness—since, as it can be divided, nothing inconceivable results. And (iii) the suposition is open to this general objection—it is a paradox that 'passion' should occur in this manner only, viz. by the bodies being split. For this theory abolishes 'alteration': but we see the same body liquid at one time and solid at another, without losing its continuity. It has suffered this change not by 'division' and composition', nor yet by 'turning' and 'intercontact' as Democritus asserts; for it has passed from the liquid to the solid state without any change of 'grouping' or 'position' in the constituents of its substance. Nor are there contained within it those 'hard' (i.e. congealed) particles 'indivisible in their bulk': on the contrary, it is liquid—and again, solid and congealed—uniformly all through. This theory, it must be added, makes growth and diminution impossible also. For if there is to be opposition (instead of the growing thing having changed as a whole, either by the admixture of something or by its own transformation), increase of size will not have resulted in any and every part.

So much, then, to establish that things generate and are generated, act and suffer action, reciprocally; and to distinguish the way in which these processes can occur from the (impossible) way in which some thinkers say they occur.

10

But we have still to explain 'combination', for that was the third of the subjects we originally proposed to discuss. Our explanation will proceed on the same method as before. We must inquire: What is 'combination', and what is that which can 'combine'? Of what things, and under what conditions, is 'combination' a property? And, further, does 'combination' exist in fact, or is it false to assert its existence?

For, according to some thinkers, it is impossible for one thing to be combined with another. They argue that (i) if both the 'combined' constituents persist unaltered, they are no more 'combined' now than they were before, but are in the same condition: while (ii) if one has been destroyed, the constituents have not been 'combined'—on the contrary, one constituent is and the other is not, whereas 'combination' demands uniformity of condition in them both: and on the same principle (iii) even if both the combining constituents have been destroyed as the result of their coalescence, they cannot 'have been combined' since they have no being at all.

What we have in this argument is, it would seem, a demand for the precise distinction of 'combination' from coming-to-be and passing-away (for it is obvious that 'combination', if it exists, must differ from these processes) and for the precise distinction of the 'combinable' from that which is such as to come-to-be and pass-away. As soon, therefore, as these distinctions are clear, the difficulties raised by the argument would be solved.

Now (i) we do not speak of the wood as 'combined' with the fire, nor of its burning as a 'combining' either of its particles with one another or of itself with the fire: what we say is that 'the fire is coming-to-be, but the wood is 'passing-away'. Similarly, we speak neither (ii) of the food as 'combining' with the body, nor (iii) of the shape as 'combining' with the wax and thus fashioning the lump. Nor can body 'combine' with white, nor (to generalize) 'properties' and 'states' with 'things': for we see them persisting unaltered. But again (iv) white and knowledge cannot be 'combined' either, nor any other of the 'adjectivals'. (Indeed, this is a blemish in the theory of those who assert that 'once upon a time all things were together and combined'. For not everything can 'combine' with everything. On the contrary, both of the constituents that are combined in the compound must originally have existed in separation: but no property can have separate existence.)

Since, however, some things are-potentially while others are-actually, the constituents combined in a compound can 'be' in a sense and yet 'not-be'. The compound may he-actually other than the constituents from which it has resulted; nevertheless each of them may still he-potentially what it was before they were combined, and both of them may survive undestroyed. (For this was the difficulty that emerged in the previous argument: and it is evident that the combining constituents not only coalesce, having formerly existed in separation, but also can again be separated out from the compound.) The constituents, therefore, neither (a) persist actually, as 'body' and 'white' persist: nor (b) are they destroyed (either one of them or both), for their 'power of action' is preserved. Hence these difficulties may be dismissed: but the problem immediately connected with them-whether combination is something relative to perception' must be set out and discussed.

When the combining constituents have been divided into parts so small, and have been juxtaposed in such a manner, that perception fails to discriminate them one from another, have they then 'been combined Or ought we to say 'No, not until any and every part of one constituent is juxtaposed to a part of the other'? The term, no doubt, is applied in the former sense: we speak, e.g. of wheat having been 'combined' with barley when each grain of the one is juxtaposed to a grain of the other. But every body is divisible and therefore, since body 'combined' with body is uniform in texture throughout, any and every part of each constituent ought to be juxtaposed to a part of the other.

No body, however, can be divided into its 'least' parts: and 'composition' is not identical with 'combination', but other than it. From these premises it clearly follows (i) that so long as the constituents are preserved in small particles, we must not speak of them as 'combined'. (For this will be a 'composition' instead of a 'blending' or 'combination': nor will every portion of the resultant exhibit the same ratio between its constituents as the whole. But we maintain that, if 'combination' has taken place, the compound must be uniform in texture throughout-any part of such a compound being the same as the whole, just as any part of water is water: whereas, if 'combination' is 'composition of the small particles', nothing of the kind will happen. On the contrary, the constituents will only be 'combined' relatively to perception: and the same thing will be 'combined' to one percipient, if his sight is not sharp, (but not to another,) while to the eye of Lynceus nothing will be 'combined'.) It clearly follows (ii) that we must not speak of the constituents as 'combined in virtue of a division such that any and every part of each is juxtaposed to a part of the other: for it is impossible for them to be thus divided. Either, then, there is no 'combination', or we have still to explain the manner in which it can take place.

Now, as we maintain, some things are such as to act and others such as to suffer action from them. Moreover, some things-viz. those Which have the same matter-'reciprocate', i.e. are such as to act upon one another and to suffer action from one another; while other things, viz. agents which have not the same matter as their patients, act without themselves suffering action. Such agents cannot 'combine'-that is why neither the art of healing nor health produces health by 'combining' with the bodies of the patients. Amongst those things, however, which are reciprocally active and passive, some are easily-divisible. Now (i) if a great quantity (or a large bulk) of one of these easily-divisible 'reciprocating' materials be brought together with a little (or with a small piece) of another, the effect produced is not 'combination', but increase of the dominant: for the other material is transformed into the dominant. (That is why a drop of wine does not 'combine' with ten thousand gallons of water: for its form is dissolved, and it is changed so as to merge in the total volume of water.) On the other hand (ii) when there is a certain equilibrium between their 'powers of action', then each of them changes out of its own nature towards the dominant: yet neither becomes the other, but both become an intermediate with properties common to both.

Thus it is clear that only those agents are 'combinable' which involve a contrariety-for these are such as to suffer action reciprocally. And, further, they combine more freely if small pieces of each of them are juxtaposed. For in that condition they change one another more easily and more quickly; whereas this effect takes a long time when agent and patient are present in bulk.

Hence, amongst the divisible susceptible materials, those whose shape is readily adaptable have a tendency to combine: for they are easily divided into small particles, since that is precisely what 'being readily adaptable in shape' implies. For instance, liquids are the most 'combinable' of all bodies-because, of all divisible materials, the liquid is most readily adaptable in shape, unless it be viscous. Viscous liquids, it is true, produce no effect except to increase the volume and bulk. But when one of the constituents is alone susceptible-or superlatively susceptible, the other being susceptible in a very slight degree-the compound resulting from their combination is either no greater in volume or only a little greater. This is what happens when tin is combined with bronze. For some things display a hesitating and ambiguous attitude towards one another-showing a slight tendency to combine and also an inclination to behave as 'receptive matter' and 'form' respectively. The behaviour of these metals is a case in point. For the tin almost vanishes, behaving as if it were an immaterial property of the bronze: having been combined, it disappears, leaving no trace except the colour it has imparted to the bronze. The same phenomenon occurs in other instances too.

It is clear, then, from the foregoing account, that 'combination' occurs, what it is, to what it is due, and what kind of thing is 'combinable'. The phenomenon depends upon the fact that some things are such as to be (a) reciprocally susceptible and (b) readily adaptable in shape, i.e. easily divisible. For such things can be 'combined' without its being necessary either that they should have been destroyed or that they should survive absolutely unaltered: and their 'combination' need not be a 'composition', nor merely 'relative to perception'. On the contrary:

anything is 'combinable' which, being readily adaptable in shape, is such as to suffer action and to act; and it is 'combinable with' another thing similarly characterized (for the 'combinable' is relative to the 'combinable'); and 'combination' is unification of the 'combinables', resulting from their 'alteration'.

Book II

1

WE have explained under what conditions 'combination', 'contact', and 'action-passion' are attributable to the things which undergo natural change. Further, we have discussed 'unqualified' coming-to-be and passing-away, and explained under what conditions they are predicable, of what subject, and owing to what cause. Similarly, we have also discussed 'alteration', and explained what 'altering' is and how it differs from coming-to-be and passing-away. But we have still to investigate the so-called 'elements' of bodies.

For the complex substances whose formation and maintenance are due to natural processes all presuppose the perceptible bodies as the condition of their coming-to-be and passing-away: but philosophers disagree in regard to the matter which underlies these perceptible bodies. Some maintain it is single, supposing it to be, e.g. Air or Fire, or an 'intermediate' between these two (but still a body with a separate existence). Others, on the contrary, postulate two or more materials-ascribing to their 'association' and 'dissociation', or to their 'alteration', the coming-to-be and passing-away of things. (Some, for instance, postulate Fire and Earth: some add Air, making three: and some, like Empedocles, reckon Water as well, thus postulating four.)

Now we may agree that the primary materials, whose change (whether it be 'association and dissociation' or a process of another kind) results in coming-to-be and passing-away, are rightly described as 'originative sources, i.e. elements'. But (i) those thinkers are in error who postulate, beside the bodies we have mentioned, a single matter-and that corporeal and separable matter. For this 'body' of theirs cannot possibly exist without a 'perceptible contrariety': this 'Boundless', which some thinkers identify with the 'original real', must be either light or heavy, either cold or hot. And (ii) what Plato has written in the Timaeus is not based on any precisely-articulated conception. For he has not stated clearly whether his 'Omnirecipient" exists in separation from the 'elements'; nor does he make any use of it. He says, indeed, that it is a substratum prior to the so-called 'elements'-underlying them, as gold underlies the things that are fashioned of gold. (And yet this comparison, if thus expressed, is itself open to criticism. Things which come-to-be and pass-away cannot be called by the name of the material out of which they have come-to-be: it is only the results of 'alteration' which retain the name of the substratum whose 'alterations' they are. However, he actually says that the truest account is to affirm that each of them is "gold".) Nevertheless he carries his analysis of the 'elements'—solids though they are—back to 'planes', and it is impossible for 'the Nurse' (i.e. the primary matter) to be identical with 'the planes'.

Our own doctrine is that although there is a matter of the perceptible bodies (a matter out of which the so-called 'elements' come-to-be), it has no separate existence, but is always bound up with a contrariety. A more precise account of these presuppositions has been given in another work': we must, however, give a detailed explanation of the primary bodies as well, since they too are similarly derived from the matter. We must reckon as an 'originative source' and as 'primary' the matter which underlies, though it is inseparable from, the contrary qualities: for the hot' is not matter for 'the cold' nor 'the cold' for 'the hot', but the substratum is matter for them both. We therefore have to recognize three 'originative sources': firstly that which potentially perceptible body, secondly the contrarieties (I mean, e.g. heat and cold), and thirdly Fire, Water, and the like. Only 'thirdly', however: for these bodies change into one another (they are not immutable as Empedocles and other thinkers assert, since 'alteration' would then have been impossible), whereas the contrarieties do not change.

Nevertheless, even so the question remains: What sorts of contrarieties, and how many of them, are to be accounted 'originative sources' of body? For all the other thinkers assume and use them without explaining why they are these or why they are just so many.

2

Since, then, we are looking for 'originative sources' of perceptible body; and since 'perceptible' is equivalent to 'tangible', and 'tangible' is that of which the perception is touch; it is clear that not all the contrarieties constitute 'forms' and 'originative sources' of body, but only those which correspond to touch. For it is in accordance with a contrariety-a contrariety, moreover, of tangible qualities-that the primary bodies are differentiated. That is why neither whiteness (and blackness), nor sweetness (and bitterness), nor (similarly) any quality belonging to the other perceptible contrarieties either, constitutes an 'element'. And yet vision is prior to touch, so that its object also is prior to the object of touch. The object of vision, however, is a quality of tangible body not qua tangible, but qua something else-qua something which may well be naturally prior to the object of touch.

Accordingly, we must segregate the tangible differences and contrarieties, and distinguish which amongst them are primary. Contrarieties correlative to touch are the following: hot-cold, dry-moist, heavy-light, hard-soft, viscous-brittle, rough-smooth, coarse-fine. Of these (i) heavy and light are neither active nor susceptible. Things are not called 'heavy' and 'light' because they act upon, or suffer action from, other things. But the 'elements' must be reciprocally active and susceptible, since they 'combine' and are transformed into one another. On the other hand (ii) hot and cold, and dry and moist, are terms, of which the first pair implies power to act and the second pair susceptibility. 'Hot' is that which 'associates' things of the same kind (for 'dissociating', which people attribute to Fire as its function, is 'associating' things of the same class, since its effect is to eliminate what is foreign), while 'cold' is that which brings together, i.e. 'associates', homogeneous and heterogeneous things alike. And moist is that which, being readily adaptable in shape, is not determinable by any limit of its own: while 'dry' is that which is readily determinable by its own limit, but not readily adaptable in shape.

From moist and dry are derived (iii) the fine and coarse, viscous and brittle, hard and soft, and the remaining tangible differences. For (a) since the moist has no determinate shape, but is readily adaptable and follows the outline of that which is in contact with it, it is characteristic of it to be 'such as to fill up'. Now 'the fine' is 'such as to fill up'. For the fine' consists of subtle particles; but that which consists of small particles is 'such as to fill up', inasmuch as it is in contact whole with whole-and 'the fine' exhibits this character in a superlative degree. Hence it is evident that the fine derives from the moist, while the coarse derives from the dry. Again (b) the viscous' derives from the moist: for 'the viscous' (e.g. oil) is a 'moist' modified in a certain way. 'The brittle', on the other hand, derives from the dry: for 'brittle' is that which is completely dry-so completely, that its solidification has actually been due to failure of moisture. Further (c) 'the soft' derives from the moist. For 'soft' is that which yields to pressure by retiring into itself, though it does not yield by total displacement as the moist does-which explains why the moist is not 'soft', although 'the soft' derives from the moist. 'The hard', on the other hand, derives from the dry: for 'hard' is that which is solidified, and the solidified is dry.

The terms 'dry' and 'moist' have more senses than one. For 'the damp', as well as the moist, is opposed to the dry: and again 'the solidified', as well as the dry, is opposed to the moist. But all these qualities derive from the dry and moist we mentioned first.' For (i) the dry is opposed to the damp: i.e. 'damp' is that which has foreign moisture on its surface ('sodden' being that which is penetrated to its core), while 'dry' is that which has lost foreign moisture. Hence it is evident that the damp will derive from the moist, and 'the dry' which is opposed to it will derive from the primary dry. Again (ii) the 'moist' and the solidified derive in the same way from the primary pair. For 'moist' is that which contains moisture of its-own deep within it ('sodden' being that which is deeply penetrated by foreign moisture), whereas 'solidified' is that which has lost this inner moisture. Hence these too derive from the primary pair, the 'solidified' from the dry and the 'solidified' from the dry the 'liquefiable' from the moist.

It is clear, then, that all the other differences reduce to the first four, but that these admit of no further reduction. For the hot is not essentially moist or dry, nor the moist essentially hot or cold: nor are the cold and the dry derivative forms, either of one another or of the hot and the moist. Hence these must be four.

3

The elementary qualities are four, and any four terms can be combined in six couples. Contraries, however, refuse to be coupled: for it is impossible for the same thing to be hot and cold, or moist and dry. Hence it is evident that the 'couplings' of the elementary qualities will be four: hot with dry and moist with hot, and again cold with dry and cold with moist. And these four couples have attached themselves to the apparently 'simple' bodies (Fire, Air, Water, and Earth) in a manner consonant with theory. For Fire is hot and dry, whereas Air is hot and moist (Air being a sort of aqueous vapour); and Water is cold and moist, while Earth is cold and dry. Thus the differences are reasonably distributed among the primary bodies, and the number of the latter is consonant with theory. For all who make the simple bodies 'elements' postulate either one, or two, or three, or four. Now (i) those who assert there is one only, and then generate everything else by condensation and rarefaction, are in effect making their 'originative sources' two, viz. the rare and the dense, or rather the hot and the cold: for it is these which are the moulding forces, while the 'one' underlies them as a 'matter'. But (ii) those who postulate two from the start-as Parmenides postulated Fire and Earth-make the intermediates (e.g. Air and Water) blends of these. The same course is followed (iii) by those who advocate three. (We may compare what Plato does in Me Divisions': for he makes 'the middle' a blend.) Indeed, there is practically no difference between those who postulate two and those who postulate three, except that the former split the middle 'element' into two, while the latter treat it as only one. But (iv) some advocate four from the start, e.g. Empedocles: yet he too draws them together so as to reduce them to the two, for he opposes all the others to Fire.

In fact, however, fire and air, and each of the bodies we have mentioned, are not simple, but blended. The 'simple' bodies are indeed similar in nature to them, but not identical with them. Thus the 'simple' body corresponding to fire is 'such-as-fire, not fire: that which corresponds to air is 'such-as-air': and so on with the rest of them. But fire is an excess of heat, just as ice is an excess of cold. For freezing and boiling are excesses of heat and cold respectively. Assuming, therefore, that ice is a freezing of moist and cold, fire analogously will be a boiling of dry and hot: a fact, by the way, which explains why nothing comes-to-be either out of ice or out of fire.

The 'simple' bodies, since they are four, fall into two pairs which belong to the two regions, each to each: for Fire and Air are forms of the body moving towards the 'limit', while Earth and Water are forms of the body which moves towards the 'centre'. Fire and Earth, moreover, are extremes and purest: Water and Air, on the contrary are intermediates and more like blends. And, further, the members of either pair are contrary to those of the other, Water being contrary to Fire and Earth to Air; for the qualities constituting Water and Earth are contrary to those that constitute Fire and Air. Nevertheless, since they are four, each of them is characterized par excellence a single quality: Earth by dry rather than by cold, Water by cold rather than by moist, Air by moist rather than by hot, and Fire by hot rather than by dry.

4

It has been established before' that the coming-to-be of the 'simple' bodies is reciprocal. At the same time, it is manifest, even on the evidence of perception, that they do come-to-be: for otherwise there would not have been 'alteration, since 'alteration' is change in respect to the qualities of the objects of touch. Consequently, we must explain (i) what is the manner of their reciprocal transformation, and (ii) whether every one of them can come to-be out of every one-or whether some can do so, but not others.

Now it is evident that all of them are by nature such as to change into one another: for coming-to-be is a change into contraries and out of contraries, and the 'elements' all involve a contrariety in their mutual relations because their distinctive qualities are contrary. For in some of them both qualities are contrary—e.g. in Fire and Water, the first of these being dry and hot, and the second moist and cold: while in others one of the qualities (though only one) is contrary—e.g. in Air and Water, the first being moist and hot, and the second moist and cold. It is evident, therefore, if we consider them in general, that every one is by nature such as to come-to-be out of every one: and when we come to consider them severally, it is not difficult to see the manner in which their transformation is effected. For, though all will result from all, both the speed and the facility of their conversion will differ in degree.

Thus (i) the process of conversion will be quick between those which have interchangeable 'complementary factors', but slow between those which have none. The reason is that it is easier for a single thing to change than for many. Air, e.g. will result from Fire if a single quality changes: for Fire, as we saw, is hot and dry while Air is hot and moist, so that there will be Air if the dry be overcome by the moist. Again, Water will result from Air if the hot be overcome by the cold: for Air, as we saw, is hot and moist while Water is cold and moist, so that, if the hot changes, there will be Water. So too, in the same manner, Earth will result from Water and Fire from Earth, since the two 'elements' in both these couples have interchangeable 'complementary factors'. For Water is moist and cold while Earth is cold and dry—so that, if the moist be overcome, there will be Earth: and again, since Fire is dry and hot while Earth is cold and dry, Fire will result from Earth if the cold pass-away.

It is evident, therefore, that the coming-to-be of the 'simple' bodies will be cyclical; and that this cyclical method of transformation is the easiest, because the consecutive 'elements' contain interchangeable 'complementary factors'. On the other hand (ii) the transformation of Fire into Water and of Air into Earth, and again of Water and Earth into Fire and Air respectively, though possible, is more difficult because it involves the change of more qualities. For if Fire is to result from Water, both the cold and the moist must pass-away: and again, both the cold and the dry must pass-away if Air is to result from Earth. So' too, if Water and Earth are to result from Fire and Air respectively-both qualities must change.

This second method of coming-to-be, then, takes a longer time. But (iii) if one quality in each of two 'elements' pass-away, the transformation, though easier, is not reciprocal. Still, from Fire plus Water there will result Earth and Air, and from Air plus Earth Fire and Water. For there will be Air, when the cold of the Water and the dry of the Fire have passed-away (since the hot of the latter and the moist of the former are left): whereas, when the hot of the Fire and the moist of the Water have passed-away, there will be Earth, owing to the survival of the dry of the Fire and the cold of the Water. So, too, in the same Way, Fire and Water will result from Air plus Earth. For there will be Water, when the hot of the Air and the dry of the Earth have passed-away (since the moist of the former and the cold of the latter are left): whereas, when the moist of the Air and the cold of the Earth have passed-away, there will be Fire, owing to the survival of the hot of the Air and the dry of the Earth-qualities essentially constitutive of Fire. Moreover, this mode of Fire's coming-to-be is confirmed by perception. For flame is par excellence Fire: but flame is burning smoke, and smoke consists of Air and Earth.

No transformation, however, into any of the 'simple' bodies can result from the passing-away of one elementary quality in each of two 'elements' when they are taken in their consecutive order, because either identical or contrary qualities are left in the pair: but no 'simple' body can be formed either out of identical, or out of contrary, qualities. Thus no 'simple' body would result, if the dry of Fire and the moist of Air were to pass-away: for the hot is left in both. On the other hand, if the hot pass-away out both, the contraries-dry and moist-are left. A similar result will occur in all the others too: for all the consecutive 'elements' contain one identical, and one contrary, quality. Hence, too, it clearly follows that, when one of the consecutive 'elements' is transformed into one, the coming-to-be is effected by the passing-away of a single quality: whereas, when two of them are transformed into a third, more than one quality must have passed-away.

We have stated that all the 'elements' come-to-be out of any one of them; and we have explained the manner in which their mutual conversion takes place. Let us nevertheless supplement our theory by the following speculations concerning them.

5

If Water, Air, and the like are a 'matter' of which the natural bodies consist, as some thinkers in fact believe, these 'elements' must be either one, or two, or more. Now they cannot all of them be one-they cannot, e.g. all be Air or Water or Fire or Earth-because 'Change is into contraries'. For if they all were Air, then (assuming Air to persist) there will be 'alteration' instead of coming-to-be. Besides, nobody supposes a single 'element' to persist, as the basis of all, in such a way that it is Water as well as Air (or any other 'element') at the same time. So there will be a certain contrariety, i.e. a differentiating quality: and the other member of this contrariety, e.g. heat, will belong to some other 'element', e.g. to Fire. But Fire will certainly not be 'hot Air'. For a change of that kind (a) is 'alteration', and (b) is not what is observed. Moreover (c) if Air is again to result out of the Fire, it will do so by the conversion of the hot into its contrary: this contrary, therefore, will belong to Air, and Air will be a cold something: hence it is impossible for Fire to be 'hot Air', since in that case the same thing will be simultaneously hot and cold. Both Fire and Air, therefore, will be something else which is the same; i.e. there will be some 'matter', other than either, common to both.

The same argument applies to all the 'elements', proving that there is no single one of them out of which they all originate. But neither is there, beside these four, some other body from which they originate-a something intermediate, e.g. between Air and Water (coarser than Air, but finer than Water), or between Air and Fire (coarser than Fire, but finer than Air). For the supposed 'intermediate' will be Air and Fire when a pair of contrasted qualities is added to it: but, since one of every two contrary qualities is a 'privation', the 'intermediate' never can exist-as some thinkers assert the 'Boundless' or the 'Environing' exists-in isolation. It is, therefore, equally and indifferently any one of the 'elements', or else it is nothing.

Since, then, there is nothing-at least, nothing perceptible-prior to these, they must be all. That being so, either they must always persist and not be transformable into one another: or they must undergo transformation-either all of them, or some only (as Plato wrote in the Timaeus).' Now it has been proved before that they must undergo reciprocal transformation. It has also been proved that the speed with which they come-to-be, one out of another, is not uniform-since the process of reciprocal transformation is relatively quick between the 'elements' with a 'complementary factor', but relatively slow between those which possess no such factor. Assuming, then, that the contrariety, in respect to which they are transformed, is one, the elements' will inevitably be two: for it is 'matter' that is the 'mean' between the two contraries, and matter is imperceptible and inseparable from them. Since, however, the 'elements' are seen to be more than two, the contrarieties must at the least be two. But the contrarieties being two, the 'elements' must be four (as they evidently are) and cannot be three: for the couplings' are four, since, though six are possible, the two in which the qualities are contrary to one another cannot occur.

These subjects have been discussed before:' but the following arguments will make it clear that, since the 'elements' are transformed into one another, it is impossible for any one of them-whether it be at the end or in the middle-to be an 'originative source' of the rest. There can be no such 'originative element' at the ends: for all of them would then be Fire or Earth, and this theory amounts to the assertion that all things are made of Fire or Earth. Nor can a 'middle-element' be such an originative source'-as some thinkers suppose that Air is transformed both into Fire and into Water, and Water both into Air and into Earth, while the 'end-elements' are not further transformed into one another. For the process must come to a stop, and cannot continue ad infinitum in a straight line in either direction, since otherwise an infinite number of contrarieties would attach to the single 'element'. Let E stand for Earth, W for Water, A for Air, and F for Fire. Then (i) since A is transformed into F and W, there will be a contrariety belonging to A F. Let these contraries be whiteness and blackness. Again (ii) since A is transformed into W, there will be another contrariety: for W is not the same as F. Let this second contrariety be dryness and moistness, D being dryness and M moistness. Now if, when A is transformed into W, the 'white' persists, Water will be moist and white: but if it does not persist, Water will be black since change is into contraries. Water, therefore, must be either white or black. Let it then be the first. On similar grounds, therefore, D (dryness) will also belong to F. Consequently F (Fire) as well as Air will be able to be transformed into Water: for it has qualities contrary to those of Water, since Fire was first taken to be black and then to be dry, while Water was moist and then showed itself white. Thus it is evident that all the 'elements' will be able to be transformed out of one another; and that, in the instances we have taken, E (Earth) also will contain the remaining two 'complementary factors', viz. the black and the moist (for these have not yet been coupled).

We have dealt with this last topic before the thesis we set out to prove. That thesis-viz. that the process cannot continue ad infinitum-will be clear from the following considerations. If Fire (which is represented by F) is not to revert, but is to be transformed in turn into some other 'element' (e.g. into Q), a new contrariety, other than those mentioned, will belong to Fire and Q: for it has been assumed that Q is not the same as any of the four, E W A and F. Let K, then, belong to F and Y to Q. Then K will belong to all four, E W A and F: for they are transformed into one another. This last point, however, we may admit, has not yet been proved: but at any rate it is clear that if Q is to be transformed in turn into yet another 'element', yet another contrariety will belong not only to Q but also to F (Fire). And, similarly, every addition of a new 'element' will carry with it the attachment of a new contrariety to the preceding elements'. Consequently, if the 'elements' are infinitely many, there will also belong to the single 'element' an infinite number of contrarieties. But if that be so, it will be impossible to define any 'element': impossible also for any to come-to-be. For if one is to result from another, it will have to pass through such a vast number of contrarieties-and indeed even more than any determinate number. Consequently (i) into some 'elements' transformation will never be effected—viz. if the intermediates are infinite in number, as they must be if the 'elements' are infinitely many: further (ii) there will not even be a transformation of Air into Fire, if the contrarieties are infinitely many: moreover (iii) all the 'elements' become one. For all the contrarieties of the 'elements' above F must belong to those below F, and vice versa: hence they will all be one.

<h1 style="text-align:center">6</h1>

As for those who agree with Empedocles that the 'elements' of body are more than one, so that they are not transformed into one another—one may well wonder in what sense it is open to them to maintain that the 'elements' are comparable. Yet Empedocles says 'For these are all not only equal...'

If it is meant that they are comparable in their amount, all the 'comparables' must possess an identical something whereby they are measured. If, e.g. one pint of Water yields ten of Air, both are measured by the same unit; and therefore both were from the first an identical something. On the other hand, suppose (ii) they are not 'comparable in their amount' in the sense that so-much of the one yields so much of the other, but comparable in 'power of action (a pint of Water, e.g. having a power of cooling equal to that of ten pints of Air); even so, they are 'comparable in their amount', though not qua 'amount' but qua Iso-much power'. There is also (iii) a third possibility. Instead of comparing their powers by the measure of their amount, they might be compared as terms in a 'correspondence': e.g. 'as x is hot, so correspondingly y is white'. But 'correspondence', though it means equality in the quantum, means similarity in a quale. Thus it is manifestly absurd that the 'simple' bodies, though they are not transformable, are comparable not merely as 'corresponding', but by a measure of their powers; i.e. that so-much Fire is comparable with many times-that-amount of Air, as being 'equally' or 'similarly' hot. For the same thing, if it be greater in amount, will, since it belongs to the same kind, have its ratio correspondingly increased.

A further objection to the theory of Empedocles is that it makes even growth impossible, unless it be increase by addition. For his Fire increases by Fire: 'And Earth increases its own frame and Ether increases Ether." These, however, are cases of addition: but it is not by addition that growing things are believed to increase. And it is far more difficult for him to account for the coming-to-be which occurs in nature. For the things which come-to-be by natural process all exhibit, in their coming-to-be, a uniformity either absolute or highly regular: while any exceptions any results which are in accordance neither with the invariable nor with the general rule are products of chance and luck. Then what is the cause determining that man comes-to-be from man, that wheat (instead of an olive) comes-to-be from wheat, either invariably or generally? Are we to say 'Bone comes-to-be if the "elements" be put together in such-and such a manner'? For, according to his own estatements, nothing comes-to-be from their 'fortuitous consilience', but only from their 'consilience' in a certain proportion. What, then, is the cause of this proportional consilience? Presumably not Fire or Earth. But neither is it Love and Strife: for the former is a cause of 'association' only, and the latter only of 'dissociation'. No: the cause in question is the essential nature of each thing-not merely to quote his words) 'a mingling and a divorce of what has been mingled'. And chance, not proportion, 'is the name given to these occurrences': for things can be 'mingled' fortuitously.

The cause, therefore, of the coming-to-be of the things which owe their existence to nature is that they are in such-and-such a determinate condition: and it is this which constitutes, the 'nature' of each thing—a 'nature' about which he says nothing. What he says, therefore, is no explanation of 'nature'. Moreover, it is this which is both 'the excellence' of each thing and its 'good': whereas he assigns the whole credit to the 'mingling'. (And yet the 'elements' at all events are 'dissociated' not by Strife, but by Love: since the 'elements' are by nature prior to the Deity, and they too are Deities.)

Again, his account of motion is vague. For it is not an adequate explanation to say that 'Love and Strife set things moving, unless the very nature of Love is a movement of this kind and the very nature of Strife a movement of that kind. He ought, then, either to have defined or to have postulated these characteristic movements, or to have demonstrated them-whether strictly or laxly or in some other fashion. Moreover, since (a) the 'simple' bodies appear to move 'naturally' as well as by compulsion, i.e. in a manner contrary to nature (fire, e.g. appears to move upwards without compulsion, though it appears to move by compulsion downwards); and since (b) what is 'natural' is contrary to that which is due to compulsion, and movement by compulsion actually occurs; it follows that 'natural movement' can also occur in fact. Is this, then, the movement that Love sets going? No: for, on the contrary, the 'natural movement' moves Earth downwards and resembles 'dissociation', and Strife rather than Love is its cause-so that in general, too, Love rather than Strife would seem to be contrary to nature. And unless Love or Strife is actually setting them in motion, the 'simple' bodies themselves have absolutely no movement or rest. But this is paradoxical: and what is more, they do in fact obviously move. For though Strife 'dissociated', it was not by Strife that the 'Ether' was borne upwards. On the contrary, sometimes he attributes its movement to something like chance ('For thus, as it ran, it happened to meet them then, though often otherwise"), while at other times he says it is the nature of Fire to be borne upwards, but 'the Ether' (to quote his words) 'sank down upon the Earth with long roots'. With such statements, too, he combines the assertion that the Order of the World is the same now, in the reign of Strife, as it was formerly in the reign of Love. What, then, is the 'first mover' of the 'elements'? What causes their motion? Presumably not Love and Strife: on the contrary, these are causes of a particular motion, if at least we assume that 'first mover' to be an originative source'.

An additional paradox is that the soul should consist of the 'elements', or that it should be one of them. How are the soul's 'alterations' to take Place? How, e.g. is the change from being musical to being unmusical, or how is memory or forgetting, to occur? For clearly, if the soul be Fire, only such modifications will happen to it as characterize Fire qua Fire: while if it be compounded out of the elements', only the corporeal modifications will occur in it. But the changes we have mentioned are none of them corporeal.

7

The discussion of these difficulties, however, is a task appropriate to a different investigation:' let us return to the 'elements' of which bodies are composed. The theories that 'there is something common to all the "elements"', and that they are reciprocally transformed', are so related that those who accept either are bound to accept the other as well. Those, on the other hand, who do not make their coming-to-be reciprocal-who refuse to suppose that any one of the 'elements' comes-to-be out of any other taken singly, except in the sense in which bricks come-to-be out of a wall-are faced with a paradox. How, on their theory, are flesh and bones or any of the other compounds to result from the 'elements' taken together?

Indeed, the point we have raised constitutes a problem even for those who generate the 'elements' out of one another. In what manner does anything other than, and beside, the 'elements' come-to-be out of them? Let me illustrate my meaning. Water can come-to-be out of Fire and Fire out of Water; for their substratum is something common to them both. But flesh too, presumably, and marrow come-to-be out of them. How, then, do such things come to-be? For (a) how is the manner of their coming-to-be to be conceived by those who maintain a theory like that of Empedocles? They must conceive it as composition-just as a wall comes-to-be out of bricks and stones: and the 'Mixture', of which they speak, will be composed of the 'elements', these being preserved in it unaltered but with their small particles juxtaposed each to each. That will be the manner, presumably, in which flesh and every other compound results from the 'elements'. Consequently, it follows that Fire and Water do not come-to-be 'out of any and every part of flesh'. For instance, although a sphere might come-to-be out of this part of a lump of wax and a pyramid out of some other part, it was nevertheless possible for either figure to have come-to-be out of either part indifferently: that is the manner of coming-to-be when 'both Fire and Water come-to-be out of any and every part of flesh'. Those, however, who maintain the theory in question, are not at liberty to conceive that 'both come-to-be out of flesh' in that manner, but only as a stone and a brick 'both come-to-be out of a wall'-viz. each out of a different place or part. Similarly (b) even for those who postulate a single matter of their 'elements' there is a certain difficulty in explaining how anything is to result from two of them taken together-e.g. from 'cold' and hot', or from Fire and Earth. For if flesh consists of both and is neither of them, nor again is a 'composition' of them in which they are preserved unaltered, what alternative is left except to identify the resultant of the two 'elements' with their matter? For the passing-away of either 'element' produces either the other or the matter.

Perhaps we may suggest the following solution. (i) There are differences of degree in hot and cold. Although, therefore, when either is fully real without qualification, the other will exist potentially; yet, when neither exists in the full completeness of its being, but both by combining destroy one another's excesses so that there exist instead a hot which (for a 'hot') is cold and a cold which (for a 'cold') is hot; then what results from these two contraries will be neither their matter, nor either of them existing in its full reality without qualification. There will result instead an 'intermediate': and this 'intermediate', according as it is potentially more hot than cold or vice versa, will possess a power-of-heating that is double or triple its power-of-cooling, or otherwise related thereto in some similar ratio. Thus all the other bodies will result from the contraries, or rather from the 'elements', in so far as these have been 'combined': while the elements' will result from the contraries, in so far as these 'exist potentially' in a special sense-not as matter 'exists potentially', but in the sense explained above. And when a thing comes-to-be in this manner, the process is 'combination'; whereas what comes-to-be in the other manner is matter. Moreover (ii) contraries also 'suffer action', in accordance with the disjunctively-articulated definition established in the early part of this work.' For the actually-hot is potentially-cold and the actually cold potentially-hot; so that hot and cold, unless they are equally balanced, are transformed into one another (and all the other contraries behave in a similar way). It is thus, then, that in the first place the 'elements' are transformed; and that (in the second place) out of the 'elements' there come-to-be flesh and bones and the like-the hot becoming cold and the cold becoming hot when they have been brought to the 'mean'. For at the 'mean' is neither hot nor cold. The 'mean', however, is of considerable extent and not indivisible. Similarly, it is qua reduced to a 'mean' condition that the dry and the moist, as well as the contraries we have used as examples, produce flesh and bone and the remaining compounds.

8

All the compound bodies—all of which exist in the region belonging to the central body—are composed of all the 'simple' bodies. For they all contain Earth because every 'simple' body is to be found specially and most abundantly in its own place. And they all contain Water because (a) the compound must possess a definite outline and Water, alone of the 'simple' bodies, is readily adaptable in shape: moreover (b) Earth has no power of cohesion without the moist. On the contrary, the moist is what holds it together; for it would fall to pieces if the moist were eliminated from it completely.

They contain Earth and Water, then, for the reasons we have given: and they contain Air and Fire, because these are contrary to Earth and Water (Earth being contrary to Air and Water to Fire, in so far as one Substance can be 'contrary' to another). Now all compounds presuppose in their coming-to-be constituents which are contrary to one another: and in all compounds there is contained one set of the contrasted extremes. Hence the other set must be contained in them also, so that every compound will include all the 'simple' bodies.

Additional evidence seems to be furnished by the food each compound takes. For all of them are fed by substances which are the same as their constituents, and all of them are fed by more substances than one. Indeed, even the plants, though it might be thought they are fed by one substance only, viz. by Water, are fed by more than one: for Earth has been mixed with the Water. That is why farmers too endeavour to mix before watering. Although food is akin to the matter, that which is fed is the 'figure'—i.e. the 'form' taken along with the matter. This fact enables us to understand why, whereas all the 'simple' bodies come-to-be out of one another, Fire is the only one of them which (as our predecessors also assert) 'is fed'. For Fire alone-or more than all the rest-is akin to the 'form' because it tends by nature to be borne towards the limit. Now each of them naturally tends to be borne towards its own place; but the 'figure'—i.e. the 'form'—Of them all is at the limits.

Thus we have explained that all the compound bodies are composed of all the 'simple' bodies.

9

Since some things are such as to come-to-be and pass-away, and since coming-to-be in fact occurs in the region about the centre, we must explain the number and the nature of the 'originative sources' of all coming-to-be alike: for a grasp of the true theory of any universal facilitates the understanding of its specific forms.

The 'originative sources', then, of the things which come-to-be are equal in number to, and identical in kind with, those in the sphere of the eternal and primary things. For there is one in the sense of 'matter', and a second in the sense of 'form': and, in addition, the third 'originative source' must be present as well. For the two first are not sufficient to bring things into being, any more than they are adequate to account for the primary things.

Now cause, in the sense of material origin, for the things which are such as to come-to-be is 'that which can be-and-not-be': and this is identical with 'that which can come-to-be-and-pass-away', since the latter, while it is at one time, at another time is not. (For whereas some things are of necessity, viz. the eternal things, others of necessity are not. And of these two sets of things, since they cannot diverge from the necessity of their nature, it is impossible for the first not to he and impossible for the second to he. Other things, however, can both be and not he.) Hence coming-to-be and passing-away must occur within the field of 'that which can be-and-not-be'. This, therefore, is cause in the sense of material origin for the things which are such as to come-to-be; while cause, in the sense of their 'end', is their 'figure' or 'form'-and that is the formula expressing the essential nature of each of them.

But the third 'originative source' must be present as well—the cause vaguely dreamed of by all our predecessors, definitely stated by none of them. On the contrary (a) some amongst them thought the nature of 'the Forms' was adequate to account for coming-to-be. Thus Socrates in the Phaedo first blames everybody else for having given no explanation; and then lays it down; that 'some things are Forms, others Participants in the Forms', and that 'while a thing is said to "be" in virtue of the Form, it is said to "come-to-be" qua sharing in," to "pass-away" qua "losing," the 'Form'. Hence he thinks that 'assuming the truth of these theses, the

Forms must be causes both of coming-to-be and of passing-away'. On the other hand (b) there were others who thought 'the matter' was adequate by itself to account for coming-to-be, since 'the movement originates from the matter'.

Neither of these theories, however, is sound. For (a) if the Forms are causes, why is their generating activity intermittent instead of perpetual and continuous-since there always are Participants as well as Forms? Besides, in some instances we see that the cause is other than the Form. For it is the doctor who implants health and the man of science who implants science, although 'Health itself' and 'Science itself' are as well as the Participants: and the same principle applies to everything else that is produced in accordance with an art. On the other hand (b) to say that 'matter generates owing to its movement' would be, no doubt, more scientific than to make such statements as are made by the thinkers we have been criticizing. For what 'alters' and transfigures plays a greater part in bringing, things into being; and we are everywhere accustomed, in the products of nature and of art alike, to look upon that which can initiate movement as the producing cause. Nevertheless this second theory is not right either.

For, to begin with, it is characteristic of matter to suffer action, i.e. to be moved: but to move, i.e. to act, belongs to a different 'power'. This is obvious both in the things that come-to-be by art and in those that come to-be by nature. Water does not of itself produce out of itself an animal: and it is the art, not the wood, that makes a bed. Nor is this their only error. They make a second mistake in omitting the more controlling cause: for they eliminate the essential nature, i.e. the 'form'. And what is more, since they remove the formal cause, they invest the forces they assign to the 'simple' bodies-the forces which enable these bodies to bring things into being-with too instrumental a character. For 'since' (as they say) 'it is the nature of the hot to dissociate, of the cold to bring together, and of each remaining contrary either to act or to suffer action', it is out of such materials and by their agency (so they maintain) that everything else comes-to-be and passes-away. Yet (a) it is evident that even Fire is itself moved, i.e. suffers action. Moreover (b) their procedure is virtually the same as if one were to treat the saw (and the various instruments of carpentry) as 'the cause' of the things that come-to-be: for the wood must be divided if a man saws, must become smooth if he planes, and so on with the remaining tools. Hence, however true it may be that Fire is active, i.e. sets things moving, there is a further point they fail to observe—viz. that Fire is inferior to the tools or instruments in the manner in which it sets things moving.

10

As to our own theory—we have given a general account of the causes in an earlier work,' we have now explained and distinguished the 'matter' and the 'form'. Further, since the change which is motion has been proved' to be eternal, the continuity of the occurrence of coming-to-be follows necessarily from what we have established: for the eternal motion, by causing 'the generator' to approach and retire, will produce coming-to-be uninterruptedly. At the same time it is clear that we were right when, in an earlier work,' we called motion (not coming-to-be) 'the primary form of change'. For it is far more reasonable that what is should cause the coming-to-be of what is not, than that what is not should cause the being of what is. Now that which is being moved is, but that which is coming-to-be is not: hence, also, motion is prior to coming-to-be.

We have assumed, and have proved, that coming-to-be and passing-away happen to things continuously; and we assert that motion causes coming-to-be. That being so, it is evident that, if the motion be single, both processes cannot occur since they are contrary to one another: for it is a law of nature that the same cause, provided it remain in the same condition, always produces the same effect, so that, from a single motion, either coming-to-be or passing-away will always result. The movements must, on the contrary, be more than one, and they must be contrasted with one another either by the sense of their motion or by its irregularity: for contrary effects demand contraries as their causes.

This explains why it is not the primary motion that causes coming-to-be and passing-away, but the motion along the inclined circle: for this motion not only possesses the necessary continuity, but includes a duality of movements as well. For if coming-to-be and passing-away are always to be continuous, there must be some body always being moved (in order that these changes may not fail) and moved with a duality of movements (in order that both changes, not one only, may result). Now the continuity of this movement is caused by the motion of the whole: but the approaching and retreating of the moving body are caused by the inclination. For the consequence of the inclination is that the body becomes alternately remote and near; and since its distance is thus unequal, its movement will be irregular. Therefore, if it generates by approaching and by its proximity, it-this very same body-destroys by retreating and becoming remote: and if it generates by many successive approaches, it also destroys by many successive retirements. For contrary effects demand contraries as their causes; and the natural processes of passing-away and coming-to-be occupy equal periods of time. Hence, too, the times-i.e. the lives-of the several kinds of living things have a number by which they are distinguished: for there is an Order controlling all things, and every time (i.e. every life) is measured by a period. Not all of them, however, are measured by the same period, but some by a smaller and others by a greater one: for to some of them the period, which is their measure, is a year, while to some it is longer and to others shorter.

And there are facts of observation in manifest agreement with our theories. Thus we see that coming-to-be occurs as the sun approaches and decay as it retreats; and we see that the two processes occupy equal times. For the durations of the natural processes of passing-away and coming-to-be are equal. Nevertheless it often happens that things pass-away in too short a time. This is due to the 'intermingling' by which the things that come-to-be and pass-away are implicated with one another. For their matter is 'irregular', i.e. is not everywhere the same: hence the processes by which they come-to-be must be 'irregular' too, i.e. some too quick and others too slow. Consequently the phenomenon in question occurs, because the 'irregular' coming-to-be of these things is the passing-away of other things.

Coming-to-be and passing-away will, as we have said, always be continuous, and will never fail owing to the cause we stated. And this continuity has a sufficient reason on our theory. For in all things, as we affirm, Nature always strives after 'the better'. Now 'being' (we have explained elsewhere the exact variety of meanings we recognize in this term) is better than 'not-being': but not all things can possess 'being', since they are too far removed from the 'originative source'. 'God therefore adopted the remaining alternative, and fulfilled the perfection of the universe by making coming-to-be uninterrupted: for the greatest possible coherence would thus be secured to existence, because that 'coming-to-be should itself come-to-be perpetually' is the closest approximation to eternal being.

The cause of this perpetuity of coming-to-be, as we have often said, is circular motion: for that is the only motion which is continuous. That, too, is why all the other things-the things, I mean, which are reciprocally transformed in virtue of their 'passions' and their 'powers of action' e.g. the 'simple' bodies imitate circular motion. For when Water is transformed into Air, Air into Fire, and the Fire back into Water, we say the coming-to-be 'has completed the circle', because it reverts again to the beginning. Hence it is by imitating circular motion that rectilinear motion too is continuous.

These considerations serve at the same time to explain what is to some people a baffling problem—viz. why the 'simple' bodies, since each them is travelling towards its own place, have not become dissevered from one another in the infinite lapse of time. The reason is their reciprocal transformation. For, had each of them persisted in its own place instead of being transformed by its neighbour, they would have got dissevered long ago. They are transformed, however, owing to the motion with its dual character: and because they are transformed, none of them is able to persist in any place allotted to it by the Order.

It is clear from what has been said (i) that coming-to-be and passing-away actually occur, (ii) what causes them, and (iii) what subject undergoes them. But (a) if there is to be movement (as we have explained elsewhere, in an earlier work') there must be something which initiates it; if there is to be movement always, there must always be something which initiates it; if the movement is to be continuous, what initiates it must be single, unmoved, ungenerated, and incapable of 'alteration'; and if the circular movements are more than one, their initiating causes must all of them, in spite of their plurality, be in some way subordinated to a single 'originative source'. Further (b) since time is continuous, movement must be continuous, inasmuch as there can be no time without movement. Time, therefore, is a 'number' of some continuous movement-a 'number', therefore, of the circular movement, as was established in the discussions at the beginning. But (c) is movement continuous because of the continuity of that which is moved, or because that in which the movement occurs (I mean, e.g. the place or the quality) is continuous? The answer must clearly be 'because that which is moved is continuous'. (For how can the quality be continuous except in virtue of the continuity of the thing to which it belongs? But if the continuity of 'that in which' contributes to make the movement continuous, this is true only of 'the place in which'; for that has 'magnitude' in a sense.) But (d) amongst continuous bodies which are moved, only that which is moved in a circle is 'continuous' in such a way that it preserves its continuity with itself throughout the movement. The conclusion therefore is that this is what produces continuous movement, viz. the body which is being moved in a circle; and its movement makes time continuous.

11

Wherever there is continuity in any process (coming-to-be or 'alteration' or any kind of change whatever) we observe consecutiveness', i.e. this coming-to-be after that without any interval. Hence we must investigate whether, amongst the consecutive members, there is any whose future being is necessary; or whether, on the contrary, every one of them may fail to come-to-be. For that some of them may fail to occur, is clear. (a) We need only appeal to the distinction between the statements 'x will be' and 'x is about to which depends upon this fact. For if it be true to say of x that it 'will be', it must at some time be true to say of it that 'it is': whereas, though it be true to say of x now that 'it is about to occur', it is quite possible for it not to come-to-be-thus a man might not walk, though he is now 'about to' walk. And (b) since (to appeal to a general principle) amongst the things which 'are' some are capable also of 'not-being', it is clear that the same ambiguous character will attach to them no less when they are coming-to-be: in other words, their coming-to-be will not be necessary.

Then are all the things that come-to-be of this contingent character? Or, on the contrary, is it absolutely necessary for some of them to come-to-be? Is there, in fact, a distinction in the field of 'coming-to-be' corresponding to the distinction, within the field of 'being', between things that cannot possibly 'not-be' and things that can 'not-be'? For instance, is it necessary that solstices shall come-to-be, i.e. impossible that they should fail to be able to occur?

Assuming that the antecedent must have come-to-be if the consequent is to be (e.g. that foundations must have come-to-be if there is to be a house: clay, if there are to be foundations), is the converse also true? If foundations have come-to-be, must a house come-to-be? The answer seems to be that the necessary nexus no longer holds, unless it is 'necessary' for the consequent (as well as for the antecedent) to come-to-be-'necessary' absolutely. If that be the case, however, 'a house must come to-be if foundations have come-to-be', as well as vice versa. For the antecedent was assumed to be so related to the consequent that, if the latter is to be, the antecedent must have come-to-be before it. If, therefore, it is necessary that the consequent should come-to-be, the antecedent also must have come-to-be: and if the antecedent has come-to-be, then the consequent also must come-to-be-not, however, because of the antecedent, but because the future being of the consequent was assumed as necessary. Hence, in any sequence, when the being of the consequent is necessary, the nexus is reciprocal-in other words, when the antecedent has come-to-be the consequent must always come-to-be too.

Now (i) if the sequence of occurrences is to proceed ad infinitum 'downwards', the coming to-be of any determinate 'this' amongst the later members of the sequence will not be absolutely, but only conditionally, necessary. For it will always be necessary that some other member shall have come-to-be before 'this' as the presupposed condition of the necessity that 'this' should come-to-be: consequently, since what is 'infinite' has no 'originative source', neither will there be in the infinite sequence any 'primary' member which will make it 'necessary' for the remaining members to come-to-be.

Nor again (ii) will it be possible to say with truth, even in regard to the members of a limited sequence, that it is 'absolutely necessary' for any one of them to come-to-be. We cannot truly say, e.g. that 'it is absolutely necessary for a house to come-to-be when foundations have been laid': for (unless it is always necessary for a house to be coming-to-be) we should be faced with the consequence that, when foundations have been laid, a thing, which need not always be, must always be. No: if its coming-to-be is to be 'necessary', it must be 'always' in its coming-to-be. For what is 'of necessity' coincides with what is 'always', since that which 'must be' cannot possibly 'not-be'. Hence a thing is eternal if its 'being' is necessary: and if it is eternal, its 'being' is necessary. And if, therefore, the 'coming-to-be' of a thing is necessary, its 'coming-to-be' is eternal; and if eternal, necessary.

It follows that the coming-to-be of anything, if it is absolutely necessary, must be cyclical—i.e. must return upon itself. For coming to-be must either be limited or not limited: and if not limited, it must be either rectilinear or cyclical. But the first of these last two alternatives is impossible if coming-to-be is to be eternal, because there could not be any 'originative source' whatever in an infinite rectilinear sequence, whether its members be taken 'downwards' (as future events) or 'upwards' (as past events). Yet coming to be must have an 'originative source' (if it is to be necessary and therefore eternal), nor can it be eternal if it is limited. Consequently it must be cyclical. Hence the nexus must be reciprocal. By this I mean that the necessary occurrence of 'this' involves the necessary occurrence of its antecedent: and conversely that, given the antecedent, it is also necessary for the consequent to come-to-be. And this reciprocal nexus will hold continuously throughout the sequence: for it makes no difference whether the reciprocal nexus, of which we are speaking, is mediated by two, or by many, members.

It is in circular movement, therefore, and in cyclical coming-to-be that the 'absolutely necessary' is to be found. In other words, if the coming-to-be of any things is cyclical, it is 'necessary' that each of them is coming-to-be and has come-to-be: and if the coming-to-be of any things is 'necessary', their coming-to-be is cyclical.

The result we have reached is logically concordant with the eternity of circular motion, i.e. the eternity of the revolution of the heavens (a fact which approved itself on other and independent evidence),' since precisely those movements which belong to, and depend upon, this eternal revolution 'come-to-be' of necessity, and of necessity 'will be'. For since the revolving body is always setting something else in motion, the movement of the things it moves must also be circular. Thus, from the being of the 'upper revolution' it follows that the sun revolves in this determinate manner; and since the sun revolves thus, the seasons in consequence come-to-be in a cycle, i.e. return upon themselves; and since they come-to-be cyclically, so in their turn do the things whose coming-to-be the seasons initiate.

Then why do some things manifestly come to-be in this cyclical fashion (as, e.g. showers and air, so that it must rain if there is to be a cloud and, conversely, there must be a cloud if it is to rain), while men and animals do not 'return upon themselves' so that the same individual comes-to-be a second time (for though your coming-to-be presupposes your father's, his coming-to-be does not presuppose yours)? Why, on the contrary, does this coming-to-be seem to constitute a rectilinear sequence?

In discussing this new problem, we must begin by inquiring whether all things 'return upon themselves' in a uniform manner; or whether, on the contrary, though in some sequences what recurs is numerically the same, in other sequences it is the same only in species. In consequence of this distinction, it is evident that those things, whose 'substance'—that which is undergoing the process-is imperishable, will be numerically, as well as specifically, the same in their recurrence: for the character of the process is determined by the character of that which undergoes it. Those things, on the other hand, whose 'substance' is perish, able (not imperishable) must 'return upon themselves' in the sense that what recurs, though specifically the same, is not the same numerically. That why, when Water comes-to-be from Air and Air from Water, the Air is the same 'specifically', not 'numerically': and if these too recur numerically the same, at any rate this does not happen with things whose 'substance' comes-to-be-whose 'substance' is such that it is essentially capable of not-being.

POETICS

Table of Contents

'Imitation' the common principle of the Arts of Poetry.

I propose to treat of Poetry in itself and of its various kinds, noting the essential quality of each; to inquire into the structure of the plot as requisite to a good poem; into the number and nature of the parts of which a poem is composed; and similarly into whatever else falls within the same inquiry. Following, then, the order of nature, let us begin with the principles which come first.

Epic poetry and Tragedy, Comedy also and Dithyrambic: poetry, and the music of the flute and of the lyre in most of their forms, are all in their general conception modes of imitation. They differ, however, from one: another in three respects,—the medium, the objects, the manner or mode of imitation, being in each case distinct.

For as there are persons who, by conscious art or mere habit, imitate and represent various objects through the medium of colour and form, or again by the voice; so in the arts above mentioned, taken as a whole, the imitation is produced by rhythm, language, or 'harmony,' either singly or combined.

Thus in the music of the flute and of the lyre, 'harmony' and rhythm alone are employed; also in other arts, such as that of the shepherd's pipe, which are essentially similar to these. In dancing, rhythm alone is used without 'harmony'; for even dancing imitates character, emotion, and action, by rhythmical movement.

There is another art which imitates by means of language alone, and that either in prose or verse—which, verse, again, may either combine different metres or consist of but one kind—but this has hitherto been without a name. For there is no common term we could apply to the mimes of Sophron and Xenarchus and the Socratic dialogues on the one hand; and, on the other, to poetic imitations in iambic, elegiac, or any similar metre. People do, indeed, add the word 'maker' or 'poet' to the name of the metre, and speak of elegiac poets, or epic (that is, hexameter) poets, as if it were not the imitation that makes the poet, but the verse that entitles them all indiscriminately to the name. Even when a treatise on medicine or natural science is brought out in verse, the name of poet is by custom given to the author; and yet Homer and Empedocles have nothing in common but the metre, so that it would be right to call the one poet, the other physicist rather than poet. On the same principle, even if a writer in his poetic imitation were to combine all metres, as Chaeremon did in his Centaur, which is a medley composed of metres of all kinds, we should bring him too under the general term poet. So much then for these distinctions.

There are, again, some arts which employ all the means above mentioned, namely, rhythm, tune, and metre. Such are Dithyrambic and Nomic poetry, and also Tragedy and Comedy; but between them the difference is, that in the first two cases these means are all employed in combination, in the latter, now one means is employed, now another.

Such, then, are the differences of the arts with respect to the medium of imitation.

The Objects of Imitation.

Since the objects of imitation are men in action, and these men must be either of a higher or a lower type (for moral character mainly answers to these divisions, goodness and badness being the distinguishing marks of moral differences), it follows that we must represent men either as better than in real life, or as worse, or as they are. It is the same in painting. Polygnotus depicted men as nobler than they are, Pauson as less noble, Dionysius drew them true to life.

Now it is evident that each of the modes of imitation above mentioned will exhibit these differences, and become a distinct kind in imitating objects that are thus distinct. Such diversities may be found even in dancing,: flute-playing, and lyre-playing. So again in language, whether prose or verse unaccompanied by music. Homer, for example, makes men better than they are; Cleophon as they are; Hegemon the Thasian, the inventor of parodies, and Nicochares, the author of the Deiliad, worse than they are. The same thing holds good of Dithyrambs and Nomes; here too one may portray different types, as Timotheus and Philoxenus differed in representing their Cyclopes. The same distinction marks off Tragedy from Comedy; for Comedy aims at representing men as worse, Tragedy as better than in actual life.

The Manner of Imitation.

There is still a third difference—the manner in which each of these objects may be imitated. For the medium being the same, and the objects the same, the poet may imitate by narration—in which case he can either take another personality as Homer does, or speak in his own person, unchanged—or he may present all his characters as living and moving before us.

These, then, as we said at the beginning, are the three differences which distinguish artistic imitation,—the medium, the objects, and the manner. So that from one point of view, Sophocles is an imitator of the same kind as Homer—for both imitate higher types of character; from another point of view, of the same kind as Aristophanes—for both imitate persons acting and doing. Hence, some say, the name of 'drama' is given to such poems, as representing action. For the same reason the Dorians claim the invention both of Tragedy and Comedy. The claim to Comedy is put forward by the Megarians,—not only by those of Greece proper, who allege that it originated under their democracy, but also by the Megarians of Sicily, for the poet Epicharmus, who is much earlier than Chionides and Magnes, belonged to that country. Tragedy too is claimed by certain Dorians of the Peloponnese. In each case they appeal to the evidence of language. The outlying villages, they say, are by them called κωμαι, by the Athenians δημι: and they assume that Comedians were so named not from κωμάζειν, 'to revel,' but because they wandered from village to village κατα κωμασ, being excluded contemptuously from the city. They add also that the Dorian word for 'doing' is δραν, and the Athenian, πραττειν.

This may suffice as to the number and nature of the various modes of imitation.

The Origin and Development of Poetry.

Poetry in general seems to have sprung from two causes, each of them lying deep in our nature. First, the instinct of imitation is implanted in man from childhood, one difference between him and other animals being that he is the most imitative of living creatures, and through imitation learns his earliest lessons; and no less universal is the pleasure felt in things imitated. We have evidence of this in the facts of experience. Objects which in themselves we view with pain, we delight to contemplate when reproduced with minute fidelity: such as the forms of the most ignoble animals and of dead bodies. The cause of this again is, that to learn gives the liveliest pleasure, not only to philosophers but to men in general; whose capacity, however, of learning is more limited. Thus the reason why men enjoy seeing a likeness is, that in contemplating it they find themselves learning or inferring, and saying perhaps, 'Ah, that is he.' For if you happen not to have seen the original, the pleasure will be due not to the imitation as such, but to the execution, the colouring, or some such other cause.

Imitation, then, is one instinct of our nature. Next, there is the instinct for 'harmony' and rhythm, metres being manifestly sections of rhythm. Persons, therefore, starting with this natural gift developed by degrees their special aptitudes, till their rude improvisations gave birth to Poetry.

Poetry now diverged in two directions, according to the individual character of the writers. The graver spirits imitated noble actions, and the actions of good men. The more trivial sort imitated the actions of meaner persons, at first composing satires, as the former did hymns to the gods and the praises of famous men. A poem of the satirical kind cannot indeed be put down to any author earlier than Homer; though many such writers probably there were. But from Homer onward, instances can be cited,—his own Margites, for example, and other similar compositions. The appropriate metre was also here introduced; hence the measure is still called the iambic or lampooning measure, being that in which people lampooned one another. Thus the older poets were distinguished as writers of heroic or of lampooning verse.

As, in the serious style, Homer is pre-eminent among poets, for he alone combined dramatic form with excellence of imitation, so he too first laid down the main lines of Comedy, by dramatising the ludicrous instead of writing personal satire. His Margites bears the same relation to Comedy that the Iliad and Odyssey do to Tragedy. But when Tragedy and Comedy came to light, the two classes of poets still followed their natural bent: the lampooners became writers of Comedy, and the Epic poets were succeeded by Tragedians, since the drama was a larger and higher form of art.

Whether Tragedy has as yet perfected its proper types or not; and whether it is to be judged in itself, or in relation also to the audience,—this raises another question. Be that as it may, Tragedy—as also Comedy—was at first mere improvisation. The one originated with the authors of the Dithyramb, the other with those of the phallic songs, which are still in use in many of our cities. Tragedy advanced by slow degrees; each new element that showed itself was in turn developed. Having passed through many changes, it found its natural form, and there it stopped.

Aeschylus first introduced a second actor; he diminished the importance of the Chorus, and assigned the leading part to the dialogue. Sophocles raised the number of actors to three, and added scene-painting. Moreover, it was not till late that the short plot was discarded for one of greater compass, and the grotesque diction of the earlier satyric form for the stately manner of Tragedy. The iambic measure then replaced the trochaic tetrameter, which was originally employed when the poetry was of the Satyric order, and had greater affinities with dancing. Once dialogue had come in, Nature herself discovered the appropriate measure. For the iambic is, of all measures, the most colloquial: we see it in the fact that conversational speech runs into iambic lines more frequently than into any other kind of verse; rarely into hexameters, and only when we drop the colloquial intonation. The additions to the number of 'episodes' or acts, and the other accessories of which tradition; tells, must be taken as already described; for to discuss them in detail would, doubtless, be a large undertaking.

Definition of the Ludicrous, and a brief sketch of the rise of Comedy.

Comedy is, as we have said, an imitation of characters of a lower type, not, however, in the full sense of the word bad, the Ludicrous being merely a subdivision of the ugly. It consists in some defect or ugliness which is not painful or destructive. To take an obvious example, the comic mask is ugly and distorted, but does not imply pain.

The successive changes through which Tragedy passed, and the authors of these changes, are well known, whereas Comedy has had no history, because it was not at first treated seriously. It was late before the Archon granted a comic chorus to a poet; the performers were till then voluntary. Comedy had already taken definite shape when comic poets, distinctively so called, are heard of. Who furnished it with masks, or prologues, or increased the number of actors,—these and other similar details remain unknown. As for the plot, it came originally from Sicily; but of Athenian writers Crates was the first who, abandoning the 'iambic' or lampooning form, generalised his themes and plots.

Epic poetry agrees with Tragedy in so far as it is an imitation in verse of characters of a higher type. They differ, in that Epic poetry admits but one kind of metre, and is narrative in form. They differ, again, in their length: for Tragedy endeavours, as far as possible, to confine itself to a single revolution of the sun, or but slightly to exceed this limit; whereas the Epic action has no limits of time. This, then, is a second point of difference; though at first the same freedom was admitted in Tragedy as in Epic poetry.

Of their constituent parts some are common to both, some peculiar to Tragedy, whoever, therefore, knows what is good or bad Tragedy, knows also about Epic poetry. All the elements of an Epic poem are found in Tragedy, but the elements of a Tragedy are not all found in the Epic poem.

Definition of Tragedy.

Of the poetry which imitates in hexameter verse, and of Comedy, we will speak hereafter. Let us now discuss Tragedy, resuming its formal definition, as resulting from what has been already said.

Tragedy, then, is an imitation of an action that is serious, complete, and of a certain magnitude; in language embellished with each kind of artistic ornament, the several kinds being found in separate parts of the play; in the form of action, not of narrative; through pity and fear effecting the proper purgation of these emotions. By 'language embellished,' I mean language into which rhythm, 'harmony,' and song enter. By 'the several kinds in separate parts,' I mean, that some parts are rendered through the medium of verse alone, others again with the aid of song.

Now as tragic imitation implies persons acting, it necessarily follows, in the first place, that Spectacular equipment will be a part of Tragedy. Next, Song and Diction, for these are the medium of imitation. By 'Diction' I mean the mere metrical arrangement of the words: as for 'Song,' it is a term whose sense every one understands.

Again, Tragedy is the imitation of an action; and an action implies personal agents, who necessarily possess certain distinctive qualities both of character and thought; for it is by these that we qualify actions themselves, and these—thought and character—are the two natural causes from which actions spring, and on actions again all success or failure depends. Hence, the Plot is the imitation of the action: for by plot I here mean the arrangement of the incidents. By Character I mean that in virtue of which we ascribe certain qualities to the agents. Thought is required wherever a statement is proved, or, it may be, a general truth enunciated. Every Tragedy, therefore, must have six parts, which parts determine its quality—namely, Plot, Character, Diction, Thought, Spectacle, Song. Two of the parts constitute the medium of imitation, one the manner, and three the objects of imitation. And these complete the list. These elements have been employed, we may say, by the poets to a man; in fact, every play contains Spectacular elements as well as Character, Plot, Diction, Song, and Thought.

But most important of all is the structure of the incidents. For Tragedy is an imitation, not of men, but of an action and of life, and life consists in action, and its end is a mode of action, not a quality. Now character determines men's qualities, but it is by their actions that they are happy or the reverse. Dramatic action, therefore, is not with a view to the representation of character: character comes in as subsidiary to the actions. Hence the incidents and the plot are the end of a tragedy; and the end is the chief thing of all. Again, without action there cannot be a tragedy; there may be without character. The tragedies of most of our modern poets fail in the rendering of character; and of poets in general this is often true. It is the same in painting; and here lies the difference between Zeuxis and Polygnotus. Polygnotus delineates character well: the style of Zeuxis is devoid of ethical quality. Again, if you string together a set of speeches expressive of character, and well finished in point of diction and thought, you will not produce the essential tragic effect nearly so well as with a play which, however deficient in these respects, yet has a plot and artistically constructed incidents. Besides which, the most powerful elements of emotional: interest in Tragedy Peripeteia or Reversal of the Situation, and Recognition scenes—are parts of the plot. A further proof is, that novices in the art attain to finish: of diction and precision of portraiture before they can construct the plot. It is the same with almost all the early poets.

The Plot, then, is the first principle, and, as it were, the soul of a tragedy: Character holds the second place. A similar fact is seen in painting. The most beautiful colours, laid on confusedly, will not give as much pleasure as the chalk outline of a portrait. Thus Tragedy is the imitation of an action, and of the agents mainly with a view to the action.

Third in order is Thought,—that is, the faculty of saying what is possible and pertinent in given circumstances. In the case of oratory, this is the function of the Political art and of the art of rhetoric: and so indeed the older poets make their characters speak the language of civic life; the poets of our time, the language of the rhetoricians. Character is that which reveals moral purpose, showing what kind of things a man chooses or avoids. Speeches, therefore, which do not make this manifest, or in which the speaker does not choose or avoid anything whatever, are not expressive of character. Thought, on the other hand, is found where something is proved to be. or not to be, or a general maxim is enunciated.

Fourth among the elements enumerated comes Diction; by which I mean, as has been already said, the expression of the meaning in words; and its essence is the same both in verse and prose.

Of the remaining elements Song holds the chief place among the embellishments.

The Spectacle has, indeed, an emotional attraction of its own, but, of all the parts, it is the least artistic, and connected least with the art of poetry. For the power of Tragedy, we may be sure, is felt even apart from representation and actors. Besides, the production of spectacular effects depends more on the art of the stage machinist than on that of the poet.

The Plot must be a Whole.

These principles being established, let us now discuss the proper structure of the Plot, since this is the first and most important thing in Tragedy.

Now, according to our definition, Tragedy is an imitation of an action that is complete, and whole, and of a certain magnitude; for there may be a whole that is wanting in magnitude. A whole is that which has a beginning, a middle, and an end. A beginning is that which does not itself follow anything by causal necessity, but after which something naturally is or comes to be. An end, on the contrary, is that which itself naturally follows some other thing, either by necessity, or as a rule, but has nothing following it. A middle is that which follows something as some other thing follows it. A well constructed plot, therefore, must neither begin nor end at haphazard, but conform to these principles.

Again, a beautiful object, whether it be a living organism or any whole composed of parts, must not only have an orderly arrangement of parts, but must also be of a certain magnitude; for beauty depends on magnitude and order. Hence a very small animal organism cannot be beautiful; for the view of it is confused, the object being seen in an almost imperceptible moment of time. Nor, again, can one of vast size be beautiful; for as the eye cannot take it all in at once, the unity and sense of the whole is lost for the spectator; as for instance if there were one a thousand miles long. As, therefore, in the case of animate bodies and organisms a certain magnitude is necessary, and a magnitude which may be easily embraced in one view; so in the plot, a certain length is necessary, and a length which can be easily embraced by the memory. The limit of length in relation to dramatic competition and sensuous presentment, is no part of artistic theory. For had it been the rule for a hundred tragedies to compete together, the performance would have been regulated by the water-clock,—as indeed we are told was formerly done. But the limit as fixed by the nature of the drama itself is this: the greater the length, the more beautiful will the piece be by reason of its size, provided that the whole be perspicuous. And to define the matter roughly, we may say that the proper magnitude is comprised within such limits, that the sequence of events, according to the law of probability or necessity, will admit of a change from bad fortune to good, or from good fortune to bad.

The Plot must be a Unity.

Unity of plot does not, as some persons think, consist in the Unity of the hero. For infinitely various are the incidents in one man's life which cannot be reduced to unity; and so, too, there are many actions of one man out of which we cannot make one action. Hence, the error, as it appears, of all poets who have composed a Heracleid, a Theseid, or other poems of the kind. They imagine that as Heracles was one man, the story of Heracles must also be a unity. But Homer, as in all else he is of surpassing merit, here too—whether from art or natural genius—seems to have happily discerned the truth. In composing the Odyssey he did not include all the adventures of Odysseus—such as his wound on Parnassus, or his feigned madness at the mustering of the host—incidents between which there was no necessary or probable connection: but he made the Odyssey, and likewise the Iliad, to centre round an action that in our sense of the word is one. As therefore, in the other imitative arts, the imitation is one when the object imitated is one, so the plot, being an imitation of an action, must imitate one action and that a whole, the structural union of the parts being such that, if any one of them is displaced or removed, the whole will be disjointed and disturbed. For a thing whose presence or absence makes no visible difference, is not an organic part of the whole.

Dramatic Unity.

It is, moreover, evident from what has been said, that it is not the function of the poet to relate what has happened, but what may happen,—what is possible according to the law of probability or necessity. The poet and the historian differ not by writing in verse or in prose. The work of Herodotus might be put into verse, and it would still be a species of history, with metre no less than without it. The true difference is that one relates what has happened, the other what may happen. Poetry, therefore, is a more philosophical and a higher thing than history: for poetry tends to express the universal, history the particular. By the universal, I mean how a person of a certain type will on occasion speak or act, according to the law of probability or necessity; and it is this universality at which poetry aims in the names she attaches to the personages. The particular is—for example—what Alcibiades did or suffered. In Comedy this is already apparent: for here the poet first constructs the plot on the lines of probability, and then inserts characteristic names;—unlike the lampooners who write about particular individuals. But tragedians still keep to real names, the reason being that what is possible is credible: what has not happened we do not at once feel sure to be possible: but what has happened is manifestly possible: otherwise it would not have happened. Still there are even some tragedies in which there are only one or two well known names, the rest being fictitious. In others, none are well known, as in Agathon's Antheus, where incidents and names alike are fictitious, and yet they give none the less pleasure. We must not, therefore, at all costs keep to the received legends, which are the usual subjects of Tragedy. Indeed, it would be absurd to attempt it; for even subjects that are known are known only to a few, and yet give pleasure to all. It clearly follows that the poet or 'maker' should be the maker of plots rather than of verses; since he is a poet because he imitates, and what he imitates are actions. And even if he chances to take an historical subject, he is none the less a poet; for there is no reason why some events that have actually happened should not conform to the law of the probable and possible, and in virtue of that quality in them he is their poet or maker.

Of all plots and actions the epeisodic are the worst. I call a plot 'epeisodic' in which the episodes or acts succeed one another without probable or necessary sequence. Bad poets compose such pieces by their own fault, good poets, to please the players; for, as they write show pieces for competition, they stretch the plot beyond its capacity, and are often forced to break the natural continuity.

But again, Tragedy is an imitation not only of a complete action, but of events inspiring fear or pity. Such an effect is best produced when the events come on us by sunrise; and the effect is heightened when, at the same time, they follow as cause and effect. The tragic wonder will thee be greater than if they happened of themselves or by accident; for even coincidences are most striking when they have an air of design. We may instance the statue of Mitys at Argos, which fell upon his murderer while he was a spectator at a festival, and killed him. Such events seem not to be due to mere chance. Plots, therefore, constructed on these principles are necessarily the best.

Definitions of Simple and Complex Plots.

Plots are either Simple or Complex, for the actions in real life, of which the plots are an imitation, obviously show a similar distinction. An action which is one and continuous in the sense above defined, I call Simple, when the change of fortune takes place without Reversal of the Situation and without Recognition.

A Complex action is one in which the change is accompanied by such Reversal, or by Recognition, or by both. These last should arise from the internal structure of the plot, so that what follows should be the necessary or probable result of the preceding action. It makes all the difference whether any given event is a case of propter hoc or post hoc.

Reversal of the Situation, Recognition, and Tragic or disastrous Incident defined and explained.

Reversal of the Situation is a change by which the action veers round to its opposite, subject always to our rule of probability or necessity. Thus in the Oedipus, the messenger comes to cheer Oedipus and free him from his alarms about his mother, but by revealing who he is, he produces the opposite effect. Again in the Lynceus, Lynceus is being led away to his death, and Danaus goes with him, meaning, to slay him; but the outcome of the preceding incidents is that Danaus is killed and Lynceus saved. Recognition, as the name indicates, is a change from ignorance to knowledge, producing love or hate between the persons destined by the poet for good or bad fortune. The best form of recognition is coincident with a Reversal of the Situation, as in the Oedipus. There are indeed other forms. Even inanimate things of the most trivial kind may in a sense be objects of recognition. Again, we may recognise or discover whether a person has done a thing or not. But the recognition which is most intimately connected with the plot and action is, as we have said, the recognition of persons. This recognition, combined, with Reversal, will produce either pity or fear; and actions producing these effects are those which, by our definition, Tragedy represents. Moreover, it is upon such situations that the issues of good or bad fortune will depend. Recognition, then, being between persons, it may happen that one person only is recognised by the other—when the latter is already known—or it may be necessary that the recognition should be on both sides. Thus Iphigenia is revealed to Orestes by the sending of the letter; but another act of recognition is required to make Orestes known to Iphigenia.

Two parts, then, of the Plot—Reversal of the Situation and Recognition—turn upon surprises. A third part is the Scene of Suffering. The Scene of Suffering is a destructive or painful action, such as death on the stage, bodily agony, wounds and the like.

The 'quantitative parts' of Tragedy defined.

The parts of Tragedy which must be treated as elements of the whole have been already mentioned. We now come to the quantitative parts, and the separate parts into which Tragedy is divided, namely, Prologue, Episode, Exode, Choric song; this last being divided into Parode and Stasimon. These are common to all plays: peculiar to some are the songs of actors from the stage and the Commoi.

The Prologue is that entire part of a tragedy which precedes the Parode of the Chorus. The Episode is that entire part of a tragedy which is between complete choric songs. The Exode is that entire part of a tragedy which has no choric song after it. Of the Choric part the Parode is the first undivided utterance of the Chorus: the Stasimon is a Choric ode without anapaests or trochaic tetrameters: the Commos is a joint lamentation of Chorus and actors. The parts of Tragedy which must be treated as elements of the whole have been already mentioned. The quantitative parts the separate parts into which it is divided—are here enumerated.

What constitutes Tragic Action.

As the sequel to what has already been said, we must proceed to consider what the poet should aim at, and what he should avoid, in constructing his plots; and by what means the specific effect of Tragedy will be produced.

A perfect tragedy should, as we have seen, be arranged not on the simple but on the complex plan. It should, moreover, imitate actions which excite pity and fear, this being the distinctive mark of tragic imitation. It follows plainly, in the first place, that the change, of fortune presented must not be the spectacle of a virtuous man brought from prosperity to adversity: for this moves neither pity nor fear; it merely shocks us. Nor, again, that of a bad man passing from adversity to prosperity: for nothing can be more alien to the spirit of Tragedy; it possesses no single tragic quality; it neither satisfies the moral sense nor calls forth pity or fear. Nor, again, should the downfall of the utter villain be exhibited. A plot of this kind would, doubtless, satisfy the moral sense, but it would inspire neither pity nor fear; for pity is aroused by unmerited misfortune, fear by the misfortune of a man like ourselves. Such an event, therefore, will be neither pitiful nor terrible. There remains, then, the character between these two extremes,—that of a man who is not eminently good and just,-yet whose misfortune is brought about not by vice or depravity, but by some error or frailty. He must be one who is highly renowned and prosperous,—a personage like Oedipus, Thyestes, or other illustrious men of such families.

A well constructed plot should, therefore, be single in its issue, rather than double as some maintain. The change of fortune should be not from bad to good, but, reversely, from good to bad. It should come about as the result not of vice, but of some great error or frailty, in a character either such as we have described, or better rather than worse. The practice of the stage bears out our view. At first the poets recounted any legend that came in their way. Now, the best tragedies are founded on the story of a few houses, on the fortunes of Alcmaeon, Oedipus, Orestes, Meleager, Thyestes, Telephus, and those others who have done or suffered something terrible. A tragedy, then, to be perfect according to the rules of art should be of this construction. Hence they are in error who censure Euripides just because he follows this principle in his plays, many of which end unhappily. It is, as we have said, the right ending. The best proof is that on the stage and in dramatic competition, such plays, if well worked out, are the most tragic in effect; and Euripides, faulty though he may be in the general management of his subject, yet is felt to be the most tragic of the poets.

In the second rank comes the kind of tragedy which some place first. Like the Odyssey, it has a double thread of plot, and also an opposite catastrophe for the good and for the bad. It is accounted the best because of the weakness of the spectators; for the poet is guided in what he writes by the wishes of his audience. The pleasure, however, thence derived is not the true tragic pleasure. It is proper rather to Comedy, where those who, in the piece, are the deadliest enemies—like Orestes and Aegisthus—quit the stage as friends at the close, and no one slays or is slain.

The tragic emotions of pity and fear should spring out of the Plot itself.

Fear and pity may be aroused by spectacular means; but they may also result from the inner structure of the piece, which is the better way, and indicates a superior poet. For the plot ought to be so constructed that, even without the aid of the eye, he who hears the tale told will thrill with horror and melt to pity at what takes place. This is the impression we should receive from hearing the story of the Oedipus. But to produce this effect by the mere spectacle is a less artistic method, and dependent on extraneous aids. Those who employ spectacular means to create a sense not of the terrible but only of the monstrous, are strangers to the purpose of Tragedy; for we must not demand of Tragedy any and every kind of pleasure, but only that which is proper to it. And since the pleasure which the poet should afford is that which comes from pity and fear through imitation, it is evident that this quality must be impressed upon the incidents.

Let us then determine what are the circumstances which strike us as terrible or pitiful.

Actions capable of this effect must happen between persons who are either friends or enemies or indifferent to one another. If an enemy kills an enemy, there is nothing to excite pity either in the act or the intention,—except so far as the suffering in itself is pitiful. So again with indifferent persons. But when the tragic incident occurs between those who are near or dear to one another—if, for example, a brother kills, or intends to kill, a brother, a son his father, a mother her son, a son his mother, or any other deed of the kind is done—these are the situations to be looked for by the poet. He may not indeed destroy the framework of the received legends—the fact, for instance, that Clytemnestra was slain by Orestes and Eriphyle by Alcmaeon but he ought to show invention of his own, and skilfully handle the traditional material. Let us explain more clearly what is meant by skilful handling.

The action may be done consciously and with knowledge of the persons, in the manner of the older poets. It is thus too that Euripides makes Medea slay her children. Or, again, the deed of horror may be done, but done in ignorance, and the tie of kinship or friendship be discovered afterwards. The Oedipus of Sophocles is an example. Here, indeed, the incident is outside the drama proper; but cases occur where it falls within the action of the play: one may cite the Alcmaeon of Astydamas, or Telegonus in the Wounded Odysseus. Again, there is a third case,—when some one is about to do an irreparable deed through ignorance, and makes the discovery before it is done. These are the only possible ways. For the deed must either be done or not done,—and that wittingly or unwittingly. But of all these ways, to be about to act knowing the persons, and then not to act, is the worst. It is shocking without being tragic, for no disaster follows. It is, therefore, never, or very rarely, found in poetry. One instance, however, is in the Antigone, where Haemon threatens to kill Creon. The next and better way is that the deed should be perpetrated. Still better, that it should be perpetrated in ignorance, and the discovery made afterwards. There is then nothing to shock us, while the discovery produces a startling effect. The last case is the best, as when in the Cresphontes Merope is about to slay her son, but, recognising who he is, spares his life. So in the Iphigenia, the sister recognises the brother just in time. Again in the Helle, the son recognises the mother when on the point of giving her up. This, then, is why a few families only, as has been already observed, furnish the subjects of tragedy. It was not art, but happy chance, that led the poets in search of subjects to impress the tragic quality upon their plots. They are compelled, therefore, to have recourse to those houses whose history contains moving incidents like these.

Enough has now been said concerning the structure of the incidents, and the right kind of plot.

The element of Character in Tragedy.

In respect of Character there are four things to be aimed at. First, and most important, it must be good. Now any speech or action that manifests moral purpose of any kind will be expressive of character: the character will be good if the purpose is good. This rule is relative to each class. Even a woman may be good, and also a slave; though the woman may be said to be an inferior being, and the slave quite worthless. The second thing to aim at is propriety. There is a type of manly valour; but valour in a woman, or unscrupulous cleverness, is inappropriate. Thirdly, character must be true to life: for this is a distinct thing from goodness and propriety, as here described. The fourth point is consistency: for though the subject of the imitation, who suggested the type, be inconsistent, still he must be consistently inconsistent. As an example of motiveless degradation of character, we have Menelaus in the Orestes: of character indecorous and inappropriate, the lament of Odysseus in the Scylla, and the speech of Melanippe: of inconsistency, the Iphigenia at Aulis,—for Iphigenia the suppliant in no way resembles her later self.

As in the structure of the plot, so too in the portraiture of character, the poet should always aim either at the necessary or the probable. Thus a person of a given character should speak or act in a given way, by the rule either of necessity or of probability; just as this event should follow that by necessary or probable sequence. It is therefore evident that the unravelling of the plot, no less than the complication, must arise out of the plot itself, it must not be brought about by the 'Deus ex Machina'—as in the Medea, or in the Return of the Greeks in the Iliad. The 'Deus ex Machina' should be employed only for events external to the drama,—for antecedent or subsequent events, which lie beyond the range of human knowledge, and which require to be reported or foretold; for to the gods we ascribe the power of seeing all things. Within the action there must be nothing irrational. If the irrational cannot be excluded, it should be outside the scope of the tragedy. Such is the irrational element in the Oedipus of Sophocles.

Again, since Tragedy is an imitation of persons who are above the common level, the example of good portrait-painters should be followed. They, while reproducing the distinctive form of the original, make a likeness which is true to life and yet more beautiful. So too the poet, in representing men who are irascible or indolent, or have other defects of character, should preserve the type and yet ennoble it. In this way Achilles is portrayed by Agathon and Homer.

These then are rules the poet should observe. Nor should he neglect those appeals to the senses, which, though not among the essentials, are the concomitants of poetry; for here too there is much room for error. But of this enough has been said in our published treatises.

Recognition: its various kinds, with examples.

What Recognition is has been already explained. We will now enumerate its kinds.

First, the least artistic form, which, from poverty of wit, is most commonly employed recognition by signs. Of these some are congenital,—such as 'the spear which the earth-born race bear on their bodies,' or the stars introduced by Carcinus in his Thyestes. Others are acquired after birth; and of these some are bodily marks, as scars; some external tokens, as necklaces, or the little ark in the Tyro by which the discovery is effected. Even these admit of more or less skilful treatment. Thus in the recognition of Odysseus by his scar, the discovery is made in one way by the nurse, in another by the swineherds. The use of tokens for the express purpose of proof—and, indeed, any formal proof with or without tokens—is a less artistic mode of recognition. A better kind is that which comes about by a turn of incident, as in the Bath Scene in the Odyssey.

Next come the recognitions invented at will by the poet, and on that account wanting in art. For example, Orestes in the Iphigenia reveals the fact that he is Orestes. She, indeed, makes herself known by the letter; but he, by speaking himself, and saying what the poet, not what the plot requires. This, therefore, is nearly allied to the fault above mentioned:—for Orestes might as well have brought tokens with him. Another similar instance is the 'voice of the shuttle' in the Tereus of Sophocles.

The third kind depends on memory when the sight of some object awakens a feeling: as in the Cyprians of Dicaeogenes, where the hero breaks into tears on seeing the picture; or again in the 'Lay of Alcinous,' where Odysseus, hearing the minstrel play the lyre, recalls the past and weeps; and hence the recognition.

The fourth kind is by process of reasoning. Thus in the Choephori: 'Some one resembling me has come: no one resembles me but Orestes: therefore Orestes has come.' Such too is the discovery made by Iphigenia in the play of Polyidus the Sophist. It was a natural reflection for Orestes to make, 'So I too must die at the altar like my sister.' So, again, in the Tydeus of Theodectes, the father says, 'I came to find my son, and I lose my own life.' So too in the Phineidae: the women, on seeing the place, inferred their fate:—'Here we are doomed to die, for here we were cast forth.' Again, there is a composite kind of recognition involving false inference on the part of one of the characters, as in the Odysseus Disguised as a Messenger. A said recognise the bow which, in fact, he had not seen; and to bring about a recognition by this means that the expectation A would recognise the bow is false inference.

But, of all recognitions, the best is that which arises from the incidents themselves, where the startling discovery is made by natural means. Such is that in the Oedipus of Sophocles, and in the Iphigenia; for it was natural that Iphigenia should wish to dispatch a letter. These recognitions alone dispense with the artificial aid of tokens or amulets. Next come the recognitions by process of reasoning.

Practical rules for the Tragic Poet.

In constructing the plot and working it out with the proper diction, the poet should place the scene, as far as possible, before his eyes. In this way, seeing everything with the utmost vividness, as if he were a spectator of the action, he will discover what is in keeping with it, and be most unlikely to overlook inconsistencies. The need of such a rule is shown by the fault found in Carcinus.

Amphiaraus was on his way from the temple. This fact escaped the observation of one who did not see the situation. On the stage, however, the piece failed, the audience being offended at the oversight.

Again, the poet should work out his play, to the best of his power, with appropriate gestures; for those who feel emotion are most convincing through natural sympathy with the characters they represent; and one who is agitated storms, one who is angry rages, with the most life-like reality. Hence poetry implies either a happy gift of nature or a strain of madness. In the one case a man can take the mould of any character; in the other, he is lifted out of his proper self.

As for the story, whether the poet takes it ready made or constructs it for himself, he should first sketch its general outline, and then fill in the episodes and amplify in detail. The general plan may be illustrated by the Iphigenia. A young girl is sacrificed; she disappears mysteriously from the eyes of those who sacrificed her; She is transported to another country, where the custom is to offer up all strangers to the goddess. To this ministry she is appointed. Some time later her own brother chances to arrive. The fact that the oracle for some reason ordered him to go there, is outside the general plan of the play. The purpose, again, of his coming is outside the action proper. However, he comes, he is seized, and, when on the point of being sacrificed, reveals who he is. The mode of recognition may be either that of Euripides or of Polyidus, in whose play he exclaims very naturally:—'So it was not my sister only, but I too, who was doomed to be sacrificed'; and by that remark he is saved.

After this, the names being once given, it remains to fill in the episodes. We must see that they are relevant to the action. In the case of Orestes, for example, there is the madness which led to his capture, and his deliverance by means of the purificatory rite. In the drama, the episodes are short, but it is these that give extension to Epic poetry. Thus the story of the Odyssey can be stated briefly. A certain man is absent from home for many years; he is jealously watched by Poseidon, and left desolate. Meanwhile his home is in a wretched plight——suitors are wasting his substance and plotting against his son. At length, tempest-tost, he himself arrives; he makes certain persons acquainted with him; he attacks the suitors with his own hand, and is himself preserved while he destroys them. This is the essence of the plot; the rest is episode.

Further rules for the Tragic Poet.

Every tragedy falls into two parts,—Complication and Unravelling or Denouement. Incidents extraneous to the action are frequently combined with a portion of the action proper, to form the Complication; the rest is the Unravelling. By the Complication I mean all that extends from the beginning of the action to the part which marks the turning-point to good or bad fortune. The Unravelling is that which extends from the beginning of the change to the end. Thus, in the

Lynceus of Theodectes, the Complication consists of the incidents presupposed in the drama, the seizure of the child, and then again * * extends from the accusation of murder to the end.

There are four kinds of Tragedy, the Complex, depending entirely on Reversal of the Situation and Recognition; the Pathetic (where the motive is passion),—such as the tragedies on Ajax and Ixion; the Ethical (where the motives are ethical),—such as the Phthiotides and the Peleus. The fourth kind is the Simple, exemplified by the Phorcides, the Prometheus, and scenes laid in Hades. The poet should endeavour, if possible, to combine all poetic elements; or failing that, the greatest number and those the most important; the more so, in face of the cavilling criticism of the day. For whereas there have hitherto been good poets, each in his own branch, the critics now expect one man to surpass all others in their several lines of excellence.

In speaking of a tragedy as the same or different, the best test to take is the plot. Identity exists where the Complication and Unravelling are the same. Many poets tie the knot well, but unravel it ill. Both arts, however, should always be mastered.

Again, the poet should remember what has been often said, and not make an Epic structure into a Tragedy—by an Epic structure I mean one with a multiplicity of plots—as if, for instance, you were to make a tragedy out of the entire story of the Iliad. In the Epic poem, owing to its length, each part assumes its proper magnitude. In the drama the result is far from answering to the poet's expectation. The proof is that the poets who have dramatised the whole story of the Fall of Troy, instead of selecting portions, like Euripides; or who have taken the whole tale of Niobe, and not a part of her story, like Aeschylus, either fail utterly or meet with poor success on the stage. Even Agathon has been known to fail from this one defect. In his Reversals of the Situation, however, he shows a marvellous skill in the effort to hit the popular taste,—to produce a tragic effect that satisfies the moral sense. This effect is produced when the clever rogue, like Sisyphus, is outwitted, or the brave villain defeated. Such an event is probable in Agathon's sense of the word: 'it is probable,' he says, 'that many things should happen contrary to probability.'

The Chorus too should be regarded as one of the actors; it should be an integral part of the whole, and share in the action, in the manner not of Euripides but of Sophocles. As for the later poets, their choral songs pertain as little to the subject of the piece as to that of any other tragedy. They are, therefore, sung as mere interludes, a practice first begun by Agathon. Yet what difference is there between introducing such choral interludes, and transferring a speech, or even a whole act, from one play to another?

Thought, or the Intellectual element, and Diction in Tragedy.

It remains to speak of Diction and Thought, the other parts of Tragedy having been already discussed. Concerning Thought, we may assume what is said in the Rhetoric, to which inquiry the subject more strictly belongs. Under Thought is included every effect which has to be produced by speech, the subdivisions being,—proof and refutation; the excitation of the feelings, such as pity, fear, anger, and the like; the suggestion of importance or its opposite. Now, it is evident that the dramatic incidents must be treated from the same points of view as the dramatic speeches, when the object is to evoke the sense of pity, fear, importance, or probability. The only difference is, that the incidents should speak for themselves without verbal exposition; while the effects aimed at in speech should be produced by the speaker, and as a result of the speech. For what were the business of a speaker, if the Thought were revealed quite apart from what he says?

Next, as regards Diction. One branch of the inquiry treats of the Modes of Utterance. But this province of knowledge belongs to the art of Delivery and to the masters of that science. It includes, for instance,—what is a command, a prayer, a statement, a threat, a question, an answer, and so forth. To know or not to know these things involves no serious censure upon the poet's art. For who can admit the fault imputed to Homer by Protagoras,—that in the words, 'Sing, goddess, of the wrath,' he gives a command under the idea that he utters a prayer? For to tell some one to do a thing or not to do it is, he says, a command. We may, therefore, pass this over as an inquiry that belongs to another art, not to poetry.

Diction, or Language in general.

Language in general includes the following parts:-Letter, Syllable, Connecting word, Noun, Verb, Inflexion or Case, Sentence or Phrase.

A Letter is an indivisible sound, yet not every such sound, but only one which can form part of a group of sounds. For even brutes utter indivisible sounds, none of which I call a letter. The sound I mean may be either a vowel, a semi-vowel, or a mute. A vowel is that which without impact of tongue or lip has an audible sound. A semi-vowel, that which with such impact has an audible sound, as S and R. A mute, that which with such impact has by itself no sound, but joined to a vowel sound becomes audible, as G and D. These are distinguished according to the form assumed by the mouth and the place where they are produced; according as they are aspirated or smooth, long or short; as they are acute, grave, or of an intermediate tone; which inquiry belongs in detail to the writers on metre.

A Syllable is a non-significant sound, composed of a mute and a vowel: for GR without A is a syllable, as also with A,—GRA. But the investigation of these differences belongs also to metrical science.

A Connecting word is a non-significant sound, which neither causes nor hinders the union of many sounds into one significant sound; it may be placed at either end or in the middle of a sentence. Or, a non-significant sound, which out of several sounds, each of them significant, is capable of forming one significant sound,—as αμθι, περι, and the like. Or, a non-significant sound, which marks the beginning, end, or division of a sentence; such, however, that it cannot correctly stand by itself at the beginning of a sentence, as μεν, ητοι, δε.

A Noun is a composite significant sound, not marking time, of which no part is in itself significant: for in double or compound words we do not employ the separate parts as if each were in itself significant. Thus in Theodorus, 'god-given,' the δωρον or 'gift' is not in itself significant.

A Verb is a composite significant sound, marking time, in which, as in the noun, no part is in itself significant. For 'man,' or 'white' does not express the idea of 'when'; but 'he walks,' or 'he has walked' does connote time, present or past.

Inflexion belongs both to the noun and verb, and expresses either the relation 'of,' 'to,' or the like; or that of number, whether one or many, as 'man' or 'men '; or the modes or tones in actual delivery, e.g. a question or a command. 'Did he go?' and 'go' are verbal inflexions of this kind.

A Sentence or Phrase is a composite significant sound, some at least of whose parts are in themselves significant; for not every such group of words consists of verbs and nouns—'the definition of man,' for example—but it may dispense even with the verb. Still it will always have some significant part, as 'in walking,' or 'Cleon son of Cleon.' A sentence or phrase may form a unity in two ways,—either as signifying one thing, or as consisting of several parts linked together. Thus the Iliad is one by the linking together of parts, the definition of man by the unity of the thing signified.

Poetic Diction.

Words are of two kinds, simple and double. By simple I mean those composed of non-significant elements, such as γη. By double or compound, those composed either of a significant and non-significant element (though within the whole word no element is significant), or of elements that are both significant. A word may likewise be triple, quadruple, or multiple in form, like so many Massilian expressions, e.g. 'Hermo-caico-xanthus who prayed to Father Zeus>.'

Every word is either current, or strange, or metaphorical, or ornamental, or newly-coined, or lengthened, or contracted, or altered.

By a current or proper word I mean one which is in general use among a people; by a strange word, one which is in use in another country. Plainly, therefore, the same word may be at once strange and current, but not in relation to the same people. The word σιγυνον, 'lance,' is to the Cyprians a current term but to us a strange one.

Metaphor is the application of an alien name by transference either from genus to species, or from species to genus, or from species to species, or by analogy, that is, proportion. Thus from genus to species, as: 'There lies my ship'; for lying at anchor is a species of lying. From species to genus, as: 'Verily ten thousand noble deeds hath Odysseus wrought'; for ten thousand is a species of large number, and is here used for a large number generally. From species to species, as: 'With blade of bronze drew away the life,' and 'Cleft the water with the vessel of unyielding bronze.' Here αρυραι, 'to draw away,' is used for ταμειν, 'to cleave,' and ταμειν again for αρυαι,—each being a species of taking away. Analogy or proportion is when the second term is to the first as the fourth to the third. We may then use the fourth for the second, or the second for the fourth. Sometimes too we qualify the metaphor by adding the term to which the proper word is relative. Thus the cup is to Dionysus as the shield to Ares. The cup may, therefore, be called 'the shield of Dionysus,' and the shield 'the cup of Ares.' Or, again, as old age is to life, so is evening to day. Evening may therefore be called 'the old age of the day,' and old age, 'the evening of life,' or, in the phrase of Empedocles, 'life's setting sun.' For some of the terms of the proportion there is at times no word in existence; still the metaphor may be used. For instance, to scatter seed is called sowing: but the action of the sun in scattering his rays is nameless. Still this process bears to the sun the same relation as sowing to the seed. Hence the expression of the poet 'sowing the god-created light.' There is another way in which this kind of metaphor may be employed. We may apply an alien term, and then deny of that term one of its proper attributes; as if we were to call the shield, not 'the cup of Ares,' but 'the wineless cup.'

A newly-coined word is one which has never been even in local use, but is adopted by the poet himself. Some such words there appear to be: as ερνυγεσ, 'sprouters,' for κερατα, 'horns,' and αρητηρ, 'supplicator,' for ιερευσ, 'priest.'

A word is lengthened when its own vowel is exchanged for a longer one, or wολησσ hen a syllable is inserted. A word is contracted when some part of it is removed. Instances of lengthening are,—π for πολεωσ, and Πηληιαδεω for Πηλειδου· of contraction, κρι, δω, and οψ, as in μια γινεται αμφοτερων οψ.

An altered word is one in which part of the ordinary form is left unchanged, and part is re-cast; as in δεξι-τερον κατα μαζον, δεξιτερον is for δεξιον.

[Nouns in themselves are either masculine, feminine, or neuter. Masculine are such as end in ν, ρ, σ, or in some letter compounded with σ,—these being two, and ξ. Feminine, such as end in vowels that are always long, namely η and ω, and—of vowels that admit of lengthening—those in α. Thus the number of letters in which nouns masculine and feminine end is the same; for ψ and ξ are equivalent to endings in σ. No noun ends in a mute or a vowel short by nature. Three only end in ι,—μηλι, κομμι, πεπερι: five end in υ. Neuter nouns end in these two latter vowels; also in ν and σ.]

How Poetry combines elevation of language with perspicuity.

The perfection of style is to be clear without being mean. The clearest style is that which uses only current or proper words; at the same time it is mean:—witness the poetry of Cleophon and of Sthenelus. That diction, on the other hand, is lofty and raised above the commonplace which employs unusual words. By unusual, I mean strange (or rare) words, metaphorical, lengthened,—anything, in short, that differs from the normal idiom. Yet a style wholly composed of such words is either a riddle or a jargon; a riddle, if it consists of metaphors; a jargon, if it consists of strange (or rare) words. For the essence of a riddle is to express true facts under impossible combinations. Now this cannot be done by any arrangement of ordinary words, but by the use of metaphor it can. Such is the riddle:—'A man I saw who on another man had glued the bronze by aid of fire,' and others of the same kind. A diction that is made up of strange (or rare) terms is a jargon. A certain infusion, therefore, of these elements is necessary to style; for the strange (or rare) word, the metaphorical, the ornamental, and the other kinds above mentioned, will raise it above the commonplace and mean, while the use of proper words will make it perspicuous. But nothing contributes more to produce a clearness of diction that is remote from commonness than the lengthening, contraction, and alteration of words. For by deviating in exceptional cases from the normal idiom, the language will gain distinction; while, at the same time, the partial conformity with usage will give perspicuity. The critics, therefore, are in error who censure these licenses of speech, and hold the author up to ridicule. Thus Eucleides, the elder, declared that it would be an easy matter to be a poet if you might lengthen syllables at will. He caricatured the practice in the very form of his diction, as in the verse: Ἐπιχαρην ειδον Μαραθωναδε Βαδιζοντα, or, ουκ αν γ εραμενοσ τον εκεινου ελλεβορον. To employ such license at all obtrusively is, no doubt, grotesque; but in any mode of poetic diction there must be moderation. Even metaphors, strange (or rare) words, or any similar forms of speech, would produce the like effect if used without propriety and with the express purpose of being ludicrous. How great a difference is made by the appropriate use of lengthening, may be seen in Epic poetry by the insertion of ordinary forms in the verse. So, again, if we take a strange (or rare) word, a metaphor, or any similar mode of expression, and replace it by the current or proper term, the truth of our observation will be manifest. For example Aeschylus and Euripides each composed the same iambic line. But the alteration of a single word by Euripides, who employed the rarer term instead of the ordinary one, makes one verse appear beautiful and the other trivial. Aeschylus in his Philoctetes says: Φαγεδαινα <δ> η μου σαρκασ ερθιει ποδοσ.

Euripides substitutes Θοιναται 'feasts on' for εσθιει 'feeds on.' Again, in the line, νυν δε μ εων ολιγιγυσ τε και ουτιδανοσ και αεικησ, the difference will be felt if we substitute the common words, νυν δε μ εων μικροσ τε και αρθενικοσ και αειδγσ. Or, if for the line, διφρον αεικελιον καταθεισ ολιγην τε τραπεισ ολιγην τε τραπεζαν, we read, διφρον μοχθηρον καταθεισ μικραν τε τραπεζαν.

Or, for ηιονεσ βοοωριν, ηιονεσκραζουριν

Again, Ariphrades ridiculed the tragedians for using phrases which no one would employ in ordinary speech: for example, δωματων απο instead of απο δωματων, ρεθεν, εγω δε νιν, Αχιλλεωσ περι instead of περι χιλλεωσ, and the like. It is precisely because such phrases are not part of the current idiom that they give distinction to the style. This, however, he failed to see.

It is a great matter to observe propriety in these several modes of expression, as also in compound words, strange (or rare) words, and so forth. But the greatest thing by far is to have a command of metaphor. This alone cannot be imparted by another; it is the mark of genius, for to make good metaphors implies an eye for resemblances.

Of the various kinds of words, the compound are best adapted to Dithyrambs, rare words to heroic poetry, metaphors to iambic. In heroic poetry, indeed, all these varieties are serviceable. But in iambic verse, which reproduces, as far as may be, familiar speech, the most appropriate words are those which are found even in prose. These are,—the current or proper, the metaphorical, the ornamental.

Concerning Tragedy and imitation by means of action this may suffice.

Epic Poetry.

As to that poetic imitation which is narrative in form and employs a single metre, the plot manifestly ought, as in a tragedy, to be constructed on dramatic principles. It should have for its subject a single action, whole and complete, with a beginning, a middle, and an end. It will thus resemble a living organism in all its unity, and produce the pleasure proper to it. It will differ in structure from historical compositions, which of necessity present not a single action, but a single period, and all that happened within that period to one person or to many, little connected together as the events may be. For as the sea-fight at Salamis and the battle with the Carthaginians in Sicily took place at the same time, but did not tend to any one result, so in the sequence of events, one thing sometimes follows another, and yet no single result is thereby produced. Such is the practice, we may say, of most poets. Here again, then, as has been already observed, the transcendent excellence of Homer is manifest. He never attempts to make the whole war of Troy the subject of his poem, though that war had a beginning and an end. It would have been too vast a theme, and not easily embraced in a single view. If, again, he had kept it within moderate limits, it must have been over-complicated by the variety of the incidents. As it is, he detaches a single portion, and admits as episodes many events from the general story of the war—such as the Catalogue of the ships and others—thus diversifying the poem. All other poets take a single hero, a single period, or an action single indeed, but with a multiplicity of parts. Thus did the author of the Cypria and of the Little Iliad. For this reason the Iliad and the Odyssey each furnish the subject of one tragedy, or, at most, of two; while the Cypria supplies materials for many, and the Little Iliad for eight—the Award of the Arms, the Philoctetes, the Neoptolemus, the Eurypylus, the Mendicant Odysseus, the Laconian Women, the Fall of Ilium, the Departure of the Fleet.

Further points of agreement with Tragedy.

Again, Epic poetry must have as many kinds as Tragedy: it must be simple, or complex, or 'ethical,' or 'pathetic.' The parts also, with the exception of song and spectacle, are the same; for it requires Reversals of the Situation, Recognitions, and Scenes of Suffering. Moreover, the thoughts and the diction must be artistic. In all these respects Homer is our earliest and sufficient model. Indeed each of his poems has a twofold character. The Iliad is at once simple and 'pathetic,' and the Odyssey complex (for Recognition scenes run through it), and at the same time 'ethical.' Moreover, in diction and thought they are supreme.

Epic poetry differs from Tragedy in the scale on which it is constructed, and in its metre. As regards scale or length, we have already laid down an adequate limit:—the beginning and the end must be capable of being brought within a single view. This condition will be satisfied by poems on a smaller scale than the old epics, and answering in length to the group of tragedies presented at a single sitting.

Epic poetry has, however, a great—a special—capacity for enlarging its dimensions, and we can see the reason. In Tragedy we cannot imitate several lines of actions carried on at one and the same time; we must confine ourselves to the action on the stage and the part taken by the players. But in Epic poetry, owing to the narrative form, many events simultaneously transacted can be presented; and these, if relevant to the subject, add mass and dignity to the poem. The Epic has here an advantage, and one that conduces to grandeur of effect, to diverting the mind of the hearer, and relieving the story with varying episodes. For sameness of incident soon produces satiety, and makes tragedies fail on the stage.

As for the metre, the heroic measure has proved its fitness by the test of experience. If a narrative poem in any other metre or in many metres were now composed, it would be found incongruous. For of all measures the heroic is the stateliest and the most massive; and hence it most readily admits rare words and metaphors, which is another point in which the narrative form of imitation stands alone. On the other hand, the iambic and the trochaic tetrameter are stirring measures, the latter being akin to dancing, the former expressive of action. Still more absurd would it be to mix together different metres, as was done by Chaeremon. Hence no one has ever composed a poem on a great scale in any other than heroic verse. Nature herself, as we have said, teaches the choice of the proper measure.

Homer, admirable in all respects, has the special merit of being the only poet who rightly appreciates the part he should take himself. The poet should speak as little as possible in his own person, for it is not this that makes him an imitator. Other poets appear themselves upon the scene throughout, and imitate but little and rarely. Homer, after a few prefatory words, at once brings in a man, or woman, or other personage; none of them wanting in characteristic qualities, but each with a character of his own.

The element of the wonderful is required in Tragedy. The irrational, on which the wonderful depends for its chief effects, has wider scope in Epic poetry, because there the person acting is not seen. Thus, the pursuit of Hector would be ludicrous if placed upon the stage—the Greeks standing still and not joining in the pursuit, and Achilles waving them back. But in the Epic poem the absurdity passes unnoticed. Now the wonderful is pleasing: as may be inferred from the fact that every one tells a story with some addition of his own, knowing that his hearers like it. It is Homer who has chiefly taught other poets the art of telling lies skilfully. The secret of it lies in a fallacy, For, assuming that if one thing is or becomes, a second is or becomes, men imagine that, if the second is, the first likewise is or becomes. But this is a false inference. Hence, where the first thing is untrue, it is quite unnecessary, provided the second be true, to add that the first is or has become. For the mind, knowing the second to be true, falsely infers the truth of the first. There is an example of this in the Bath Scene of the Odyssey.

Accordingly, the poet should prefer probable impossibilities to improbable possibilities. The tragic plot must not be composed of irrational parts. Everything irrational should, if possible, be excluded; or, at all events, it should lie outside the action of the play (as, in the Oedipus, the hero's ignorance as to the manner of Laius' death); not within the drama,—as in the Electra, the messenger's account of the Pythian games; or, as in the Mysians, the man who has come from Tegea to Mysia and is still speechless. The plea that otherwise the plot would have been ruined, is ridiculous; such a plot should not in the first instance be constructed. But once the irrational has been introduced and an air of likelihood imparted to it, we must accept it in spite of the absurdity. Take even the irrational incidents in the Odyssey, where Odysseus is left upon the shore of Ithaca. How intolerable even these might have been would be apparent if an inferior poet were to treat the subject. As it is, the absurdity is veiled by the poetic charm with which the poet invests it.

The diction should be elaborated in the pauses of the action, where there is no expression of character or thought. For, conversely, character and thought are merely obscured by a diction that is over brilliant.

Critical Objections brought against Poetry, and the principles on which they are to be answered.

With respect to critical difficulties and their solutions, the number and nature of the sources from which they may be drawn may be thus exhibited.

The poet being an imitator, like a painter or any other artist, must of necessity imitate one of three objects,—things as they were or are, things as they are said or thought to be, or things as they ought to be. The vehicle of expression is language,—either current terms or, it may be, rare words or metaphors. There are also many modifications of language, which we concede to the poets. Add to this, that the standard of correctness is not the same in poetry and politics, any more than in poetry and any other art. Within the art of poetry itself there are two kinds of faults, those which touch its essence, and those which are accidental. If a poet has chosen to imitate something, through want of capacity, the error is inherent in the poetry. But if the failure is due to a wrong choice if he has represented a horse as throwing out both his off legs at once, or introduced technical inaccuracies in medicine, for example, or in any other art the error is not essential to the poetry. These are the points of view from which we should consider and answer the objections raised by the critics.

First as to matters which concern the poet's own art. If he describes the impossible, he is guilty of an error; but the error may be justified, if the end of the art be thereby attained (the end being that already mentioned), if, that is, the effect of this or any other part of the poem is thus rendered more striking. A case in point is the pursuit of Hector. If, however, the end might have been as well, or better, attained without violating the special rules of the poetic art, the error is not justified: for every kind of error should, if possible, be avoided.

Again, does the error touch the essentials of the poetic art, or some accident of it? For example,—not to know that a hind has no horns is a less serious matter than to paint it inartistically.

Further, if it be objected that the description is not true to fact, the poet may perhaps reply,—'But the objects are as they ought to be': just as Sophocles said that he drew men as they ought to be; Euripides, as they are. In this way the objection may be met. If, however, the representation be of neither kind, the poet may answer,—This is how men say the thing is.' This applies to tales about the gods. It may well be that these stories are not higher than fact nor yet true to fact:

they are, very possibly, what Xenophanes says of them. But anyhow, 'this is what is said.' Again, a description may be no better than the fact: 'still, it was the fact'; as in the passage about the arms: 'Upright upon their butt-ends stood the spears.' This was the custom then, as it now is among the Illyrians.

Again, in examining whether what has been said or done by some one is poetically right or not, we must not look merely to the particular act or saying, and ask whether it is poetically good or bad. We must also consider by whom it is said or done, to whom, when, by what means, or for what end; whether, for instance, it be to secure a greater good, or avert a greater evil.

Other difficulties may be resolved by due regard to the usage of language. We may note a rare word, as in ουρηασ μεν πρωτον, where the poet perhaps employs ουρηασ not in the sense of mules, but of sentinels. So, again, of Dolon: 'ill-favoured indeed he was to look upon.' It is not meant that his body was ill-shaped, but that his face was ugly; for the Cretans use the word ευειδεσ, 'well-favoured,' to denote a fair face. Again, ζωροτερον δε κεραιε, 'mix the drink livelier,' does not mean `mix it stronger' as for hard drinkers, but 'mix it quicker.'

Sometimes an expression is metaphorical, as 'Now all gods and men were sleeping through the night,'—while at the same time the poet says: 'Often indeed as he turned his gaze to the Trojan plain, he marvelled at the sound of flutes and pipes.' 'All' is here used metaphorically for 'many,' all being a species of many. So in the verse,—'alone she hath no part . .,' οιη, 'alone,' is metaphorical; for the best known may be called the only one.

Again, the solution may depend upon accent or breathing. Thus Hippias of Thasos solved the difficulties in the lines,—διδομεν(διδομεν)δε οι, and το μεν ου(ου)καταπυθεται ομβρω.

Or again, the question may be solved by punctuation, as in Empedocles,—'Of a sudden things became mortal that before had learnt to be immortal, and things unmixed before mixed.'

Or again, by ambiguity of meaning,—as παρωχηκεν δε πλεω νυξ, where the word πλεω is ambiguous.

Or by the usage of language. Thus any mixed drink is called οινοσ, 'wine.' Hence Ganymede is said 'to pour the wine to Zeus,' though the gods do not drink wine. So too workers in iron are called χαλκεασ, or workers in bronze. This, however, may also be taken as a metaphor.

Again, when a word seems to involve some inconsistency of meaning, we should consider how many senses it may bear in the particular passage. For example: 'there was stayed the spear of bronze'—we should ask in how many ways we may take 'being checked there.' The true mode of interpretation is the precise opposite of what Glaucon mentions. Critics, he says, jump at certain groundless conclusions; they pass adverse judgment and then proceed to reason on it; and, assuming that the poet has said whatever they happen to think, find fault if a thing is inconsistent with their own fancy. The question about Icarius has been treated in this fashion. The critics imagine he was a Lacedaemonian. They think it strange, therefore, that Telemachus should not have met him when he went to Lacedaemon. But the Cephallenian story may perhaps be the true one. They allege that Odysseus took a wife from among themselves, and that her father was Icadius not Icarius. It is merely a mistake, then, that gives plausibility to the objection.

In general, the impossible must be justified by reference to artistic requirements, or to the higher reality, or to received opinion. With respect to the requirements of art, a probable impossibility is to be preferred to a thing improbable and yet possible. Again, it may be impossible that there should be men such as Zeuxis painted. 'Yes,' we say, 'but the impossible is the higher thing; for the ideal type must surpass the reality.' To justify the irrational, we appeal to what is commonly said to be. In addition to which, we urge that the irrational sometimes does not violate reason; just as 'it is probable that a thing may happen contrary to probability.'

Things that sound contradictory should be examined by the same rules as in dialectical refutation whether the same thing is meant, in the same relation, and in the same sense. We should therefore solve the question by reference to what the poet says himself, or to what is tacitly assumed by a person of intelligence.

The element of the irrational, and, similarly, depravity of character, are justly censured when there is no inner necessity for introducing them. Such is the irrational element in the introduction of Aegeus by Euripides and the badness of Menelaus in the Orestes.

Thus, there are five sources from which critical objections are drawn. Things are censured either as impossible, or irrational, or morally hurtful, or contradictory, or contrary to artistic correctness. The answers should be sought under the twelve heads above mentioned.

A general estimate of the comparative worth of Epic Poetry and Tragedy.

The question may be raised whether the Epic or Tragic mode of imitation is the higher. If the more refined art is the higher, and the more refined in every case is that which appeals to the better sort of audience, the art which imitates anything and everything is manifestly most unrefined. The audience is supposed to be too dull to comprehend unless something of their own is thrown in by the performers, who therefore indulge in restless movements. Bad flute-players twist and twirl, if they have to represent 'the quoit-throw,' or hustle the coryphaeus when they perform the 'Scylla.' Tragedy, it is said, has this same defect. We may compare the opinion that the older actors entertained of their successors. Mynniscus used to call Callippides 'ape' on account of the extravagance of his action, and the same view was held of Pindarus. Tragic art, then, as a whole, stands to Epic in the same relation as the younger to the elder actors. So we are told that Epic poetry is addressed to a cultivated audience, who do not need gesture; Tragedy, to an inferior public. Being then unrefined, it is evidently the lower of the two.

Now, in the first place, this censure attaches not to the poetic but to the histrionic art; for gesticulation may be equally overdone in epic recitation, as by Sosi-stratus, or in lyrical competition, as by Mnasitheus the Opuntian. Next, all action is not to be condemned any more than all dancing—but only that of bad performers. Such was the fault found in Callippides, as also in others of our own day, who are censured for representing degraded women. Again, Tragedy like Epic poetry produces its effect even without action; it reveals its power by mere reading. If, then, in all other respects it is superior, this fault, we say, is not inherent in it.

And superior it is, because it has all the epic elements—it may even use the epic metre—with the music and spectacular effects as important accessories; and these produce the most vivid of pleasures. Further, it has vividness of impression in reading as well as in representation. Moreover, the art attains its end within narrower limits; for the concentrated effect is more pleasurable than one which is spread over a long time and so diluted. What, for example, would be the effect of the Oedipus of Sophocles, if it were cast into a form as long as the Iliad? Once more, the Epic imitation has less unity; as is shown by this, that any Epic poem will furnish subjects for several tragedies. Thus if the story adopted by the poet has a strict unity, it must either be concisely told and appear truncated; or, if it conform to the Epic canon of length, it must seem weak and watery. if, I mean, the poem is constructed out of several actions, like the Iliad and the Odyssey, which have many such parts, each with a certain magnitude of its own. Yet these poems are as perfect as possible in structure; each is, in the highest degree attainable, an imitation of a single action.

If, then, Tragedy is superior to Epic poetry in all these respects, and, moreover, fulfils its specific function better as an art for each art ought to produce, not any chance pleasure, but the pleasure proper to it, as already stated it plainly follows that Tragedy is the higher art, as attaining its end more perfectly.

Thus much may suffice concerning Tragic and Epic poetry in general; their several kinds and parts, with the number of each and their differences; the causes that make a poem good or bad; the objections of the critics and the answers to these objections.

The Nicomachean Ethics

Table of Contents

Book I: The Good for Man

Chapter 1:All human activities aim at some good: some goods subordinate to others.

EVERY art and every inquiry, and similarly every action and pursuit, is thought to aim at some good; and for this reason the good has rightly been declared to be that at which all things aim. But a certain difference is found among ends; some are activities, others are products apart from the activities that produce them. Where there are ends apart from the actions, it is the nature of the products to be better than the activities. Now, as there are many actions, arts, and sciences, their ends also are many; the end of the medical art is health, that of shipbuilding a vessel, that of strategy victory, that of economics wealth. But where such arts fall under a single capacity—as bridle-making and the other arts concerned with the equipment of horses fall under the art of riding, and this and every military action under strategy, in the same way other arts fall under yet others—in all of these the ends of the master arts are to be preferred to all the subordinate ends; for it is for the sake of the former that the latter are pursued. It makes no difference whether the activities themselves are the ends of the actions, or something else apart from the activities, as in the case of the sciences just mentioned.

Chapter 2: The science of the good for man is politics.

If, then, there is some end of the things we do, which we desire for its own sake (everything else being desired for the sake of this), and if we do not choose everything for the sake of something else (for at that rate the process would go on to infinity, so that our desire would be empty and vain), clearly this must be the good and the chief good. Will not the knowledge of it, then, have a great influence on life? Shall we not, like archers who have a mark to aim at, be more likely to hit upon what is right? If so, we must try, in outline at least, to determine what it is, and of which of the sciences or capacities it is the object. It would seem to belong to the most authoritative art and that which is most truly the master art. And politics appears to be of this nature; for it is this that ordains which of the sciences should be studied in a state, and which each class of citizens should learn and up to what point they should learn them; and we see even the most highly esteemed of capacities to fall under this, e.g. strategy, economics, rhetoric; now, since politics uses the rest of the sciences, and since, again, it legislates as to what we are to do and what we are to abstain from, the end of this science must include those of the others, so that this end must be the good for man. For even if the end is the same for a single man and for a state, that of the state seems at all events something greater and more complete whether to attain or to preserve; though it is worth while to attain the end merely for one man, it is finer and more godlike to attain it for a nation or for city-states. These, then, are the ends at which our inquiry aims, since it is political science, in one sense of that term.

Chapter 3: We must not expect more precision than the subject-matter admits. The student should have reached years of discretion.

Our discussion will be adequate if it has as much clearness as the subject-matter admits of, for precision is not to be sought for alike in all discussions, any more than in all the products of the crafts. Now fine and just actions, which political science investigates, admit of much variety and fluctuation of opinion, so that they may be thought to exist only by convention, and not by nature. And goods also give rise to a similar fluctuation because they bring harm to many people; for before now men have been undone by reason of their wealth, and others by reason of their courage. We must be content, then, in speaking of such subjects and with such premises to indicate the truth roughly and in outline, and in speaking about things which are only for the most part true and with premises of the same kind to reach conclusions that are no better. In the same spirit, therefore, should each type of statement be received; for it is the mark of an educated man to look for precision in each class of things just so far as the nature of the subject admits; it is evidently equally foolish to accept probable reasoning from a mathematician and to demand from a rhetorician scientific proofs.

Now each man judges well the things he knows, and of these he is a good judge. And so the man who has been educated in a subject is a good judge of that subject, and the man who has received an all-round education is a good judge in general. Hence a young man is not a proper hearer of lectures on political science; for he is inexperienced in the actions that occur in life, but its discussions start from these and are about these; and, further, since he tends to follow his passions, his study will be vain and unprofitable, because the end aimed at is not knowledge but action. And it makes no difference whether he is young in years or youthful in character; the defect does not depend on time, but on his living, and pursuing each successive object, as passion directs. For to such persons, as to the incontinent, knowledge brings no profit; but to those who desire and act in accordance with a rational principle knowledge about such matters will be of great benefit.

These remarks about the student, the sort of treatment to be expected, and the purpose of the inquiry, may be taken as our preface.

Chapter 4: It [the good] is generally agreed to be happiness, but there are various views as to what happiness is. What is required at the start is an unreasoned conviction about the facts, such as is produced by a good upbringing.

Let us resume our inquiry and state, in view of the fact that all knowledge and every pursuit aims at some good, what it is that we say political science aims at and what is the highest of all goods achievable by action. Verbally there is very general agreement; for both the general run of men and people of superior refinement say that it is happiness, and identify living well and doing well with being happy; but with regard to what happiness is they differ, and the many do not give the same account as the wise. For the former think it is some plain and obvious thing, like pleasure, wealth, or honour; they differ, however, from one another—and often even the same man identifies it with different things, with health when he is ill, with wealth when he is poor; but, conscious of their ignorance, they admire those who proclaim some great ideal that is above their comprehension. Now some thought that apart from these many goods there is another which is self-subsistent and causes the goodness of all these as well. To examine all the opinions that have been held were perhaps somewhat fruitless; enough to examine those that are most prevalent or that seem to be arguable.

Let us not fail to notice, however, that there is a difference between arguments from and those to the first principles. For Plato, too, was right in raising this question and asking, as he used to do, 'are we on the way from or to the first principles?' There is a difference, as there is in a race-course between the course from the judges to the turning-point and the way back. For, while we must begin with what is known, things are objects of knowledge in two senses — some to us, some without qualification. Presumably, then, we must begin with things known to us. Hence any one who is to listen intelligently to lectures about what is noble and just, and generally, about the subjects of political science must have been brought up in good habits. For the fact is the starting-point, and if this is sufficiently plain to him, he will not at the start need the reason as well; and the man who has been well brought up has or can easily get startingpoints. And as for him who neither has nor can get them, let him hear the words of Hesiod:

Far best is he who knows all things himself;
Good, he that hearkens when men counsel right;
But he who neither knows, nor lays to heart
Another's wisdom, is a useless wight.

Chapter 5: Discussion of the popular views that the good is pleasure, honour, wealth; a fourth kind of life, that of contemplation, deferred for future discussion.

Let us, however, resume our discussion from the point at which we digressed. To judge from the lives that men lead, most men, and men of the most vulgar type, seem (not without some ground) to identify the good, or happiness, with pleasure; which is the reason why they love the life of enjoyment. For there are, we may say, three prominent types of life—that just mentioned, the political, and thirdly the contemplative life. Now the mass of mankind are evidently quite slavish in their tastes, preferring a life suitable to beasts, but they get some ground for their view from the fact that many of those in high places share the tastes of Sardanapallus. A consideration of the prominent types of life shows that people of superior refinement and of active disposition identify happiness with honour; for this is, roughly speaking, the end of the political life. But it seems too superficial to be what we are looking for, since it is thought to depend on those who bestow honour rather than on him who receives it, but the good we divine to be something proper to a man and not easily taken from him. Further, men seem to pursue honour in order that they may be assured of their goodness; at least it is by men of practical wisdom that they seek to be honoured, and among those who know them, and on the ground of their virtue; clearly, then, according to them, at any rate, virtue is better. And perhaps one might even suppose this to be, rather than honour, the end of the political life. But even this appears somewhat incomplete; for possession of virtue seems actually compatible with being asleep, or with lifelong inactivity, and, further, with the greatest sufferings and misfortunes; but a man who was living so no one would call happy, unless he were maintaining a thesis at all costs. But enough of this; for the subject has been sufficiently treated even in the current discussions. Third comes the contemplative life, which we shall consider later.

The life of money-making is one undertaken under compulsion, and wealth is evidently not the good we are seeking; for it is merely useful and for the sake of something else. And so one might rather take the aforenamed objects to be ends; for they are loved for themselves. But it is evident that not even these are ends; yet many arguments have been thrown away in support of them. Let us leave this subject, then.

Chapter 6: Discussion of the philosophical view that there is an Idea of good.

We had perhaps better consider the universal good and discuss thoroughly what is meant by it, although such an inquiry is made an uphill one by the fact that the Forms have been introduced by friends of our own. Yet it would perhaps be thought to be better, indeed to be our duty, for the sake of maintaining the truth even to destroy what touches us closely, especially as we are philosophers or lovers of wisdom; for, while both are dear, piety requires us to honour truth above our friends.

The men who introduced this doctrine did not posit Ideas of classes within which they recognized priority and posteriority (which is the reason why they did not maintain the existence of an Idea embracing all numbers); but the term 'good' is used both in the category of substance and in that of quality and in that of relation, and that which is per se, i.e. substance, is prior in nature to the relative (for the latter is like an off shoot and accident of being); so that there could not be a common Idea set over all these goods. Further, since 'good' has as many senses as 'being' (for it is predicated both in the category of substance, as of God and of reason, and in quality, i.e. of the virtues, and in quantity, i.e. of that which is moderate, and in relation, i.e. of the useful, and in time, i.e. of the right opportunity, and in place, i.e. of the right locality and the like), clearly it cannot be something universally present in all cases and single; for then it could not have been predicated in all the categories but in one only. Further, since of the things answering to one Idea there is one science, there would have been one science of all the goods; but as it is there are many sciences even of the things that fall under one category, e.g. of opportunity, for opportunity in war is studied by strategics and in disease by medicine, and the moderate in food is studied by medicine and in exercise by the science of gymnastics. And one might ask the question, what in the world they mean by 'a thing itself', is (as is the case) in 'man himself' and in a particular man the account of man is one and the same. For in so far as they are man, they will in no respect differ; and if this is so, neither will 'good itself' and particular goods, in so far as they are good. But again it will not be good any the more for being eternal, since that which lasts long is no whiter than that which perishes in a day. The Pythagoreans seem to give a more plausible account of the good, when they place the one in the column of goods; and it is they that Speusippus seems to have followed.

But let us discuss these matters elsewhere; an objection to what we have said, however, may be discerned in the fact that the Platonists have not been speaking about all goods, and that the goods that are pursued and loved for themselves are called good by reference to a single Form, while those which tend to produce or to preserve these somehow or to prevent their contraries are called so by reference to these, and in a secondary sense. Clearly, then, goods must be spoken of in two ways, and some must be good in themselves, the others by reason of these. Let us separate, then, things good in themselves from things useful, and consider whether the former are called good by reference to a single Idea. What sort of goods would one call good in themselves? Is it those that are pursued even when isolated from others, such as intelligence, sight, and certain pleasures and honours? Certainly, if we pursue these also for the sake of something else, yet one would place them among things good in themselves. Or is nothing other than the Idea of good good in itself? In that case the Form will be empty. But if the things we have named are also things good in themselves, the account of the good will have to appear as something identical in them all, as that of whiteness is identical in snow and in white lead. But of honour, wisdom, and pleasure, just in respect of their goodness, the accounts are distinct and diverse. The good, therefore, is not some common element answering to one Idea.

But what then do we mean by the good? It is surely not like the things that only chance to have the same name. Are goods one, then, by being derived from one good or by all contributing to one good, or are they rather one by analogy? Certainly as sight is in the body, so is reason in the soul, and so on in other cases. But perhaps these subjects had better be dismissed for the present; for perfect precision about them would be more appropriate to another branch of philosophy. And similarly with regard to the Idea; even if there is some one good which is universally predicable of goods or is capable of separate and independent existence, clearly it could not be achieved or attained by man; but we are now seeking something attainable. Perhaps, however, some one might think it worth while to recognize this with a view to the goods that are attainable and achievable; for having this as a sort of pattern we shall know better the goods that are good for us, and if we know them shall attain them. This argument has some plausibility, but seems to clash with the procedure of the sciences; for all of these, though they aim at some good and seek to supply the deficiency of it, leave on one side the knowledge of the good. Yet that all the exponents of the arts should be ignorant of, and should not even seek, so great an aid is not probable. It is hard, too, to see how a weaver or a carpenter will be benefited in regard to his own craft by knowing this 'good itself', or how the man who has viewed the Idea itself will be a better doctor or general thereby. For a doctor seems not even to study health in this way, but the health of man, or perhaps rather the health of a particular man; it is individuals that he is healing. But enough of these topics.

Chapter 7: The good must be something final and self-sufficient. Definition of happiness reached by considering the characteristic

function of man.

Let us again return to the good we are seeking, and ask what it can be. It seems different in different actions and arts; it is different in medicine, in strategy, and in the other arts likewise. What then is the good of each? Surely that for whose sake everything else is done. In medicine this is health, in strategy victory, in architecture a house, in any other sphere something else, and in every action and pursuit the end; for it is for the sake of this that all men do whatever else they do. Therefore, if there is an end for all that we do, this will be the good achievable by action, and if there are more than one, these will be the goods achievable by action.

So the argument has by a different course reached the same point; but we must try to state this even more clearly. Since there are evidently more than one end, and we choose some of these (e.g. wealth, flutes, and in general instruments) for the sake of something else, clearly not all ends are final ends; but the chief good is evidently something final. Therefore, if there is only one final end, this will be what we are seeking, and if there are more than one, the most final of these will be what we are seeking. Now we call that which is in itself worthy of pursuit more final than that which is worthy of pursuit for the sake of something else, and that which is never desirable for the sake of something else more final than the things that are desirable both in themselves and for the sake of that other thing, and therefore we call final without qualification that which is always desirable in itself and never for the sake of something else.

Now such a thing happiness, above all else, is held to be; for this we choose always for self and never for the sake of something else, but honour, pleasure, reason, and every virtue we choose indeed for themselves (for if nothing resulted from them we should still choose each of them), but we choose them also for the sake of happiness, judging that by means of them we shall be happy. Happiness, on the other hand, no one chooses for the sake of these, nor, in general, for anything other than itself.

From the point of view of self-sufficiency the same result seems to follow; for the final good is thought to be self-sufficient. Now by self-sufficient we do not mean that which is sufficient for a man by himself, for one who lives a solitary life, but also for parents, children, wife, and in general for his friends and fellow citizens, since man is born for citizenship. But some limit must be set to this; for if we extend our requirement to ancestors and descendants and friends' friends we are in for an infinite series. Let us examine this question, however, on another occasion; the self-sufficient we now define as that which when isolated makes life desirable and lacking in nothing; and such we think happiness to be; and further we think it most desirable of all things, without being counted as one good thing among others—if it were so counted it would clearly be made more desirable by the addition of even the least of goods; for that which is added becomes an excess of goods, and of goods the greater is always more desirable. Happiness, then, is something final and self-sufficient, and is the end of action.

Presumably, however, to say that happiness is the chief good seems a platitude, and a clearer account of what it is still desired. This might perhaps be given, if we could first ascertain the function of man. For just as for a flute-player, a sculptor, or an artist, and, in general, for all things that have a function or activity, the good and the 'well' is thought to reside in the function, so would it seem to be for man, if he has a function. Have the carpenter, then, and the tanner certain functions or activities, and has man none? Is he born without a function? Or as eye, hand, foot, and in general each of the parts evidently has a function, may one lay it down that man similarly has a function apart from all these? What then can this be? Life seems to be common even to plants, but we are seeking what is peculiar to man. Let us exclude, therefore, the life of nutrition and growth. Next there would be a life of perception, but it also seems to be common even to the horse, the ox, and every animal. There remains, then, an active life of the element that has a rational principle; of this, one part has such a principle in the sense of being obedient to one, the other in the sense of possessing one and exercising thought. And, as 'life of the rational element' also has two meanings, we must state that life in the sense of activity is what we mean; for this seems to be the more proper sense of the term. Now if the function of man is an activity of soul which follows or implies a rational principle, and if we say 'so-and-so' and 'a good so-and-so' have a function which is the same in kind, e.g. a lyre, and a good lyre-player, and so without qualification in all cases, eminence in respect of goodness being idded to the name of the function (for the function of a lyre-player is to play the lyre, and that of a good lyre-player is to do so well): if this is the case, and we state the function of man to be a certain kind of life, and this to be an activity or actions of the soul implying a rational principle, and the function of a good man to be the good and noble performance of these, and if any action is well performed when it is performed in accordance with the appropriate excellence: if this is the case, human good turns out to be activity of soul in accordance with virtue, and if there are more than one virtue, in accordance with the best and most complete.

But we must add 'in a complete life.' For one swallow does not make a summer, nor does one day; and so too one day, or a short time, does not make a man blessed and happy.

Let this serve as an outline of the good; for we must presumably first sketch it roughly, and then later fill in the details. But it would seem that any one is capable of carrying on and articulating what has once been well outlined, and that time is a good discoverer or partner in such a work; to which facts the advances of the arts are due; for any one can add what is lacking. And we must also remember what has been said before, and not look for precision in all things alike, but in each class of things such precision as accords with the subject-matter, and so much as is appropriate to the inquiry. For a carpenter and a geometer investigate the right angle in different ways; the former does so in so far as the right angle is useful for his work, while the latter inquires what it is or what sort of thing it is; for he is a spectator of the truth. We must act in the same way, then, in all other matters as well, that our main task may not be subordinated to minor questions. Nor must we demand the cause in all matters alike; it is enough in some cases that the fact be well established, as in the case of the first principles; the fact is the primary thing or first principle. Now of first principles we see some by induction, some by perception, some by a certain habituation, and others too in other ways. But each set of principles we must try to investigate in the natural way, and we must take pains to state them definitely, since they have a great influence on what follows. For the beginning is thought to be more than half of the whole, and many of the questions we ask are cleared up by it.

Chapter 8: This definition is confirmed by current beliefs about happiness.

We must consider it, however, in the light not only of our conclusion and our premises, but also of what is commonly said about it; for with a true view all the data harmonize, but with a false one the facts soon clash. Now goods have been divided into three classes, and some are described as external, others as relating to soul or to body; we call those that relate to soul most properly and truly goods, and psychical actions and activities we class as relating to soul. Therefore our account must be sound, at least according to this view, which is an old one and agreed on by philosophers. It is correct also in that we identify the end with certain actions and activities; for thus it falls among goods of the soul and not among external goods. Another belief which harmonizes with our account is that the happy man lives well and does well; for we have practically defined happiness as a sort of good life and good action. The characteristics that are looked for in happiness seem also, all of them, to belong to what we have defined happiness as being. For some identify happiness with virtue, some with practical wisdom, others with a kind of philosophic wisdom, others with these, or one of these, accompanied by pleasure or not without pleasure; while others include also external prosperity. Now some of these views have been held by many men and men of old, others by a few eminent persons; and it is not probable that either of these should be entirely mistaken, but rather that they should be right in at least some one respect or even in most respects.

With those who identify happiness with virtue or some one virtue our account is in harmony; for to virtue belongs virtuous activity. But it makes, perhaps, no small difference whether we place the chief good in possession or in use, in state of mind or in activity. For the state of mind may exist without producing any good result, as in a man who is asleep or in some other way quite inactive, but the activity cannot; for one who has the activity will of necessity be acting, and acting well. And as in the Olympic Games it is not the most beautiful and the strongest that are crowned but those who compete (for it is some of these that are victorious), so those who act win, and rightly win, the noble and good things in life.

Their life is also in itself pleasant. For pleasure is a state of soul, and to each man that which he is said to be a lover of is pleasant; e.g. not only is a horse pleasant to the lover of horses, and a spectacle to the lover of sights, but also in the same way just acts are pleasant to the lover of justice and in general virtuous acts to the lover of virtue. Now for most men their pleasures are in conflict with one another because these are not by nature pleasant, but the lovers of what is noble find pleasant the things that are by nature pleasant; and virtuous actions are such, so that these are pleasant for such men as well as in their own nature. Their life, therefore, has no further need of pleasure as a sort of adventitious charm, but has its pleasure in itself. For, besides what we have said, the man who does not rejoice in noble actions is not even good; since no one would call a man just who did not enjoy acting justly, nor any man liberal who did not enjoy liberal actions; and similarly in all other cases. If this is so, virtuous actions must be in themselves pleasant. But they are also good and noble, and have each of these attributes in the highest degree, since the good man judges well about these attributes; his judgement is such as we have described. Happiness then is the best, noblest, and most pleasant thing in the world, and these attributes are not severed as in the inscription at Delos —

Most noble is that which is justest, and best is health;
But pleasantest is it to win what we love.

For all these properties belong to the best activities; and these, or one—the best—of these, we identify with happiness.

Yet evidently, as we said, it needs the external goods as well; for it is impossible, or not easy, to do noble acts without the proper equipment. In many actions we use friends and riches and political power as instruments; and there are some things the lack of which takes the lustre from happiness, as good birth, goodly children, beauty; for the man who is very ugly in appearance or ill-born or solitary and childless is not very likely to be happy, and perhaps a man would be still less likely if he had thoroughly bad children or friends or had lost good children or friends by death. As we said, then, happiness seems to need this sort of prosperity in addition; for which reason some identify happiness with good fortune, though others identify it with virtue.

Chapter 9: Is happiness acquired by learning or habituation, or sent by God or by chance?

For this reason also the question is asked, whether happiness is to be acquired by learning or by habituation or some other sort of training, or comes in virtue of some divine providence or again by chance. Now if there is any gift of the gods to men, it is reasonable that happiness should be god-given, and most surely god-given of all human things inasmuch as it is the best. But this question would perhaps be more appropriate to another inquiry; happiness seems, however, even if it is not god-sent but comes as a result of virtue and some process of learning or training, to be among the most godlike things; for that which is the prize and end of virtue seems to be the best thing in the world, and something godlike and blessed.

It will also on this view be very generally shared; for all who are not maimed as regards their potentiality for virtue may win it by a certain kind of study and care. But if it is better to be happy thus than by chance, it is reasonable that the facts should be so, since everything that depends on the action of nature is by nature as good as it can be, and similarly everything that depends on art or any rational cause, and especially if it depends on the best of all causes. To entrust to chance what is greatest and most noble would be a very defective arrangement.

The answer to the question we are asking is plain also from the definition of happiness; for it has been said to be a virtuous activity of soul, of a certain kind. Of the remaining goods, some must necessarily pre-exist as conditions of happiness, and others are naturally co-operative and useful as instruments. And this will be found to agree with what we said at the outset; for we stated the end of political science to be the best end, and political science spends most of its pains on making the citizens to be of a certain character, viz. good and capable of noble acts.

It is natural, then, that we call neither ox nor horse nor any other of the animals happy; for none of them is capable of sharing in such activity. For this reason also a boy is not happy; for he is not yet capable of such acts, owing to his age; and boys who are called happy are being congratulated by reason of the hopes we have for them. For there is required, as we said, not only complete virtue but also a complete life, since many changes occur in life, and all manner of chances, and the most prosperous may fall into great misfortunes in old age, as is told of Priam in the Trojan Cycle; and one who has experienced such chances and has ended wretchedly no one calls happy.

Chapter 10: Should no man be called happy while he lives?

Must no one at all, then, be called happy while he lives; must we, as Solon says, see the end? Even if we are to lay down this doctrine, is it also the case that a man is happy when he is dead? Or is not this quite absurd, especially for us who say that happiness is an activity? But if we do not call the dead man happy, and if Solon does not mean this, but that one can then safely call a man blessed as being at last beyond evils and misfortunes, this also affords matter for discussion; for both evil and good are thought to exist for a dead man, as much as for one who is alive but not aware of them; e.g. honours and dishonours and the good or bad fortunes of children and in general of descendants. And this also presents a problem; for though a man has lived happily up to old age and has had a death worthy of his life, many reverses may befall his descendants—some of them may be good and attain the life they deserve, while with others the opposite may be the case; and clearly too the degrees of relationship between them and their ancestors may vary indefinitely. It would be odd, then, if the dead man were to share in these changes and become at one time happy, at another wretched; while it would also be odd if the fortunes of the descendants did not for some time have some effect on the happiness of their ancestors.

But we must return to our first difficulty; for perhaps by a consideration of it our present problem might be solved. Now if we must see the end and only then call a man happy, not as being happy but as having been so before, surely this is a paradox, that when he is happy the attribute that belongs to him is not to be truly predicated of him because we do not wish to call living men happy, on account of the changes that may befall them, and because we have assumed happiness to be something permanent and by no means easily changed, while a single man may suffer many turns of fortune's wheel. For clearly if we were to keep pace with his fortunes, we should often call the same man happy and again wretched, making the happy man out to be chameleon and insecurely based. Or is this keeping pace with his fortunes quite wrong? Success or failure in life does not depend on these, but human life, as we said, needs these as mere additions, while virtuous activities or their opposites are what constitute happiness or the reverse.

The question we have now discussed confirms our definition. For no function of man has so much permanence as virtuous activities (these are thought to be more durable even than knowledge of the sciences), and of these themselves the most valuable are more durable because those who are happy spend their life most readily and most continuously in these; for this seems to be the reason why we do not forget them. The attribute in question, then, will belong to the happy man, and he will be happy throughout his life; for always, or by preference to everything else, he will be engaged in virtuous action and contemplation, and he will bear the chances of life most nobly and altogether decorously, if he is 'truly good' and 'foursquare beyond reproach'.

Now many events happen by chance, and events differing in importance; small pieces of good fortune or of its opposite clearly do not weigh down the scales of life one way or the other, but a multitude of great events if they turn out well will make life happier (for not only are they themselves such as to add beauty to life, but the way a man deals with them may be noble and good), while if they turn out ill they crush and maim happiness; for they both bring pain with them and hinder many activities. Yet even in these nobility shines through, when a man bears with resignation many great misfortunes, not through insensibility to pain but through nobility and greatness of soul.

If activities are, as we said, what gives life its character, no happy man can become miserable; for he will never do the acts that are hateful and mean. For the man who is truly good and wise, we think, bears all the chances life becomingly and always makes the best of circumstances, as a good general makes the best military use of the army at his command and a good shoemaker makes the best shoes out of the hides that are given him; and so with all other craftsmen. And if this is the case, the happy man can never become miserable; though he will not reach blessedness, if he meet with fortunes like those of Priam.

Nor, again, is he many-coloured and changeable; for neither will he be moved from his happy state easily or by any ordinary misadventures, but only by many great ones, nor, if he has had many great misadventures, will he recover his happiness in a short time, but if at all, only in a long and complete one in which he has attained many splendid successes.

When then should we not say that he is happy who is active in accordance with complete virtue and is sufficiently equipped with external goods, not for some chance period but throughout a complete life? Or must we add 'and who is destined to live thus and die as befits his life'? Certainly the future is obscure to us, while happiness, we claim, is an end and something in every way final. If so, we shall call happy those among living men in whom these conditions are, and are to be, fulfilled—but happy men. So much for these questions.

Chapter 11: Do the fortunes of the living affect the dead?

That the fortunes of descendants and of all a man's friends should not affect his happiness at all seems a very unfriendly doctrine, and one opposed to the opinions men hold; but since the events that happen are numerous and admit of all sorts of difference, and some come more near to us and others less so, it seems a long—nay, an infinite—task to discuss each in detail; a general outline will perhaps suffice. If, then, as some of a man's own misadventures have a certain weight and influence on life while others are, as it were, lighter, so too there are differences among the misadventures of our friends taken as a whole, and it makes a difference whether the various suffering befall the living or the dead (much more even than whether lawless and terrible deeds are presupposed in a tragedy or done on the stage), this difference also must be taken into account; or rather, perhaps, the fact that doubt is felt whether the dead share in any good or evil. For it seems, from these considerations, that even if anything whether good or evil penetrates to them, it must be something weak and negligible, either in itself or for them, or if not, at least it must be such in degree and kind as not to make happy those who are not happy nor to take away their blessedness from those who are. The good or bad fortunes of friends, then, seem to have some effects on the dead, but effects of such a kind and degree as neither to make the happy unhappy nor to produce any other change of the kind.

Chapter 12: Virtue is praiseworthy, but happiness is above praise.

These questions having been definitely answered, let us consider whether happiness is among the things that are praised or rather among the things that are prized; for clearly it is not to be placed among potentialities. Everything that is praised seems to be praised because it is of a certain kind and is related somehow to something else; for we praise the just or brave man and in general both the good man and virtue itself because of the actions and functions involved, and we praise the strong man, the good runner, and so on, because he is of a certain kind and is related in a certain way to something good and important. This is clear also from the praises of the gods; for it seems absurd that the gods should be referred to our standard, but this is done because praise involves a reference, to something else. But if if praise is for things such as we have described, clearly what applies to the best things is not praise, but something greater and better, as is indeed obvious; for what we do to the gods and the most godlike of men is to call them blessed and happy. And so too with good things; no one praises happiness as he does justice, but rather calls it blessed, as being something more divine and better.

Eudoxus also seems to have been right in his method of advocating the supremacy of pleasure; he thought that the fact that, though a good, it is not praised indicated it to be better than the things that are praised, and that this is what God and the good are; for by reference to these all other things are judged. Praise is appropriate to virtue, for as a result of virtue men tend to do noble deeds, but encomia are bestowed on acts, whether of the body or of the soul. But perhaps nicety in these matters is more proper to those who have made a study of encomia; to us it is clear from what has been said that happiness is among the things that are prized and perfect. It seems to be so also from the fact that it is a first principle; for it is for the sake of this that we all do all that we do, and the first principle and cause of goods is, we claim, something prized and divine.

Chapter 13: Division of the faculties, and resultant division of virtue into intellectual and moral.

Since happiness is an activity of soul in accordance with perfect virtue, we must consider the nature of virtue; for perhaps we shall thus see better the nature of happiness. The true student of politics, too, is thought to have studied virtue above all things; for he wishes to make his fellow citizens good and obedient to the laws. As an example of this we have the lawgivers of the Cretans and the Spartans, and any others of the kind that there may have been. And if this inquiry belongs to political science, clearly the pursuit of it will be in accordance with our original plan. But clearly the virtue we must study is human virtue; for the good we were seeking was human good and the happiness human happiness. By human virtue we mean not that of the body but that of the soul; and happiness also we call an activity of soul. But if this is so, clearly the student of politics must know somehow the facts about soul, as the man who is to heal the eyes or the body as a whole must know about the eyes or the body; and all the more since politics is more prized and better than medicine; but even among doctors the best educated spend much labour on acquiring knowledge of the body. The student of politics, then, must study the soul, and must study it with these objects in view, and do so just to the extent which is sufficient for the questions we are discussing; for further precision is perhaps something more laborious than our purposes require.

Some things are said about it, adequately enough, even in the discussions outside our school, and we must use these; e.g. that one element in the soul is irrational and one has a rational principle. Whether these are separated as the parts of the body or of anything divisible are, or are distinct by definition but by nature inseparable, like convex and concave in the circumference of a circle, does not affect the present question.

Of the irrational element one division seems to be widely distributed, and vegetative in its nature, I mean that which causes nutrition and growth; for it is this kind of power of the soul that one must assign to all nurslings and to embryos, and this same power to fullgrown creatures; this is more reasonable than to assign some different power to them. Now the excellence of this seems to be common to all species and not specifically human; for this part or faculty seems to function most in sleep, while goodness and badness are least manifest in sleep (whence comes the saying that the happy are not better off than the wretched for half their lives; and this happens naturally enough, since sleep is an inactivity of the soul in that respect in which it is called good or bad), unless perhaps to a small extent some of the movements actually penetrate to the soul, and in this respect the dreams of good men are better than those of ordinary people. Enough of this subject, however; let us leave the nutritive faculty alone, since it has by its nature no share in human excellence.

There seems to be also another irrational element in the soul-one which in a sense, however, shares in a rational principle. For we praise the rational principle of the continent man and of the incontinent, and the part of their soul that has such a principle, since it urges them aright and towards the best objects; but there is found in them also another element naturally opposed to the rational principle, which fights against and resists that principle. For exactly as paralysed limbs when we intend to move them to the right turn on the contrary to the left, so is it with the soul; the impulses of incontinent people move in contrary directions. But while in the body we see that which moves astray, in the soul we do not. No doubt, however, we must none the less suppose that in the soul too there is something contrary to the rational principle, resisting and opposing it. In what sense it is distinct from the other elements does not concern us. Now even this seems to have a share in a rational principle, as we said; at any rate in the continent man it obeys the rational principle and presumably in the temperate and brave man it is still more obedient; for in him it speaks, on all matters, with the same voice as the rational principle.

Therefore the irrational element also appears to be two-fold. For the vegetative element in no way shares in a rational principle, but the appetitive and in general the desiring element in a sense shares in it, in so far as it listens to and obeys it; this is the sense in which we speak of 'taking account' of one's father or one's friends, not that in which we speak of 'accounting for a mathematical property. That the irrational element is in some sense persuaded by a rational principle is indicated also by the giving of advice and by all reproof and exhortation. And if this element also must be said to have a rational principle, that which has a rational principle (as well as that which has not) will be twofold, one subdivision having it in the strict sense and in itself, and the other having a tendency to obey as one does one's father.

Virtue too is distinguished into kinds in accordance with this difference; for we say that some of the virtues are intellectual and others moral, philosophic wisdom and understanding and practical wisdom being intellectual, liberality and temperance moral. For in speaking about a man's character we do not say that he is wise or has understanding but that he is good-tempered or temperate; yet we praise the wise man also with respect to his state of mind; and of states of mind we call those which merit praise virtues.

Book II: Moral Virtue

Chapter I: It [moral virtue], like the arts, is acquired by repetition of the corresponding acts.

VIRTUE, then, being of two kinds, intellectual and moral, intellectual virtue in the main owes both its birth and its growth to teaching (for which reason it requires experience and time), while moral virtue comes about as a result of habit, whence also its name (ethike) is one that is formed by a slight variation from the word ethos (habit). From this it is also plain that none of the moral virtues arises in us by nature; for nothing that exists by nature can form a habit contrary to its nature. For instance the stone which by nature moves downwards cannot be habituated to move upwards, not even if one tries to train it by throwing it up ten thousand times; nor can fire be habituated to move downwards, nor can anything else that by nature behaves in one way be trained to behave in another. Neither by nature, then, nor contrary to nature do the virtues arise in us; rather we are adapted by nature to receive them, and are made perfect by habit.

Again, of all the things that come to us by nature we first acquire the potentiality and later exhibit the activity (this is plain in the case of the senses; for it was not by often seeing or often hearing that we got these senses, but on the contrary we had them before we used them, and did not come to have them by using them); but the virtues we get by first exercising them, as also happens in the case of the arts as well. For the things we have to learn before we can do them, we learn by doing them, e.g. men become builders by building and lyreplayers by playing the lyre; so too we become just by doing just acts, temperate by doing temperate acts, brave by doing brave acts.

This is confirmed by what happens in states; for legislators make the citizens good by forming habits in them, and this is the wish of every legislator, and those who do not effect it miss their mark, and it is in this that a good constitution differs from a bad one.

Again, it is from the same causes and by the same means that every virtue is both produced and destroyed, and similarly every art; for it is from playing the lyre that both good and bad lyre-players are produced. And the corresponding statement is true of builders and of all the rest; men will be good or bad builders as a result of building well or badly. For if this were not so, there would have been no need of a teacher, but all men would have been born good or bad at their craft. This, then, is the case with the virtues also; by doing the acts that we do in our transactions with other men we become just or unjust, and by doing the acts that we do in the presence of danger, and being habituated to feel fear or confidence, we become brave or cowardly. The same is true of appetites and feelings of anger; some men become temperate and good-tempered, others self-indulgent and irascible, by behaving in one way or the other in the appropriate circumstances. Thus, in one word, states of character arise out of like activities. This is why the activities we exhibit must be of a certain kind; it is because the states of character correspond to the differences between these. It makes no small difference, then, whether we form habits of one kind or of another from our very youth; it makes a very great difference, or rather all the difference.

Chapter 2: These acts cannot be prescribed exactly, but must avoid excess and defect.

Since, then, the present inquiry does not aim at theoretical knowledge like the others (for we are inquiring not in order to know what virtue is, but in order to become good, since otherwise our inquiry would have been of no use), we must examine the nature of actions, namely how we ought to do them; for these determine also the nature of the states of character that are produced, as we have said. Now, that we must act according to the right rule is a common principle and must be assumed—it will be discussed later, i.e. both what the right rule is, and how it is related to the other virtues. But this must be agreed upon beforehand, that the whole account of matters of conduct must be given in outline and not precisely, as we said at the very beginning that the accounts we demand must be in accordance with the subject-matter; matters concerned with conduct and questions of what is good for us have no fixity, any more than matters of health. The general account being of this nature, the account of particular cases is yet more lacking in exactness; for they do not fall under any art or precept but the agents themselves must in each case consider what is appropriate to the occasion, as happens also in the art of medicine or of navigation.

But though our present account is of this nature we must give what help we can. First, then, let us consider this, that it is the nature of such things to be destroyed by defect and excess, as we see in the case of strength and of health (for to gain light on things imperceptible we must use the evidence of sensible things); both excessive and defective exercise destroys the strength, and similarly drink or food which is above or below a certain amount destroys the health, while that which is proportionate both produces and increases and preserves it. So too is it, then, in the case of temperance and courage and the other virtues. For the man who flies from and fears everything and does not stand his ground against anything becomes a coward, and the man who fears nothing at all but goes to meet every danger becomes rash; and similarly the man who indulges in every pleasure and abstains from none becomes self-indulgent, while the man who shuns every pleasure, as boors do, becomes in a way insensible; temperance and courage, then, are destroyed by excess and defect, and preserved by the mean.

But not only are the sources and causes of their origination and growth the same as those of their destruction, but also the sphere of their actualization will be the same; for this is also true of the things which are more evident to sense, e.g. of strength; it is produced by taking much food and undergoing much exertion, and it is the strong man that will be most able to do these things. So too is it with the virtues; by abstaining from pleasures we become temperate, and it is when we have become so that we are most able to abstain from them; and similarly too in the case of courage; for by being habituated to despise things that are terrible and to stand our ground against them we become brave, and it is when we have become so that we shall be most able to stand our ground against them.

Chapter 3: Pleasure in doing virtuous acts is a sign that the virtuous disposition has been acquired: a variety of considerations show the essential connexion of moral virtue with pleasure and pain.

We must take as a sign of states of character the pleasure or pain that ensues on acts; for the man who abstains from bodily pleasures and delights in this very fact is temperate, while the man who is annoyed at it is self-indulgent, and he who stands his ground against things that are terrible and delights in this or at least is not pained is brave, while the man who is pained is a coward. For moral excellence is concerned with pleasures and pains; it is on account of the pleasure that we do bad things, and on account of the pain that we abstain from noble ones. Hence we ought to have been brought up in a particular way from our very youth, as Plato says, so as both to delight in and to be pained by the things that we ought; for this is the right education.

Again, if the virtues are concerned with actions and passions, and every passion and every action is accompanied by pleasure and pain, for this reason also virtue will be concerned with pleasures and pains. This is indicated also by the fact that punishment is inflicted by these means; for it is a kind of cure, and it is the nature of cures to be effected by contraries.

Again, as we said but lately, every state of soul has a nature relative to and concerned with the kind of things by which it tends to be made worse or better; but it is by reason of pleasures and pains that men become bad, by pursuing and avoiding these—either the pleasures and pains they ought not or when they ought not or as they ought not, or by going wrong in one of the other similar ways that may be distinguished. Hence men even define the virtues as certain states of impassivity and rest; not well, however, because they speak absolutely, and do not say 'as one ought' and 'as one ought not' and 'when one ought or ought not', and the other things that may be added. We assume, then, that this kind of excellence tends to do what is best with regard to pleasures and pains, and vice does the contrary.

The following facts also may show us that virtue and vice are concerned with these same things. There being three objects of choice and three of avoidance, the noble, the advantageous, the pleasant, and their contraries, the base, the injurious, the painful, about all of these the good man tends to go right and the bad man to go wrong, and especially about pleasure; for this is common to the animals, and also it accompanies all objects of choice; for even the noble and the advantageous appear pleasant.

Again, it has grown up with us all from our infancy; this is why it is difficult to rub off this passion, engrained as it is in our life. And we measure even our actions, some of us more and others less, by the rule of pleasure and pain. For this reason, then, our whole inquiry must be about these; for to feel delight and pain rightly or wrongly has no small effect on our actions.

Again, it is harder to fight with pleasure than with anger, to use Heraclitus' phrase', but both art and virtue are always concerned with what is harder; for even the good is better when it is harder. Therefore for this reason also the whole concern both of virtue and of political science is with pleasures and pains; for the man who uses these well will be good, he who uses them badly bad.

That virtue, then, is concerned with pleasures and pains, and that by the acts from which it arises it is both increased and, if they are done differently, destroyed, and that the acts from which it arose are those in which it actualizes itself—let this be taken as said.

Chapter 4: The actions that produce moral virtue are not good in the same sense as those that flow from it: the latter must fulfil certain conditions not necessary in the case of the arts.

The question might be asked,; what we mean by saying that we must become just by doing just acts, and temperate by doing temperate acts; for if men do just and temperate acts, they are already just and temperate, exactly as, if they do what is in accordance with the laws of grammar and of music, they are grammarians and musicians.

Or is this not true even of the arts? It is possible to do something that is in accordance with the laws of grammar, either by chance or at the suggestion of another. A man will be a grammarian, then, only when he has both done something grammatical and done it grammatically; and this means doing it in accordance with the grammatical knowledge in himself.

Again, the case of the arts and that of the virtues are not similar; for the products of the arts have their goodness in themselves, so that it is enough that they should have a certain character, but if the acts that are in accordance with the virtues have themselves a certain character it does not follow that they are done justly or temperately. The agent also must be in a certain condition when he does them; in the first place he must have knowledge, secondly he must choose the acts, and choose them for their own sakes, and thirdly his action must proceed from a firm and unchangeable character. These are not reckoned in as conditions of the possession of the arts, except the bare knowledge; but as a condition of the possession of the virtues knowledge has little or no weight, while the other conditions count not for a little but for everything, i.e. the very conditions which result from often doing just and temperate acts.

Actions, then, are called just and temperate when they are such as the just or the temperate man would do; but it is not the man who does these that is just and temperate, but the man who also does them as just and temperate men do them. It is well said, then, that it is by doing just acts that the just man is produced, and by doing temperate acts the temperate man; without doing these no one would have even a prospect of becoming good.

But most people do not do these, but take refuge in theory and think they are being philosophers and will become good in this way, behaving somewhat like patients who listen attentively to their doctors, but do none of the things they are ordered to do. As the latter will not be made well in body by such a course of treatment, the former will not be made well in soul by such a course of philosophy.

Chapter 5: Its [moral virtue's] genus: it is a state of character, not a passion nor a faculty.

Next we must consider what virtue is. Since things that are found in the soul are of three kinds—passions, faculties, states of character, virtue must be one of these. By passions I mean appetite, anger, fear, confidence, envy, joy, friendly feeling, hatred, longing, emulation, pity, and in general the feelings that are accompanied by pleasure or pain; by faculties the things in virtue of which we are said to be capable of feeling these, e.g. of becoming angry or being pained or feeling pity; by states of character the things in virtue of which we stand well or badly with reference to the passions, e.g. with reference to anger we stand badly if we feel it violently or too weakly, and well if we feel it moderately; and similarly with reference to the other passions.

Now neither the virtues nor the vices are passions, because we are not called good or bad on the ground of our passions, but are so called on the ground of our virtues and our vices, and because we are neither praised nor blamed for our passions (for the man who feels fear or anger is not praised, nor is the man who simply feels anger blamed, but the man who feels it in a certain way), but for our virtues and our vices we are praised or blamed.

Again, we feel anger and fear without choice, but the virtues are modes of choice or involve choice. Further, in respect of the passions we are said to be moved, but in respect of the virtues and the vices we are said not to be moved but to be disposed in a particular way.

For these reasons also they are not faculties; for we are neither called good nor bad, nor praised nor blamed, for the simple capacity of feeling the passions; again, we have the faculties by nature, but we are not made good or bad by nature; we have spoken of this before. If, then, the virtues are neither passions nor faculties, all that remains is that they should be states of character.

Thus we have stated what virtue is in respect of its genus.

Chapter 6: Its differentia: it is a disposition to choose the mean.

We must, however, not only describe virtue as a state of character, but also say what sort of state it is. We may remark, then, that every virtue or excellence both brings into good condition the thing of which it is the excellence and makes the work of that thing be done well; e.g. the excellence of the eye makes both the eye and its work good; for it is by the excellence of the eye that we see well. Similarly the excellence of the horse makes a horse both good in itself and good at running and at carrying its rider and at awaiting the attack of the enemy. Therefore, if this is true in every case, the virtue of man also will be the state of character which makes a man good and which makes him do his own work well.

How this is to happen we have stated already, but it will be made plain also by the following consideration of the specific nature of virtue. In everything that is continuous and divisible it is possible to take more, less, or an equal amount, and that either in terms of the thing itself or relatively to us; and the equal is an intermediate between excess and defect. By the intermediate in the object I mean that which is equidistant from each of the extremes, which is one and the same for all men; by the intermediate relatively to us that which is neither too much nor too little—and this is not one, nor the same for all. For instance, if ten is many and two is few, six is the intermediate, taken in terms of the object; for it exceeds and is exceeded by an equal amount; this is intermediate according to arithmetical proportion. But the intermediate relatively to us is not to be taken so; if ten pounds are too much for a particular person to eat and two too little, it does not follow that the trainer will order six pounds; for this also is perhaps too much for the person who is to take it, or too little—too little for Milo, too much for the beginner in athletic exercises. The same is true of running and wrestling. Thus a master of any art avoids excess and defect, but seeks the intermediate and chooses this—the intermediate not in the object but relatively to us.

If it is thus, then, that every art does its work well—by looking to the intermediate and judging its works by this standard (so that we often say of good works of art that it is not possible either to take away or to add anything, implying that excess and defect destroy the goodness of works of art, while the mean preserves it; and good artists, as we say, look to this in their work), and if, further, virtue is more exact and better than any art, as nature also is, then virtue must have the quality of aiming at the intermediate. I mean moral virtue; for it is this that is concerned with passions and actions, and in these there is excess, defect, and the intermediate. For instance, both fear and confidence and appetite and anger and pity and in general pleasure and pain may be felt both too much and too little, and in both cases not well; but to feel them at the right times, with reference to the right objects, towards the right people, with the right motive, and in the right way, is what is both intermediate and best, and this is characteristic of virtue. Similarly with regard to actions also there is excess, defect, and the intermediate. Now virtue is concerned with passions and actions, in which excess is a form of failure, and so is defect, while the intermediate is praised and is a form of success; and being praised and being successful are both characteristics of virtue. Therefore virtue is a kind of mean, since, as we have seen, it aims at what is intermediate.

Again, it is possible to fail in many ways (for evil belongs to the class of the unlimited, as the Pythagoreans conjectured, and good to that of the limited), while to succeed is possible only in one way (for which reason also one is easy and the other difficult—to miss the mark easy, to hit it difficult); for these reasons also, then, excess and defect are characteristic of vice, and the mean of virtue;

For men are good in but one way, but bad in many.

Virtue, then, is a state of character concerned with choice, lying in a mean, i.e. the mean relative to us, this being determined by a rational principle, and by that principle by which the man of practical wisdom would determine it. Now it is a mean between two vices, that which depends on excess and that which depends on defect; and again it is a mean because the vices respectively fall short of or exceed what is right in both passions and actions, while virtue both finds and chooses that which is intermediate. Hence in respect of its substance and the definition which states its essence virtue is a mean, with regard to what is best and right an extreme.

But not every action nor every passion admits of a mean; for some have names that already imply badness, e.g. spite, shamelessness, envy, and in the case of actions adultery, theft, murder; for all of these and suchlike things imply by their names that they are themselves bad, and not the excesses or deficiencies of them. It is not possible, then, ever to be right with regard to them; one must always be wrong. Nor does goodness or badness with regard to such things depend on committing adultery with the right woman, at the right time, and in the right way, but simply to do any of them is to go wrong. It would be equally absurd, then, to expect that in unjust, cowardly, and voluptuous action there should be a mean, an excess, and a deficiency; for at that rate there would be a mean of excess and of deficiency, an excess of excess, and a deficiency of deficiency. But as there is no excess and deficiency of temperance and courage because what is intermediate is in a sense an extreme, so too of the actions we have mentioned there is no mean nor any excess and deficiency, but however they are done they are wrong; for in general there is neither a mean of excess and deficiency, nor excess and deficiency of a mean.

Chapter 7: This proposition illustrated by reference to the particular virtues.

We must, however, not only make this general statement, but also apply it to the individual facts. For among statements about conduct those which are general apply more widely, but those which are particular are more genuine, since conduct has to do with individual cases, and our statements must harmonize with the facts in these cases. We may take these cases from our table. With regard to feelings of fear and confidence courage is the mean; of the people who exceed, he who exceeds in fearlessness has no name (many of the states have no name), while the man who exceeds in confidence is rash, and he who exceeds in fear and falls short in confidence is a coward. With regard to pleasures and pains—not all of them, and not so much with regard to the pains—the mean is temperance, the excess self-indulgence. Persons deficient with regard to the pleasures are not often found; hence such persons also have received no name. But let us call them 'insensible'.

With regard to giving and taking of money the mean is liberality, the excess and the defect prodigality and meanness. In these actions people exceed and fall short in contrary ways; the prodigal exceeds in spending and falls short in taking, while the mean man exceeds in taking and falls short in spending. (At present we are giving a mere outline or summary, and are satisfied with this; later these states will be more exactly determined.) With regard to money there are also other dispositions—a mean, magnificence (for the magnificent man differs from the liberal man; the former deals with large sums, the latter with small ones), an excess, tastelessness and vulgarity, and a deficiency, niggardliness; these differ from the states opposed to liberality, and the mode of their difference will be stated later. With regard to honour and dishonour the mean is proper pride, the excess is known as a sort of 'empty vanity', and the deficiency is undue humility; and as we said liberality was related to magnificence, differing from it by dealing with small sums, so there is a state similarly related to proper pride, being concerned with small honours while that is concerned with great. For it is possible to desire honour as one ought, and more than one ought, and less, and the man who exceeds in his desires is called ambitious, the man who falls short unambitious, while the intermediate person has no name. The dispositions also are nameless, except that that of the ambitious man is called ambition. Hence the people who are at the extremes lay claim to the middle place; and we ourselves sometimes call the intermediate person ambitious and sometimes unambitious, and sometimes praise the ambitious man and sometimes the unambitious. The reason of our doing this will be stated in what follows; but now let us speak of the remaining states according to the method which has been indicated.

With regard to anger also there is an excess, a deficiency, and a mean. Although they can scarcely be said to have names, yet since we call the intermediate person good-tempered let us call the mean good temper; of the persons at the extremes let the one who exceeds be called irascible, and his vice irascibility, and the man who falls short an inirascible sort of person, and the deficiency inirascibility.

There are also three other means, which have a certain likeness to one another, but differ from one another: for they are all concerned with intercourse in words and actions, but differ in that one is concerned with truth in this sphere, the other two with pleasantness; and of this one kind is exhibited in giving amusement, the other in all the circumstances of life. We must therefore speak of these too, that we may the better see that in all things the mean is praise-worthy, and the extremes neither praiseworthy nor right, but worthy of blame. Now most of these states also have no names, but we must try, as in the other cases, to invent names ourselves so that we may be clear and easy to follow. With regard to truth, then, the intermediate is a truthful sort of person and the mean may be called truthfulness, while the pretence which exaggerates is boastfulness and the person characterized by it a boaster, and that which understates is mock modesty and the person characterized by it mock-modest. With regard to pleasantness in the giving of amusement the intermediate person is ready-witted and the disposition ready wit, the excess is buffoonery and the person characterized by it a buffoon, while the man who falls short is a sort of boor and his state is boorishness. With regard to the remaining kind of pleasantness, that which is exhibited in life in general, the man who is pleasant in the right way is friendly and the mean is friendliness, while the man who exceeds is an obsequious person if he has no end in view, a flatterer if he is aiming at his own advantage, and the man who falls short and is unpleasant in all circumstances is a quarrelsome and surly sort of person.

There are also means in the passions and concerned with the passions; since shame is not a virtue, and yet praise is extended to the modest man. For even in these matters one man is said to be intermediate, and another to exceed, as for instance the bashful man who is ashamed of everything; while he who falls short or is not ashamed of anything at all is shameless, and the intermediate person is modest. Righteous indignation is a mean between envy and spite, and these states are concerned with the pain and pleasure that are felt at the fortunes of our neighbours; the man who is characterized by righteous indignation is pained at undeserved good fortune, the envious man, going beyond him, is pained at all good fortune, and the spiteful man falls so far short of being pained that he even rejoices. But these states there will be an opportunity of describing elsewhere; with regard to justice, since it has not one simple meaning, we shall, after describing the other states, distinguish its two kinds and say how each of them is a mean; and similarly we shall treat also of the rational virtues.

Chapter 8: The extremes are opposed to each other and the

mean.

There are three kinds of disposition, then, two of them vices, involving excess and deficiency respectively, and one a virtue, viz. the mean, and all are in a sense opposed to all; for the extreme states are contrary both to the intermediate state and to each other, and the intermediate to the extremes; as the equal is greater relatively to the less, less relatively to the greater, so the middle states are excessive relatively to the deficiencies, deficient relatively to the excesses, both in passions and in actions. For the brave man appears rash relatively to the coward, and cowardly relatively to the rash man; and similarly the temperate man appears self-indulgent relatively to the insensible man, insensible relatively to the self-indulgent, and the liberal man prodigal relatively to the mean man, mean relatively to the prodigal. Hence also the people at the extremes push the intermediate man each over to the other, and the brave man is called rash by the coward, cowardly by the rash man, and correspondingly in the other cases.

These states being thus opposed to one another, the greatest contrariety is that of the extremes to each other, rather than to the intermediate; for these are further from each other than from the intermediate, as the great is further from the small and the small from the great than both are from the equal. Again, to the intermediate some extremes show a certain likeness, as that of rashness to courage and that of prodigality to liberality; but the extremes show the greatest unlikeness to each other; now contraries are defined as the things that are furthest from each other, so that things that are further apart are more contrary.

To the mean in some cases the deficiency, in some the excess is more opposed; e.g. it is not rashness, which is an excess, but cowardice, which is a deficiency, that is more opposed to courage, and not insensibility, which is a deficiency, but self-indulgence, which is an excess, that is more opposed to temperance. This happens from two reasons, one being drawn from the thing itself; for because one extreme is nearer and liker to the intermediate, we oppose not this but rather its contrary to the intermediate. E.g. since rashness is thought liker and nearer to courage, and cowardice more unlike, we oppose rather the latter to courage; for things that are further from the intermediate are thought more contrary to it. This, then, is one cause, drawn from the thing itself; another is drawn from ourselves; for the things to which we ourselves more naturally tend seem more contrary to the intermediate. For instance, we ourselves tend more naturally to pleasures, and hence are more easily carried away towards self-indulgence than towards propriety. We describe as contrary to the mean, then, rather the directions in which we more often go to great lengths; and therefore self-indulgence, which is an excess, is the more contrary to temperance.

Chapter 9: The mean is hard to attain, and is grasped by perception, not by reasoning.

That moral virtue is a mean, then, and in what sense it is so, and that it is a mean between two vices, the one involving excess, the other deficiency, and that it is such because its character is to aim at what is intermediate in passions and in actions, has been sufficiently stated. Hence also it is no easy task to be good. For in everything it is no easy task to find the middle, e.g. to find the middle of a circle is not for every one but for him who knows; so, too, any one can get angry—that is easy—or give or spend money; but to do this to the right person, to the right extent, at the right time, with the right motive, and in the right way, that is not for every one, nor is it easy; wherefore goodness is both rare and laudable and noble.

Hence he who aims at the intermediate must first depart from what is the more contrary to it, as Calypso advises —

Hold the ship out beyond that surf and spray.

For of the extremes one is more erroneous, one less so; therefore, since to hit the mean is hard in the extreme, we must as a second best, as people say, take the least of the evils; and this will be done best in the way we describe. But we must consider the things towards which we ourselves also are easily carried away; for some of us tend to one thing, some to another; and this will be recognizable from the pleasure and the pain we feel. We must drag ourselves away to the contrary extreme; for we shall get into the intermediate state by drawing well away from error, as people do in straightening sticks that are bent.

Now in everything the pleasant or pleasure is most to be guarded against; for we do not judge it impartially. We ought, then, to feel towards pleasure as the elders of the people felt towards Helen, and in all circumstances repeat their saying; for if we dismiss pleasure thus we are less likely to go astray. It is by doing this, then, (to sum the matter up) that we shall best be able to hit the mean.

But this is no doubt difficult, and especially in individual cases; for or is not easy to determine both how and with whom and on what provocation and how long one should be angry; for we too sometimes praise those who fall short and call them good-tempered, but sometimes we praise those who get angry and call them manly. The man, however, who deviates little from goodness is not blamed, whether he do so in the direction of the more or of the less, but only the man who deviates more widely; for he does not fail to be noticed. But up to what point and to what extent a man must deviate before he becomes blameworthy it is not easy to determine by reasoning, any more than anything else that is perceived by the senses; such things depend on particular facts, and the decision rests with perception. So much, then, is plain, that the intermediate state is in all things to be praised, but that we must incline sometimes towards the excess, sometimes towards the deficiency; for so shall we most easily hit the mean and what is right.

Book III: Moral Virtue

Chapter I: Praise and blame attach to voluntary actions, i.e. actions done (1) not under compulsion, and (2) with knowledge of the circumstances.

SINCE virtue is concerned with passions and actions, and on voluntary passions and actions praise and blame are bestowed, on those that are involuntary pardon, and sometimes also pity, to distinguish the voluntary and the involuntary is presumably necessary for those who are studying the nature of virtue, and useful also for legislators with a view to the assigning both of honours and of punishments. Those things, then, are thought involuntary, which take place under compulsion or owing to ignorance; and that is compulsory of which the moving principle is outside, being a principle in which nothing is contributed by the person who is acting or is feeling the passion, e.g. if he were to be carried somewhere by a wind, or by men who had him in their power.

But with regard to the things that are done from fear of greater evils or for some noble object (e.g. if a tyrant were to order one to do something base, having one's parents and children in his power, and if one did the action they were to be saved, but otherwise would be put to death), it may be debated whether such actions are involuntary or voluntary. Something of the sort happens also with regard to the throwing of goods overboard in a storm; for in the abstract no one throws goods away voluntarily, but on condition of its securing the safety of himself and his crew any sensible man does so. Such actions, then, are mixed, but are more like voluntary actions; for they are worthy of choice at the time when they are done, and the end of an action is relative to the occasion. Both the terms, then, 'voluntary' and 'involuntary', must be used with reference to the moment of action. Now the man acts voluntarily; for the principle that moves the instrumental parts of the body in such actions is in him, and the things of which the moving principle is in a man himself are in his power to do or not to do. Such actions, therefore, are voluntary, but in the abstract perhaps involuntary; for no one would choose any such act in itself.

For such actions men are sometimes even praised, when they endure something base or painful in return for great and noble objects gained; in the opposite case they are blamed, since to endure the greatest indignities for no noble end or for a trifling end is the mark of an inferior person. On some actions praise indeed is not bestowed, but pardon is, when one does what he ought not under pressure which overstrains human nature and which no one could withstand. But some acts, perhaps, we cannot be forced to do, but ought rather to face death after the most fearful sufferings; for the things that 'forced' Euripides Alcmaeon to slay his mother seem absurd. It is difficult sometimes to determine what should be chosen at what cost, and what should be endured in return for what gain, and yet more difficult to abide by our decisions; for as a rule what is expected is painful, and what we are forced to do is base, whence praise and blame are bestowed on those who have been compelled or have not.

What sort of acts, then, should be called compulsory? We answer that without qualification actions are so when the cause is in the external circumstances and the agent contributes nothing. But the things that in themselves are involuntary, but now and in return for these gains are worthy of choice, and whose moving principle is in the agent, are in themselves involuntary, but now and in return for these gains voluntary. They are more like voluntary acts; for actions are in the class of particulars, and the particular acts here are voluntary. What sort of things are to be chosen, and in return for what, it is not easy to state; for there are many differences in the particular cases.

But if some one were to say that pleasant and noble objects have a compelling power, forcing us from without, all acts would be for him compulsory; for it is for these objects that all men do everything they do. And those who act under compulsion and unwillingly act with pain, but those who do acts for their pleasantness and nobility do them with pleasure; it is absurd to make external circumstances responsible, and not oneself, as being easily caught by such attractions, and to make oneself responsible for noble acts but the pleasant objects responsible for base acts. The compulsory, then, seems to be that whose moving principle is outside, the person compelled contributing nothing.

Everything that is done by reason of ignorance is not voluntary; it is only what produces pain and repentance that is involuntary. For the man who has done something owing to ignorance, and feels not the least vexation at his action, has not acted voluntarily, since he did not know what he was doing, nor yet involuntarily, since he is not pained. Of people, then, who act by reason of ignorance he who repents is thought an involuntary agent, and the man who does not repent may, since he is different, be called a not voluntary agent; for, since he differs from the other, it is better that he should have a name of his own.

Acting by reason of ignorance seems also to be different from acting in ignorance; for the man who is drunk or in a rage is thought to act as a result not of ignorance but of one of the causes mentioned, yet not knowingly but in ignorance.

Now every wicked man is ignorant of what he ought to do and what he ought to abstain from, and it is by reason of error of this kind that men become unjust and in general bad; but the term 'involuntary' tends to be used not if a man is ignorant of what is to his advantage—for it is not mistaken purpose that causes involuntary action (it leads rather to wickedness), nor ignorance of the universal (for that men are blamed), but ignorance of particulars, i.e. of the circumstances of the action and the objects with which it is concerned. For it is on these that both pity and pardon depend, since the person who is ignorant of any of these acts involuntarily.

Perhaps it is just as well, therefore, to determine their nature and number. A man may be ignorant, then, of who he is, what he is doing, what or whom he is acting on, and sometimes also what (e.g. what instrument) he is doing it with, and to what end (e.g. he may think his act will conduce to some one's safety), and how he is doing it (e.g. whether gently or violently). Now of all of these no one could be ignorant unless he were mad, and evidently also he could not be ignorant of the agent; for how could he not know himself? But of what he is doing a man might be ignorant, as for instance people say 'it slipped out of their mouths as they were speaking', or 'they did not know it was a secret', as Aeschylus said of the mysteries, or a man might say he 'let it go off when he merely wanted to show its working', as the man did with the catapult. Again, one might think one's son was an enemy, as Merope did, or that a pointed spear had a button on it, or that a stone was pumicestone; or one might give a man a draught to save him, and really kill him; or one might want to touch a man, as people do in sparring, and really wound him. The ignorance may relate, then, to any of these things, i.e. of the circumstances of the action, and the man who was ignorant of any of these is thought to have acted involuntarily, and especially if he was ignorant on the most important points; and these are thought to be the circumstances of the action and its end. Further, the doing of an act that is called involuntary in virtue of ignorance of this sort must be painful and involve repentance.

Since that which is done under compulsion or by reason of ignorance is involuntary, the voluntary would seem to be that of which the moving principle is in the agent himself, he being aware of the particular circumstances of the action. Presumably acts done by reason of anger or appetite are not rightly called involuntary. For in the first place, on that showing none of the other animals will act voluntarily, nor will children; and secondly, is it meant that we do not do voluntarily any of the acts that are due to appetite or anger, or that we do the noble acts voluntarily and the base acts involuntarily? Is not this absurd, when one and the same thing is the cause? But it would surely be odd to describe as involuntary the things one ought to desire; and we ought both to be angry at certain things and to have an appetite for certain things, e.g. for health and for learning. Also what is involuntary is thought to be painful, but what is in accordance with appetite is thought to be pleasant. Again, what is the difference in respect of involuntariness between errors committed upon calculation and those committed in anger? Both are to be avoided, but the irrational passions are thought not less human than reason is, and therefore also the actions which proceed from anger or appetite are the man's actions. It would be odd, then, to treat them as involuntary.

Chapter 2: Moral virtue implies that the action is done (3) by choice; the object of choice is the result of previous deliberation.

Both the voluntary and the involuntary having been delimited, we must next discuss choice; for it is thought to be most closely bound up with virtue and to discriminate characters better than actions do.

Choice, then, seems to be voluntary, but not the same thing as the voluntary; the latter extends more widely. For both children and the lower animals share in voluntary action, but not in choice, and acts done on the spur of the moment we describe as voluntary, but not as chosen.

Those who say it is appetite or anger or wish or a kind of opinion do not seem to be right. For choice is not common to irrational creatures as well, but appetite and anger are. Again, the incontinent man acts with appetite, but not with choice; while the continent man on the contrary acts with choice, but not with appetite. Again, appetite is contrary to choice, but not appetite to appetite. Again, appetite relates to the pleasant and the painful, choice neither to the painful nor to the pleasant.

Still less is it anger; for acts due to anger are thought to be less than any others objects of choice.

But neither is it wish, though it seems near to it; for choice cannot relate to impossibles, and if any one said he chose them he would be thought silly; but there may be a wish even for impossibles, e.g. for immortality. And wish may relate to things that could in no way be brought about by one's own efforts, e.g. that a particular actor or athlete should win in a competition; but no one chooses such things, but only the things that he thinks could be brought about by his own efforts. Again, wish relates rather to the end, choice to the means; for instance, we wish to be healthy, but we choose the acts which will make us healthy, and we wish to be happy and say we do, but we cannot well say we choose to be so; for, in general, choice seems to relate to the things that are in our own power.

For this reason, too, it cannot be opinion; for opinion is thought to relate to all kinds of things, no less to eternal things and impossible things than to things in our own power; and it is distinguished by its falsity or truth, not by its badness or goodness, while choice is distinguished rather by these.

Now with opinion in general perhaps no one even says it is identical. But it is not identical even with any kind of opinion; for by choosing what is good or bad we are men of a certain character, which we are not by holding certain opinions. And we choose to get or avoid something good or bad, but we have opinions about what a thing is or whom it is good for or how it is good for him; we can hardly be said to opine to get or avoid anything. And choice is praised for being related to the right object rather than for being rightly related to it, opinion for being truly related to its object. And we choose what we best know to be good, but we opine what we do not quite know; and it is not the same people that are thought to make the best choices and to have the best opinions, but some are thought to have fairly good opinions, but by reason of vice to choose what they should not. If opinion precedes choice or accompanies it, that makes no difference; for it is not this that we are considering, but whether it is identical with some kind of opinion.

What, then, or what kind of thing is it, since it is none of the things we have mentioned? It seems to be voluntary, but not all that is voluntary to be an object of choice. Is it, then, what has been decided on by previous deliberation? At any rate choice involves a rational principle and thought. Even the name seems to suggest that it is what is chosen before other things.

Chapter 3: The nature of deliberation and its objects: choice is the deliberate desire of things in our own power.

Do we deliberate about everything, and is everything a possible subject of deliberation, or is deliberation impossible about some things? We ought presumably to call not what a fool or a madman would deliberate about, but what a sensible man would deliberate about, a subject of deliberation. Now about eternal things no one deliberates, e.g. about the material universe or the incommensurability of the diagonal and the side of a square. But no more do we deliberate about the things that involve movement but always happen in the same way, whether of necessity or by nature or from any other cause, e.g. the solstices and the risings of the stars; nor about things that happen now in one way, now in another, e.g. droughts and rains; nor about chance events, like the finding of treasure. But we do not deliberate even about all human affairs; for instance, no Spartan deliberates about the best constitution for the Scythians. For none of these things can be brought about by our own efforts.

We deliberate about things that are in our power and can be done; and these are in fact what is left. For nature, necessity, and chance are thought to be causes, and also reason and everything that depends on man. Now every class of men deliberates about the things that can be done by their own efforts. And in the case of exact and self-contained sciences there is no deliberation, e.g. about the letters of the alphabet (for we have no doubt how they should be written); but the things that are brought about by our own efforts, but not always in the same way, are the things about which we deliberate, e.g. questions of medical treatment or of money-making. And we do so more in the case of the art of navigation than in that of gymnastics, inasmuch as it has been less exactly worked out, and again about other things in the same ratio, and more also in the case of the arts than in that of the sciences; for we have more doubt about the former. Deliberation is concerned with things that happen in a certain way for the most part, but in which the event is obscure, and with things in which it is indeterminate. We call in others to aid us in deliberation on important questions, distrusting ourselves as not being equal to deciding.

We deliberate not about ends but about means. For a doctor does not deliberate whether he shall heal, nor an orator whether he shall persuade, nor a statesman whether he shall produce law and order, nor does any one else deliberate about his end. They assume the end and consider how and by what means it is to be attained; and if it seems to be produced by several means they consider by which it is most easily and best produced, while if it is achieved by one only they consider how it will be achieved by this and by what means this will be achieved, till they come to the first cause, which in the order of discovery is last. For the person who deliberates seems to investigate and analyse in the way described as though he were analysing a geometrical construction (not all investigation appears to be deliberation—for instance mathematical investigations—but all deliberation is investigation), and what is last in the order of analysis seems to be first in the order of becoming. And if we come on an impossibility, we give up the search, e.g. if we need money and this cannot be got; but if a thing appears possible we try to do it. By 'possible' things I mean things that might be brought about by our own efforts; and these in a sense include things that can be brought about by the efforts of our friends, since the moving principle is in ourselves. The subject of investigation is sometimes the instruments, sometimes the use of them; and similarly in the other cases—sometimes the means, sometimes the mode of using it or the means of bringing it about. It seems, then, as has been said, that man is a moving principle of actions; now deliberation is about the things to be done by the agent himself, and actions are for the sake of things other than themselves. For the end cannot be a subject of deliberation, but only the means; nor indeed can the particular facts be a subject of it, as whether this is bread or has been baked as it should; for these are matters of perception. If we are to be always deliberating, we shall have to go on to infinity.

The same thing is deliberated upon and is chosen, except that the object of choice is already determinate, since it is that which has been decided upon as a result of deliberation that is the object of choice. For every one ceases to inquire how he is to act when he has brought the moving principle back to himself and to the ruling part of himself; for this is what chooses. This is plain also from the ancient constitutions, which Homer represented; for the kings announced their choices to the people. The object of choice being one of the things in our own power which is desired after deliberation, choice will be deliberate desire of things in our own power; for when we have decided as a result of deliberation, we desire in accordance with our deliberation.

We may take it, then, that we have described choice in outline, and stated the nature of its objects and the fact that it is concerned with means.

Chapter 4: The object of rational wish is the end, i.e. the good or the apparent good.

That wish is for the end has already been stated; some think it is for the good, others for the apparent good. Now those who say that the good is the object of wish must admit in consequence that that which the man who does not choose aright wishes for is not an object of wish (for if it is to be so, it must also be good; but it was, if it so happened, bad); while those who say the apparent good is the object of wish must admit that there is no natural object of wish, but only what seems good to each man. Now different things appear good to different people, and, if it so happens, even contrary things.

If these consequences are unpleasing, are we to say that absolutely and in truth the good is the object of wish, but for each person the apparent good; that that which is in truth an object of wish is an object of wish to the good man, while any chance thing may be so the bad man, as in the case of bodies also the things that are in truth wholesome are wholesome for bodies which are in good condition, while for those that are diseased other things are wholesome—or bitter or sweet or hot or heavy, and so on; since the good man judges each class of things rightly, and in each the truth appears to him? For each state of character has its own ideas of the noble and the pleasant, and perhaps the good man differs from others most by seeing the truth in each class of things, being as it were the norm and measure of them. In most things the error seems to be due to pleasure; for it appears a good when it is not. We therefore choose the pleasant as a good, and avoid pain as an evil.

Chapter 5: We are responsible for bad as well as for good actions.

The end, then, being what we wish for, the means what we deliberate about and choose, actions concerning means must be according to choice and voluntary. Now the exercise of the virtues is concerned with means. Therefore virtue also is in our own power, and so too vice. For where it is in our power to act it is also in our power not to act, and vice versa; so that, if to act, where this is noble, is in our power, not to act, which will be base, will also be in our power, and if not to act, where this is noble, is in our power, to act, which will be base, will also be in our power. Now if it is in our power to do noble or base acts, and likewise in our power not to do them, and this was what being good or bad meant, then it is in our power to be virtuous or vicious.

The saying that 'no one is voluntarily wicked nor involuntarily happy' seems to be partly false and partly true; for no one is involuntarily happy, but wickedness is voluntary. Or else we shall have to dispute what has just been said, at any rate, and deny that man is a moving principle or begetter of his actions as of children. But if these facts are evident and we cannot refer actions to moving principles other than those in ourselves, the acts whose moving principles are in us must themselves also be in our power and voluntary.

Witness seems to be borne to this both by individuals in their private capacity and by legislators themselves; for these punish and take vengeance on those who do wicked acts (unless they have acted under compulsion or as a result of ignorance for which they are not themselves responsible), while they honour those who do noble acts, as though they meant to encourage the latter and deter the former. But no one is encouraged to do the things that are neither in our power nor voluntary; it is assumed that there is no gain in being persuaded not to be hot or in pain or hungry or the like, since we shall experience these feelings none the less. Indeed, we punish a man for his very ignorance, if he is thought responsible for the ignorance, as when penalties are doubled in the case of drunkenness; for the moving principle is in the man himself, since he had the power of not getting drunk and his getting drunk was the cause of his ignorance. And we punish those who are ignorant of anything in the laws that they ought to know and that is not difficult, and so too in the case of anything else that they are thought to be ignorant of through carelessness; we assume that it is in their power not to be ignorant, since they have the power of taking care.

But perhaps a man is the kind of man not to take care. Still they are themselves by their slack lives responsible for becoming men of that kind, and men make themselves responsible for being unjust or self-indulgent, in the one case by cheating and in the other by spending their time in drinking bouts and the like; for it is activities exercised on particular objects that make the corresponding character. This is plain from the case of people training for any contest or action; they practise the activity the whole time. Now not to know that it is from the exercise of activities on particular objects that states of character are produced is the mark of a thoroughly senseless person. Again, it is irrational to suppose that a man who acts unjustly does not wish to be unjust or a man who acts self-indulgently to be self-indulgent. But if without being ignorant a man does the things which will make him unjust, he will be unjust voluntarily. Yet it does not follow that if he wishes he will cease to be unjust and will be just. For neither does the man who is ill become well on those terms. We may suppose a case in which he is ill voluntarily, through living incontinently and disobeying his doctors. In that case it was then open to him not to be ill, but not now, when he has thrown away his chance, just as when you have let a stone go it is too late to recover it; but yet it was in your power to throw it, since the moving principle was in you. So, too, to the unjust and to the self-indulgent man it was open at the beginning not to become men of this kind, and so they are unjust and selfindulgent voluntarily; but now that they have become so it is not possible for them not to be so.

But not only are the vices of the soul voluntary, but those of the body also for some men, whom we accordingly blame; while no one blames those who are ugly by nature, we blame those who are so owing to want of exercise and care. So it is, too, with respect to weakness and infirmity; no one would reproach a man blind from birth or by disease or from a blow, but rather pity him, while every one would blame a man who was blind from drunkenness or some other form of self-indulgence. Of vices of the body, then, those in our own power are blamed, those not in our power are not. And if this be so, in the other cases also the vices that are blamed must be in our own power.

Now some one may say that all men desire the apparent good, but have no control over the appearance, but the end appears to each man in a form answering to his character. We reply that if each man is somehow responsible for his state of mind, he will also be himself somehow responsible for the appearance; but if not, no one is responsible for his own evildoing, but every one does evil acts through ignorance of the end, thinking that by these he will get what is best, and the aiming at the end is not self-chosen but one must be born with an eye, as it were, by which to judge rightly and choose what is truly good, and he is well endowed by nature who is well endowed with this. For it is what is greatest and most noble, and what we cannot get or learn from another, but must have just such as it was when given us at birth, and to be well and nobly endowed with this will be perfect and true excellence of natural endowment. If this is true, then, how will virtue be more voluntary than vice? To both men alike, the good and the bad, the end appears and is fixed by nature or however it may be, and it is by referring everything else to this that men do whatever they do.

Whether, then, it is not by nature that the end appears to each man such as it does appear, but something also depends on him, or the end is natural but because the good man adopts the means voluntarily virtue is voluntary, vice also will be none the less voluntary; for in the case of the bad man there is equally present that which depends on himself in his actions even if not in his end. If, then, as is asserted, the virtues are voluntary (for we are ourselves somehow partly responsible for our states of character, and it is by being persons of a certain kind that we assume the end to be so and so), the vices also will be voluntary; for the same is true of them.

With regard to the virtues in general we have stated their genus in outline, viz. that they are means and that they are states of character, and that they tend, and by their own nature, to the doing of the acts by which they are produced, and that they are in our power and voluntary, and act as the right rule prescribes. But actions and states of character are not voluntary in the same way; for we are masters of our actions from the beginning right to the end, if we know the particular facts, but though we control the beginning of our states of character the gradual progress is not obvious any more than it is in illnesses; because it was in our power, however, to act in this way or not in this way, therefore the states are voluntary.

Let us take up the several virtues, however, and say which they are and what sort of things they are concerned with and how they are concerned with them; at the same time it will become plain how many they are. And first let us speak of courage.

Chapter 6: Courage concerned with the feelings of fear and confidence—strictly speaking, with the fear of death in battle.

That it is a mean with regard to feelings of fear and confidence has already been made evident; and plainly the things we fear are terrible things, and these are, to speak without qualification, evils; for which reason people even define fear as expectation of evil. Now we fear all evils, e.g. disgrace, poverty, disease, friendlessness, death, but the brave man is not thought to be concerned with all; for to fear some things is even right and noble, and it is base not to fear them—e.g. disgrace; he who fears this is good and modest, and he who does not is shameless. He is, however, by some people called brave, by a transference of the word to a new meaning; for he has in him something which is like the brave man, since the brave man also is a fearless person. Poverty and disease we perhaps ought not to fear, nor in general the things that do not proceed from vice and are not due to a man himself. But not even the man who is fearless of these is brave. Yet we apply the word to him also in virtue of a similarity; for some who in the dangers of war are cowards are liberal and are confident in face of the loss of money. Nor is a man a coward if he fears insult to his wife and children or envy or anything of the kind; nor brave if he is confident when he is about to be flogged. With what sort of terrible things, then, is the brave man concerned? Surely with the greatest; for no one is more likely than he to stand his ground against what is awe-inspiring. Now death is the most terrible of all things; for it is the end, and nothing is thought to be any longer either good or bad for the dead. But the brave man would not seem to be concerned even with death in all circumstances, e.g. at sea or in disease. In what circumstances, then? Surely in the noblest. Now such deaths are those in battle; for these take place in the greatest and noblest danger. And these are correspondingly honoured in city-states and at the courts of monarchs. Properly, then, he will be called brave who is fearless in face of a noble death, and of all emergencies that involve death; and the emergencies of war are in the highest degree of this kind. Yet at sea also, and in disease, the brave man is fearless, but not in the same way as the seaman; for he has given up hope of safety, and is disliking the thought of death in this shape, while they are hopeful because of their experience. At the same time, we show courage in situations where there is the opportunity of showing prowess or where death is noble; but in these forms of death neither of these conditions is fulfilled.

Chapter 7: The motive of courage is the sense of honour: characteristics of the opposite vices, cowardice and rashness.

What is terrible is not the same for all men; but we say there are things terrible even beyond human strength. These, then, are terrible to every one—at least to every sensible man; but the terrible things that are not beyond human strength differ in magnitude and degree, and so too do the things that inspire confidence. Now the brave man is as dauntless as man may be. Therefore, while he will fear even the things that are not beyond human strength, he will face them as he ought and as the rule directs, for honour's sake; for this is the end of virtue. But it is possible to fear these more, or less, and again to fear things that are not terrible as if they were. Of the faults that are committed one consists in fearing what one should not, another in fearing as we should not, another in fearing when we should not, and so on; and so too with respect to the things that inspire confidence. The man, then, who faces and who fears the right things and from the right motive, in the right way and from the right time, and who feels confidence under the corresponding conditions, is brave; for the brave man feels and acts according to the merits of the case and in whatever way the rule directs. Now the end of every activity is conformity to the corresponding state of character. This is true, therefore, of the brave man as well as of others. But courage is noble. Therefore the end also is noble; for each thing is defined by its end. Therefore it is for a noble end that the brave man endures and acts as courage directs.

Of those who go to excess he who exceeds in fearlessness has no name (we have said previously that many states of character have no names), but he would be a sort of madman or insensible person if he feared nothing, neither earthquakes nor the waves, as they say the Celts do not; while the man who exceeds in confidence about what really is terrible is rash. The rash man, however, is also thought to be boastful and only a pretender to courage; at all events, as the brave man is with regard to what is terrible, so the rash man wishes to appear; and so he imitates him in situations where he can. Hence also most of them are a mixture of rashness and cowardice; for, while in these situations they display confidence, they do not hold their ground against what is really terrible. The man who exceeds in fear is a coward; for he fears both what he ought not and as he ought not, and all the similar characterizations attach to him. He is lacking also in confidence; but he is more conspicuous for his excess of fear in painful situations. The coward, then, is a despairing sort of person; for he fears everything. The brave man, on the other hand, has the opposite disposition; for confidence is the mark of a hopeful disposition. The coward, the rash man, and the brave man, then, are concerned with the same objects but are differently disposed towards them; for the first two exceed and fall short, while the third holds the middle, which is the right, position; and rash men are precipitate, and wish for dangers beforehand but draw back when they are in them, while brave men are keen in the moment of action, but quiet beforehand.

As we have said, then, courage is a mean with respect to things that inspire confidence or fear, in the circumstances that have been stated; and it chooses or endures things because it is noble to do so, or because it is base not to do so. But to die to escape from poverty or love or anything painful is not the mark of a brave man, but rather of a coward; for it is softness to fly from what is troublesome, and such a man endures death not because it is noble but to fly from evil.

Chapter 8: Five kinds of courage improperly so called.

Courage, then, is something of this sort, but the name is also applied to five other kinds.

First comes the courage of the citizen-soldier; for this is most like true courage. Citizen-soldiers seem to face dangers because of the penalties imposed by the laws and the reproaches they would otherwise incur, and because of the honours they win by such action; and therefore those peoples seem to be bravest among whom cowards are held in dishonour and brave men in honour. This is the kind of courage that Homer depicts, e.g. in Diomede and in Hector:

First will Polydamas be to heap reproach on me then; and

For Hector one day 'mid the Trojans shall utter his vaulting harangue:

Afraid was Tydeides, and fled from my face.

This kind of courage is most like to that which we described earlier, because it is due to virtue; for it is due to shame and to desire of a noble object (i.e. honour) and avoidance of disgrace, which is ignoble. One might rank in the same class even those who are compelled by their rulers; but they are inferior, inasmuch as they do what they do not from shame but from fear, and to avoid not what is disgraceful but what is painful; for their masters compel them, as Hector does:

But if I shall spy any dastard that cowers far from the fight,
Vainly will such an one hope to escape from the dogs.

And those who give them their posts, and beat them if they retreat, do the same, and so do those who draw them up with trenches or something of the sort behind them; all of these apply compulsion. But one ought to be brave not under compulsion but because it is noble to be so.

(2) Experience with regard to particular facts is also thought to be courage; this is indeed the reason why Socrates thought courage was knowledge. Other people exhibit this quality in other dangers, and professional soldiers exhibit it in the dangers of war; for there seem to be many empty alarms in war, of which these have had the most comprehensive experience; therefore they seem brave, because the others do not know the nature of the facts. Again, their experience makes them most capable in attack and in defence, since they can use their arms and have the kind that are likely to be best both for attack and for defence; therefore they fight like armed men against unarmed or like trained athletes against amateurs; for in such contests too it is not the bravest men that fight best, but those who are strongest and have their bodies in the best condition. Professional soldiers turn cowards, however, when the danger puts too great a strain on them and they are inferior in numbers and equipment; for they are the first to fly, while citizen-forces die at their posts, as in fact happened at the temple of Hermes. For to the latter flight is disgraceful and death is preferable to safety on those terms; while the former from the very beginning faced the danger on the assumption that they were stronger, and when they know the facts they fly, fearing death more than disgrace; but the brave man is not that sort of person.

(3) Passion also is sometimes reckoned as courage; those who act from passion, like wild beasts rushing at those who have wounded them, are thought to be brave, because brave men also are passionate; for passion above all things is eager to rush on danger, and hence Homer's 'put strength into his passion' and 'aroused their spirit and passion and 'hard he breathed panting' and 'his blood boiled'. For all such expressions seem to indicate the stirring and onset of passion. Now brave men act for honour's sake, but passion aids them; while wild beasts act under the influence of pain; for they attack because they have been wounded or because they are afraid, since if they are in a forest they do not come near one. Thus they are not

brave because, driven by pain and passion, they rush on danger without foreseeing any of the perils, since at that rate even asses would be brave when they are hungry; for blows will not drive them from their food; and lust also makes adulterers do many daring things. (Those creatures are not brave, then, which are driven on to danger by pain or passion.) The 'courage' that is due to passion seems to be the most natural, and to be courage if choice and motive be added.

Men, then, as well as beasts, suffer pain when they are angry, and are pleased when they exact their revenge; those who fight for these reasons, however, are pugnacious but not brave; for they do not act for honour's sake nor as the rule directs, but from strength of feeling; they have, however, something akin to courage.

(4) Nor are sanguine people brave; for they are confident in danger only because they have conquered often and against many foes. Yet they closely resemble brave men, because both are confident; but brave men are confident for the reasons stated earlier, while these are so because they think they are the strongest and can suffer nothing. (Drunken men also behave in this way; they become sanguine). When their adventures do not succeed, however, they run away; but it was the mark of a brave man to face things that are, and seem, terrible for a man, because it is noble to do so and disgraceful not to do so. Hence also it is thought the mark of a braver man to be fearless and undisturbed in sudden alarms than to be so in those that are foreseen; for it must have proceeded more from a state of character, because less from preparation; acts that are foreseen may be chosen by calculation and rule, but sudden actions must be in accordance with one's state of character.

(5) People who are ignorant of the danger also appear brave, and they are not far removed from those of a sanguine temper, but are inferior inasmuch as they have no self-reliance while these have. Hence also the sanguine hold their ground for a time; but those who have been deceived about the facts fly if they know or suspect that these are different from what they supposed, as happened to the Argives when they fell in with the Spartans and took them for Sicyonians.

We have, then, described the character both of brave men and of those who are thought to be brave.

Chapter 9: Relation of courage to pain and pleasure.

Though courage is concerned with feelings of confidence and of fear, it is not concerned with both alike, but more with the things that inspire fear; for he who is undisturbed in face of these and bears himself as he should towards these is more truly brave than the man who does so towards the things that inspire confidence. It is for facing what is painful, then, as has been said, that men are called brave. Hence also courage involves pain, and is justly praised; for it is harder to face what is painful than to abstain from what is pleasant.

Yet the end which courage sets before it would seem to be pleasant, but to be concealed by the attending circumstances, as happens also in athletic contests; for the end at which boxers aim is pleasant—the crown and the honours—but the blows they take are distressing to flesh and blood, and painful, and so is their whole exertion; and because the blows and the exertions are many the end, which is but small, appears to have nothing pleasant in it. And so, if the case of courage is similar, death and wounds will be painful to the brave man and against his will, but he will face them because it is noble to do so or because it is base not to do so. And the more he is possessed of virtue in its entirety and the happier he is, the more he will be pained at the thought of death; for life is best worth living for such a man, and he is knowingly losing the greatest goods, and this is painful. But he is none the less brave, and perhaps all the more so, because he chooses noble deeds of war at that cost. It is not the case, then, with all the virtues that the exercise of them is pleasant, except in so far as it reaches its end. But it is quite possible that the best soldiers may be not men of this sort but those who are less brave but have no other good; for these are ready to face danger, and they sell their life for trifling gains.

So much, then, for courage; it is not difficult to grasp its nature in outline, at any rate, from what has been said.

Chapter 10: Temperance is limited to certain pleasures of touch.

After courage let us speak of temperance; for these seem to be the virtues of the irrational parts. We have said that temperance is a mean with regard to pleasures (for it is less, and not in the same way, concerned with pains); self-indulgence also is manifested in the same sphere. Now, therefore, let us determine with what sort of pleasures they are concerned. We may assume the distinction between bodily pleasures and those of the soul, such as love of honour and love of learning; for the lover of each of these delights in that of which he is a lover, the body being in no way affected, but rather the mind; but men who are concerned with such pleasures are called neither temperate nor self-indulgent. Nor, again, are those who are concerned with the other pleasures that are not bodily; for those who are fond of hearing and telling stories and who spend their days on anything that turns up are called gossips, but not self-indulgent, nor are those who are pained at the loss of money or of friends.

Temperance must be concerned with bodily pleasures, but not all even of these; for those who delight in objects of vision, such as colours and shapes and painting, are called neither temperate nor self-indulgent; yet it would seem possible to delight even in these either as one should or to excess or to a deficient degree.

And so too is it with objects of hearing; no one calls those who delight extravagantly in music or acting self-indulgent, nor those who do so as they ought temperate.

Nor do we apply these names to those who delight in odour, unless it be incidentally; we do not call those self-indulgent who delight in the odour of apples or roses or incense, but rather those who delight in the odour of unguents or of dainty dishes; for self-indulgent people delight in these because these remind them of the objects of their appetite. And one may see even other people, when they are hungry, delighting in the smell of food; but to delight in this kind of thing is the mark of the self-indulgent man; for these are objects of appetite to him.

Nor is there in animals other than man any pleasure connected with these senses, except incidentally. For dogs do not delight in the scent of hares, but in the eating of them, but the scent told them the hares were there; nor does the lion delight in the lowing of the ox, but in eating it; but he perceived by the lowing that it was near, and therefore appears to delight in the lowing; and similarly he does not delight because he sees 'a stag or a wild goat', but because he is going to make a meal of it. Temperance and self-indulgence, however, are concerned with the kind of pleasures that the other animals share in, which therefore appear slavish and brutish; these are touch and taste. But even of taste they appear to make little or no use; for the business of taste is the discriminating of flavours, which is done by winetasters and people who season dishes; but they hardly take pleasure in making these discriminations, or at least self-indulgent people do not, but in the actual enjoyment, which in all cases comes through touch, both in the case of food and in that of drink and in that of sexual intercourse. This is why a certain gourmand prayed that his throat might become longer than a crane's, implying that it was the contact that he took pleasure in. Thus the sense with which self-indulgence is connected is the most widely shared of the senses; and self-indulgence would seem to be justly a matter of reproach, because it attaches to us not as men but as animals. To delight in such things, then, and to love them above all others, is brutish. For even of the pleasures of touch the most liberal have been eliminated, e.g. those produced in the gymnasium by rubbing and by the consequent heat; for the contact characteristic of the self-indulgent man does not affect the whole body but only certain parts.

Chapter 11: Characteristics of temperance and its opposites, self-indulgence and 'insensibility'.

Of the appetites some seem to be common, others to be peculiar to individuals and acquired, e.g. the appetite for food is natural, since every one who is without it craves for food or drink, and sometimes for both, and for love also (as Homer says) if he is young and lusty; but not every one craves for this or that kind of nourishment or love, nor for the same things. Hence such craving appears to be our very own. Yet it has of course something natural about it; for different things are pleasant to different kinds of people, and some things are more pleasant to every one than chance objects. Now in the natural appetites few go wrong, and only in one direction, that of excess; for to eat or drink whatever offers itself till one is surfeited is to exceed the natural amount, since natural appetite is the replenishment of one's deficiency. Hence these people are called belly-gods, this implying that they fill their belly beyond what is right. It is people of entirely slavish character that become like this. But with regard to the pleasures peculiar to individuals many people go wrong and in many ways. For while the people who are 'fond of so and so' are so called because they delight either in the wrong things, or more than most people do, or in the wrong way, the self-indulgent exceed in all three ways; they both delight in some things that they ought not to delight in (since they are hateful), and if one ought to delight in some of the things they delight in, they do so more than one ought and than most men do.

Plainly, then, excess with regard to pleasures is self-indulgence and is culpable; with regard to pains one is not, as in the case of courage, called temperate for facing them or self-indulgent for not doing so, but the selfindulgent man is so called because he is pained more than he ought at not getting pleasant things (even his pain being caused by pleasure), and the temperate man is so called because he is not pained at the absence of what is pleasant and at his abstinence from it.

The self-indulgent man, then, craves for all pleasant things or those that are most pleasant, and is led by his appetite to choose these at the cost of everything else; hence he is pained both when he fails to get them and when he is merely craving for them (for appetite involves pain); but it seems absurd to be pained for the sake of pleasure. People who fall short with regard to pleasures and delight in them less than they should are hardly found; for such insensibility is not human. Even the other animals distinguish different kinds of food and enjoy some and not others; and if there is any one who finds nothing pleasant and nothing more attractive than anything else, he must be something quite different from a man; this sort of person has not received a name because he hardly occurs. The temperate man occupies a middle position with regard to these objects. For he neither enjoys the things that the self-indulgent man enjoys most—but rather dislikes them—nor in general the things that he should not, nor anything of this sort to excess, nor does he feel pain or craving when they are absent, or does so only to a moderate degree, and not more than he should, nor when he should not, and so on; but the things that, being pleasant, make for health or for good condition, he will desire moderately and as he should, and also other pleasant things if they are not hindrances to these ends, or contrary to what is noble, or beyond his means. For he who neglects these conditions loves such pleasures more than they are worth, but the temperate man is not that sort of person, but the sort of person that the right rule prescribes.

Chapter 12: Self-indulgence more voluntary than cowardice: comparison of the self-indulgent man to the spoilt child.

Self-indulgence is more like a voluntary state than cowardice. For the former is actuated by pleasure, the latter by pain, of which the one is to be chosen and the other to be avoided; and pain upsets and destroys the nature of the person who feels it, while pleasure does nothing of the sort. Therefore self-indulgence is more voluntary. Hence also it is more a matter of reproach; for it is easier to become accustomed to its objects, since there are many things of this sort in life, and the process of habituation to them is free from danger, while with terrible objects the reverse is the case. But cowardice would seem to be voluntary in a different degree from its particular manifestations; for it is itself painless, but in these we are upset by pain, so that we even throw down our arms and disgrace ourselves in other ways; hence our acts are even thought to be done under compulsion. For the self-indulgent man, on the other hand, the particular acts are voluntary (for he does them with craving and desire), but the whole state is less so; for no one craves to be self-indulgent.

The name self-indulgence is applied also to childish faults; for they bear a certain resemblance to what we have been considering. Which is called after which, makes no difference to our present purpose; plainly, however, the later is called after the earlier. The transference of the name seems not a bad one; for that which desires what is base and which develops quickly ought to be kept in a chastened condition, and these characteristics belong above all to appetite and to the child, since children in fact live at the beck and call of appetite, and it is in them that the desire for what is pleasant is strongest. If, then, it is not going to be obedient and subject to the ruling principle, it will go to great lengths; for in an irrational being the desire for pleasure is insatiable even if it tries every source of gratification, and the exercise of appetite increases its innate force, and if appetites are strong and violent they even expel the power of calculation. Hence they should be moderate and few, and should in no way oppose the rational principle—and this is what we call an obedient and chastened state—and as the child should live according to the direction of his tutor, so the appetitive element should live according to rational principle. Hence the appetitive element in a temperate man should harmonize with the rational principle; for the noble is the mark at which both aim, and the temperate man craves for the things be ought, as he ought, as when he ought; and when he ought; and this is what rational principle directs.

Here we conclude our account of temperance.

Book IV: Moral Virtue

Chapter I: Liberality, prodigality, meanness.

LET us speak next of liberality. It seems to be the mean with regard to wealth; for the liberal man is praised not in respect of military matters, nor of those in respect of which the temrate man is praised, nor of judicial decisions, but with regard to the giving and taking of wealth, and especially in respect of giving. Now by 'wealth' we mean all the things whose value is measured by money. Further, prodigality and meanness are excesses and defects with regard to wealth; and meanness we always impute to those who care more than they ought for wealth, but we sometimes apply the word 'prodigality' in a complex sense; for we call those men prodigals who are incontinent and spend money on self-indulgence. Hence also they are thought the poorest characters; for they combine more vices than one. Therefore the application of the word to them is not its proper use; for a 'prodigal' means a man who has a single evil quality, that of wasting his substance; since a prodigal is one who is being ruined by his own fault, and the wasting of substance is thought to be a sort of ruining of oneself, life being held to depend on possession of substance.

This, then, is the sense in which we take the word 'prodigality'. Now the things that have a use may be used either well or badly; and riches is a useful thing; and everything is used best by the man who has the virtue concerned with it; riches, therefore, will be used best by the man who has the virtue concerned with wealth; and this is the liberal man. Now spending and giving seem to be the using of wealth; taking and keeping rather the possession of it. Hence it is more the mark of the liberal man to give to the right people than to take from the right sources and not to take from the wrong. For it is more characteristic of virtue to do good than to have good done to one, and more characteristic to do what is noble than not to do what is base; and it is not hard to see that giving implies doing good and doing what is noble, and taking implies having good done to one or not acting basely. And gratitude is felt towards him who gives, not towards him who does not take, and praise also is bestowed more on him. It is easier, also, not to take than to give; for men are apter to give away their own too little than to take what is another's. Givers, too, are called liberal; but those who do not take are not praised for liberality but rather for justice; while those who take are hardly praised at all. And the liberal are almost the most loved of all virtuous characters, since they are useful; and this depends on their giving.

Now virtuous actions are noble and done for the sake of the noble. Therefore the liberal man, like other virtuous men, will give for the sake of the noble, and rightly; for he will give to the right people, the right amounts, and at the right time, with all the other qualifications that accompany right giving; and that too with pleasure or without pain; for that which is virtuous is pleasant or free from pain—least of all will it be painful. But he who gives to the wrong people or not for the sake of the noble but for some other cause, will be called not liberal but by some other name. Nor is he liberal who gives with pain; for he would prefer the wealth to the noble act, and this is not characteristic of a liberal man. But no more will the liberal man take from wrong sources; for such taking is not characteristic of the man who sets no store by wealth. Nor will he be a ready asker; for it is not characteristic of a man who confers benefits to accept them lightly. But he will take from the right sources, e.g. from his own possessions, not as something noble but as a necessity, that he may have something to give. Nor will he neglect his own property, since he wishes by means of this to help others. And he will refrain from giving to anybody and everybody, that he may have something to give to the right people, at the right time, and where it is noble to do so. It is highly characteristic of a liberal man also to go to excess in giving, so that he leaves too little for himself; for it is the nature of a liberal man not to look to himself. The term 'liberality' is used relatively to a man's substance; for liberality resides not in the multitude of the gifts but in the state of character of the giver, and this is relative to the giver's substance. There is therefore nothing to prevent the man who gives less from being the more liberal man, if he has less to give those are thought to be more liberal who have not made their wealth but inherited it; for in the first place they have no experience of want, and secondly all men are fonder of their own productions, as are parents and poets. It is not easy for the liberal man to be rich, since he is not apt either at taking or at keeping, but at giving away, and does not value wealth for its own sake but as a means to giving. Hence comes the charge that is brought against fortune, that those who deserve riches most get it least. But it is not unreasonable that it should turn out so; for he cannot have wealth, any more than anything else, if he does not take pains to have it. Yet he will not give to the wrong people nor at the wrong time, and so on; for he would no longer be acting in accordance with liberality, and if he spent on these objects he would have nothing to spend on the right objects. For, as has been said, he is liberal who spends according to his substance and on the right objects; and he who exceeds is prodigal. Hence we do not call despots prodigal; for it is thought not easy for them to give and spend beyond the amount of their possessions. Liberality, then, being a mean with regard to giving and taking of wealth, the liberal man will both give and spend the right amounts and on the right objects, alike in small things and in great, and that with pleasure; he will also take the right amounts and from the right sources. For, the virtue being a mean with regard to both, he will do both as he ought; since this sort of taking accompanies proper giving, and that which is not of this sort is contrary to it, and accordingly the giving and taking that accompany each other are present together in the same man, while the contrary kinds evidently are not. But if he happens to spend in a manner contrary to what is right and noble, he will be pained, but moderately and as he ought; for it is the mark of virtue both to be pleased and to be pained at the right objects and in the right way. Further, the liberal man is easy to deal with in money matters; for he can be got the better of, since he sets no store by money, and is more annoyed if he has not spent something that he ought than pained if he has spent something that he ought not, and does not agree with the saying of Simonides.

The prodigal errs in these respects also; for he is neither pleased nor pained at the right things or in the right way; this will be more evident as we go on. We have said that prodigality and meanness are excesses and deficiencies, and in two things, in giving and in taking; for we include spending under giving. Now prodigality exceeds in giving and not taking, while meanness falls short in giving, and exceeds in taking, except in small things.

The characteristics of prodigality are not often combined; for it is not easy to give to all if you take from none; private persons soon exhaust their substance with giving, and it is to these that the name of prodigals is applied—though a man of this sort would seem to be in no small degree better than a mean man. For he is easily cured both by age and by poverty, and thus he may move towards the middle state. For he has the characteristics of the liberal man, since he both gives and refrains from taking, though he does neither of these in the right manner or well. Therefore if he were brought to do so by habituation or in some other way, he would be liberal; for he will then give to the right people, and will not take from the wrong sources. This is why he is thought to have not a bad character; it is not the mark of a wicked or ignoble man to go to excess in giving and not taking, but only of a foolish one. The man who is prodigal in this way is thought much better than the mean man both for the aforesaid reasons and because he benefits many while the other benefits no one, not even himself.

But most prodigal people, as has been said, also take from the wrong sources, and are in this respect mean. They become apt to take because they wish to spend and cannot do this easily; for their possessions soon run short. Thus they are forced to provide means from some other source. At the same time, because they care nothing for honour, they take recklessly and from any source; for they have an appetite for giving, and they do not mind how or from what source. Hence also their giving is not liberal; for it is not noble, nor does it aim at nobility, nor is it done in the right way; sometimes they make rich those who should be poor, and will give nothing to people of respectable character, and much to flatterers or those who provide them with some other pleasure. Hence also most of them are self-indulgent; for they spend lightly and waste money on their indulgences, and incline towards pleasures because they do not live with a view to what is noble.

The prodigal man, then, turns into what we have described if he is left untutored, but if he is treated with care he will arrive at the intermediate and right state. But meanness is both incurable (for old age and every disability is thought to make men mean) and more innate in men than prodigality; for most men are fonder of getting money than of giving. It also extends widely, and is multiform, since there seem to be many kinds of meanness.

For it consists in two things, deficiency in giving and excess in taking, and is not found complete in all men but is sometimes divided; some men go to excess in taking, others fall short in giving. Those who are called by such names as 'miserly', 'close', 'stingy', all fall short in giving, but do not covet the possessions of others nor wish to get them. In some this is due to a sort of honesty and avoidance of what is disgraceful (for some seem, or at least profess, to hoard their money for this reason, that they may not some day be forced to do something disgraceful; to this class belong the cheeseparer and every one of the sort; he is so called from his excess of unwillingness to give anything); while others again keep their hands off the property of others from fear, on the ground that it is not easy, if one takes the property of others oneself, to avoid having one's own taken by them; they are therefore content neither to take nor to give.

Others again exceed in respect of taking by taking anything and from any source, e.g. those who ply sordid trades, pimps and all such people, and those who lend small sums and at high rates. For all of these take more than they ought and from wrong sources. What is common to them is evidently sordid love of gain; they all put up with a bad name for the sake of gain, and little gain at that. For those who make great gains but from wrong sources, and not the right gains, e.g. despots when they sack cities and spoil temples, we do not call mean but rather wicked, impious, and unjust. But the gamester and the footpad (and the highwayman) belong to the class of the mean, since they have a sordid love of gain. For it is for gain that both of them ply their craft and endure the disgrace of it, and the one faces the greatest dangers for the sake of the booty, while the other makes gain from his friends, to whom he ought to be giving. Both, then, since they are willing to make gain from wrong sources, are sordid lovers of gain; therefore all such forms of taking are mean.

And it is natural that meanness is described as the contrary of liberality; for not only is it a greater evil than prodigality, but men err more often in this direction than in the way of prodigality as we have described it.

So much, then, for liberality and the opposed vices.

Chapter 2: Magnificence, vulgarity, niggardliness.

It would seem proper to discuss magnificence next. For this also seems to be a virtue concerned with wealth; but it does not like liberality extend to all the actions that are concerned with wealth, but only to those that involve expenditure; and in these it surpasses liberality in scale. For, as the name itself suggests, it is a fitting expenditure involving largeness of scale. But the scale is relative; for the expense of equipping a trireme is not the same as that of heading a sacred embassy. It is what is fitting, then, in relation to the agent, and to the circumstances and the object. The man who in small or middling things spends according to the merits of the case is not called magnificent (e.g. the man who can say 'many a gift I gave the wanderer'), but only the man who does so in great things. For the magnificent man is liberal, but the liberal man is not necessarily magnificent. The deficiency of this state of character is called niggardliness, the excess vulgarity, lack of taste, and the like, which do not go to excess in the amount spent on right objects, but by showy expenditure in the wrong circumstances and the wrong manner; we shall speak of these vices later.

The magnificent man is like an artist; for he can see what is fitting and spend large sums tastefully. For, as we said at the begining, a state of character is determined by its activities and by its objects. Now the expenses of the magnificent man are large and fitting. Such, therefore, are also his results; for thus there will be a great expenditure and one that is fitting to its result. Therefore the result should be worthy of the expense, and the expense should be worthy of the result, or should even exceed it. And the magnificent man will spend such sums for honour's sake; for this is common to the virtues. And further he will do so gladly and lavishly; for nice calculation is a niggardly thing. And he will consider how the result can be made most beautiful and most becoming rather than for how much it can be produced and how it can be produced most cheaply. It is necessary, then, that the magnificent man be also liberal. For the liberal man also will spend what he ought and as he ought; and it is in these matters that the greatness implied in the name of the magnificent man—his bigness, as it were—is manifested, since liberality is concerned with these matters; and at an equal expense he will produce a more magnificent work of art. For a possession and a work of art have not the same excellence. The most valuable possession is that which is worth most, e.g. gold, but the most valuable work of art is that which is great and beautiful (for the contemplation of such a work inspires admiration, and so does magnificence); and a work has an excellence—viz. magnificence—which involves magnitude. Magnificence is an attribute of expenditures of the kind which we call honourable, e.g. those connected with the gods—votive offerings, buildings, and sacrifices—and similarly with any form of religious worship, and all those that are proper objects of public-spirited ambition, as when people think they ought to equip a chorus or a trireme, or entertain the city, in a brilliant way. But in all cases, as has been said, we have regard to the agent as well and ask who he is and what means he has; for the expenditure should be worthy of his means, and suit not only the result but also the producer. Hence a poor man cannot be magnificent, since he has not the means with which to spend large sums fittingly; and he who tries is a fool, since he spends beyond what can be expected of him and what is proper, but it is right expenditure that is virtuous. But great expenditure is becoming to those who have suitable means to start with, acquired by their own efforts or from ancestors or connexions, and to people of high birth or reputation, and so on; for all these things bring with them greatness and prestige. Primarily, then, the magnificent man is of this sort, and magnificence is shown in expenditures of this sort, as has been said; for these are the greatest and most honourable. Of private occasions of expenditure the most suitable are those that take place once for all, e.g. a wedding or anything of the kind, or anything that interests the whole city or the people of position in it, and also the receiving of foreign guests and the sending of them on their way, and gifts and counter-gifts; for the magnificent man spends not on himself but on public objects, and gifts bear some resemblance to votive offerings. A magnificent man will also furnish his house suitably to his wealth (for even a house is a sort of public ornament), and will spend by preference on those works that are lasting (for these are the most beautiful), and on every class of things he will spend what is becoming; for the same things are not suitable for gods and for men, nor in a temple and in a tomb. And since each expenditure may be great of its kind, and what is most magnificent absolutely is great expenditure on a great object, but what is magnificent here is what is great in these circumstances, and greatness in the work differs from greatness in the expense (for the most beautiful ball or bottle is magnificent as a gift to a child, but the price of it is small and mean),—therefore it is characteristic of the magnificent man, whatever kind of result he is producing, to produce it magnificently (for such a result is not easily surpassed) and to make it worthy of the expenditure.

Such, then, is the magnificent man; the man who goes to excess and is vulgar exceeds, as has been said, by spending beyond what is right. For on small objects of expenditure he spends much and displays a tasteless showiness; e.g. he gives a club dinner on the scale of a wedding banquet, and when he provides the chorus for a comedy he brings them on to the stage in purple, as they do at Megara. And all such things he will do not for honour's sake but to show off his wealth, and because he thinks he is admired for these things, and where he ought to spend much he spends little and where little, much. The niggardly man on the other hand will fall short in everything, and after spending the greatest sums will spoil the beauty of the result for a trifle, and whatever he is doing he will hesitate and consider how he may spend least, and lament even that, and think he is doing everything on a bigger scale than he ought.

These states of character, then, are vices; yet they do not bring disgrace because they are neither harmful to one's neighbour nor very unseemly.

Chapter 3: Pride, vanity, humility.

Pride seems even from its name to be concerned with great things; what sort of great things, is the first question we must try to answer. It makes no difference whether we consider the state of character or the man characterized by it. Now the man is thought to be proud who thinks himself worthy of great things, being worthy of them; for he who does so beyond his deserts is a fool, but no virtuous man is foolish or silly. The proud man, then, is the man we have described. For he who is worthy of little and thinks himself worthy of little is temperate, but not proud; for pride implies greatness, as beauty implies a goodsized body, and little people may be neat and well-proportioned but cannot be beautiful. On the other hand, he who thinks himself worthy of great things, being unworthy of them, is vain; though not every one who thinks himself worthy of more than he really is worthy of in vain. The man who thinks himself worthy of worthy of less than he is really worthy of is unduly humble, whether his deserts be great or moderate, or his deserts be small but his claims yet smaller. And the man whose deserts are great would seem most unduly humble, for what would he have done if they had been less? The proud man, then, is an extreme in respect of the greatness of his claims, but a mean in respect of the rightness of them; for he claims what is accordance with his merits, while the others go to excess or fall short.

If, then, he deserves and claims great things, and above all the great things, he will be concerned with one thing in particular. Desert is relative to external goods; and the greatest of these, we should say, is that which we render to the gods, and which people of position most aim at, and which is the prize appointed for the noblest deeds; and this is honour; that is surely the greatest of external goods. Honours and dishonours, therefore, are the objects with respect to which the proud man is as he should be. And even apart from argument it is with honour that proud men appear to be concerned; for it is honour that they chiefly claim, but in accordance with their deserts. The unduly humble man falls short both in comparison with his own merits and in comparison with the proud man's claims. The vain man goes to excess in comparison with his own merits, but does not exceed the proud man's claims.

Now the proud man, since he deserves most, must be good in the highest degree; for the better man always deserves more, and the best man most. Therefore the truly proud man must be good. And greatness in every virtue would seem to be characteristic of a proud man. And it would be most unbecoming for a proud man to fly from danger, swinging his arms by his sides, or to wrong another; for to what end should he do disgraceful acts, he to whom nothing is great? If we consider him point by point we shall see the utter absurdity of a proud man who is not good. Nor, again, would he be worthy of honour if he were bad; for honour is the prize of virtue, and it is to the good that it is rendered. Pride, then, seems to be a sort of crown of the virtues; for it makes them greater, and it is not found without them. Therefore it is hard to be truly proud; for it is impossible without nobility and goodness of character. It is chiefly with honours and dishonours, then, that the proud man is concerned; and at honours that are great and conferred by good men he will be moderately Pleased, thinking that he is coming by his own or even less than his own; for there can be no honour that is worthy of perfect virtue, yet he will at any rate accept it since they have nothing greater to bestow on him; but honour from casual people and on trifling grounds he will utterly despise, since it is not this that he deserves, and dishonour too, since in his case it cannot be just. In the first place, then, as has been said, the proud man is concerned with honours; yet he will also bear himself with moderation towards wealth and power and all good or evil fortune, whatever may befall him, and will be neither over-joyed by good fortune nor over-pained by evil. For not even towards honour does he bear himself as if it were a very great thing. Power and wealth are desirable for the sake of honour (at least those who have them wish to get honour by means of them); and for him to whom even honour is a little thing the others must be so too. Hence proud men are thought to be disdainful.

The goods of fortune also are thought to contribute towards pride. For men who are well-born are thought worthy of honour, and so are those who enjoy power or wealth; for they are in a superior position, and everything that has a superiority in something good is held in greater honour. Hence even such things make men prouder; for they are honoured by some for having them; but in truth the good man alone is to be honoured; he, however, who has both advantages is thought the more worthy of honour. But those who without virtue have such goods are neither justified in making great claims nor entitled to the name of 'proud'; for these things imply perfect virtue. Disdainful and insolent, however, even those who have such goods become. For without virtue it is not easy to bear gracefully the goods of fortune; and, being unable to bear them, and thinking themselves superior to others, they despise others and themselves do what they please. They imitate the proud man without being like him, and this they do where they can; so they do not act virtuously, but they do despise others. For the proud man despises justly (since he thinks truly), but the many do so at random.

He does not run into trifling dangers, nor is he fond of danger, because he honours few things; but he will face great dangers, and when he is in danger he is unsparing of his life, knowing that there are conditions on which life is not worth having. And he is the sort of man to confer benefits, but he is ashamed of receiving them; for the one is the mark of a superior, the other of an inferior. And he is apt to confer greater benefits in return; for thus the original benefactor besides being paid will incur a debt to him, and will be the gainer by the transaction. They seem also to remember any service they have done, but not those they have received (for he who receives a service is inferior to him who has done it, but the proud man wishes to be superior), and to hear of the former with pleasure, of the latter with displeasure; this, it seems, is why Thetis did not mention to Zeus the services she had done him, and why the Spartans did not recount their services to the Athenians, but those they had received. It is a mark of the proud man also to ask for nothing or scarcely anything, but to give help readily, and to be dignified towards people who enjoy high position and good fortune, but unassuming towards those of the middle class; for it is a difficult and lofty thing to be superior to the former, but easy to be so to the latter, and a lofty bearing over the former is no mark of ill-breeding, but among humble people it is as vulgar as a display of strength against the weak. Again, it is characteristic of the proud man not to aim at the things commonly held in honour, or the things in which others excel; to be sluggish and to hold back except where great honour or a great work is at stake, and to be a man of few deeds, but of great and notable ones. He must also be open in his hate and in his love (for to conceal one's feelings, i.e. to care less for truth than for what people will think, is a coward's part), and must speak and act openly; for he is free of speech because he is contemptuous, and he is given to telling the truth, except when he speaks in irony to the vulgar. He must be unable to make his life revolve round another, unless it be a friend; for this is slavish, and for this reason all flatterers are servile and people lacking in self-respect are flatterers. Nor is he given to admiration; for nothing to him is great. Nor is he mindful of wrongs; for it is not the part of a proud man to have a long memory, especially for wrongs, but rather to overlook them. Nor is he a gossip; for he will speak neither about himself nor about another, since he cares not to be praised nor for others to be blamed; nor again is he given to praise; and for the same reason he is not an evil-speaker, even about his enemies, except from haughtiness. With regard to necessary or small matters he is least of all me given to lamentation or the asking of favours; for it is the part of one who takes such matters seriously to behave so with respect to them. He is one who will possess beautiful and profitless things rather than profitable and useful ones; for this is more proper to a character that suffices to itself.

Further, a slow step is thought proper to the proud man, a deep voice, and a level utterance; for the man who takes few things seriously is not likely to be hurried, nor the man who thinks nothing great to be excited, while a shrill voice and a rapid gait are the results of hurry and excitement.

Such, then, is the proud man; the man who falls short of him is unduly humble, and the man who goes beyond him is vain. Now even these are not thought to be bad (for they are not malicious), but only mistaken. For the unduly humble man, being worthy of good things, robs himself of what he deserves, and to have something bad about him from the fact that he does not think himself worthy of good things, and seems also not to know himself; else he would have desired the things he was worthy of, since these were good. Yet such people are not thought to be fools, but rather unduly retiring. Such a reputation, however, seems actually to make them worse; for each class of people aims at what corresponds to its worth, and these people stand back even from noble actions and undertakings, deeming themselves unworthy, and from external goods no less. Vain people, on the other hand, are fools and ignorant of themselves, and that manifestly; for, not being worthy of them, they attempt honourable undertakings, and then are found out; and tetadorn themselves with clothing and outward show and such things, and wish their strokes of good fortune to be made public, and speak about them as if they would be honoured for them. But undue humility is more opposed to pride than vanity is; for it is both commoner and worse.

Pride, then, is concerned with honour on the grand scale, as has been said.

Chapter 4: Ambition, unambitiousness, and the mean between them.

There seems to be in the sphere of honour also, as was said in our first remarks on the subject, a virtue which would appear to be related to pride as liberality is to magnificence. For neither of these has anything to do with the grand scale, but both dispose us as is right with regard to middling and unimportant objects; as in getting and giving of wealth there is a mean and an excess and defect, so too honour may be desired more than is right, or less, or from the right sources and in the right way. We blame both the ambitious man as am at honour more than is right and from wrong sources, and the unambitious man as not willing to be honoured even for noble reasons. But sometimes we praise the ambitious man as being manly and a lover of what is noble, and the unambitious man as being moderate and self-controlled, as we said in our first treatment of the subject. Evidently, since 'fond of such and such an object' has more than one meaning, we do not assign the term 'ambition' or 'love of honour' always to the same thing, but when we praise the quality we think of the man who loves honour more than most people, and when we blame it we think of him who loves it more than is right. The mean being without a name, the extremes seem to dispute for its place as though that were vacant by default. But where there is excess and defect, there is also an intermediate; now men desire honour both more than they should and less;

therefore it is possible also to do so as one should; at all events this is the state of character that is praised, being an unnamed mean in respect of honour. Relatively to ambition it seems to be unambitiousness, and relatively to unambitiousness it seems to be ambition, while relatively to both severally it seems in a sense to be both together. This appears to be true of the other virtues also. But in this case the extremes seem to be contradictories because the mean has not received a name.

Chapter 5:Good temper, irascibility, inirascibility.

Good temper is a mean with respect to anger; the middle state being unnamed, and the extremes almost without a name as well, we place good temper in the middle position, though it inclines towards the deficiency, which is without a name. The excess might called a sort of 'irascibility'. For the passion is anger, while its causes are many and diverse.

The man who is angry at the right things and with the right people, and, further, as he ought, when he ought, and as long as he ought, is praised. This will be the good-tempered man, then, since good temper is praised. For the good-tempered man tends to be unperturbed and not to be led by passion, but to be angry in the manner, at the things, and for the length of time, that the rule dictates; but he is thought to err rather in the direction of deficiency; for the good-tempered man is not revengeful, but rather tends to make allowances.

The deficiency, whether it is a sort of 'inirascibility' or whatever it is, is blamed. For those who are not angry at the things they should be angry at are thought to be fools, and so are those who are not angry in the right way, at the right time, or with the right persons; for such a man is thought not to feel things nor to be pained by them, and, since he does not get angry, he is thought unlikely to defend himself; and to endure being insulted and put up with insult to one's friends is slavish.

The excess can be manifested in all the points that have been named (for one can be angry with the wrong persons, at the wrong things, more than is right, too quickly, or too long); yet all are not found in the same person. Indeed they could not; for evil destroys even itself, and if it is complete becomes unbearable. Now hot-tempered people get angry quickly and with the wrong persons and at the wrong things and more than is right, but their anger ceases quickly—which is the best point about them. This happens to them because they do not restrain their anger but retaliate openly owing to their quickness of temper, and then their anger ceases. By reason of excess choleric people are quick-tempered and ready to be angry with everything and on every occasion; whence their name. Sulky people are hard to appease, and retain their anger long; for they repress their passion. But it ceases when they retaliate; for revenge relieves them of their anger, producing in them pleasure instead of pain. If this does not happen they retain their burden; for owing to its not being obvious no one even reasons with them, and to digest one's anger in oneself takes time. Such people are most troublesome to themselves and to their dearest friends. We call bad-tempered those who are angry at the wrong things, more than is right, and longer, and cannot be appeased until they inflict vengeance or punishment.

To good temper we oppose the excess rather than the defect; for not only is it commoner since revenge is the more human), but bad-tempered people are worse to live with.

What we have said in our earlier treatment of the subject is plain also from what we are now saying; viz. that it is not easy to define how, with whom, at what, and how long one should be angry, and at what point right action ceases and wrong begins. For the man who strays a little from the path, either towards the more or towards the less, is not blamed; since sometimes we praise those who exhibit the deficiency, and call them good-tempered, and sometimes we call angry people manly, as being capable of ruling. How far, therefore, and how a man must stray before he becomes blameworthy, it is not easy to state in words; for the decision depends on the particular facts and on perception. But so much at least is plain, that the middle state is praiseworthy—that in virtue of which we are angry with the right people, at the right things, in the right way, and so on, while the excesses and defects are blameworthy—slightly so if they are present in a low degree, more if in a higher degree, and very much if in a high degree. Evidently, then, we must cling to the middle state—-Enough of the states relative to anger.

Chapter 6: Friendliness, obsequiousness, churlishness.

In gatherings of men, in social life and the interchange of words and deeds, some men are thought to be obsequious, viz. those who to give pleasure praise everything and never oppose, but think it their duty 'to give no pain to the people they meet'; while those who, on the contrary, oppose everything and care not a whit about giving pain are called churlish and contentious. That the states we have named are culpable is plain enough, and that the middle state is laudable—that in virtue of which a man will put up with, and will resent, the right things and in the right way; but no name has been assigned to it, though it most resembles friendship. For the man who corresponds to this middle state is very much what, with affection added, we call a good friend. But the state in question differs from friendship in that it implies no passion or affection for one's associates; since it is not by reason of loving or hating that such a man takes everything in the right way, but by being a man of a certain kind. For he will behave so alike towards those he knows and those he does not know, towards intimates and those who are not so, except that in each of these cases he will behave as is befitting; for it is not proper to have the same care for intimates and for strangers, nor again is it the same conditions that make it right to give pain to them. Now we have said generally that he will associate with people in the right way; but it is by reference to what is honourable and expedient that he will aim at not giving pain or at contributing pleasure. For he seems to be concerned with the pleasures and pains of social life; and wherever it is not honourable, or is harmful, for him to contribute pleasure, he will refuse, and will choose rather to give pain; also if his acquiescence in another's action would bring disgrace, and that in a high degree, or injury, on that other, while his opposition brings a little pain, he will not acquiesce but will decline. He will associate differently with people in high station and with ordinary people, with closer and more distant acquaintances, and so too with regard to all other differences, rendering to each class what is befitting, and while for its own sake he chooses to contribute pleasure, and avoids the giving of pain, he will be guided by the consequences, if these are greater, i.e. honour and expediency. For the sake of a great future pleasure, too, he will inflict small pains.

The man who attains the mean, then, is such as we have described, but has not received a name; of those who contribute pleasure, the man who aims at being pleasant with no ulterior object is obsequious, but the man who does so in order that he may get some advantage in the direction of money or the things that money buys is a flatterer; while the man who quarrels with everything is, as has been said, churlish and contentious. And the extremes seem to be contradictory to each other because the mean is without a name.

Chapter 7: Truthfulness, boastfulness, mock-modesty.

The mean opposed to boastfulness is found in almost the same sphere; and this also is without a name. It will be no bad plan to describe these states as well; for we shall both know the facts about character better if we go through them in detail, and we shall be convinced that the virtues are means if we see this to be so in all cases. In the field of social life those who make the giving of pleasure or pain their object in associating with others have been described; let us now describe those who pursue truth or falsehood alike in words and deeds and in the claims they put forward. The boastful man, then, is thought to be apt to claim the things that bring glory, when he has not got them, or to claim more of them than he has, and the mock-modest man on the other hand to disclaim what he has or belittle it, while the man who observes the mean is one who calls a thing by its own name, being truthful both in life and in word, owning to what he has, and neither more nor less. Now each of these courses may be adopted either with or without an object. But each man speaks and acts and lives in accordance with his character, if he is not acting for some ulterior object. And falsehood is in itself mean and culpable, and truth noble and worthy of praise. Thus the truthful man is another case of a man who, being in the mean, is worthy of praise, and both forms of untruthful man are culpable, and particularly the boastful man.

Let us discuss them both, but first of all the truthful man. We are not speaking of the man who keeps faith in his agreements, i.e. in the things that pertain to justice or injustice (for this would belong to another virtue), but the man who in the matters in which nothing of this sort is at stake is true both in word and in life because his character is such. But such a man would seem to be as a matter of fact equitable. For the man who loves truth, and is truthful where nothing is at stake, will still more be truthful where something is at stake; he will avoid falsehood as something base, seeing that he avoided it even for its own sake; and such a man is worthy of praise. He inclines rather to understate the truth; for this seems in better taste because exaggerations are wearisome.

He who claims more than he has with no ulterior object is a contemptible sort of fellow (otherwise he would not have delighted in falsehood), but seems futile rather than bad; but if he does it for an object, he who does it for the sake of reputation or honour is (for a boaster) not very much to be blamed, but he who does it for money, or the things that lead to money, is an uglier character (it is not the capacity that makes the boaster, but the purpose; for it is in virtue of his state of character and by being a man of a certain kind that he is boaster); as one man is a liar because he enjoys the lie itself, and another because he desires reputation or gain. Now those who boast for the sake of reputation claim such qualities as will praise or congratulation, but those whose object is gain claim qualities which are of value to one's neighbours and one's lack of which is not easily detected, e.g. the powers of a seer, a sage, or a physician. For this reason it is such things as these that most people claim and boast about; for in them the above-mentioned qualities are found.

Mock-modest people, who understate things, seem more attractive in character; for they are thought to speak not for gain but to avoid parade; and here too it is qualities which bring reputation that they disclaim, as Socrates used to do. Those who disclaim trifling and obvious qualities are called humbugs and are more contemptible; and sometimes this seems to be boastfulness, like the Spartan dress; for both excess and great deficiency are boastful. But those who use understatement with moderation and understate about matters that do not very much force themselves on our notice seem attractive. And it is the boaster that seems to be opposed to the truthful man; for he is the worse character.

Chapter 8: Ready wit, buffoonery, boorishness.

Since life includes rest as well as activity, and in this is included leisure and amusement, there seems here also to be a kind of intercourse which is tasteful; there is such a thing as saying — and again listening to—what one should and as one should. The kind of people one is speaking or listening to will also make a difference. Evidently here also there is both an excess and a deficiency as compared with the mean. Those who carry humour to excess are thought to be vulgar buffoons, striving after humour at all costs, and aiming rather at raising a laugh than at saying what is becoming and at avoiding pain to the object of their fun; while those who can neither make a joke themselves nor put up with those who do are thought to be boorish and unpolished. But those who joke in a tasteful way are called ready-witted, which implies a sort of readiness to turn this way and that; for such sallies are thought to be movements of the character, and as bodies are discriminated by their movements, so too are characters. The ridiculous side of things is not far to seek, however, and most people delight more than they should in amusement and in jestinly. and so even buffoons are called ready-witted because they are found attractive; but that they differ from the ready-witted man, and to no small extent, is clear from what has been said.

To the middle state belongs also tact; it is the mark of a tactful man to say and listen to such things as befit a good and well-bred man; for there are some things that it befits such a man to say and to hear by way of jest, and the well-bred man's jesting differs from that of a vulgar man, and the joking of an educated man from that of an uneducated. One may see this even from the old and the new comedies; to the authors of the former indecency of language was amusing, to those of the latter innuendo is more so; and these differ in no small degree in respect of propriety. Now should we define the man who jokes well by his saying what is not unbecoming to a well-bred man, or by his not giving pain, or even giving delight, to the hearer? Or is the latter definition, at any rate, itself indefinite, since different things are hateful or pleasant to different people? The kind of jokes he will listen to will be the same; for the kind he can put up with are also the kind he seems to make. There are, then, jokes he will not make; for the jest is a sort of abuse, and there are things that lawgivers forbid us to abuse; and they should, perhaps, have forbidden us even to make a jest of such. The refined and well-bred man, therefore, will be as we have described, being as it were a law to himself.

Such, then, is the man who observes the mean, whether he be called tactful or ready-witted. The buffoon, on the other hand, is the slave of his sense of humour, and spares neither himself nor others if he can raise a laugh, and says things none of which a man of refinement would say, and to some of which he would not even listen. The boor, again, is useless for such social intercourse; for he contributes nothing and finds fault with everything. But relaxation and amusement are thought to be a necessary element in life.

The means in life that have been described, then, are three in number, and are all concerned with an interchange of words and deeds of some kind. They differ, however, in that one is concerned with truth; and the other two with pleasantness. Of those concerned with pleasure, one is displayed in jests, the other in the general social intercourse of life.

Chapter 9: Shame, bashfulness, shamelessness.

Shame should not be described as a virtue; for it is more like a feeling than a state of character. It is defined, at any rate, as a kind of fear of dishonour, and produces an effect similar to that produced by fear of danger; for people who feel disgraced blush, and those who fear death turn pale. Both, therefore, seem to be in a sense bodily conditions, which is thought to be characteristic of feeling rather than of a state of character.

The feeling is not becoming to every age, but only to youth. For we think young people should be prone to the feeling of shame because they live by feeling and therefore commit many errors, but are restrained by shame; and we praise young people who are prone to this feeling, but an older person no one would praise for being prone to the sense of disgrace, since we think he should not do anything that need cause this sense. For the sense of disgrace is not even characteristic of a good man, since it is consequent on bad actions (for such actions should not be done; and if some actions are disgraceful in very truth and others only according to common opinion, this makes no difference; for neither class of actions should be done, so that no disgrace should be felt); and it is a mark of a bad man even to be such as to do any disgraceful action. To be so constituted as to feel disgraced if one does such an action, and for this reason to think oneself good, is absurd; for it is for voluntary actions that shame is felt, and the good man will never voluntarily do bad actions. But shame may be said to be conditionally a good thing; if a good man does such actions, he will feel disgraced; but the virtues are not subject to such a qualification. And if shamelessness—not to be ashamed of doing base actions—is bad, that does not make it good to be ashamed of doing such actions. Continence too is not virtue, but a mixed sort of state; this will be shown later. Now, however, let us discuss justice.

Book V: Moral Virtue

Chapter 1: The just as the lawful (universal justice) and the just as the fair and equal (particular justice): the former considered.

WITH regards to justice and injustice we must (1) consider what kind of actions they are concerned with, (2) what sort of mean justice is, and (3) between what extremes the just act is intermediate. Our investigation shall follow the same course as the preceding discussions.

We see that all men mean by justice that kind of state of character which makes people disposed to do what is just and makes them act justly and wish for what is just; and similarly by injustice that state which makes them act unjustly and wish for what is unjust. Let us too, then, lay this down as a general basis. For the same is not true of the sciences and the faculties as of states of character. A faculty or a science which is one and the same is held to relate to contrary objects, but a state of character which is one of two contraries does not produce the contrary results; e.g. as a result of health we do not do what is the opposite of healthy, but only what is healthy; for we say a man walks healthily, when he walks as a healthy man would.

Now often one contrary state is recognized from its contrary, and often states are recognized from the subjects that exhibit them; for (A) if good condition is known, bad condition also becomes known, and (B) good condition is known from the things that are in good condition, and they from it. If good condition is firmness of flesh, it is necessary both that bad condition should be flabbiness of flesh and that the wholesome should be that which causes firmness in flesh. And

416

Aristotle

it follows for the most part that if one contrary is ambiguous the other also will be ambiguous; e.g. if 'just' is so, that 'unjust' will be so too.

Now 'justice' and 'injustice' seem to be ambiguous, but because their different meanings approach near to one another the ambiguity escapes notice and is not obvious as it is, comparatively, when the meanings are far apart, e.g. (for here the difference in outward form is great) as the ambiguity in the use of kleis for the collar-bone of an animal and for that with which we lock a door. Let us take as a starting-point, then, the various meanings of 'an unjust man'. Both the lawless man and the grasping and unfair man are thought to be unjust, so that evidently both the law-abiding and the fair man will be just. The just, then, is the lawful and the fair, the unjust the unlawful and the unfair.

Since the unjust man is grasping, he must be concerned with goods—not all goods, but those with which prosperity and adversity have to do, which taken absolutely are always good, but for a particular person are not always good. Now men pray for and pursue these things; but they should not, but should pray that the things that are good absolutely may also be good for them, and should choose the things that are good for them. The unjust man does not always choose the greater, but also the less—in the case of things bad absolutely; but because the lesser evil is itself thought to be in a sense good, and graspingness is directed at the good, therefore he is thought to be grasping. And he is unfair; for this contains and is common to both.

Since the lawless man was seen to be unjust and the law-abiding man just, evidently all lawful acts are in a sense just acts; for the acts laid down by the legislative art are lawful, and each of these, we say, is just. Now the laws in their enactments on all subjects aim at the common advantage either of all or of the best or of those who hold power, or something of the sort; so that in one sense we call those acts just that tend to produce and preserve happiness and its components for the political society. And the law bids us do both the acts of a brave man (e.g. not to desert our post nor take to flight nor throw away our arms), and those of a temperate man (e.g. not to commit adultery nor to gratify one's lust), and those of a good-tempered man (e.g. not to strike another nor to speak evil), and similarly with regard to the other virtues and forms of wickedness, commanding some acts and forbidding others; and the rightly framed law does this rightly, and the hastily conceived one less well. This form of justice, then, is complete virtue, but not absolutely, but in relation to our neighbour. And therefore justice is often thought to be the greatest of virtues, and 'neither evening nor morning star' is so wonderful; and proverbially 'in justice is every virtue comprehended'. And it is complete virtue in its fullest sense, because it is the actual exercise of complete virtue. It is complete because he who possesses it can exercise his virtue not only in himself but towards his neighbour also; for many men can exercise virtue in their own affairs, but not in their relations to their neighbour. This is why the saying of Bias is thought to be true, that 'rule will show the man'; for a ruler is necessarily in relation to other men and a member of a society. For this same reason justice, alone of the virtues, is thought to be 'another's good', because it is related to our neighbour; for it does what is advantageous to another, either a ruler or a copartner. Now the worst man is he who exercises his wickedness both towards himself and towards his friends, and the best man is not he who exercises his virtue towards himself but he who exercises it towards another; for this is a difficult task. Justice in this sense, then, is not part of virtue but virtue entire, nor is the contrary injustice a part of vice but vice entire. What the difference is between virtue and justice in this sense is plain from what we have said; they are the same but their essence is not the same; what, as a relation to one's neighbour, is justice is, as a certain kind of state without qualification, virtue.

Chapter 2: The latter considered: divided into distributive and rectificatory justice.

But at all events what we are investigating is the justice which is a part of virtue; for there is a justice of this kind, as we maintain. Similarly it is with injustice in the particular sense that we are concerned.

That there is such a thing is indicated by the fact that while the man who exhibits in action the other forms of wickedness acts wrongly indeed, but not graspingly (e.g. the man who throws away his shield through cowardice or speaks harshly through bad temper or fails to help a friend with money through meanness), when a man acts graspingly he often exhibits none of these vices,—no, nor all together, but certainly wickedness of some kind (for we blame him) and injustice. There is, then, another kind of injustice which is a part of injustice in the wide sense, and a use of the word 'unjust' which answers to a part of what is unjust in the wide sense of 'contrary to the law'. Again if one man commits adultery for the sake of gain and makes money by it, while another does so at the bidding of appetite though he loses money and is penalized for it, the latter would be held to be self-indulgent rather than grasping, but the former is unjust, but not self-indulgent; evidently, therefore, he is unjust by reason of his making gain by his act. Again, all other unjust acts are ascribed invariably to some particular kind of wickedness, e.g. adultery to self-indulgence, the desertion of a comrade in battle to cowardice, physical violence to anger; but if a man makes gain, his action is ascribed to no form of wickedness but injustice. Evidently, therefore, there is apart from injustice in the wide sense another, 'particular', injustice which shares the name and nature of the first, because its definition falls within the same genus; for the significance of both consists in a relation to one's neighbour, but the one is concerned with honour or money or safety—or that which includes all these, if we had a single name for it—and its motive is the pleasure that arises from gain; while the other is concerned with all the objects with which the good man is concerned.

It is clear, then, that there is more than one kind of justice, and that there is one which is distinct from virtue entire; we must try to grasp its genus and differentia.

The unjust has been divided into the unlawful and the unfair, and the just into the lawful and the fair. To the unlawful answers the afore-mentioned sense of injustice. But since unfair and the unlawful are not the same, but are different as a part is from its whole (for all that is unfair is unlawful, but not all that is unlawful is unfair), the unjust and injustice in the sense of the unfair are not the same as but different from the former kind, as part from whole; for injustice in this sense is a part of injustice in the wide sense, and similarly justice in the one sense of justice in the other. Therefore we must speak also about particular justice and particular and similarly about the just and the unjust. The justice, then, which answers to the whole of virtue, and the corresponding injustice, one being the exercise of virtue as a whole, and the other that of vice as a whole, towards one's neighbour, we may leave on one side. And how the meanings of 'just' and 'unjust' which answer to these are to be distinguished is evident; for practically the majority of the acts commanded by the law are those which are prescribed from the point of view of virtue taken as a whole; for the law bids us practise every virtue and forbids us to practise any vice. And the things that tend to produce virtue taken as a whole are those of the acts prescribed by the law which have been prescribed with a view to education for the common good. But with regard to the education of the individual as such, which makes him without qualification a good man, we must determine later whether this is the function of the political art or of another; for perhaps it is not the same to be a good man and a good citizen of any state taken at random.

Of particular justice and that which is just in the corresponding sense, (A) one kind is that which is manifested in distributions of honour or money or the other things that fall to be divided among those who have a share in the constitution (for in these it is possible for one man to have a share either unequal or equal to that of another), and (B) one is that which plays a rectifying part in transactions between man and man. Of this there are two divisions; of transactions (1) some are voluntary and (2) others involuntary—voluntary such transactions as sale, purchase, loan for consumption, pledging, loan for use, depositing, letting (they are called voluntary because the origin of these transactions is voluntary), while of the involuntary (a) some are clandestine, such as theft, adultery, poisoning, procuring, enticement of slaves, assassination, false witness, and (b) others are violent, such as assault, imprisonment, murder, robbery with violence, mutilation, abuse, insult.

Chapter 3: Distributive justice, in accordance with geometrical proportion.

(A) We have shown that both the unjust man and the unjust act are unfair or unequal; now it is clear that there is also an intermediate between the two unequals involved in either case. And this is the equal; for in any kind of action in which there's a more and a less there is also what is equal. If, then, the unjust is unequal, just is equal, as all men suppose it to be, even apart from argument. And since the equal is intermediate, the just will be an intermediate. Now equality implies at least two things. The just, then, must be both intermediate and equal and relative (i.e. for certain persons). And since the equal is intermediate it must be between certain things (which are respectively greater and less); equal, it involves two things; qua just, it is for certain people. The just, therefore, involves at least four terms; for the persons for whom it is in fact just are two, and the things in which it is manifested, the objects distributed, are two. And the same equality will exist between the persons and between the things concerned; for as the latter the things concerned are related, so are the former; if they are not equal, they will not have what is equal, but this is the origin of quarrels and complaints—when either equals have and are awarded unequal shares, or unequals equal shares. Further, this is plain from the fact that awards should be 'according to merit'; for all men agree that what is just in distribution must be according to merit in some sense, though they do not all specify the same sort of merit, but democrats identify it with the status of freeman, supporters of oligarchy with wealth (or with noble birth), and supporters of aristocracy with excellence.

The just, then, is a species of the proportionate (proportion being not a property only of the kind of number which consists of abstract units, but of number in general). For proportion is equality of ratios, and involves four terms at least (that discrete proportion involves four terms is plain, but so does continuous proportion, for it uses one term as two and mentions it twice; e.g. 'as the line A is to the line B, so is the line B to the line C'; the line B, then, has been mentioned twice, so that if the line B be assumed twice, the proportional terms will be four); and the just, too, involves at least four terms, and the ratio between one pair is the same as that between the other pair; for there is a similar distinction between the persons and between the things. As the term A, then, is to B, so will C be to D, and therefore, alternando, as A is to C, B will be to D. Therefore also the whole is in the same ratio to the whole; and this coupling the distribution effects, and, if the terms are so combined, effects justly. The conjunction, then, of the term A with C and of B with D is what is just in distribution, and this species of the just is intermediate, and the unjust is what violates the proportion; for the proportional is intermediate, and the just is proportional. (Mathematicians call this kind of proportion geometrical; for it is in geometrical proportion that it follows that the whole is to the whole as either part is to the corresponding part.) This proportion is not continuous; for we cannot get a single term standing for a person and a thing.

This, then, is what the just is—the proportional; the unjust is what violates the proportion. Hence one term becomes too great, the other too small, as indeed happens in practice; for the man who acts unjustly has too much, and the man who is unjustly treated too little, of what is good. In the case of evil the reverse is true; for the lesser evil is reckoned a good in comparison with the greater evil, since the lesser evil is rather to be chosen than the greater, and what is worthy of choice is good, and what is worthier of choice a greater good.

This, then, is one species of the just.

Chapter 4: Rectificatory justice, in accordance with arithmetical progression.

(B) The remaining one is the rectificatory, which arises in connexion with transactions both voluntary and involuntary. This form of the just has a different specific character from the former. For the justice which distributes common possessions is always in accordance with the kind of proportion mentioned above (for in the case also in which the distribution is made from the common funds of a partnership it will be according to the same ratio which the funds put into the business by the partners bear to one another); and the injustice opposed to this kind of justice is that which violates the proportion. But the justice in transactions between man and man is a sort of equality indeed, and the injustice a sort of inequality; not according to that kind of proportion, however, but according to arithmetical proportion. For it makes no difference whether a good man has defrauded a bad man or a bad man a good one, nor whether it is a good or a bad man that has committed adultery; the law looks only to the distinctive character of the injury, and treats the parties as equal, if one is in the wrong and the other is being wronged, and if one inflicted injury and the other has received it. Therefore, this kind of injustice being an inequality, the judge tries to equalize it; for in the case also in which one has received and the other has inflicted a wound, or one has slain and the other been slain, the suffering and the action have been unequally distributed; but the judge tries to equalize by means of the penalty, taking away from the gain of the assailant. For the term 'gain' is applied generally to such cases, even if it be not a term appropriate to certain cases, e.g. to the person who inflicts a wound and 'loss' to the sufferer; at all events when the suffering has been estimated, the one is called loss and the other gain. Therefore the equal is intermediate between the greater and the less, but the gain and the loss are respectively greater and less in contrary ways; more of the good and less of the evil are gain, and the contrary is loss; intermediate between them is, as we saw, equal, which we say is just; therefore corrective justice will be the intermediate between loss and gain. This is why, when people dispute, they take refuge in the judge; and to go to the judge is to go to justice; for the nature of the judge is to be a sort of animate justice; and they seek the judge as an intermediate, and in some states they call judges mediators, on the assumption that if they get what is intermediate they will get what is just. The just, then, is an intermediate, since the judge is so. Now the judge restores equality; it is as though there were a line divided into unequal parts, and he took away that by which the greater segment exceeds the half, and added it to the smaller segment. And when the whole has been equally divided, then they say they have 'their own'—i.e. when they have got what is equal. The equal is intermediate between the greater and the lesser line according to arithmetical proportion. It is for this reason also that it is called just (sikaion), because it is a division into two equal parts (sicha), just as if one were to call it sichaion; and the judge (sikastes) is one who bisects (sichastes). For when something is subtracted from one of two equals and added to the other, the other is in excess by these two; since if what was taken from the one had not been added to the other, the latter would have been in excess by one only. It therefore exceeds the intermediate by one, and the intermediate exceeds by one that from which something was taken. By this, then, we shall recognize both what we must subtract from that which has more, and what we must add to that which has less; we must add to the latter that by which the intermediate exceeds it, and subtract from the greatest that by which it exceeds the intermediate. Let the lines AA', BB', CC' be equal to one another; from the line AA' let the segment AE have been subtracted, and to the line CC' let the segment CD have been added, so that the whole line DCC' exceeds the line EA' by the segment CD and the segment CF; therefore it exceeds the line BB' by the segment CD. (See diagram.)

These names, both loss and gain, have come from voluntary exchange; for to have more than one's own is called gaining, and to have less than one's original share is called losing, e.g. in buying and selling and in all other matters in which the law has left people free to make their own terms; but when they get neither more nor less but just what belongs to themselves, they say that they have their own and that they neither lose nor gain.

Therefore the just is intermediate between a sort of gain and a sort of loss, viz. those which are involuntary; it consists in having an equal amount before and after the transaction.

Chapter 5: Justice in exchange, reciprocity in accordance with proportion.

Some think that reciprocity is without qualification just, as the Pythagoreans said; for they defined justice without qualification as reciprocity. Now 'reciprocity' fits neither distributive nor rectificatory justice—yet people want even the justice of Rhadamanthus to mean this:

Should a man suffer what he did, right justice would be done — for in many cases reciprocity and rectificatory justice are not in accord; e.g. (1) if an official has inflicted a wound, he should not be wounded in return, and if some one has wounded an official, he ought not to be wounded only but punished in addition. Further (2) there is a great difference between a voluntary and an involuntary act. But in associations for exchange this sort of justice does hold men together—reciprocity in accordance with a proportion and not on the basis of precisely equal return. For it is by proportionate requital that the city holds together. Men seek to return either evil for evil—and if they cana not do so, think their position mere slavery—or good for good—and if they cannot do so there is no exchange, but it is by exchange that they hold together. This is why they give a prominent place to the temple of the Graces—to promote the requital of services; for this is characteristic of grace—we should serve in return one who has shown grace to us, and should another time take the initiative in showing it.

Aristotle

Now proportionate return is secured by cross-conjunction. Let A be a builder, B a shoemaker, C a house, D a shoe. The builder, then, must get from the shoemaker the latter's work, and must himself give him in return his own. If, then, first there is proportionate equality of goods, and then reciprocal action takes place, the result we mention will be effected. If not, the bargain is not equal, and does not hold; for there is nothing to prevent the work of the one being better than that of the other; they must therefore be equated. (And this is true of the other arts also; for they would have been destroyed if what the patient suffered had not been just what the agent did, and of the same amount and kind.) For it is not two doctors that associate for exchange, but a doctor and a farmer, or in general people who are different and unequal; but these must be equated. This is why all things that are exchanged must be somehow comparable. It is for this end that money has been introduced, and it becomes in a sense an intermediate; for it measures all things, and therefore the excess and the defect—how many shoes are equal to a house or to a given amount of food. The number of shoes exchanged for a house (or for a given amount of food) must therefore correspond to the ratio of builder to shoemaker. For if this be not so, there will be no exchange and no intercourse. And this proportion will not be effected unless the goods are somehow equal. All goods must therefore be measured by some one thing, as we said before. Now this unit is in truth demand, which holds all things together (for if men did not need one another's goods at all, or did not need them equally, there would be either no exchange or not the same exchange); but money has become by convention a sort of representative of demand; and this is why it has the name 'money' (nomisma)—because it exists not by nature but by law (nomos) and it is in our power to change it and make it useless. There will, then, be reciprocity when the terms have been equated so that as farmer is to shoemaker, the amount of the shoemaker's work is to that of the farmer's work for which it exchanges. But we must not bring them into a figure of proportion when they have already exchanged (otherwise one extreme will have both excesses), but when they still have their own goods. Thus they are equals and associates just because this equality can be effected in their case. Let A be a farmer, C food, B a shoemaker, D his product equated to C. If it had not been possible for reciprocity to be thus effected, there would have been no association of the parties. That demand holds things together as a single unit is shown by the fact that when men do not need one another, i.e. when neither needs the other or one does not need the other, they do not exchange, as we do when some one wants what one has oneself, e.g. when people permit the exportation of corn in exchange for wine. This equation therefore must be established. And for the future exchange—that if we do not need a thing now we shall have it if ever we do need it—money is as it were our surety; for it must be possible for us to get what we want by bringing the money. Now the same thing happens to money itself as to goods—it is not always worth the same; yet it tends to be steadier. This is why all goods must have a price set on them; for then there will always be exchange, and if so, association of man with man. Money, then, acting as a measure, makes goods commensurate and equates them; for neither would there have been association if there were not exchange, nor exchange if there were not equality, nor equality if there were not commensurability. Now in truth it is impossible that things differing so much should become commensurate, but with reference to demand they may become so sufficiently. There must, then, be a unit, and that fixed by agreement (for which reason it is called money); for it is this that makes all things commensurate, since all things are measured by money. Let A be a house, B ten minae, C a bed. A is half of B, if the house is worth five minae or equal to them; the bed, C, is a tenth of B; it is plain, then, how many beds are equal to a house, viz. five. That exchange took place thus before there was money is plain; for it makes no difference whether it is five beds that exchange for a house, or the money value of five beds.

We have now defined the unjust and the just. These having been marked off from each other, it is plain that just action is intermediate between acting unjustly and being unjustly treated; for the one is to have too much and the other to have too little. Justice is a kind of mean, but not in the same way as the other virtues, but because it relates to an intermediate amount, while injustice relates to the extremes. And justice is that in virtue of which the just man is said to be a doer, by choice, of that which is just, and one who will distribute either between himself and another or between two others not so as to give more of what is desirable to himself and less to his neighbour (and conversely with what is harmful), but so as to give what is equal in accordance with proportion; and similarly in distributing between two other persons. Injustice on the other hand is similarly related to the unjust, which is excess and defect, contrary to proportion, of the useful or hurtful. For which reason injustice is excess and defect, viz. because it is productive of excess and defect—in one's own case excess of what is in its own nature useful and defect of what is hurtful, while in the case of others it is as a whole like what it is in one's own case, but proportion may be violated in either direction. In the unjust act to have too little is to be unjustly treated; to have too much is to act unjustly.

Let this be taken as our account of the nature of justice and injustice, and similarly of the just and the unjust in general.

Chapter 6: Political justice and analogous kinds of justice.

Since acting unjustly does not necessarily imply being unjust, we must ask what sort of unjust acts imply that the doer is unjust with respect to each type of injustice, e.g. a thief, an adulterer, or a brigand. Surely the answer does not turn on the difference between these types. For a man might even lie with a woman knowing who she was, but the origin of his might be not deliberate choice but passion. He acts unjustly, then, but is not unjust; e.g. a man is not a thief, yet he stole, nor an adulterer, yet he committed adultery; and similarly in all other cases.

Now we have previously stated how the reciprocal is related to the just; but we must not forget that what we are looking for is not only what is just without qualification but also political justice. This is found among men who share their life with a view to selfsufficiency, men who are free and either proportionately or arithmetically equal, so that between those who do not fulfil this condition there is no political justice but justice in a special sense and by analogy. For justice exists only between men whose mutual relations are governed by law; and law exists for men between whom there is injustice; for legal justice is the discrimination of the just and the unjust. And between men between whom there is injustice there is also unjust action (though there is not injustice between all between whom there is unjust action), and this is assigning too much to oneself of things good in themselves and too little of things evil in themselves. This is why we do not allow a man to rule, but rational principle, because a man behaves thus in his own interests and becomes a tyrant. The magistrate on the other hand is the guardian of justice, and, if of justice, then of equality also. And since he is assumed to have no more than his share, if he is just (for he does not assign to himself more of what is good in itself, unless such a share is proportional to his merits—so that it is for others that he labours, and it is for this reason that men, as we stated previously, say that justice is 'another's good'), therefore a reward must be given him, and this is honour and privilege; but those for whom such things are not enough become tyrants.

The justice of a master and that of a father are not the same as the justice of citizens, though they are like it; for there can be no injustice in the unqualified sense towards thing that are one's own, but a man's chattel, and his child until it reaches a certain age and sets up for itself, are as it were part of himself, and no one chooses to hurt himself (for which reason there can be no injustice towards oneself). Therefore the justice or injustice of citizens is not manifested in these relations; for it was as we saw according to law, and between people naturally subject to law, and these as we saw' are people who have an equal share in ruling and being ruled. Hence justice can more truly be manifested towards a wife than towards children and chattels, for the former is household justice; but even this is different from political justice.

Chapter 7: Natural and legal justice.

Of political justice part is natural, part legal, natural, that which everywhere has the same force and does not exist by people's thinking this or that; legal, that which is originally indifferent, but when it has been laid down is not indifferent, e.g. that a prisoner's ransom shall be a mina, or that a goat and not two sheep shall be sacrificed, and again all the laws that are passed for particular cases, e.g. that sacrifice shall be made in honour of Brasidas, and the provisions of decrees. Now some think that all justice is of this sort, because that which is by nature is unchangeable and has everywhere the same force (as fire burns both here and in Persia), while they see change in the things recognized as just. This, however, is not true in this unqualified way, but is true in a sense; or rather, with the gods it is perhaps not true at all, while with us there is something that is just even by nature, yet all of it is changeable; but still some is by nature, some not by nature. It is evident which sort of thing, among things capable of being otherwise, is by nature, and which is not but is legal and conventional, assuming that both are equally changeable. And in all other things the same distinction will apply; by nature the right hand is stronger, yet it is possible that all men should come to be ambidextrous. The things which are just by virtue of convention and expediency are like measures; for wine and corn measures are not everywhere equal, but larger in wholesale and smaller in retail markets. Similarly, the things which are just not by nature but by human enactment are not everywhere the same, since constitutions also are not the same, though there is but one which is everywhere by nature the best. Of things just and lawful each is related as the universal to its particulars; for the things that are done are many, but of them each is one, since it is universal.

There is a difference between the act of injustice and what is unjust, and between the act of justice and what is just; for a thing is unjust by nature or by enactment; and this very thing, when it has been done, is an act of injustice, but before it is done is not yet that but is unjust. So, too, with an act of justice (though the general term is rather 'just action', and 'act of justice' is applied to the correction of the act of injustice).

Each of these must later be examined separately with regard to the nature and number of its species and the nature of the things with which it is concerned.

Chapter 8: The scale of degrees of wrongdoing.

Acts just and unjust being as we have described them, a man acts unjustly or justly whenever he does such acts voluntarily; when involuntarily, he acts neither unjustly nor justly except in an incidental way; for he does things which happen to be just or unjust. Whether an act is or is not one of injustice (or of justice) is determined by its voluntariness or involuntariness; for when it is voluntary it is blamed, and at the same time is then an act of injustice; so that there will be things that are unjust but not yet acts of injustice, if voluntariness be not present as well. By the voluntary I mean, as has been said before, any of the things in a man's own power which he does with knowledge, i.e. not in ignorance either of the person acted on or of the instrument used or of the end that will be attained (e.g. whom he is striking, with what, and to what end), each such act being done not incidentally nor under compulsion (e.g. if A takes B's hand and therewith strikes C, B does not act voluntarily; for the act was not in his own power). The person struck may be the striker's father, and the striker may know that it is a man or one of the persons present, but not know that it is his father; a similar distinction may be made in the case of the end, and with regard to the whole action. Therefore that which is done in ignorance, or though not done in ignorance is not in the agent's power, or is done under compulsion, is involuntary (for many natural processes, even, we knowingly both perform and experience, none of which is either voluntary or involuntary; e.g. growing old or dying). But in the case of unjust and just acts alike the injustice or justice may be only incidental; for a man might return a deposit unwillingly and from fear, and then he must not be said either to do what is just or to act justly, except in an incidental way. Similarly the man who under compulsion and unwillingly fails to return the deposit must be said to act unjustly, and to do what is unjust, only incidentally. Of voluntary acts we do some by choice, others not by choice; by choice those which we do after deliberation, not by choice those which we do without previous deliberation. Thus there are three kinds of injury in transactions between man and man; those done in ignorance are mistakes when the person acted on, the act, the instrument, or the end that will be attained is other than the agent supposed; the agent thought either that he was not hiting any one or that he was not hitting with this missile or not hitting this person or to this end, but a result followed other than that which he thought likely (e.g. he threw not with intent to wound but only to prick), or the person hit or the missile was other than he supposed. Now when (1) the injury takes place contrary to reasonable expectation, it is a misadventure. When (2) it is not contrary to reasonable expectation, but does not imply vice, it is a mistake (for a man makes a mistake when the fault originates in him, but is the victim of accident when the origin lies outside him). When (3) he acts with knowledge but not after deliberation, it is an act of injustice—e.g. the acts due to anger or to other passions necessary or natural to man; for when men do such harmful and mistaken acts they act unjustly, and the acts are acts of injustice, but this does not imply that the doers are unjust or wicked; for the injury is not due to vice. But when (4) a man acts from choice, he is an unjust man and a vicious man.

Hence acts proceeding from anger are rightly judged not to be done of malice aforethought; for it is not the man who acts in anger but he who enraged him that starts the mischief. Again, the matter in dispute is not whether the thing happened or not, but its justice; for it is apparent injustice that occasions rage. For they do not dispute about the occurrence of the act—as in commercial transactions where one of the two parties must be vicious—unless they do so owing to forgetfulness; but, agreeing about the fact, they dispute on which side justice lies (whereas a man who has deliberately injured another cannot help knowing that he has done so), so that the one thinks he is being treated unjustly and the other disagrees.

But if a man harms another by choice, he acts unjustly; and these are the acts of injustice which imply that the doer is an unjust man, provided that the act violates proportion or equality. Similarly, a man is just when he acts justly by choice; but he acts justly if he merely acts voluntarily.

Of involuntary acts some are excusable, others not. For the mistakes which men make not only in ignorance but also from ignorance are excusable, while those which men do not from ignorance but (though they do them in ignorance) owing to a passion which is neither natural nor such as man is liable to, are not excusable.

Chapter 9: Can a man be voluntarily treated unjustly? Is it the distributor or the recipient that is guilty of injustice in distribution? Justice not so easy as it might seem, because it is not a way of acting but an inner disposition.

Assuming that we have sufficiently defined the suffering and doing of injustice, it may be asked (1) whether the truth is expressed in Euripides' paradoxical words:

I slew my mother, that's my tale in brief.

Were you both willing, or unwilling both?

Is it truly possible to be willingly treated unjustly, or is all suffering of injustice the contrary involuntary, as all unjust action is voluntary? And is all suffering of injustice of the latter kind or else all of the former, or is it sometimes voluntary, sometimes involuntary? So, too, with the case of being justly treated; all just action is voluntary, so that it is reasonable that there should be a similar opposition in either case—that both being unjustly and being justly treated should be either alike voluntary or alike involuntary. But it would be thought paradoxical even in the case of being justly treated, if it were always voluntary; for some are unwillingly treated justly. (2) One might raise this question also, whether every one who has suffered what is unjust is being unjustly treated, or on the other hand it is with suffering as with acting. In action and in passivity alike it is possible to partake of justice incidentally, and similarly (it is plain) of injustice; for to do what is unjust is not the same as to act unjustly, nor to suffer what is unjust as to be treated unjustly, and similarly in the case of acting justly and being justly treated; for it is impossible to be unjustly treated if the other does not act unjustly, or justly treated unless he acts justly. Now if to act unjustly is simply to harm some one voluntarily, and 'voluntarily' means 'knowing the person acted on, the instrument, and the manner of one's acting', and the incontinent man voluntarily harms himself, not only will he voluntarily be unjustly treated but it will be possible to treat oneself unjustly. (This also is one of the questions in doubt, whether a man can treat himself unjustly.) Again, a man may voluntarily, owing to incontinence, be harmed by another who acts voluntarily, so that it would be possible to be voluntarily treated unjustly. Or is our definition incorrect; must we to 'harming another, with knowledge both of the person acted on, of the instrument, and of the manner' add 'contrary to the wish of the person acted on'? Then a man may be voluntarily harmed and voluntarily suffer what is unjust, but no one is voluntarily treated unjustly; for no one wishes to be unjustly treated, not even the incontinent man. He acts contrary to his wish; for no one wishes for what he does not think to be good, but the incontinent man does do things that he does not think he ought to do. Again, one who gives what is his own, as Homer says Glaucus gave Diomede

Armour of gold for brazen, the price of a hundred beeves for nine,

is not unjustly treated; for though to give is in his power, to be unjustly treated is not, but there must be some one to treat him unjustly. It is plain, then, that being unjustly treated is not voluntary.

Of the questions we intended to discuss two still remain for discussion; (3) whether it is the man who has assigned to another more than his share that acts unjustly, or he who has the excessive share, and (4) whether it is possible to treat oneself unjustly. The questions are connected; for if the former alternative is possible and the distributor acts unjustly and not the man who has the excessive share, then if a man assigns more to another than to himself, knowingly and voluntarily, he treats himself unjustly; which is what modest people seem to do, since the virtuous man tends to take less than his share. Or does this statement too need qualification? For (a) he perhaps gets more than his share of some other good, e.g. of honour or of intrinsic nobility. (b) The question is solved by applying the distinction we applied to unjust action; for he suffers nothing contrary to his own wish, so that he is not unjustly treated as far as this goes, but at most only suffers harm.

It is plain too that the distributor acts unjustly, but not always the man who has the excessive share; for it is not he to whom what is unjust appertains that acts unjustly, but he to whom it appertains to do the unjust act voluntarily, i.e. the person in whom lies the origin of the action, and this lies in the distributor, not in the receiver. Again, since the word 'do' is ambiguous, and there is a sense in which lifeless things, or a hand, or a servant who obeys an order, may be said to slay, he who gets an excessive share does not act unjustly, though he 'does' what is unjust.

Again, if the distributor gave his judgement in ignorance, he does not act unjustly in respect of legal justice, and his judgement is not unjust in this sense, but in a sense it is unjust (for legal justice and primordial justice are different); but if with knowledge he judged unjustly, he is himself aiming at an excessive share either of gratitude or of revenge. As much, then, as if he were to share in the plunder, the man who has judged unjustly for these reasons has got too much; the fact that what he gets is different from what he distributes makes no difference, for even if he awards land with a view to sharing in the plunder he gets not land but money.

Men think that acting unjustly is in their power, and therefore that being just is easy. But it is not; to lie with one's neighbour's wife, to wound another, to deliver a bribe, is easy and in our power, but to do these things as a result of a certain state of character is neither easy nor in our power. Similarly to know what is just and what is unjust requires, men think, no great wisdom, because it is not hard to understand the matters dealt with by the laws (though these are not the things that are just, except incidentally); but how actions must be done and distributions effected in order to be just, to know this is a greater achievement than knowing what is good for the health; though even there, while it is easy to know that honey, wine, hellebore, cautery, and the use of the knife are so, to know how, to whom, and when these should be applied with a view to producing health, is no less an achievement than that of being a physician. Again, for this very reason men think that acting unjustly is characteristic of the just man no less than of the unjust, because he would be not less but even more capable of doing each of these unjust acts; for he could lie with a woman or wound a neighbour; and the brave man could throw away his shield and turn to flight in this direction or in that. But to play the coward or to act unjustly consists not in doing these things, except incidentally, but in doing them as the result of a certain state of character, just as to practise medicine and healing consists not in applying or not applying the knife, in using or not using medicines, but in doing so in a certain way.

Just acts occur between people who participate in things good in themselves and can have too much or too little of them; for some beings (e.g. presumably the gods) cannot have too much of them, and to others, those who are incurably bad, not even the smallest share in them is beneficial but all such goods are harmful, while to others they are beneficial up to a point; therefore justice is essentially something human.

Chapter 10: Equity, a corrective of legal justice.

Our next subject is equity and the equitable (to epiekes), and their respective relations to justice and the just. For on examination they appear to be neither absolutely the same nor generically different; and while we sometime praise what is equitable and the equitable man (so that we apply the name by way of praise even to instances of the other virtues, instead of 'good' meaning by epieikestebon that a thing is better), at other times, when we reason it out, it seems strange if the equitable, being something different from the just, is yet praiseworthy; for either the just or the equitable is not good, if they are different; or, if both are good, they are the same.

These, then, are pretty much the considerations that give rise to the problem about the equitable; they are all in a sense correct and not opposed to one another; for the equitable, though it is better than one kind of justice, yet is just, and it is not as being a different class of thing that it is better than the just. The same thing, then, is just and equitable, and while both are good the equitable is superior. What creates the problem is that the equitable is just, but not the legally just but a correction of legal justice. The reason is that all law is universal but about some things it is not possible to make a universal statement which shall be correct. In those cases, then, in which it is necessary to speak universally, but not possible to do so correctly, the law takes the usual case, though it is not ignorant of the possibility of error. And it is none the less correct; for the error is in the law nor in the legislator but in the nature of the thing, since the matter of practical affairs is of this kind from the start. When the law speaks universally, then, and a case arises on it which is not covered by the universal statement, then it is right, where the legislator fails us and has erred by oversimplicity, to correct the omission—to say what the legislator himself would have said had he been present, and would have put into his law if he had known. Hence the equitable is just, and better than one kind of justice—not better than absolute justice but better than the error that arises from the absoluteness of the statement. And this is the nature of the equitable, a correction of law where it is defective owing to its universality. In fact this is the reason why all things are not determined by law, that about some things it is impossible to lay down a law, so that a decree is needed. For when the thing is indefinite the rule also is indefinite, like the leaden rule used in making the Lesbian moulding; the rule adapts itself to the shape of the stone and is not rigid, and so too the decree is adapted to the facts.

It is plain, then, what the equitable is, and that it is just and is better than one kind of justice. It is evident also from this who the equitable man is; the man who chooses and does such acts, and is no stickler for his rights in a bad sense but tends to take less than his share though he has the law oft his side, is equitable, and this state of character is equity, which is a sort of justice and not a different state of character.

Chapter 11: Can a man treat himself unjustly?

Whether a man can treat himself unjustly or not, is evident from what has been said. For (a) one class of just acts are those acts in accordance with any virtue which are prescribed by the law; e.g. the law does not expressly permit suicide, and what it does not expressly permit it forbids. Again, when a man in violation of the law harms another (otherwise than in retaliation) voluntarily, he acts unjustly, and a voluntary agent is one who knows both the person he is affecting by his action and the instrument he is using; and he who through anger voluntarily stabs himself does this contrary to the right rule of life, and this the law does not allow; therefore he is acting unjustly. But towards whom? Surely towards the state, not towards himself. For he suffers voluntarily, but no one is voluntarily treated unjustly. This is also the reason why the state punishes; a certain loss of civil rights attaches to the man who destroys himself, on the ground that he is treating the state unjustly.

Further (b) in that sense of 'acting unjustly' in which the man who 'acts unjustly' is unjust only and not bad all round, it is not possible to treat oneself unjustly (this is different from the former sense; the unjust man in one sense of the term is wicked in a particularized way just as the coward is, not in the sense of being wicked all round, so that his 'unjust act' does not manifest wickedness in general). For (i) that would imply the possibility of the same thing's having been subtracted from and added to the same thing at the same time; but this is impossible—the just and the unjust always involve more than one person. Further, (ii) unjust action is voluntary and done by choice, and takes the initiative (for the man who because he has suffered does the same in return is not thought to act unjustly); but if a man harms himself he suffers and does the same things at the same time. Further, (iii) if a man could treat himself unjustly, he could be voluntarily treated unjustly. Besides, (iv) no one acts unjustly without committing particular acts of injustice; but no one can commit adultery with his own wife or housebreaking on his own house or theft on his own property,

In general, the question 'can a man treat himself unjustly?' is solved also by the distinction we applied to the question 'can a man be voluntarily treated unjustly?'

(It is evident too that both are bad, being unjustly treated and acting unjustly; for the one means having less and the other having more than the intermediate amount, which plays the part here that the healthy does in the medical art, and that good condition does in the art of bodily training. But still acting unjustly is the worse, for it involves vice and is blameworthy—involves vice which is either of the complete and unqualified kind or almost so (we must admit the latter alternative, because not all voluntary unjust action implies injustice as a state of character), while being unjustly treated does not involve vice and injustice in oneself. In itself, then, being unjustly treated is less bad, but there is nothing to prevent its being incidentally a greater evil. But theory cares nothing for this; it calls pleurisy a more serious mischief than a stumble; yet the latter may become incidentally the more serious, if the fall due to it leads to your being taken prisoner or put to death the enemy.)

Metaphorically and in virtue of a certain resemblance there is a justice, not indeed between a man and himself, but between certain parts of him; yet not every kind of justice but that of master and servant or that of husband and wife. For these are the ratios in which the part of the soul that has a rational principle stands to the irrational part; and it is with a view to these parts that people also think a man can be unjust to himself, viz. because these parts are liable to suffer something contrary to their respective desires; there is therefore thought to be a mutual justice between them as between ruler and ruled.

Let this be taken as our account of justice and the other, i.e. the other moral, virtues.

Book VI: Intellectual Virtue

Chapter 1: Reasons for studying intellectual virtue: intellect divided into the contemplative and the calculative.

SINCE we have previously said that one ought to choose that which is intermediate, not the excess nor the defect, and that the intermediate is determined by the dictates of the right rule, let us discuss the nature of these dictates. In all the states of character we have mentioned, as in all other matters, there is a mark to which the man who has the rule looks, and heightens or relaxes his activity accordingly, and there is a standard which determines the mean states which we say are intermediate between excess and defect, being in accordance with the right rule. But such a statement, though true, is by no means clear; for not only here but in all other pursuits which are objects of knowledge it is indeed true to say that we must not exert ourselves nor relax our efforts too much nor too little, but to an intermediate extent and as the right rule dictates; but if a man had only this knowledge he would be none the wiser e.g. we should not know what sort of medicines to apply to our body if some one were to say 'all those which the medical art prescribes, and which agree with the practice of one who possesses the art'. Hence it is necessary with regard to the states of the soul also not only that this true statement should be made, but also that it should be determined what is the right rule and what is the standard that fixes it.

We divided the virtues of the soul and a said that some are virtues of character and others of intellect. Now we have discussed in detail the moral virtues; with regard to the others let us express our view as follows, beginning with some remarks about the soul. We said before that there are two parts of the soul—that which grasps a rule or rational principle, and the irrational; let us now draw a similar distinction within the part which grasps a rational principle. And let it be assumed that there are two parts which grasp a rational principle—one by which we contemplate the kind of things whose originative causes are invariable, and one by which we contemplate variable things; for where objects differ in kind the part of the soul answering to each of the two is different in kind, since it is in virtue of a certain likeness and kinship with their objects that they have the knowledge they have. Let one of these parts be called the scientific and the other the calculative; for to deliberate and to calculate are the same thing, but no one deliberates about the invariable. Therefore the calculative is one part of the faculty which grasps a rational principle. We must, then, learn what is the best state of each of these two parts; for this is the virtue of each.

Chapter 2: The object of the former is truth, that of the latter truth corresponding with right desire.

The virtue of a thing is relative to its proper work. Now there are three things in the soul which control action and truth—sensation, reason, desire.

Of these sensation originates no action; this is plain from the fact that the lower animals have sensation but no share in action.

What affirmation and negation are in thinking, pursuit and avoidance are in desire; so that since moral virtue is a state of character concerned with choice, and choice is deliberate desire, therefore both the reasoning must be true and the desire right, if the choice is to be good, and the latter must pursue just what the former asserts. Now this kind of intellect and of truth is practical; of the intellect which is contemplative, not practical nor productive, the good and the bad state are truth and falsity respectively (for this is the work of everything intellectual); while of the part which is practical and intellectual the good state is truth in agreement with right desire.

The origin of action—its efficient, not its final cause—is choice, and that of choice is desire and reasoning with a view to an end. This is why choice cannot

exist either without reason and intellect or without a moral state; for good action and its opposite cannot exist without a combination of intellect and character. Intellect itself, however, moves nothing, but only the intellect which aims at an end and is practical; for this rules the productive intellect, as well, since every one who makes makes for an end, and that which is made is not an end in the unqualified sense (but only an end in a particular relation, and the end of a particular operation)—only that which is done is that; for good action is an end, and desire aims at this. Hence choice is either desiderative reason or ratiocinative desire, and such an origin of action is a man. (It is to be noted that nothing that is past is an object of choice, e.g. no one chooses to have sacked Troy; for no one deliberates about the past, but about what is future and capable of being otherwise, while what is past is not capable of not having taken place; hence Agathon is right in saying
For this alone is lacking even to God,
To make undone things that have once been done.)

The work of both the intellectual parts, then, is truth. Therefore the states that are most strictly those in respect of which each of these parts will reach truth are the virtues of the two parts.

Chapter 3: Science—demonstrative knowledge of the necessary and eternal.

Let us begin, then, from the beginning, and discuss these states once more. Let it be assumed that the states by virtue of which the soul possesses truth by way of affirmation or denial are five in number, i.e. art, scientific knowledge, practical wisdom, philosophic wisdom, intuitive reason; we do not include judgement and opinion because in these we may be mistaken.

Now what scientific knowledge is, if we are to speak exactly and not follow mere similarities, is plain from what follows. We all suppose that what we know is not even capable of being otherwise; of things capable of being otherwise we do not know, when they have passed outside our observation, whether they exist or not. Therefore the object of scientific knowledge is of necessity. Therefore it is eternal; for things that are of necessity in the unqualified sense are all eternal; and things that are eternal are ungenerated and imperishable. Again, every science is thought to be capable of being taught, and its object of being learned. And all teaching starts from what is already known, as we maintain in the Analytics also; for it proceeds sometimes through induction and sometimes by syllogism. Now induction is the starting-point which knowledge even of the universal presupposes, while syllogism proceeds from universals. There are therefore starting-points from which syllogism proceeds, which are not reached by syllogism; it is therefore by induction that they are acquired. Scientific knowledge is, then, a state of capacity to demonstrate, and has the other limiting characteristics which we specify in the Analytics, for it is when a man believes in a certain way and the starting-points are known to him that he has scientific knowledge, since if they are not better known to him than the conclusion, he will have his knowledge only incidentally.

Let this, then, be taken as our account of scientific knowledge.

Chapter 4: Art—knowledge of how to make things.

In the variable are included both things made and things done; making and acting are different (for their nature we treat even the discussions outside our school as reliable); so that the reasoned state of capacity to act is different from the reasoned state of capacity to make. Hence too they are not included one in the other; for neither is acting making nor is making acting. Now since architecture is an art and is essentially a reasoned state of capacity to make, and there is neither any art that is not such a state nor any such state that is not an art, art is identical with a state of capacity to make, involving a true course of reasoning. All art is concerned with coming into being, i.e. with contriving and considering how something may come into being which is capable of either being or not being, and whose origin is in the maker and not in the thing made; for art is concerned neither with things that are, or come into being, by necessity, nor with things that do so in accordance with nature (since these have their origin in themselves). Making and acting being different, art must be a matter of making, not of acting. And in a sense chance and art are concerned with the same objects; as Agathon says, 'art loves chance and chance loves art'. Art, then, as has been is a state concerned with making, involving a true course of reasoning, and lack of art on the contrary is a state concerned with making, involving a false course of reasoning; both are concerned with the variable.

Chapter 5: Practical wisdom—knowledge of how to secure the ends of human life.

Regarding practical wisdom we shall get at the truth by considering who are the persons we credit with it. Now it is thought to be the mark of a man of practical wisdom to be able to deliberate well about what is good and expedient for himself, not in some particular respect, e.g. about what sorts of thing conduce to health or to strength, but about what sorts of thing conduce to the good life in general. This is shown by the fact that we credit men with practical wisdom in some particular respect when they have calculated well with a view to some good end which is one of those that are not the object of any art. It follows that in the general sense also the man who is capable of deliberating has practical wisdom. Now no one deliberates about things that are invariable, nor about things that it is impossible for him to do. Therefore, since scientific knowledge involves demonstration, but there is no demonstration of things whose first principles are variable (for all such things might actually be otherwise), and since it is impossible to deliberate about things that are of necessity, practical wisdom cannot be scientific knowledge nor art; not science because that which can be done is capable of being otherwise, not art because action and making are different kinds of thing. The remaining alternative, then, is that it is a true and reasoned state of capacity to act with regard to the things that are good or bad for man. For while making has an end other than itself, action cannot; for good action itself is its end. It is for this reason that we think Pericles and men like him have practical wisdom, viz. because they can see what is good for themselves and what is good for men in general; we consider that those can do this who are good at managing households or states. (This is why we call temperance (sophrosune) by this name; we imply that it preserves one's practical wisdom (sozousa tan phronsin). Now what it preserves is a judgement of the kind we have described. For it is not any and every judgement that pleasant and painful objects destroy and pervert, e.g. the judgement that the triangle has or has not its angles equal to two right angles, but only judgements about what is to be done. For the originating causes of the things that are done consist in the end at which they are aimed; but the man who has been ruined by pleasure or pain forthwith fails to see any such originating cause—to see that for the sake of this or because of this he ought to choose and do whatever he chooses and does; for vice is destructive of the originating cause of action.) Practical wisdom, then, must be a reasoned and true state of capacity to act with regard to human goods. But further, while there is such a thing as excellence in art, there is no such thing as excellence in practical wisdom; and in art he who errs willingly is preferable, but in practical wisdom, as in the virtues, he is the reverse. Plainly, then, practical wisdom is a virtue and not an art. There being two parts of the soul that can follow a course of reasoning, it must be the virtue of one of the two, i.e. of that part which forms opinions; for opinion is about the variable and so is practical wisdom. But yet it is not only a reasoned state; this is shown by the fact that a state of that sort may forgotten but practical wisdom cannot.

Chapter 6: Intuitive reason—knowledge of the principles from which science proceeds.

Scientific knowledge is judgement about things that are universal and necessary, and the conclusions of demonstration, and all scientific knowledge, follow from first principles (for scientific knowledge involves apprehension of a rational ground). This being so, the first principle from which what is scientifically known follows cannot be an object of scientific knowledge, of art, or of practical wisdom; for that which can be scientifically known can be demonstrated, and art and practical wisdom deal with things that are variable. Nor are these first principles the objects of philosophic wisdom, for it is a mark of the philosopher to have demonstration about some things. If, then, the states of mind by which we have truth and are never deceived about things invariable or even variable are scientific knowlededge, practical wisdom, philosophic wisdom, and intuitive reason, and it cannot be any of the three (i.e. practical wisdom, scientific knowledge, or philosophic wisdom), the remaining alternative is that it is intuitive reason that grasps the first principles.

Chapter 7: Philosophic wisdom—the union of intuitive reason and science.

Wisdom (1) in the arts we ascribe to their most finished exponents, e.g. to Phidias as a sculptor and to Polyclitus as a maker of portrait-statues, and here we mean nothing by wisdom except excellence in art; but (2) we think that some people are wise in general, not in some particular field or in any other limited respect, as Homer says in the Margites,
Him did the gods make neither a digger nor yet a ploughman
Nor wise in anything else.
Therefore wisdom must plainly be the most finished of the forms of knowledge. It follows that the wise man must not only know what follows from the first principles, but must also possess truth about the first principles. Therefore wisdom must be intuitive reason combined with scientific knowledge—scientific knowledge of the highest objects which has received as it were its proper completion.

Of the highest objects, we say; for it would be strange to think that the art of politics, or practical wisdom, is the best knowledge, since man is not the best thing in the world. Now if what is healthy or good is different for men and for fishes, but what is white or straight is always the same, any one would say that what is wise is the same but what is practically wise is different; for it is to that which observes well the various matters concerning itself that one ascribes practical wisdom, and it is to this that one will entrust such matters. This is why we say that some even of the lower animals have practical wisdom, viz. those which are found to have a power of foresight with regard to their own life. It is evident also that philosophic wisdom and the art of politics cannot be the same; for if the state of mind concerned with a man's own interests is to be called philosophic wisdom, there will be many philosophic wisdoms; there will not be one concerned with the good of all animals (any more than there is one art of medicine for all existing things), but a different philosophic wisdom about the good of each species.

But if the argument be that man is the best of the animals, this makes no difference; for there are other things much more divine in their nature even than man, e.g., most conspicuously, the bodies of which the heavens are framed. From what has been said it is plain, then, that philosophic wisdom is scientific knowledge, combined with intuitive reason, of the things that are highest by nature. This is why we say Anaxagoras, Thales, and men like them have philosophic but not practical wisdom, when we see them ignorant of what is to their own advantage, and why we say that they know things that are remarkable, admirable, difficult, and divine, but useless; viz. because it is not human goods that they seek.

Practical wisdom on the other hand is concerned with things human and things about which it is possible to deliberate; for we say this is above all the work of the man of practical wisdom, to deliberate well, but no one deliberates about things invariable, nor about things which have not an end, and that a good that can be brought about by action. The man who is without qualification good at deliberating is the man who is capable of aiming in accordance with calculation at the best for man of things attainable by action. Nor is practical wisdom concerned with universals only—it must also recognize the particulars; for it is practical, and practice is concerned with particulars. This is why some who do not know, and especially those who have experience, are more practical than others who know; for if a man knew that light meats are digestible and wholesome, but did not know which sorts of meat are light, he would not produce health, but the man who knows that chicken is wholesome is more likely to produce health.

Now practical wisdom is concerned with action; therefore one should have both forms of it, or the latter in preference to the former. But of practical as of philosophic wisdom there must be a controlling kind.

Chapter 8: Relations between practical wisdom and political science.

Political wisdom and practical wisdom are the same state of mind, but their essence is not the same. Of the wisdom concerned with the city, the practical wisdom which plays a controlling part is legislative wisdom, while that which is related to this as particulars to their universal is known by the general name 'political wisdom'; this has to do with action and deliberation, for a decree is a thing to be carried out in the form of an individual act. This is why the exponents of this art are alone said to 'take part in politics'; for these alone 'do things' as manual labourers 'do things'.

Practical wisdom also is identified especially with that form of it which is concerned with a man himself—with the individual; and this is known by the general name 'practical wisdom'; of the other kinds one is called household management, another legislation, the third politics, and of the latter one part is called deliberative and the other judicial. Now knowing what is good for oneself will be one kind of knowledge, but it is very different from the other kinds; and the man who knows and concerns himself with his own interests is thought to have practical wisdom, while politicians are thought to be busybodies; hence the word of Euripides,

But how could I be wise, who might at ease,
Numbered among the army's multitude,
Have had an equal share?
For those who aim too high and do too much.

Those who think thus seek their own good, and consider that one ought to do so. From this opinion, then, has come the view that such men have practical wisdom; yet perhaps one's own good cannot exist without household management, nor without a form of government. Further, how one should order one's own affairs is not clear and needs inquiry.

What has been said is confirmed by the fact that while young men become geometricians and mathematicians and wise in matters like these, it is thought that a young man of practical wisdom cannot be found. The cause is that such wisdom is concerned not only with universals but with particulars, which become familiar from experience, but a young man has no experience, for it is length of time that gives experience; indeed one might ask this question too, why a boy may become a mathematician, but not a philosopher or a physicist. It is because the objects of mathematics exist by abstraction, while the first principles of these other subjects come from experience, and because young men have no conviction about the latter but merely use the proper language, while the essence of mathematical objects is plain enough to them?

Further, error in deliberation may be either about the universal or about the particular; we may fall to know either that all water that weighs heavy is bad, or that this particular water weighs heavy.

That practical wisdom is not scientific knowledge is evident; for it is, as has been said, concerned with the ultimate particular fact, since the thing to be done is of this nature. It is opposed, then, to intuitive reason; for intuitive reason is of the limiting premisses, for which no reason can be given, while practical wisdom is concerned with the ultimate particular, which is the object not of scientific knowledge but of perception—not the perception of qualities peculiar to one sense but a perception akin to that by which we perceive that the particular figure before us is a triangle; for in that direction as well as in that of the major premiss there will be a limit. But this is rather perception than practical wisdom, though it is another kind of perception than that of the qualities peculiar to each sense.

Chapter 9: Goodness in deliberation, how related to practical wisdom.

There is a difference between inquiry and deliberation; for deliberation is inquiry into a particular kind of thing. We must grasp the nature of excellence in deliberation as well whether it is a form of scientific knowledge, or opinion, or skill in conjecture, or some other kind of thing. Scientific knowledge it is not; for men do not inquire about the things they know about, but good deliberation is a kind of deliberation, and he who deliberates inquires and calculates. Nor is it skill in conjecture; for this both involves no reasoning and is something that is quick in its operation, while men deliberate a long time, and they say that one should carry out quickly the conclusions of one's deliberation, but should deliberate slowly. Again, readiness of mind is different from excellence in deliberation; it is a sort of skill in conjecture. Nor again is excellence in deliberation opinion of any sort. But since the man who deliberates badly makes a mistake, while he who deliberates well does so correctly, excellence in deliberation is clearly a kind of correctness, but neither of knowledge nor of opinion; for there is no such thing as correctness of knowledge (since there is no such thing as error of knowledge), and correctness of opinion is truth; and at the same time everything that is an object of opinion is already determined. But again excellence in deliberation involves reasoning. The remaining alternative, then, is that it is correctness of thinking; for this is not yet assertion, since, while even opinion is not inquiry but has reached the stage of assertion, the man who is deliberating, whether he does so well or ill, is searching for something and calculating.

But excellence in deliberation is a certain correctness of deliberation; hence we must first inquire what deliberation is and what it is about. And, there being more than one kind of correctness, plainly excellence in deliberation is not any and every kind; for (1) the incontinent man and the bad man, if he is clever, will reach as a result of his calculation what he sets before himself, so that he will have deliberated correctly, but he will have got for himself a great evil. Now to have deliberated well is thought to be a good thing; for it is this kind of correctness of deliberation that is excellence in deliberation, viz. that which tends to attain what is good. But (2) it is possible to attain even good by a false syllogism, and to attain what one ought to do but not by the right means, the middle term being false; so that this too is not yet excellence in deliberation this state in virtue of which one attains what one ought but not by the right means. Again (3) it is possible to attain it by long deliberation while another man attains it quickly. Therefore in the former case we have not yet got excellence in deliberation, which is rightness with regard to the expedient—rightness in respect both of the end, the manner, and the time. (4) Further it is possible to have deliberated well either in the unqualified sense or with reference to a particular end. Excellence in deliberation in the unqualified sense, then, is that which succeeds with reference to what is the end in the unqualified sense, and excellence in deliberation in a particular sense is that which succeeds relatively to a particular end. If, then, it is characteristic of men of practical wisdom to have deliberated well, excellence in deliberation will be correctness with regard to what conduces to the end of which practical wisdom is the true apprehension.

Chapter 10: Understanding—the critical quality answering to the imperative quality practical wisdom.

Understanding, also, and goodness of understanding, in virtue of which men are said to be men of understanding or of good understanding, are neither entirely the same as opinion or scientific knowledge (for at that rate all men would have been men of understanding), nor are they one of the particular sciences, such as medicine, the science of things connected with health, or geometry, the science of spatial magnitudes. For understanding is neither about things that are always and are unchangeable, nor about any and every one of the things that come into being, but about things which may become subjects of questioning and deliberation. Hence it is about the same objects as practical wisdom; but understanding and practical wisdom are not the same. For practical wisdom issues commands, since its end is what ought to be done or not to be done; but understanding only judges. (Understanding is identical with goodness of understanding, men of understanding with men of good understanding.) Now understanding is neither the having nor the acquiring of practical wisdom; but as learning is called understanding when it means the exercise of the faculty of knowledge, so 'understanding' is applicable to the exercise of the faculty of opinion for the purpose of judging of what some one else says about matters with which practical wisdom is concerned—and of judging soundly; for 'well' and 'soundly' are the same thing. And from this has come the use of the name 'understanding' in virtue of which men are said to be 'of good understanding', viz. from the application of the word to the grasping of scientific truth; for we often call such grasping understanding.

Chapter 11: Judgement—right discrimination of the equitable: the place of intuition in morals.

What is called judgement, in virtue of which men are said to 'be sympathetic judges' and to 'have judgement', is the right discrimination of the equitable. This is shown by the fact that we say the equitable man is above all others a man of sympathetic judgement, and identify equity with sympathetic judgement about certain facts. And sympathetic judgement is judgement which discriminates what is equitable and does so correctly; and correct judgement is that which judges what is true.

Now all the states we have considered converge, as might be expected, to the same point; for when we speak of judgement and understanding and practical wisdom and intuitive reason we credit the same people with possessing judgement and having reached years of reason and with having practical wisdom and understanding. For all these faculties deal with ultimates, i.e. with particulars; and being a man of understanding and of good or sympathetic judgement consists in being able judge about the things with which practical wisdom is concerned; for the equities are common to all good men in relation to other men. Now all things which have to be done are included among particulars or ultimates; for not only must the man of practical wisdom know particular facts, but understanding and judgement are also concerned with things to be done, and these are ultimates. And intuitive reason is concerned with the ultimates in both directions; for both the first terms and the last are objects of intuitive reason and not of argument, and the intuitive reason which is presupposed by demonstrations grasps the unchangeable and first terms, while the intuitive reason involved in practical reasonings grasps the last and variable fact, i.e. the minor premiss. For these variable facts are the starting-points for the apprehension of the end, since the universals are reached from the particulars; of these therefore we must have perception, and this perception is intuitive reason.

This is why these states are thought to be natural endowments—why, while no one is thought to be a philosopher by nature, people are thought to have by nature judgement, understanding, and intuitive reason. This is shown by the fact that we think our powers correspond to our time of life, and that a particular age brings with it intuitive reason and judgement; this implies that nature is the cause. (Hence intuitive reason is both beginning and end; for demonstrations are from these and about these.) Therefore we ought to attend to the undemonstrated sayings and opinions of experienced and older people or of people of practical wisdom not less than to demonstrations; for because experience has given them an eye they see aright.

We have stated, then, what practical and philosophic wisdom are, and with what each of them is concerned, and we have said that each is the virtue of a different part of the soul.

Chapter 12: What is the use of philosophic and of practical wisdom? Philosophic wisdom is the formal cause of happiness; practical wisdom is what ensures the taking of proper means to the proper ends desired by moral virtue.

Difficulties might be raised as to the utility of these qualities of mind. For (1) philosophic wisdom will contemplate none of the things that will make a man happy (for it is not concerned with any coming into being), and though practical wisdom has this merit, for what purpose do we need it? Practical wisdom is the quality of mind concerned with things just and noble and good for man, but these are the things which it is the mark of a good man to do, and we are none the more able to act for knowing them if the virtues are states of character, just as we are none the better able to act for knowing the things that are healthy and sound, in the sense not of producing but of issuing from the state of health; for we are none the more able to act for having the art of medicine or of gymnastics. But (2) if we are to say that a man should have practical wisdom not for the sake of knowing moral truths but for the sake of becoming good, practical wisdom will be of no use to those who are good; again it is of no use to those who have not virtue; for it will make no difference whether they have practical wisdom themselves or obey others who have it, and it would be enough for us to do what we do in the case of health; though we wish to become healthy, yet we do not learn the art of medicine. (3) Besides this, it would be thought strange if practical wisdom, being inferior to philosophic wisdom, is to be put in authority over it, as seems to be implied by the fact that the art which produces anything rules and issues commands about that thing.

These, then, are the questions we must discuss; so far we have only stated the difficulties.

(1) Now first let us say that in themselves these states must be worthy of choice because they are the virtues of the two parts of the soul respectively, even if neither of them produce anything.

(2) Secondly, they do produce something, not as the art of medicine produces health, however, but as health produces health; so does philosophic wisdom produce happiness; for, being a part of virtue entire, by being possessed and by actualizing itself it makes a man happy.

(3) Again, the work of man is achieved only in accordance with practical wisdom as well as with moral virtue; for virtue makes us aim at the right mark, and practical wisdom makes us take the right means. (Of the fourth part of the soul—the nutritive—there is no such virtue; for there is nothing which it is in its power to do or not to do.)

(4) With regard to our being none the more able to do because of our practical wisdom what is noble and just, let us begin a little further back, starting with the following principle. As we say that some people who do just acts are not necessarily just, i.e. those who do the acts ordained by the laws either unwillingly or owing to ignorance or for some other reason and not for the sake of the acts themselves (though, to be sure, they do what they should and all the things that the good man ought), so is it, it seems, that in order to be good one must be in a certain state when one does the several acts, i.e. one must do them as a result of choice and for the sake of the acts themselves. Now virtue makes the choice right, but the question of the things which should naturally be done to carry out our choice belongs not to virtue but to another faculty. We must devote our attention to these matters and give a clearer statement about them. There is a faculty which is called cleverness; and this is such as to be able to do the things that tend towards the mark we have set before ourselves, and to hit it. Now if the mark be noble, the cleverness is laudable, but if the mark be bad, the cleverness is mere smartness; hence we call even men of practical wisdom clever or smart. Practical wisdom is not the faculty, but it does not exist without this faculty. And this eye of the soul acquires its formed state not without the aid of virtue, as has been said and is plain; for the syllogisms which deal with acts to be done are things which involve a starting-point, viz. 'since the end, i.e. what is best, is of such and such a nature', whatever it may be (let it for the sake of argument be what we please), and this is not evident except to the good man; for wickedness perverts us and causes us to be deceived about the starting-points of action. Therefore it is evident that it is impossible to be practically wise without being good.

Chapter 13: Relation of practical wisdom to natural virtue, moral virtue, and the right rule.

We must therefore consider virtue also once more; for virtue too is similarly related; as practical wisdom is to cleverness—not the same, but like it—so is natural virtue to virtue in the strict sense. For all men think that each type of character belongs to its possessors in some sense by nature; for from the very moment of birth we are just or fitted for selfcontrol or brave or have the other moral qualities; but yet we seek something else as that which is good in the strict sense—we seek for the presence of such qualities in another way. For both children and brutes have the natural dispositions to these qualities, but without reason these are evidently hurtful. Only we seem to see this much, that, while one may be led astray by them, as a strong body which moves without sight may stumble badly because of its lack of sight, still, if a man once acquires reason, that makes a difference in action; and his state, while still like what it was, will then be virtue in the strict sense. Therefore, as in the part of us which forms opinions there are two types, cleverness and practical wisdom, so too in the moral part there are two types, natural virtue and virtue in the strict sense, and of these the latter involves practical wisdom. This is why some say that all the virtues are forms of practical wisdom, and why Socrates in one respect was on the right track while in another he went astray; in thinking that all the virtues were forms of practical wisdom he was wrong, but in saying they implied practical wisdom he was right. This is confirmed by the fact that even now all men, when they define virtue, after naming the state of character and its objects add 'that (state) which is in accordance with the right rule'; now the right rule is that which is in accordance with practical wisdom. All men, then, seem somehow to divine that this kind of state is virtue, viz. that which is in accordance with practical wisdom. But we must go a little further. For it is not merely the state in accordance with the right rule, but the state that implies the presence of the right rule, that is virtue; and practical wisdom is a right rule about such matters. Socrates, then, thought the virtues were rules or rational principles (for he thought they were, all of them, forms of scientific knowledge), while we think they involve a rational principle.

It is clear, then, from what has been said, that it is not possible to be good in the strict sense without practical wisdom, nor practically wise without moral virtue. But in this way we may also refute the dialectical argument whereby it might be contended that the virtues exist in separation from each other; the same man, it might be said, is not best equipped by nature for all the virtues, so that he will have already acquired one when he has not yet acquired another. This is possible in respect of the natural virtues, but not in respect of those in respect of which a man is called without qualification good; for with the presence of the one quality, practical wisdom, will be given all the virtues. And it is plain that, even if it were of no practical value, we should have needed it because it is the virtue of the part of us in question; plain too that the choice will not be right without practical wisdom any more than without virtue; for the one deter, mines the end and the other makes us do the things that lead to the end.

But again it is not supreme over philosophic wisdom, i.e. over the superior part of us, any more than the art of medicine is over health; for it does not use it but provides for its coming into being; it issues orders, then, for its sake, but not to it. Further, to maintain its supremacy would be like saying that the art of politics rules the gods because it issues orders about all the affairs of the state.

Book VII: Continence and Incontinence

Chapter 1: Six varieties of character: method of treatment: current opinions.

LET us now make a fresh beginning and point out that of moral states to be avoided there are three kinds—vice, incontinence, brutishness. The contraries of two of these are evident, — one we call virtue, the other continence; to brutishness it would be most fitting to oppose superhuman virtue, a heroic and divine kind of virtue, as Homer has represented Priam saying of Hector that he was very good,

For he seemed not, he,

The child of a mortal man, but as one that of God's seed came.

Therefore if, as they say, men become gods by excess of virtue, of this kind must evidently be the state opposed to the brutish state; for as a brute has no vice or virtue, so neither has a god; his state is higher than virtue, and that of a brute is a different kind of state from vice.

Now, since it is rarely that a godlike man is found—to use the epithet of the Spartans, who when they admire any one highly call him a 'godlike man'—so too the brutish type is rarely found among men; it is found chiefly among barbarians, but some brutish qualities are also produced by disease or deformity; and we also call by this evil name those men who go beyond all ordinary standards by reason of vice. Of this kind of disposition, however, we must later make some mention, while we have discussed vice before we must now discuss incontinence and softness (or effeminacy), and continence and endurance; for we must treat each of the two neither as identical with virtue or wickedness, nor as a different genus. We must, as in all other cases, set the observed facts before us and, after first discussing the difficulties, go on to prove, if possible, the truth of all the common opinions about these affections of the mind, or, failing this, of the greater number and the most authoritative; for if we both refute the objections and leave the common opinions undisturbed, we shall have proved the case sufficiently.

Now (1) both continence and endurance are thought to be included among things good and praiseworthy, and both incontinence and soft, ness among things bad and blameworthy; and the same man is thought to be continent and ready to abide by the result of his calculations, or incontinent and ready to abandon them. And (2) the incontinent man, knowing that what he does is bad, does it as a result of passion, while the continent man, knowing that his appetites are bad, refuses on account of his rational principle to follow them (3) The temperate man all men call continent and disposed to endurance, while the continent man some maintain to be always temperate but others do not; and some call the self-indulgent man incontinent and the incontinent man selfindulgent indiscriminately, while others distinguish them. (4) The man of practical wisdom, they sometimes say, cannot be incontinent, while sometimes they say that some who are practically wise and clever are incontinent. Again (5) men are said to be incontinent even with respect to anger, honour, and gain—These, then, are the things that are said.

Chapter 2: Contradictions involved in these opinions.

Now we may ask (1) how a man who judges rightly can behave incontinently. That he should behave so when he has knowledge, some say is impossible; for it would be strange—so Socrates thought—if when knowledge was in a man something else could master it and drag it about like a slave. For Socrates was entirely opposed to the view in question, holding that there is no such thing as incontinence; no one, he said, when he judges acts against what he judges best—people act so only by reason of ignorance. Now this view plainly contradicts the observed facts, and we must inquire about what happens to such a man; if he acts by reason of ignorance, what is the manner of his ignorance? For that the man who behaves incontinently does not, before he gets into this state, think he ought to act so, is evident. But there are some who concede certain of Socrates' contentions but not others; that nothing is stronger than knowledge they admit, but not that on one acts contrary to what has seemed to him the better course, and therefore they say that the incontinent man has not knowledge when he is mastered by his pleasures, but opinion. But if it is opinion and not knowledge, if it is not a strong conviction that resists but a weak one, as in men who hesitate, we sympathize with their failure to stand by such convictions against strong appetites; but we do not sympathize with wickedness, nor with any of the other blameworthy states. Is it then practical wisdom whose resistance is mastered? That is the strongest of all states. But this is absurd; the same man will be at once practically wise and incontinent, but no one would say that it is the part of a practically wise man to do willingly the basest acts. Besides, it has been shown before that the man of practical wisdom is one who will act (for he is a man concerned with the individual facts) and who has the other virtues.

(2) Further, if continence involves having strong and bad appetites, the temperate man will not be continent nor the continent man temperate; for a temperate man will have neither excessive nor bad appetites. But the continent man must; for if the appetites are good, the state of character that restrains us from following them is bad, so that not all continence will be good; while if they are weak and not bad, there is nothing admirable in resisting them, and if they are weak and bad, there is nothing great in resisting these either.

(3) Further, if continence makes a man ready to stand by any and every opinion, it is bad, i.e. if it makes him stand even by a false opinion; and if incontinence makes a man apt to abandon any and every opinion, there will be a

good incontinence, of which Sophocles' Neoptolemus in the Philoctetes will be an instance; for he is to be praised for not standing by what Odysseus persuaded him to do, because he is pained at telling a lie.

(4) Further, the sophistic argument presents a difficulty; the syllogism arising from men's wish to expose paradoxical results arising from an opponent's view, in order that they may be admired when they succeed, is one that puts us in a difficulty (for thought is bound fast when it will not rest because the conclusion does not satisfy it, and cannot advance because it cannot refute the argument). There is an argument from which it follows that folly coupled with incontinence is virtue; for a man does the opposite of what he judges, owing to incontinence, but judges what is good to be evil and something that he should not do, and consequence he will do what is good and not what is evil.

(5) Further, he who on conviction does and pursues and chooses what is pleasant would be thought to be better than one who does so as a result not of calculation but of incontinence; for he is easier to cure since he may be persuaded to change his mind. But to the incontinent man may be applied the proverb 'when water chokes, what is one to wash it down with?' If he had been persuaded of the rightness of what he does, he would have desisted when he was persuaded to change his mind; but now he acts in spite of his being persuaded of something quite different.

(6) Further, if incontinence and continence are concerned with any and every kind of object, who is it that is incontinent in the unqualified sense? No one has all the forms of incontinence, but we say some people are incontinent without qualification.

Chapter 3: Solution of the problem, in what sense the incontinent man acts against knowledge.

Of some such kind are the difficulties that arise; some of these points must be refuted and the others left in possession of the field; for the solution of the difficulty is the discovery of the truth. (1) We must consider first, then, whether incontinent people act knowingly or not, and in what sense knowingly; then (2) with what sorts of object the incontinent and the continent man may be said to be concerned (i.e. whether with any and every pleasure and pain or with certain determinate kinds), and whether the continent man and the man of endurance are the same or different; and similarly with regard to the other matters germane to this inquiry. The starting-point of our investigation is (a) the question whether the continent man and the incontinent are differentiated by their objects or by their attitude, i.e. whether the incontinent man is incontinent simply by being concerned with such and such objects, or, instead, by his attitude, or, instead of that, by both these things; (b) the second question is whether incontinence and continence are concerned with any and every object or not. The man who is incontinent in the unqualified sense is neither concerned with any and every object, but with precisely those with which the self-indulgent man is concerned, nor is he characterized by being simply related to these (for then his state would be the same as self-indulgence), but by being related to them in a certain way. For the one is led on in accordance with his own choice, thinking that he ought always to pursue the present pleasure; while the other does not think so, but yet pursues it.

(1) As for the suggestion that it is true opinion and not knowledge against which we act incontinently, that makes no difference to the argument; for some people when in a state of opinion do not hesitate, but think they know exactly. If, then, the notion is that owing to their weak conviction those who have opinion are more likely to act against their judgement than those who know, we answer that there need be no difference between knowledge and opinion in this respect; for some men are no less convinced of what they think than others of what they know; as is shown by the of Heraclitus. But (a), since we use the word 'know' in two senses (for both the man who has knowledge but is not using it and he who is using it are said to know), it will make a difference whether, when a man does what he should not, he has the knowledge but is not exercising it, or is exercising it; for the latter seems strange, but not the former.

(b) Further, since there are two kinds of premisses, there is nothing to prevent a man's having both premisses and acting against his knowledge, provided that he is using only the universal premiss and not the particular; for it is particular acts that have to be done. And there are also two kinds of universal term; one is predicable of the agent, the other of the object; e.g. 'dry food is good for every man', and 'I am a man', or 'such and such food is dry'; but whether 'this food is such and such', of this the incontinent man either has not or is not exercising the knowledge. There will, then, be, firstly, an enormous difference between these manners of knowing, so that to know in one way when we act incontinently would not seem anything strange, while to know in the other way would be extraordinary.

And further (c) the possession of knowledge in another sense than those just named is something that happens to men; for within the case of having knowledge but not using it we see a difference of state, admitting of the possibility of having knowledge in a sense and yet not having it, as in the instance of a man asleep, mad, or drunk. But now this is just the condition of men under the influence of passions; for outbursts of anger and sexual appetites and some other such passions, it is evident, actually alter our bodily condition, and in some men even produce fits of madness. It is plain, then, that incontinent people must be said to be in a similar condition to men asleep, mad, or drunk. The fact that men use the language that flows from knowledge proves nothing; for even men under the influence of these passions utter scientific proofs and verses of Empedocles, and those who have just begun to learn a science can string together its phrases, but do not yet know it; for it has to become part of themselves, and that takes time; so that we must suppose that the use of language by men in an incontinent state means no more than its utterance by actors on the stage. (d) Again, we may also view the cause as follows with reference to the facts of human nature. The one opinion is universal, the other is concerned with the particular facts, and here we come to something within the sphere of perception; when a single opinion results from the two, the soul must in one type of case affirm the conclusion, while in the case of opinions concerned with production it must immediately act (e.g. if 'everything sweet ought to be tasted', and 'this is sweet', in the sense of being one of the particular sweet things, the man who can act and is not prevented must at the same time actually act accordingly). When, then, the universal opinion is present in us forbidding us to taste, and there is also the opinion that 'everything sweet is pleasant', and that 'this is sweet' (now this is the opinion that is active), and when appetite happens to be present in us, the one opinion bids us avoid the object, but appetite leads us towards it (for it can move each of our bodily parts); so that it turns out that a man behaves incontinently under the influence (in a sense) of a rule and an opinion, and of one not contrary in itself, but only incidentally—for the appetite is contrary, not the opinion—to the right rule. It also follows that this is the reason why the lower animals are not incontinent, viz. because they have no universal judgement but only imagination and memory of particulars.

The explanation of how the ignorance is dissolved and the incontinent man regains his knowledge, is the same as in the case of the man drunk or asleep and is not peculiar to this condition; we must go to the students of natural science for it. Now, the last premiss both being an opinion about a perceptible object, and being what determines our actions this a man either has not when he is in the state of passion, or has it in the sense in which having knowledge did not mean knowing but only talking, as a drunken man may utter the verses of Empedocles. And because the last term is not universal nor equally an object of scientific knowledge with the universal term, the position that Socrates sought to establish actually seems to result; for it is not in the presence of what is thought to be knowledge proper that the affection of incontinence arises (nor is it this that is 'dragged about' as a result of the state of passion), but in that of perceptual knowledge.

This must suffice as our answer to the question of action with and without knowledge, and how it is possible to behave incontinently with knowledge.

Chapter 4: Solution of the problem, what is the sphere of incontinence: its proper and its extended sense distinguished.

(2) We must next discuss whether there is any one who is incontinent without qualification, or all men who are incontinent are so in a particular sense, and if there is, with what sort of objects he is concerned. That both continent persons and persons of endurance, and incontinent and soft persons, are concerned with pleasures and pains, is evident.

Now of the things that produce pleasure some are necessary, while others are worthy of choice in themselves but admit of excess, the bodily causes of pleasure being necessary (by such I mean both those concerned with food and those concerned with sexual intercourse, i.e. the bodily matters with which we defined self-indulgence and temperance as being concerned), while the others are not necessary but worthy of choice in themselves (e.g. victory, honour, wealth, and good and pleasant things of this sort). This being so, (a) those who go to excess with reference to the latter, contrary to the right rule which is in themselves, are not called incontinent simply, but incontinent with the qualification 'in respect of money, gain, honour, or anger',—not simply incontinent, on the ground that they are different from incontinent people and are called incontinent by reason of a resemblance. (Compare the case of Anthropos (Man), who won a contest at the Olympic games; in his case the general definition of man differed little from the definition peculiar to him, but yet it was different.) This is shown by the fact that incontinence either without qualification or in respect of some particular bodily pleasure is blamed not only as a fault but as a kind of vice, while none of the people who are incontinent in these other respects is so blamed.

But (b) of the people who are incontinent with respect to bodily enjoyments, with which we say the temperate and the self-indulgent man are concerned, he who pursues the excesses of things pleasant—and shuns those of things painful, of hunger and thirst and heat and cold and all the objects of touch and taste—not by choice but contrary to his choice and his judgement, is called incontinent, not with the qualification 'in respect of this or that', e.g. of anger, but just simply. This is confirmed by the fact that men are called 'soft' with regard to these pleasures, but not with regard to any of the others. And for this reason we group together the incontinent and the self-indulgent, the continent and the temperate man—but not any of these other types—because they are concerned somehow with the same pleasures and pains; but though these are concerned with the same objects, they are not similarly related to them, but some of them make a deliberate choice while the others do not.

This is why we should describe as self-indulgent rather the man who without appetite or with but a slight appetite pursues the excesses of pleasure and avoids moderate pains, than the man who does so because of his strong appetites; for what would the former do, if he had in addition a vigorous appetite, and a violent pain at the lack of the 'necessary' objects?

Now of appetites and pleasures some belong to the class of things generically noble and good—for some pleasant things are by nature worthy of choice, while others are contrary to these, and others are intermediate, to adopt our previous distinction—e.g. wealth, gain, victory, honour. And with reference to all objects whether of this or of the intermediate kind men are not blamed for being affected by them, for desiring and loving them, but for doing so in a certain way, i.e. for going to excess. (This is why all those who contrary to the rule either are mastered by or pursue one of the objects which are naturally noble and good, e.g. those who busy themselves more than they ought about honour or about children and parents, (are not wicked); for these too are good, and those who busy themselves about them are praised; but yet there is an excess even in them—if like Niobe one were to fight even against the gods, or were to be as much devoted to one's father as Satyrus nicknamed 'the filial', who was thought to be very silly on this point.) There is no wickedness, then, with regard to these objects, for the reason named, viz. because each of them is by nature a thing worthy of choice for its own sake; yet excesses in respect of them are bad and to be avoided. Similarly there is no incontinence with regard to them; for incontinence is not only to be avoided but is also a thing worthy of blame; but owing to a similarity in the state of feeling people apply the name incontinence, adding in each case what it is in respect of, as we may describe as a bad doctor or a bad actor one whom we should not call bad, simply. As, then, in this case we do not apply the term without qualification because each of these conditions is no shadness but only analogous to it, so it is clear that in the other case also that alone must be taken to be incontinence and continence which is concerned with the same objects as temperance and self-indulgence, but we apply the term to anger by virtue of a resemblance; and this is why we say with a qualification 'incontinent in respect of anger' as we say 'incontinent in respect of honour, or of gain'.

Chapter 5: Incontinence in its extended sense includes a brutish and a morbid form.

(1) Some things are pleasant by nature, and of these (a) some are so without qualification, and (b) others are so with reference to particular classes either of animals or of men; while (2) others are not pleasant by nature, but (a) some of them become so by reason of injuries to the system, and (b) others by reason of acquired habits, and (c) others by reason of originally bad natures. This being so, it is possible with regard to each of the latter kinds to discover similar states of character to those recognized with regard to the former; I mean (A) the brutish states, as in the case of the female who, they say, rips open pregnant women and devours the infants, or of the things in which some of the tribes about the Black Sea that have gone savage are said to delight—in raw meat or in human flesh, or in lending their children to one another to feast upon—or of the story told of Phalaris.

These states are brutish, but (B) others arise as a result of disease (or, in some cases, of madness, as with the man who sacrificed and ate his mother, or with the slave who ate the liver of his fellow), and others are morbid states (C) resulting from custom, e.g. the habit of plucking out the hair or of gnawing the nails, or even coals or earth, and in addition to these paederasty; for these arise in some by nature and in others, as in those who have been the victims of lust from childhood, from habit.

Now those in whom nature is the cause of such a state no one would call incontinent, any more than one would apply the epithet to women because of the passive part they play in copulation; nor would one apply it to those who are in a morbid condition as a result of habit. To have these various types of habit is beyond the limits of vice, as brutishness is too; for a man who has them to master or be mastered by them is not simple (continence or) incontinence but that which is so by analogy, as the man who is in this condition in respect of fits of anger is to be called incontinent in respect of that feeling but not incontinent simply. For every excessive state whether of folly, of cowardice, of self-indulgence, or of bad temper, is either brutish or morbid; the man who is by nature apt to fear everything, even the squeak of a mouse, is cowardly with a brutish cowardice, while the man who feared a weasel did so in consequence of disease; and of foolish people those who by nature are thoughtless and live by their senses alone are brutish, like some races of the distant barbarians, while those who are so as a result of disease (e.g. of epilepsy) or of madness are morbid. Of these characteristics it is possible to have some only at times, and not to be mastered by them. e.g. Phalaris may have restrained a desire to eat the flesh of a child or an appetite for unnatural sexual pleasure; but it is also possible to be mastered, not merely to have the feelings. Thus, as the wickedness which is on the human level is called wickedness simply, while that which is not is called wickedness not simply but with the qualification 'brutish' or 'morbid', in the same way it is plain that some incontinence is brutish and some morbid, while only that which corresponds to human self-indulgence is incontinence simply.

That incontinence and continence, then, are concerned only with the same objects as selfindulgence and temperance and that what is concerned with other objects is a type distinct from incontinence, and called incontinence by a metaphor and not simply, is plain.

Chapter 6: Incontinence in respect of anger less disgraceful than incontinence proper.

That incontinence in respect of anger is less disgraceful than that in respect of the appetites is what we will now proceed to see. (1) Anger seems to listen to argument to some extent, but to mishear it, as do hasty servants who run out before they have heard the whole of what one says, and then muddle the order, or as dogs bark if there is but a knock at the door, before looking to see if it is a friend; so anger by reason of the warmth and hastiness of its nature, though it hears, does not hear an order, and springs to take revenge. For argument or imagination informs us that we have been insulted or slighted, and anger, reasoning as it were that anything like this must be fought against, boils up straightway; while appetite, if argument or perception merely says that an object is pleasant, springs to the enjoyment of it. Therefore anger obeys the argument in a sense, but appetite does not. It is therefore more disgraceful; for the man who is incontinent in respect of anger is in a sense conquered by argument, while the other is conquered by appetite and not by argument.

(2) Further, we pardon people more easily for following natural desires, since we pardon them more easily for following such appetites as are common to all men,

and in so far as they are common; now anger and bad temper are more natural than the appetites for excess, i.e. for unnecessary objects. Take for instance the man who defended himself on the charge of striking his father by saying 'yes, but he struck his father, and he struck his, and' (pointing to his child) 'this boy will strike me when he is a man; it runs in the family'; or the man who when he was being dragged along by his son bade him stop at the doorway, since he himself had dragged his father only as far as that.

(3) Further, those who are more given to plotting against others are more criminal. Now a passionate man is not given to plotting, nor is anger itself—it is open; but the nature of appetite is illustrated by what the poets call Aphrodite, 'guile-weaving daughter of Cyprus', and by Homer's words about her 'embroidered girdle':

And the whisper of wooing is there,
Whose subtlety stealeth the wits of the wise, how prudent soe'er.

Therefore if this form of incontinence is more criminal and disgraceful than that in respect of anger, it is both incontinence without qualification and in a sense vice.

(4) Further, no one commits wanton outrage with a feeling of pain, but every one who acts in anger acts with pain, while the man who commits outrage acts with pleasure. If, then, those acts at which it is most just to be angry are more criminal than others, the incontinence which is due to appetite is the more criminal; for there is no wanton outrage involved in anger.

Plainly, then, the incontinence concerned with appetite is more disgraceful than that concerned with anger, and continence and incontinence are concerned with bodily appetites and pleasures; but we must grasp the differences among the latter themselves. For, as has been said at the beginning, some are human and natural both in kind and in magnitude, others are brutish, and others are due to organic injuries and diseases. Only with the first of these are temperance and self-indulgence concerned; this is why we call the lower animals neither temperate nor self-indulgent except by a metaphor, and only if some one race of animals exceeds another as a whole in wantonness, destructiveness, and omnivorous greed; these have no power of choice or calculation, but they are departures from the natural norm, as, among men, madmen are. Now brutishness is a less evil than vice, though more alarming; for it is not that the better part has been perverted, as in man,—they have no better part. Thus it is like comparing a lifeless thing with a living in respect of badness; for the badness of that which has no originative source of movement is always less hurtful, and reason is an originative source. Thus it is like comparing injustice in the abstract with an unjust man. Each is in some sense worse; for a bad man will do ten thousand times as much evil as a brute.

Chapter 7: Softness and endurance: two forms of incontinence—weakness and impetuosity.

With regard to the pleasures and pains and appetites and aversions arising through touch and taste, to which both self-indulgence and temperance were formerly narrowed down, it possible to be in such a state as to be defeated even by those of them which most people master, or to master even those by which most people are defeated; among these possibilities, those relating to pleasures are incontinence and continence, those relating to pains softness and endurance. The state of most people is intermediate, even if they lean more towards the worse states.

Now, since some pleasures are necessary while others are not, and are necessary up to a point while the excesses of them are not, nor the deficiencies, and this is equally true of appetites and pains, the man who pursues the excesses of things pleasant, or pursues to excess necessary objects, and does so by choice, for their own sake and not at all for the sake of any result distinct from them, is self-indulgent; for such a man is of necessity unlikely to repent, and therefore incurable, since a man who cannot repent cannot be cured. The man who is deficient in his pursuit of them is the opposite of self-indulgent; the man who is intermediate is temperate. Similarly, there is the man who avoids bodily pains not because he is defeated by them but by choice. (Of those who do not choose such acts, one kind of man is led to them as a result of the pleasure involved, another because he avoids the pain arising from the appetite, so that these types differ from one another. Now any one would think worse of a man with no appetite or with weak appetite were he to do something disgraceful, than if he did it under the influence of powerful appetite, and worse of him if he struck a blow not in anger than if he did it in anger; for what would he have done if he had been strongly affected? This is why the self-indulgent man is worse than the incontinent.) of the states named, then, the latter is rather a kind of softness; the former is self-indulgence. While to the incontinent man is opposed the continent, to the soft is opposed the man of endurance; for endurance consists in resisting, while continence consists in conquering, and resisting and conquering are different, as not being beaten is different from winning; this is why continence is also more worthy of choice than endurance. Now the man who is defective in respect of resistance to the things which most men both resist and resist successfully is soft and effeminate; for effeminacy too is a kind of softness; such a man trails his cloak to avoid the pain of lifting it, and plays the invalid without thinking himself wretched, though the man he imitates is a wretched man.

The case is similar with regard to continence and incontinence. For if a man is defeated by violent and excessive pleasures or pains, there is nothing wonderful in that; indeed we are ready to pardon him if he has resisted, as Theodectes' Philoctetes does when bitten by the snake, or Carcinus' Cercyon in the Alope, and as people who try to restrain their laughter burst out into a guffaw, as happened to Xenophantus. But it is surprising if a man is defeated by and cannot resist pleasures or pains which most men can hold out against, when this is not due to heredity or disease, like the softness that is hereditary with the kings of the Scythians, or that which distinguishes the female sex from the male.

The lover of amusement, too, is thought to be self-indulgent, but is really soft. For amusement is a relaxation, since it is a rest from work; and the lover of amusement is one of the people who go to excess in this.

Of incontinence one kind is impetuosity, another weakness. For some men after deliberating fail, owing to their emotion, to stand by the conclusions of their deliberation, others because they have not deliberated are led by their emotion; since some men (just as people who first tickle others are not tickled themselves), if they have first perceived and seen what is coming and have first roused themselves and their calculative faculty, are not defeated by their emotion, whether it be pleasant or painful. It is keen and excitable people that suffer especially from the impetuous form of incontinence; for the former by reason of their quickness and the latter by reason of the violence of their passions do not await the argument, because they are apt to follow their imagination.

Chapter 8: Self-indulgence worse than incontinence.

The self-indulgent man, as was said, is not apt to repent; for he stands by his choice; but incontinent man is likely to repent. This is why the position is not as it was expressed in the formulation of the problem, but the selfindulgent man is incurable and the incontinent man curable; for wickedness is like a disease such as dropsy or consumption, while incontinence is like epilepsy; the former is a permanent, the latter an intermittent badness. And generally incontinence and vice are different in kind; vice is unconscious of itself, incontinence is not (of incontinent men themselves, those who become temporarily beside themselves are better than those who have the rational principle but do not abide by it, since the latter are defeated by a weaker passion, and do not act without previous deliberation like the others); for the incontinent man is like the people who get drunk quickly and on little wine, i.e. on less than most people.

Evidently, then, incontinence is not vice (though perhaps it is so in a qualified sense); for incontinence is contrary to choice while vice is in accordance with choice; not but what they are similar in respect of the actions they lead to; as in the saying of Demodocus about the Milesians, 'the Milesians are not without sense, but they do the things that senseless people do', so too incontinent people are not criminal, but they will do criminal acts.

Now, since the incontinent man is apt to pursue, not on conviction, bodily pleasures that are excessive and contrary to the right rule, while the self-indulgent man is convinced because he is the sort of man to pursue them, it is on the contrary the former that is easily persuaded to change his mind, while the latter is not. For virtue and vice respectively preserve and destroy the first principle, and in actions the final cause is the first principle, as the hypotheses are in mathematics; neither in that case is it argument that teaches the first principles, nor is it so here—virtue either natural or produced by habituation is what teaches right opinion about the first principle. Such a man as this, then, is temperate; his contrary is the self-indulgent.

But there is a sort of man who is carried away as a result of passion and contrary to the right rule—a man whom passion masters so that he does not act according to the right rule, but does not master to the extent of making him ready to believe that he ought to pursue such pleasures without reserve; this is the incontinent man, who is better than the self-indulgent man, and not bad without qualification; for the best thing in him, the first principle, is preserved. And contrary to him is another kind of man, he who abides by his convictions and is not carried away, at least as a result of passion. It is evident from these considerations that the latter is a good state and the former a bad one.

Chapter 9: Relation of continence to obstinancy, incontinence, 'insensibility', temperence.

Is the man continent who abides by any and every rule and any and every choice, or the man who abides by the right choice, and is he incontinent who abandons any and every choice and any and every rule, or he who abandons the rule that is not false and the choice that is right; this is how we put it before in our statement of the problem. Or is it incidentally any and every choice but per se the true rule and the right choice by which the one abides and the other does not? If any one chooses or pursues this for the sake of that, per se he pursues and chooses the latter, but incidentally the former. But when we speak without qualification we mean what is per se. Therefore in a sense the one abides by, and the other abandons, any and every opinion; but without qualification, the true opinion.

There are some who are apt to abide by their opinion, who are called strong-headed, viz. those who are hard to persuade in the first instance and are not easily persuaded to change; these have in them something like the continent man, as the prodigal is in a way like the liberal man and the rash man like the confident man; but they are different in many respects. For it is to passion and appetite that the one will not yield, since on occasion the continent man will be easy to persuade; but it is to argument that the others refuse to yield, for they do form appetites and many of them are led by their pleasures. Now the people who are strong-headed are the opinionated, the ignorant, and the boorish—the opinionated being influenced by pleasure and pain; for they delight in the victory they gain if they are not persuaded to change, and are pained if their decisions become null and void as decrees sometimes do; so that they are liker the incontinent than the continent man.

But there are some who fail to abide by their resolutions, not as a result of incontinence, e.g. Neoptolemus in Sophocles' Philoctetes; yet it was for the sake of pleasure that he did not stand fast—but a noble pleasure; for telling the truth was noble to him, but he had been persuaded by Odysseus to tell the lie. For not every one who does anything for the sake of pleasure is either self-indulgent or bad or incontinent, but he who does it for a disgraceful pleasure.

Since there is also a sort of man who takes less delight than he should in bodily things, and does not abide by the rule, he who is intermediate between him and the incontinent man is the continent man; for the incontinent man fails to abide by the rule because he delights too much in them, and this man because he delights in them too little; while the continent man abides by the rule and does not change on either account. Now if continence is good, both the contrary states must be bad, as they actually appear to be; but because the other extreme is seen in few people and seldom, as temperance is thought to be contrary only to self-indulgence, so is continence to incontinence.

Since many names are applied analogically, it is by analogy that we have come to speak of the 'continence' the temperate man; for both the continent man and the temperate man are such as to do nothing contrary to the rule for the sake of the bodily pleasures, but the former has and the latter has not bad appetites, and the latter is such as not to feel pleasure contrary to the rule, while the former is such as to feel pleasure but not to be led by it. And the incontinent and the self-indulgent man are also like another; they are different, but both pursue bodily pleasures—the latter, however, also thinking that he ought to do so, while the former does not think this.

Chapter 10: Practical wisdom is not compatible with incontinence, but cleverness is.

Nor can the same man have practical wisdom and be incontinent; for it has been shown' that a man is at the same time practically wise, and good in respect of character. Further, a man has practical wisdom not by knowing only but by being able to act; but the incontinent man is unable to act—there is, however, nothing to prevent a clever man from being incontinent; this is why it is sometimes actually thought that some people have practical wisdom but are incontinent, viz. because cleverness and practical wisdom differ in the way we have described in our first discussions, and are near together in respect of their reasoning, but differ in respect of their purpose—nor yet is the incontinent man like the man who knows and is contemplating a truth, but like the man who is asleep or drunk. And he acts willingly (for he acts in a sense with knowledge both of what he does and of the end to which he does it), but is not wicked, since his purpose is good; so that he is half-wicked. And he is not a criminal; for he does not act of malice aforethought; of the two types of incontinent man the one does not abide by the conclusions of his deliberation, while the excitable man does not deliberate at all. And thus the incontinent man like a city which passes all the right decrees and has good laws, but makes no use of them, as in Anaxandrides' jesting remark,

The city willed it, that cares nought for laws;

but the wicked man is like a city that uses its laws, but has wicked laws to use.

Now incontinence and continence are concerned with that which is in excess of the state characteristic of most men; for the continent man abides by his resolutions more and the incontinent man less than most men can.

Of the forms of incontinence, that of excitable people is more curable than that of those who deliberate but do not abide by their decisions, and those who are incontinent through habituation are more curable than those in whom incontinence is innate; for it is easier to change a habit than to change one's nature; even habit is hard to change just because it is like nature, as Evenus says:

I say that habit's but a long practice, friend,

And this becomes men's nature in the end.

We have now stated what continence, incontinence, endurance, and softness are, and how these states are related to each other.

Chapter 11: Three views hostile to pleasure, and the arguments for them.

The study of pleasure and pain belongs to the province of the political philosopher; for he is the architect of the end, with a view to which we call one thing bad and another good without qualification. Further, it is one of our necessary tasks to consider them; for not only did we lay it down that moral virtue and vice are concerned with pains and pleasures, but most people say that happiness involves pleasure; this is why the blessed man is called by a name derived from a word meaning enjoyment.

Now (1) some people think that no pleasure is a good, either in itself or incidentally, since the good and pleasure are not the same; (2) others think that some pleasures are good but that most are bad. (3) Again there is a third view, that even if all pleasures are good, yet the best thing in the world cannot be pleasure. (1) The reasons given for the view that pleasure is not a good at all are (a) that every pleasure is a perceptible process to a natural state, and that no process is of the same kind as its end, e.g. no process of building of the same kind as a house. (b) A temperate man avoids pleasures. (c) A man of practical wisdom pursues what is free from pain, not what is pleasant. (d) The pleasures are a hindrance to thought, and the more so the more one delights in them, e.g. in sexual pleasure; for no one could think of anything while absorbed in this. (e) There is no art of pleasure; but every good is the product of some art. (f) Children and the brutes pursue pleasures. (2) The reasons for the view that not all pleasures are good are that (a) there are pleasures that are actually base and objects of reproach, and (b) there are harmful pleasures; for some pleasant things are unhealthy. (3) The reason for the view that the best thing in the world is not pleasure is that pleasure is not an end but a process.

Chapter 12: Discussion of the view that pleasure is not a good.

These are pretty much the things that are said. That it does not follow from these grounds that pleasure is not a good, or even the chief good, is plain from the following considerations. (A) (a) First, since that which is good may be so in either of two senses (one thing good simply and another good for a particular person), natural constitutions and states of being, and therefore also the corresponding movements and processes, will be correspondingly divisible. Of those which are thought to be bad some will be bad if taken without qualification but not bad for a particular person, but worthy of his choice, and some will not be worthy of choice even for a particular person, but only at a particular time and for a short period, though not without qualification; while others are not even pleasures, but seem to be so, viz. all those which involve pain and whose end is curative, e.g. the processes that go on in sick persons.

(b) Further, one kind of good being activity and another being state, the processes that restore us to our natural state are only incidentally pleasant; for that matter the activity at work in the appetites for them is the activity of so much of our state and nature as has remained unimpaired; for there are actually pleasures that involve no pain or appetite (e.g. those of contemplation), the nature in such a case not being defective at all. That the others are incidental is indicated by the fact that men do not enjoy the same pleasant objects when their nature is in its settled state as they do when it is being replenished, but in the former case they enjoy the things that are pleasant without qualification, in the latter the contraries of these as well; for then they enjoy even sharp and bitter things, none of which is pleasant either by nature or without qualification. The states they produce, therefore, are not pleasures naturally or without qualification; for as pleasant things differ, so do the pleasures arising from them.

(c) Again, it is not necessary that there should be something else better than pleasure, as some say the end is better than the process; for leasures are not processes nor do they all involve process—they are activities and ends; nor do they arise when we are becoming something, but when we are exercising some faculty; and not all pleasures have an end different from themselves, but only the pleasures of persons who are being led to the perfecting of their nature. This is why it is not right to say that pleasure is perceptible process, but it should rather be called activity of the natural state, and instead of 'perceptible' 'unimpeded'. It is thought by some people to be process just because they think it is in the strict sense good; for they think that activity is process, which it is not.

(B) The view that pleasures are bad because some pleasant things are unhealthy is like saying that healthy things are bad because some healthy things are bad for money-making; both are bad in the respect mentioned, but they are not bad for that reason—indeed, thinking itself is sometimes injurious to health.

Neither practical wisdom nor any state of being is impeded by the pleasure arising from it; it is foreign pleasures that impede, for the pleasures arising from thinking and learning will make us think and learn all the more.

(C) The fact that no pleasure is the product of any art arises naturally enough; there is no art of any other activity either, but only of the corresponding faculty; though for that matter the arts of the perfumer and the cook are thought to be arts of pleasure.

(D) The arguments based on the grounds that the temperate man avoids pleasure and that the man of practical wisdom pursues the painless life, and that children and the brutes pursue pleasure, are all refuted by the same consideration. We have pointed out in what sense pleasures are good without qualification and in what sense some are not good; now both the brutes and children pursue pleasures of the latter kind (and the man of practical wisdom pursues tranquil freedom from that kind), viz. those which imply appetite and pain, i.e. the bodily pleasures (for it is these that are of this nature) and the excesses of them, in respect of which the self-indulgent man is self-indulent. This is why the temperate man avoids these pleasures; for even he has pleasures of his own.

Chapter 13: Discussion of the view that pleasure is not the chief good.

But further (E) it is agreed that pain is bad and to be avoided; for some pain is without qualification bad, and other pain is bad because it is in some respect an impediment to us. Now the contrary of that which is to be avoided, qua something to be avoided and bad, is good. Pleasure, then, is necessarily a good. For the answer of Speusippus, that pleasure is contrary both to pain and to good, as the greater is contrary both to the less and to the equal, is not successful; since he would not say that pleasure is essentially just a species of evil.

And (F) if certain pleasures are bad, that does not prevent the chief good from being some pleasure, just as the chief good may be some form of knowledge though certain kinds of knowledge are bad. Perhaps it is even necessary, if each disposition has unimpeded activities, that, whether the activity (if unimpeded) of all our dispositions or that of some one of them is happiness, this should be the thing most worthy of our choice; and this activity is pleasure. Thus the chief good would be some pleasure, though most pleasures might perhaps be bad without qualification. And for this reason all men think that the happy life is pleasant and weave pleasure into their ideal of happiness—and reasonably too; for no activity is perfect when it is impeded, and happiness is a perfect thing; this is why the happy man needs the goods of the body and external goods, i.e. those of fortune, viz. in order that he may not be impeded in these ways. Those who say that the victim on the rack or the man who falls into great misfortunes is happy if he is good, are, whether they mean to or not, talking nonsense. Now because we need fortune as well as other things, some people think good fortune the same thing as happiness; but it is not that, for even good fortune itself when in excess is an impediment, and perhaps should then be no longer called good fortune; for its limit is fixed by reference to happiness.

And indeed the fact that all things, both brutes and men, pursue pleasure is an indication of its being somehow the chief good:

No voice is wholly lost that many peoples...

But since no one nature or state either is or is thought the best for all, neither do all pursue the same pleasure; yet all pursue pleasure. And perhaps they actually pursue not the pleasure they think they pursue nor that which they would say they pursue, but the same pleasure; for all things have by nature something divine in them. But the bodily pleasures have appropriated the name both because we oftenest steer our course for them and because all men share in them; thus because they alone are familiar, men think there are no others.

It is evident also that if pleasure, i.e. the activity of our faculties, is not a good, it will not be the case that the happy man lives a pleasant life; for to what end should he need pleasure, if it is not a good but the happy man may even live a painful life? For pain is neither an evil nor a good, if pleasure is not; why then should he avoid it? Therefore, too, the life of the good man will not be pleasanter than that of any one else, if his activities are not more pleasant.

Chapter 14: Discussion of the view that most pleasures are bad, and of the tendency to identify bodily pleasures with pleasure in general.

(G) With regard to the bodily pleasures, those who say that some pleasures are very much to be chosen, viz. the noble pleasures, but not the bodily pleasures, i.e. those with which the self-indulgent man is concerned, must consider why, then, the contrary pains are bad. For the contrary of bad is good. Are the necessary pleasures good in the sense in which even that which is not bad is good? Or are they good up to a point? Is it that where you have states and processes of which there cannot be too much, there cannot be too much of the corresponding pleasure, and that where there can be too much of the one there can be too much of the other also? Now there can be too much of bodily goods, and the bad man is bad by virtue of pursuing the excess, not by virtue of pursuing the necessary pleasures (for all men enjoy in some way or other both dainty foods and wines and sexual intercourse, but not all men do so as they ought). The contrary is the case with pain; for he does not avoid the excess of it, he avoids it altogether; and this is peculiar to him, for the alternative to excess of pleasure is not pain, except to the man who pursues this excess.

Since we should state not only the truth, but also the cause of error—for this contributes towards producing conviction, since when a reasonable explanation is given of why the false view appears true, this tends to produce belief in the true view—therefore we must state why the bodily pleasures appear the more worthy of choice. (a) Firstly, then, it is because they expel pain; owing to the excesses of pain that men experience, they pursue excessive and in general bodily pleasure as being a cure for the pain. Now curative agencies produce intense feeling—which is the reason why they are pursued—because they show up against the contrary pain. (Indeed pleasure is thought not to be good for these two reasons, as has been said, viz. that (a) some of them are activities belonging to a bad nature—either congenital, as in the case of a brute, or due to habit, i.e. those of bad men; while (b) others are meant to cure a defective nature, and it is better to be in a healthy state than to be getting into it, but these arise during the process of being made perfect and are therefore only incidentally good.) (b) Further, they are pursued because of their violence by those who cannot enjoy other pleasures. (At all events they go out of their way to manufacture thirsts somehow for themselves. When these are harmless, the practice is irreproachable; when they are hurtful, it is bad.) For they have nothing else to enjoy, and, besides, a neutral state is painful to many people because of their nature. For the animal nature is always in travail, as the students of natural science also testify, saying that sight and hearing are painful; but we have become used to this, as they maintain. Similarly, while, in youth, people are, owing to the growth that is going on, in a situation like that of drunken men, and youth is pleasant, on the other hand people of excitable nature always need relief; for even their body is ever in torment owing to its special composition, and they are always under the influence of violent desire; but pain is driven out both by the contrary pleasure, and by any chance pleasure if it be strong; and for these reasons they become self-indulgent and bad. But the pleasures that do not involve pains do not admit of excess; and these are among the things pleasant by nature and not incidentally. By things pleasant incidentally I mean those that act as cures (for because as a result people are cured, through some action of the part that remains healthy, for this reason the process is thought pleasant); by things naturally pleasant I mean those that stimulate the action of the healthy nature.

There is no one thing that is always pleasant, because our nature is not simple but there is another element in us as well, inasmuch as we are perishable creatures, so that if the one element does something, this is unnatural to the other nature, and when the two elements are evenly balanced, what is done seems neither painful nor pleasant; for if the nature of anything were simple, the same action would always be most pleasant to it. This is why God always enjoys a single and simple pleasure; for there is not only an activity of movement but an activity of immobility, and pleasure is found more in rest than in movement. But 'change in all things is sweet', as the poet says, because of some vice; for as it is the vicious man that is changeable, so the nature that needs change is vicious; for it is not simple nor good.

We have now discussed continence and incontinence, and pleasure and pain, both what each is and in what sense some of them are good and others bad; it remains to speak of friendship.

Book VIII: Friendship

Chapter 1: Friendship both necessary and noble: main questions about it.

AFTER what we have said, a discussion of friendship would naturally follow, since it is a virtue or implies virtue, and is besides most necessary with a view to living. For without friends no one would choose to live, though he had all other goods; even rich men and those in possession of office and of dominating power are thought to need friends most of all; for what is the use of such prosperity without the opportunity of beneficence, which is exercised chiefly and in its most laudable form towards friends? Or how can prosperity be guarded and preserved without friends? The greater it is, the more exposed is it to risk. And in poverty and in other misfortunes men think friends are the only refuge. It helps the young, too, to keep from error; it aids older people by ministering to their needs and supplementing the activities that are failing from weakness; those in the prime of life it stimulates to noble actions—'two going together'—for with friends men are more able both to think and to act. Again, parent seems by nature to feel it for offspring and offspring for parent, not only among men but among birds and among most animals; it is felt mutually by members of the same race, and especially by men, whence we praise lovers of their fellowmen. We may even in our travels how near and dear every man is to every other. Friendship seems too to hold states together, and lawgivers to care more for it than for justice; for unanimity seems to be something like friendship, and this they aim at most of all, and expel faction as their worst enemy; and when men are friends they have no need of justice, while when they are just they need friendship as well, and the truest form of justice is thought to be a friendly quality.

But it is not only necessary but also noble; for we praise those who love their friends, and it is thought to be a fine thing to have many friends; and again we think it is the same people that are good men and are friends.

Not a few things about friendship are matters of debate. Some define it as a kind of likeness and say like people are friends, whence come the sayings 'like to like', 'birds of a feather flock together', and so on; others on the contrary say 'two of a trade never agree'. On this very question they inquire for deeper and more physical causes, Euripides saying that 'parched earth loves the rain, and stately heaven when filled with rain loves to fall to earth', and Heraclitus that 'it is what opposes that helps' and 'from different tones comes the fairest tune' and 'all things are produced through strife'; while Empedocles, as well as others, expresses the opposite view that like aims at like. The physical problems we may leave alone (for they do not belong to the present inquiry); let us examine those which are human and involve character and feeling, e.g. whether friendship can arise between any two people or people cannot be friends if they are wicked, and whether there is one species of friendship or more than one. Those who think there is only one because it admits of degrees have relied on an inadequate indication; for even things different in species admit of degree. We have discussed this matter previously.

Chapter 2: Three objects of love: implications of friendship.

The kinds of friendship may perhaps be cleared up if we first come to know the object of love. For not everything seems to be loved but only the lovable, and this is good, pleasant, or useful; but it would seem to be that by which some good or pleasure is produced that is useful, so that it is the good and the useful that are lovable as ends. Do men love, then, the good, or what is good for them? These sometimes clash. So too with regard to the pleasant. Now it is thought that each loves what is good for himself, and that the good is without qualification lovable, and what is good for each man is lovable for him; but each man loves not what is good for him but what seems good. This however will make no difference; we shall just have to say that this is 'that which seems lovable'. Now there are three grounds on which people love; of the love of lifeless objects we do not use the word 'friendship'; for it is not mutual love, nor is there a wishing of good to the other (for it would surely be ridiculous to wish wine well; if one wishes anything for it, it is that it may keep, so that one may have it oneself); but to a friend we say we ought to wish what is good for his sake. But to those who thus wish good we ascribe only goodwill, if the wish is not reciprocated; goodwill when it is reciprocal being friendship. Or must we add 'when it is recognized'? For many people have goodwill to those whom they have not seen but judge to be good or useful; and one of these might return this feeling. These people seem to bear goodwill to each other; but how could one call them friends when they do not know their mutual feelings? To be friends, then, the must be mutually recognized as bearing goodwill and wishing well to each other for one of the aforesaid reasons.

Chapter 3: Three corresponding kinds of friendship: superiority of friendship whose motive is the good.

Now these reasons differ from each other in kind; so, therefore, do the corresponding forms of love and friendship. There are therefore three kinds of friendship, equal in number to the things that are lovable; for with respect to each there is a mutual and recognized love, and those who love each other wish well to each other in that respect in which they love one another. Now those who love each other for their utility do not love each other for themselves but in virtue of some good which they get from each other. So too with those who love for the sake of pleasure; it is not for their character that men love ready-witted people, but because they find them pleasant. Therefore those who love for the sake of utility love for the sake of what is good for themselves, and those who love for the sake of pleasure do so for the sake of what is pleasant to themselves, and not in so far as the other is the person loved but in so far as he is useful or pleasant. And thus these friendships are only incidental; for it is not as being the man he is that the loved person is loved, but as providing some good or pleasure. Such friendships, then, are easily dissolved, if the parties do not remain like themselves; for if the one party is no longer pleasant or useful the other ceases to love him.

Now the useful is not permanent but is always changing. Thus when the motive of the friendship is done away, the friendship is dissolved, inasmuch as it existed only for the ends in question. This kind of friendship seems to exist chiefly between old people (for at that age people pursue not the pleasant but the useful) and, of those who are in their prime or young, between those who pursue utility. And such people do not live much with each other either; for sometimes they do not even find each other pleasant; therefore they do not need such companionship unless they are useful to each other; for they are pleasant to each other only in so far as they rouse in each other hopes of something good to come. Among such friendships people also class the friendship of a host and guest. On the other hand the friendship of young people seems to aim at pleasure; for they live under the guidance of emotion, and pursue above all what is pleasant to themselves and what is immediately before them; but with increasing age their pleasures become different. This is why they quickly become friends and quickly cease to be so; their friendship changes with the object that is found pleasant, and such pleasure alters quickly. Young people are amorous too; for the greater part of the friendship of love depends on emotion and aims at pleasure; this is why they fall in love and quickly fall out of love, changing often within a single day. But these people do wish to spend their days and lives together; for it is thus that they attain the purpose of their friendship.

Perfect friendship is the friendship of men who are good, and alike in virtue; for these wish well alike to each other qua good, and they are good themselves. Now those who wish well to their friends for their sake are most truly friends; for they do this by reason of own nature and not incidentally; therefore their friendship lasts as long as they are good—and goodness is an enduring thing. And each is good without qualification and to his friend, for the good are both good without qualification and useful to each other. So too they are pleasant; for the good are pleasant both without qualification and to each other, since to each his own activities and others like them are pleasurable, and the actions of the good are the same or like. And such a friendship is as might be expected permanent, since there meet in it all the qualities that friends should have. For all friendship is for the sake of good or of pleasure—good or pleasure either in the abstract or such as will be enjoyed by him who has the friendly feeling—and is based on a certain resemblance; and to a friendship of good men all the qualities we have named belong in virtue of the nature of the friends themselves; for in the case of this kind of friendship the other qualities also are alike in both friends, and that which is good without qualification is also without qualification pleasant, and these are the most lovable qualities. Love and friendship therefore are found most and in their best form between such men.

But it is natural that such friendships should be infrequent; for such men are rare. Further, such friendship requires time and familiarity; as the proverb says, men cannot know each other till they have 'eaten salt together'; nor can they admit each other to friendship or be friends till each has been found lovable and been trusted by each. Those who quickly show the marks of friendship to each other wish to be friends, but are not friends unless they both are lovable and know the fact; for a wish for friendship may arise quickly, but friendship does not.

Chapter 4: Contrast between the best and the inferior kinds.

This kind of friendship, then, is perfect both in respect of duration and in all other respects, and in it each gets from each in all respects the same as, or something like what, he gives; which is what ought to happen between friends. Friendship for the sake of pleasure bears a resemblance to this kind; for good people too are pleasant to each other. So too does friendship for the sake of utility; for the good are also useful to each other. Among men of these inferior sorts too, friendships are most permanent when the friends get the same thing from each other (e.g. pleasure), and not only that but also from the same source, as happens between readywitted people, not as happens between lover and beloved. For these do not take pleasure in the same things, but the one in seeing the beloved and the other in receiving attentions from his lover; and when the bloom of youth is passing the friendship sometimes passes too (for the one finds no pleasure in the sight of the other, and the other gets no attentions from the first); but many lovers on the other hand are constant, if familiarity has led them to love each other's characters, these being alike. But those who exchange not pleasure but utility in their amour are both less truly friends and less constant. Those who are friends for the sake of utility part when the advantage is at an end; for they were lovers not of each other but of profit.

For the sake of pleasure or utility, then, even bad men may be friends of each other, or good men of bad, or one who is neither good nor bad may be a friend to any sort of person, but for their own sake clearly only good men can be friends; for bad men do not delight in each other unless some advantage come of the relation.

The friendship of the good too and this alone is proof against slander; for it is not easy to trust any one talk about a man who has long been tested by oneself; and it is among good men that trust and the feeling that 'he would never wrong me' and all the other things that are demanded in true friendship are found. In the other kinds of friendship, however, there is nothing to prevent these evils arising. For men apply the name of friends even to those whose motive is utility, in which sense states are said to be friendly (for the alliances of states seem to aim at advantage), and to those who love each other for the sake of pleasure, in which sense children are called friends. Therefore we too ought perhaps to call such people friends, and say that there are several kinds of friendship—firstly and in the proper sense that of good men qua good, and by analogy the other kinds; for it is in virtue of something good and something akin to what is found in true friendship that they are friends, since even the pleasant is good for the lovers of pleasure. But these two kinds of friendship are not often united, nor do the same people become friends for the sake of utility and of pleasure; for things that are only incidentally connected are not often coupled together.

Friendship being divided into these kinds, bad men will be friends for the sake of pleasure or of utility, being in this respect like each other, but good men will be friends for their own sake, i.e. in virtue of their goodness. These, then, are friends without qualification; the others are friends incidentally and through a resemblance to these.

Chapter 5: The state of friendship distinguished from the activity of friendship and from the feeling of friendliness.

As in regard to the virtues some men are called good in respect of a state of character, others in respect of an activity, so too in the case of friendship; for those who live together delight in each other and confer benefits on each other, but those who are asleep or locally separated are not performing, but are disposed to perform, the activities of friendship; distance does not break off the friendship absolutely, but only the activity of it. But if the absence is lasting, it seems actually to make men forget their friendship; hence the saying 'out of sight, out of mind'. Neither old people nor sour people seem to make friends easily; for there is little that is pleasant in them, and no one can spend his days with one whose company is painful, or not pleasant, since nature seems above all to avoid the painful and to aim at the pleasant. Those, however, who approve of each other but do not live together seem to be well-disposed rather than actual friends. For there is nothing so characteristic of friends as living together (since while it people who are in need that desire benefits, even those who are supremely happy desire to spend their days together; for solitude suits such people least of all); but people cannot live together if they are not pleasant and do not enjoy the same things, as friends who are companions seem to do.

The truest friendship, then, is that of the good, as we have frequently said; for that which is without qualification good or pleasant seems to be lovable and desirable, and for each person that which is good or pleasant to him; and the good man is lovable and desirable to the good man for both these reasons. Now it looks as if love were a feeling, friendship a state of character; for love may be felt just as much towards lifeless things, but mutual love involves choice and choice springs from a state of character; and men wish well to those whom they love, for their sake, not as a result of feeling but as a result of a state of character. And in loving a friend men love what is good for themselves; for the good man in becoming a friend becomes a good to his friend. Each, then, both loves what is good for himself, and makes an equal return in goodwill and in pleasantness; for friendship is said to be equality, and both of these are found most in the friendship of the good.

Chapter 6: Various relations between the three kinds.

Between sour and elderly people friendship arises less readily, inasmuch as they are less good-tempered and enjoy companionship less; for these are thou to be the greatest marks of friendship productive of it. This is why, while men become friends quickly, old men do not; it is because men do not become friends with those in whom they do not delight; and similarly sour people do not quickly make friends either. But such men may bear goodwill to each other; for they wish one another well and aid one another in need; but they are hardly friends because they do not spend their days together nor delight in each other, and these are thought the greatest marks of friendship.

One cannot be a friend to many people in the sense of having friendship of the perfect type with them, just as one cannot be in love with many people at once (for love is a sort of excess of feeling, and it is the nature of such only to be felt towards one person); and it is not easy for many people at the same time to please the same person very greatly, or perhaps even to be good in his eyes. One must, too, acquire some experience of the other person and become familiar with him, and that is very hard. But with a view to utility or pleasure it is possible that many people should please one; for many people are useful or pleasant, and these services take little time.

Of these two kinds that which is for the sake of pleasure is the more like friendship, when both parties get the same things from each other and delight in each other or in the things, as in the friendships of the young; for generosity is more found in such friendships. Friendship based on utility is for the commercially minded. People who are supremely happy, too, have no need of useful friends, but do need pleasant friends; for they wish to live with some one and, though they can endure for a short time what is painful, no one could put up with it continuously, nor even with the Good itself if it were painful to him; this is why they look out for friends who are pleasant. Perhaps they should look out for friends who, being pleasant, are also good, and good for them too; for so they will have all the characteristics that friends should have.

People in positions of authority seem to have friends who fall into distinct classes; some people are useful to them and others are pleasant, but the same people are rarely both; for they seek neither those whose pleasantness is accompanied by virtue nor those whose utility is with a view to noble objects, but in their desire for pleasure they seek for ready-witted people, and their other friends they choose as being clever at doing what they are told, and these characteristics are rarely combined. Now we have said that the good man is at the same time pleasant and useful; but such a man does not become the friend of one who surpasses him in station, unless he is surpassed also in virtue; if this is not so, he does not establish equality by being proportionally exceeded in both respects. But people who surpass him in both respects are not so easy to find.

However that may be, the aforesaid friendships involve equality; for the friends get the same things from one another and wish the same things for one another, or exchange one thing for another, e.g. pleasure for utility; we have said, however, that they are both less truly friendships and less permanent.

But it is from their likeness and their unlikeness to the same thing that they are thought both to be and not to be friendships. It is by their likeness to the friendship of virtue that they seem to be friendships (for one of them involves pleasure and the other utility, and these characteristics belong to the friendship of virtue as well); while it is because the friendship of virtue is proof against slander and permanent, while these quickly change (besides differing from the former in many other respects), that they appear not to be friendships; i.e. it is because of their unlikeness to the friendship of virtue.

Chapter 7: In unequal friendships a proportion must be maintained.

But there is another kind of friendship, viz. that which involves an inequality between the parties, e.g. that of father to son and in general of elder to younger, that of man to wife and in general that of ruler to subject. And these friendships differ also from each other; for it is not the same that exists between parents and children and between rulers and subjects, nor is even that of father to son the same as that of son to father, nor that of husband to wife the same as that of wife to husband. For the virtue and the function of each of these is different, and so are the reasons for which they love; the love and the friendship are therefore different also. Each party, then, neither gets the same from the other, nor ought to seek it; but when children render to parents what they ought to render to those who brought them into the world, and parents render what they should to their children, the friendship of such persons will be abiding and excellent. In all friendships implying inequality the love also should be proportional, i.e. the better should be more loved than he loves, and so should the more useful, and similarly in each of the other cases; for when the love is in proportion to the merit of the parties, then in a sense arises equality, which is certainly held to be characteristic of friendship.

But equality does not seem to take the same form in acts of justice and in friendship; for in acts of justice what is equal in the primary sense is that which is in proportion to merit, while quantitative equality is secondary, but in friendship quantitative equality is primary and proportion to merit secondary. This becomes clear if there is a great interval in respect of virtue or vice or wealth or anything else between the parties; for then they are no longer friends, and do not even expect to be so. And this is most manifest in the case of the gods; for they surpass us most decisively in all good things. But it is clear also in the case of kings; for with them, too, men who are much their inferiors do not expect to be friends; nor do men of no account expect to be friends with the best or wisest men. In such cases it is not possible to define exactly up to what point friends can remain friends; for much can be taken away and friendship remain, but when one party is removed to a great distance, as God is, the possibility of friendship ceases. This is in fact the origin of the question whether friends really wish for their friends the greatest goods, e.g. that of being gods; since in that case their friends will no longer be friends to them, and therefore will not be good things for them (for friends are good things). The answer is that if we were right in saying that friend wishes good to friend for his sake, his friend must remain the sort of being he is, whatever that may be; therefore it is for him oily so long as he remains a man that he will wish the greatest goods. But perhaps not all the greatest goods; for it is for himself most of all that each man wishes what is good.

Chapter 8: Loving is more of the essence of friendship than being loved.

Most people seem, owing to ambition, to wish to be loved rather than to love; which is why most men love flattery; for the flatterer is a friend in an inferior position, or pretends to be such and to love more than he is loved; and being loved seems to be akin to being honoured, and this is what most people aim at. But it seems to be not for its own sake that people choose honour, but incidentally. For most people enjoy being honoured by those in positions of authority because of their hopes (for they think that if they want anything they will get it from them; and therefore they delight in honour as a token of favour to come); while those who desire honour from good men, and men who know, are aiming at confirming their own opinion of themselves; they delight in honour, therefore, because they believe in their own goodness on the strength of the judgement of those who speak about them. In being loved, on the other hand, people delight for its own sake; whence it would seem to be better than being honoured, and friendship to be desirable in itself. But it seems to lie in loving rather than in being loved, as is indicated by the delight mothers take in loving; for some mothers hand over their children to be brought up, and so long as they know their fate they love them and do not seek to be loved in return (if they cannot have both), but seem to be satisfied if they see them prospering; and they themselves love their children even if these owing to their ignorance give them nothing of a mother's due. Now since friendship depends more on loving, and it is those who love their friends that are praised, loving seems to be the characteristic virtue of friends, so that it is only those in whom this is found in due measure that are lasting friends, and only their friendship that endures.

It is in this way more than any other that even unequals can be friends; they can be equalized. Now equality and likeness are friendship, and especially the likeness of those who are like in virtue; for being steadfast in themselves they hold fast to each other, and neither ask nor give base services, but (one may say) even prevent them; for it is characteristic of good men neither to go wrong themselves nor to let their friends do so. But wicked men have no steadfastness (for they do not remain even like to themselves), but become friends for a short time because they delight in each other's wickedness. Friends who are useful or pleasant last longer; i.e. as long as they provide each other with enjoyments or advantages. Friendship for utility's sake seems to be that which most easily exists between contraries, e.g. between poor and rich, between ignorant and learned; for what a man actually lacks he aims at, and one gives something else in return. But under this head, too, might bring lover and beloved, beautiful and ugly. This is why lovers sometimes seem ridiculous, when they demand to be loved as they love; if they are equally lovable their claim can perhaps be justified, but when they have nothing lovable about them it is ridiculous. Perhaps, however, contrary does not even aim at contrary by its own nature, but only incidentally, the desire being for what is intermediate; for that is what is good, e.g. it is good for the dry not to become wet but to come to the intermediate state, and similarly with the hot and in all other cases. These subjects we may dismiss; for they are indeed somewhat foreign to our inquiry.

Chapter 9: Parallelism of friendship and justice: the state comprehends all lesser communities.

Friendship and justice seem, as we have said at the outset of our discussion, to be concerned with the same objects and exhibited between the same persons. For in every community there is thought to be some form of justice, and friendship too; at least men address as friends their fellow-voyagers and fellowsoldiers, and so too those associated with them in any other kind of community. And the extent of their association is the extent of their friendship, as it is the extent to which justice exists between them. And the proverb 'what friends have is common property' expresses the truth; for friendship depends on community. Now brothers and comrades have all things in common, but the others to whom we have referred have definite things in common—some more things, others fewer; for of friendships, too, some are more and others less truly friendships. And the claims of justice differ too; the duties of parents to children, and those of brothers to each other are not the same, nor those of comrades and those of fellow-citizens, and so, too, with the other kinds of friendship. There is a difference, therefore, also between the acts that are unjust towards each of these classes of associates, and the injustice increases by being exhibited towards those who are friends in a fuller sense; e.g. it is a more terrible thing to defraud a comrade than a fellow-citizen, more terrible not to help a brother than a stranger, and more terrible to wound a father than any one else. And the demands of justice also seem to increase with the intensity of the friendship, which implies that friendship and justice exist between the same persons and have an equal extension.

Now all forms of community are like parts of the political community; for men journey together with a view to some particular advantage, and to provide something that they need for the purposes of life; and it is for the sake of advantage that the political community too seems both to have come together originally and to endure, for this is what legislators aim at, and they call just that which is to the common advantage. Now the other communities aim at advantage bit by bit, e.g. sailors at what is advantageous on a voyage with a view to making money or something of the kind, fellow-soldiers at what is advantageous in war, whether it is wealth or victory or the taking of a city that they seek, and members of tribes and demes act similarly (Some communities seem to arise for the sake or pleasure, viz. religious guilds and social clubs; for these exist respectively for the sake of offering sacrifice and of companionship. But all these seem to fall under the political community; for it aims not at present advantage but at what is advantageous for life as a whole), offering sacrifices and arranging gatherings for the purpose, and assigning honours to the gods, and providing pleasant relaxations for themselves. For the ancient sacrifices and gatherings seem to take place after the harvest as a sort of firstfruits, because it was at these seasons that people had most leisure. All the communities, then, seem to be parts of the political community; and the particular kinds friendship will correspond to the particular kinds of community.

Chapter 10: Classification of constitutions: analogies with family relations.

There are three kinds of constitution, and an equal number of deviation-forms—perversions, as it were, of them. The constitutions are monarchy, aristocracy, and thirdly that which is based on a property qualification, which it seems appropriate to call timocratic, though most people are wont to call it polity. The best of these is monarchy, the worst timocracy. The deviation from monarchy is tyrany; for both are forms of one-man rule, but there is the greatest difference between them; the tyrant looks to his own advantage, the king to that of his subjects. For a man is not a king unless he is sufficient to himself and excels his subjects in all good things; and such a man needs nothing further; therefore he will not look to his own interests but to those of his subjects; for a king who is not like that would be a mere titular king. Now tyranny is the very contrary of this; the tyrant pursues his own good. And it is clearer in the case of tyranny that it is the worst deviation-form; but it is the contrary of the best that is worst. Monarchy passes over into tyranny; for tyranny is the evil form of one-man rule and the bad king becomes a tyrant. Aristocracy passes over into oligarchy by the badness of the rulers, who distribute contrary to equity what belongs to the city—all or most of the good things to themselves, and office always to the same people, paying most regard to wealth; thus the rulers are few and are bad men instead of the most worthy. Timocracy passes over into democracy; for these are coterminous, since it is the ideal even of timocracy to be the rule of the majority, and all who have the property qualification count as equal. Democracy is the least bad of the deviations; for in its case the form of constitution is but a slight deviation. These then are the changes to which constitutions are most subject; for these are the smallest and easiest transitions.

One may find resemblances to the constitutions and, as it were, patterns of them even in households. For the association of a father with his sons bears the form of monarchy, since the father cares for his children; and this is why Homer calls Zeus 'father'; it is the ideal of monarchy to be paternal rule. But among the Persians the rule of the father is tyrannical; they use their sons as slaves. Tyrannical too is the rule of a master over slaves; for it is the advantage of the master that is brought about in it. Now this seems to be a correct form of government, but the Persian type is perverted; for the modes of rule appropriate to different relations are diverse. The association of man and wife seems to be aristocratic; for the man rules in accordance with his worth, and in those matters in which a man should rule, but the matters that befit a woman he hands over to her. If the man rules in everything the relation passes over into oligarchy; for in doing so he is not acting in accordance with their respective worth, and not ruling in virtue of his superiority. Sometimes, however, women rule, because they are heiresses; so their rule is not in virtue of excellence but due to wealth and power, as in oligarchies. The association of brothers is like timocracy; for they are equal, except in so far as they differ in age; hence if they differ much in age, the friendship is no longer of the fraternal type. Democracy is found chiefly in masterless dwellings (for here every one is on an equality), and in those in which the ruler is weak and every one has licence to do as he pleases.

Chapter 11: Corresponding forms of friendship, and of justice.

Each of the constitutions may be seen to involve friendship just in so far as it involves justice. The friendship between a king and his subjects depends on an excess of benefits conferred; for he confers benefits on his subjects if being a good man he cares for them with a view to their well-being, as a shepherd does for his sheep (whence Homer called Agamemnon 'shepherd of the peoples'). Such too is the friendship of a father, though this exceeds the other in the greatness of the benefits conferred; for he is responsible for the existence of his children, which is thought the greatest good, and for their nurture and upbringing.

These things are ascribed to ancestors as well. Further, by nature a father tends to rule over his sons, ancestors over descendants, a king over his subjects. These friendships imply superiority of one party over the other, which is why ancestors are honoured. The justice therefore that exists between persons so related is not the same on both sides but is in every case proportioned to merit; for that is true of the friendship as well. The friendship of man and wife, again, is the same that is found in an aristocracy; for it is in accordance with virtue the better gets more of what is good, and each gets what befits him; and so, too, with the justice in these relations. The friendship of brothers is like that of comrades; for they are equal and of like age, and such persons are for the most part like in their feelings and their character. Like this, too, is the friendship appropriate to timocratic government; for in such a constitution the ideal is for the citizens to be equal and fair; therefore rule is taken in turn, and on equal terms; and the friendship appropriate here will correspond.

But in the deviation-forms, as justice hardly exists, so too does friendship. It exists least in the worst form; in tyranny there is little or no friendship. For where there is nothing common to ruler and ruled, there is not friendship either, since there is not justice; e.g. between craftsman and tool, soul and body, master and slave; the latter in each case is benefited by that which uses it, but there is no friendship nor justice towards lifeless things. But neither is there friendship towards a horse or an ox, nor to a slave qua slave. For there is nothing common to the two parties; the slave is a living tool and the tool a lifeless slave. Qua slave then, one cannot be friends with him. But qua man one can; for there seems to be some justice between any man and any other who can share in a system of law or be a party to an agreement; therefore there can also be friendship with him in so far as he is a man. Therefore while in tyrannies friendship and justice hardly exist, in democracies they exist more fully; for where the citizens are equal they have much in common.

Chapter 12: Various forms of friendship between relations.

Every form of friendship, then, involves association, as has been said. One might, however, mark off from the rest both the friendship of kindred and that of comrades. Those of fellow-citizens, fellow-tribesmen, fellow-voyagers, and the like are more like mere friendships of association; for they seem to rest on a sort of compact. With them we might class the friendship of host and guest. The friendship of kinsmen itself, while it seems to be of many kinds, appears to depend in every case on parental friendship; for parents love their children as being a part of themselves, and children their parents as being something originating from them. Now (1) arents know their offspring better than there children know that they are their children, and (2) the originator feels his offspring to be his own more than the offspring do their begetter; for the product belongs to the producer (e.g. a tooth or hair or anything else to him whose it is), but the producer does not belong to the product, or belongs in a less degree. And (3) the length of time produces the same result; parents love their children as soon as these are born, but children love their parents only after time has elapsed and they have acquired understanding or the power of discrimination by the senses. From these considerations it is also plain why mothers love more than fathers do. Parents, then, love their children as themselves (for their issue are by virtue of their separate existence a sort of other selves), while children love their parents as being born of them, and brothers love each other as being born of the same parents; for their identity with them makes them identical with each other (which is the reason why people talk of 'the same blood', 'the same stock', and so on). They are, therefore, in a sense the same thing, though in separate individuals. Two things that contribute greatly to friendship are a common upbringing and similarity of age; for 'two of an age take to each other', and people brought up together tend to be comrades; whence the friendship of brothers is akin to that of comrades. And cousins and other kinsmen are bound up together by derivation from brothers, viz. by being derived from the same parents. They come to be closer together or farther apart by virtue of the nearness or distance of the original ancestor.

The friendship of children to parents, and of men to gods, is a relation to them as to something good and superior; for they have conferred the greatest benefits, since they are the causes of their being and of their nourishment, and of their education from their birth; and this kind of friendship possesses pleasantness and utility also, more than that of strangers, inasmuch as their life is lived more in common. The friendship of brothers has the characteristics found in that of comrades (and especially when these are good), and in general between people who are like each other, inasmuch as they belong more to each other and start with a love for each other from their very birth, and inasmuch as those born of the same parents and brought up together and similarly educated are more akin in character; and the test of time has been applied most fully and convincingly in their case.

Between other kinsmen friendly relations are found in due proportion. Between man and wife friendship seems to exist by nature; for man is naturally inclined to form couples—even more than to form cities, inasmuch as the household is earlier and more necessary than the city, and reproduction is more common to man with the animals. With the other animals the union extends only to this point, but human beings live together not only for the sake of reproduction but also for the various purposes of life; for from the start the functions are divided, and those of man and woman are different; so they help each other by throwing their peculiar gifts into the common stock. It is for these reasons that both utility and pleasure seem to be found in this kind of friendship. But this friendship may be based also on virtue, if the parties are good; for each has its own virtue and they will delight in the fact. And children seem to be a bond of union (which is the reason why childless people part more easily); for children are a good common to both and what is common holds them together.

How man and wife and in general friend and friend ought mutually to behave seems to be the same question as how it is just for them to behave; for a man does not seem to have the same duties to a friend, a stranger, a comrade, and a schoolfellow.

Chapter 13: Principles of interchange of services (a) in friendship between equals.

There are three kinds of friendship, as we said at the outset of our inquiry, and in respect of each some are friends on an equality and others by virtue of a superiority (for not only can equally good men become friends but a better man can make friends with a worse, and similarly in friendships of pleasure or utility the friends may be equal or unequal in the benefits they confer). This being so, equals must effect the required equalization on a basis of equality in love and in all other respects, while unequals must render what is in proportion to their superiority or inferiority. Complaints and reproaches arise either only or chiefly in the friendship of utility, and this is only to be expected. For those who are friends on the ground of virtue are anxious to do well by each other (since that is a mark of virtue and of friendship), and between men who are emulating each other in this there cannot be complaints or quarrels; no one is offended by a man who loves him and does well by him—if he is a person of nice feeling he takes his revenge by doing well by the other. And the man who excels the other in the services he renders will not complain of his friend, since he gets what he aims at; for each man desires what is good. Nor do complaints arise much even in friendships of pleasure; for both get at the same time what they desire, if they enjoy spending their time together; and even a man who complained of another for not affording him pleasure would seem ridiculous, since it is in his power not to spend his days with him.

But the friendship of utility is full of complaints; for as they use each other for their own interests they always want to get the better of the bargain, and think they have got less than they should, and blame their partners because they do not get all they 'want and deserve'; and those who do well by others cannot help them as much as those whom they benefit want.

Now it seems that, as justice is of two kinds, one unwritten and the other legal, one kind of friendship of utility is moral and the other legal. And so complaints arise most of all when men do not dissolve the relation in the spirit of the same type of friendship in which they contracted it. The legal type is that which is on fixed terms; its purely commercial variety is on the basis of immediate payment, while the more liberal variety allows time but stipulates for a definite quid pro quo. In this variety the debt is clear and not ambiguous, but in the postponement it contains an element of friendliness; and so some states do not allow suits arising out of such agreements, but think men who have bargained on a basis of credit ought to accept the consequences. The moral type is not on fixed terms; it makes a gift, or does whatever it does, as to a friend; but one expects to receive as much or more, as having not given but lent; and if a man is worse off when the relation is dissolved than he was when it was contracted he will complain. This happens because all or most men, while they wish for what is noble, choose what is advantageous; now it is noble to do well by another without a view to repayment, but it is the receiving of benefits that is advantageous. Therefore if we can we should return the equivalent of what we have received (for we must not make a man our friend against his will; we must recognize that we were mistaken at the first and took a benefit from a person we should not have taken it from—since it was not from a friend, nor from one who did it just for the sake of acting so—and we must settle up just as if we had been benefited on fixed terms). Indeed, one would agree to repay if one could (if one could not, even the giver would not have expected one to do so); therefore if it is possible we must repay. But at the outset we must consider the man by whom we are being benefited and on what terms he is acting, in order that we may accept the benefit on these terms, or else decline it.

It is disputable whether we ought to measure a service by its utility to the receiver and make the return with a view to that, or by the benevolence of the giver. For those who have received say they have received from their benefactors what meant little to the latter and what they might have got from others—minimizing the service; while the givers, on the contrary, say it was the biggest thing they had, and what could not have been got from others, and that it was given in times of danger or similar need. Now if the friendship is one that aims at utility, surely the advantage to the receiver is the measure. For it is he that asks for the service, and the other man helps him on the assumption that he will receive the equivalent; so the assistance has been precisely as great as the advantage to the receiver, and therefore he must return as much as he has received, or even more (for that would be nobler). In friendships based on virtue on the other hand, complaints do not arise, but the purpose of the doer is a sort of measure; for in purpose lies the essential element of virtue and character.

Chapter 14: (b) In friendship between unequals.

Differences arise also in friendships based on superiority; for each expects to get more out of them, but when this happens the friendship is dissolved. Not only does the better man think he ought to get more, since more should be assigned to a good man, but the more useful similarly expects this; they say a useless man should not get as much as they should, since it becomes an act of public service and not a friendship if the proceeds of the friendship do not answer to the worth of the benefits conferred. For they think that, as in a commercial partnership those who put more in get more out, so it should be in friendship. But the man who is in a state of need and inferiority makes the opposite claim; they think it is the part of a good friend to help those who are in need, what, they say, is the use of being the friend of a good man or a powerful man, if one is to get nothing out of it?

At all events it seems that each party is justified in his claim, and that each should get more out of the friendship than the other—not more of the same thing, however, but the superior more honour and the inferior more gain; for honour is the prize of virtue and of beneficence, while gain is the assistance required by inferiority.

It seems to be so in constitutional arrangements also; the man who contributes nothing good to the common stock is not honoured; for what belongs to the public is given to the man who benefits the public, and honour does belong to the public. It is not possible to get wealth from the common stock and at the same time honour. For no one puts up with the smaller share in all things; therefore to the man who loses in wealth they assign honour and to the man who is willing to be paid, wealth, since the proportion to merit equalizes the parties and preserves the friendship, as we have said. This then is also the way in which we should associate with unequals; the man who is benefited in respect of wealth or virtue must give honour in return, repaying what he can. For friendship asks a man to do what he can, not what is proportional to the merits of the case; since that cannot always be done, e.g. in honours paid to the gods or to parents; for no one could ever return to them the equivalent of what he gets, but the man who serves them to the utmost of his power is thought to be a good man. This is why it would not seem open to a man to disown his father (though a father may disown his son); being in debt, he should repay, but there is nothing by doing which a son will have done the equivalent of what he has received, so that he is always in debt. But creditors can remit a debt; and a father can therefore do so too. At the same time it is thought that presumably no one would repudiate a son who was not far gone in wickedness; for apart from the natural friendship of father and son it is human nature not to reject a son's assistance. But the son, if he is wicked, will naturally avoid aiding his father, or not be zealous about it; for most people wish to get benefits, but avoid doing them, as a thing unprofitable——So much for these questions.

Book IX: Friendship

Chapter 1: (c) In friendship in which the motives on the two sides are different.

IN all friendships between dissimilars it is, as we have said, proportion that equalizes the parties and preserves the friendship; e.g. in the political form of friendship the shoemaker gets a return for his shoes in proportion to his worth, and the weaver and all other craftsmen do the same. Now here a common measure has been provided in the form of money, and therefore everything is referred to this and measured by this; but in the friendship of lovers sometimes the lover complains that his excess of love is not met by love in return though perhaps there is nothing lovable about him), while often the beloved complains that the lover who formerly promised everything now performs nothing. Such incidents happen when the lover loves the beloved for the sake of pleasure while the beloved loves the lover for the sake of utility, and they do not both possess the qualities expected of them. If these be the objects of the friendship it is dissolved when they do not get the things that formed the motives of their love; for each did not love the other person himself but the qualities he had, and these were not enduring; that is why the friendships also are transient. But the love of characters, as has been said, endures because it is self-dependent. Differences arise when what they get is something different and not what they desire; for it is like getting nothing at all when we do not get what we aim at; compare the story of the person who made promises to a lyre-player, promising him the more, the better he sang, but in the morning, when the other demanded the fulfilment of his promises, said that he had given pleasure for pleasure. Now if this had been what each wanted, all would have been well; but if the one wanted enjoyment but the other gain, and the one has what he wants while the other has not, the terms of the association will not have been properly fulfilled; for what each in fact wants is what he attends to, and it is for the sake of that that that he will give what he has.

But who is to fix the worth of the service; he who makes the sacrifice or he who has got the advantage? At any rate the other seems to leave it to him. This is what they say Protagoras used to do; whenever he taught anything whatsoever, he bade the learner assess the value of the knowledge, and accepted the amount so fixed. But in such matters some men approve of the saying 'let a man have his fixed reward'. Those who get the money first and then do none of the things they said they would, owing to the extravagance of their promises, naturally find themselves the objects of complaint; for they do not fulfil what they agreed to. The sophists are perhaps compelled to do this because no one would give money for the things they do know. These people then, if they do not do what they have been paid for, are naturally made the objects of complaint.

But where there is no contract of service, those who give up something for the sake of the other party cannot (as we have said) be complained of (for that is the nature of the friendship of virtue), and the return to them must be made on the basis of their purpose (for it is purpose that is the characteristic thing in a friend and in virtue). And so too, it seems, should one make a return to those with whom one has studied philosophy; for their worth cannot be measured against money, and they can get no honour which will balance their services, but still it is perhaps enough, as it is with the gods and with one's parents, to give them what one can.

If the gift was not of this sort, but was made with a view to a return, it is no doubt preferable that the return made should be one that seems fair to both parties, but if this cannot be achieved, it would seem not only necessary that the person who gets the first service should fix the reward, but also just; for if the other gets in return the equivalent of the advantage the beneficiary has received, or the price lie would have paid for the pleasure, he will have got what is fair as from the other.

We see this happening too with things put up for sale, and in some places there are laws providing that no actions shall arise out of voluntary contracts, on the assumption that one should settle with a person to whom one has given credit, in the spirit in which one bargained with him. The law holds that it is more just that the person to whom credit was given should fix the terms than that the person who gave credit should do so. For most things are not assessed at the same value by those who have them and those who want them; each class values highly what is its own and what it is offering; yet the return is made on the terms fixed by the receiver. But no doubt the receiver should assess a thing not at what it seems worth when he has it, but at what he assessed it at before he had it.

Chapter 2: Conflict of obligations.

A further problem is set by such questions as, whether one should in all things give the preference to one's father and obey him, or whether when one is ill one should trust a doctor, and when one has to elect a general should elect a man of military skill; and similarly whether one should render a service by preference to a friend or to a good man, and should show gratitude to a benefactor or oblige a friend, if one cannot do both.

All such questions are hard, are they not, to decide with precision? For they admit of many variations of all sorts in respect both of the magnitude of the service and of its nobility necessity. But that we should not give the preference in all things to the same person is plain enough; and we must for the most part return benefits rather than oblige friends, as we must pay back a loan to a creditor rather than make one to a friend. But perhaps even this is not always true; e.g. should a man who has been ransomed out of the hands of brigands ransom his ransomer in return, whoever he may be (or pay him if he has not been captured but demands payment) or should he ransom his father? It would seem that he should ransom his father in preference even to himself. As we have said, then, generally the debt should be paid, but if the gift is exceedingly noble or exceedingly necessary, one should defer to these considerations. For sometimes it is not even fair to return the equivalent of what one has received, when the one man has done a service to one whom he knows to be good, while the other makes a return to one whom he believes to be bad. For that matter, one should sometimes not lend in return to one who has lent to oneself; for the one person lent to a good man, expecting to recover his loan, while the other has no hope of recovering from one who is believed to be bad. Therefore if the facts really are so, the demand is not fair; and if they are not, but people think they are, they would be held to be doing nothing strange in refusing. As we have often pointed out, then, discussions about feelings and actions have just as much definiteness as their subject-matter.

That we should not make the same return to every one, nor give a father the preference in everything, as one does not sacrifice everything to Zeus, is plain enough; but since we ought to render different things to parents, brothers, comrades, and benefactors, we ought to render to each class what is appropriate and becoming. And this is what people seem in fact to do; to marriages they invite their kinsfolk; for these have a part in the family and therefore in the doings that affect the family; and at funerals also they think that kinsfolk, before all others, should meet, for the same reason. And it would be thought that in the matter of food we should help our parents before all others, since we owe our own nourishment to them, and it is more honourable to help in this respect the authors of our being even before ourselves; and honour too one should give to one's parents as one does to the gods, but not any and every honour; for that matter one should not give the same honour to one's father and one's mother, nor again should one give them the honour due to a philosopher or to a general, but the honour due to a father, or again to a mother. To all older persons, too, one should give honour appropriate to their age, by rising to receive them and finding seats for them and so on; while to comrades and brothers one should allow freedom of speech and common use of all things. To kinsmen, too, and fellow-tribesmen and fellow-citizens and to every other class one should always try to assign what is appropriate, and to compare the claims of each class with respect to nearness of relation and to virtue or usefulness. The comparison is easier when the persons belong to the same class, and more laborious when they are different. Yet we must not on that account shrink from the task, but decide the question as best we can.

Chapter 3: Occasions of breaking off friendship.

Another question that arises is whether friendships should or should not be broken off when the other party does not remain the same. Perhaps we may say that there is nothing strange in breaking off a friendship based on utility or pleasure, when our friends no longer have these attributes. For it was of these attributes that we were the friends; and when these have failed it is reasonable to love no longer. But one might complain of another if, when he loved us for our usefulness or pleasantness, he pretended to love us for our character. For, as we said at the outset, most differences arise between friends when they are not friends in the spirit in which they think they are. So when a man has deceived himself and has thought he was being loved for his character, when the other person was doing nothing of the kind, he must blame himself; when he has been deceived by the pretences of the other person, it is just that he should complain against his deceiver; he will complain with more justice than one does against people who counterfeit the currency, inasmuch as the wrongdoing is concerned with something more valuable.

But if one accepts another man as good, and he turns out badly and is seen to do so, must one still love him? Surely it is impossible, since not everything can be loved, but only what is good. What is evil neither can nor should be loved, for it is not one's duty to be a lover of evil, nor to become like what is bad; and we have said that like is dear like. Must the friendship, then, be forthwith broken off? Or is this not so in all cases, but only when one's friends are incurable in their wickedness? If they are capable of being reformed one should rather come to the assistance of their character or their property, inasmuch as this is better and more characteristic of friendship. But a man who breaks off such a friendship would seem to be doing nothing strange; for it was not to a man of this sort that he was a friend; when his friend has changed, therefore, and he is unable to save him, he gives him up.

But if one friend remained the same while the other became better and far outstripped him in virtue, should the latter treat the former as a friend? Surely he cannot. When the interval is great this becomes most plain, e.g. in the case of childish friendships; if one friend remained a child in intellect while the other became a fully developed man, how could they be friends when they neither approved of the same things nor delighted in and were pained by the same things? For not even with regard to each other will their tastes agree, and without this (as we saw) they cannot be friends; for they cannot live together. But we have discussed these matters.

Should he, then, behave no otherwise towards him than he would if he had never been his friend? Surely he should keep a remembrance of their former intimacy, and as we think we ought to oblige friends rather than strangers, so to those who have been our friends we ought to make some allowance for our former friendship, when the breach has not been due to excess of wickedness.

Chapter 4: Friendship is based on self-love.

Friendly relations with one's neighbours, and the marks by which friendships are defined, seem to have proceeded from a man's relations to himself. For (1) we define a friend as one who wishes and does what is good, or seems so, for the sake of his friend, or (2) as one who wishes his friend to exist and live, for his sake; which mothers do to their children, and friends do who have come into conflict. And (3) others define him as one who lives with and (4) has the same tastes as another, or (5) one who grieves and rejoices with his friend; and this too is found in mothers most of all. It is by some one of these characterstics that friendship too is defined.

Now each of these is true of the good man's relation to himself (and of all other men in so far as they think themselves good; virtue and the good man seem, as has been said, to be the measure of every class of things). For his opinions are harmonious, and he desires the same things with all his soul; and therefore he wishes for himself what is good and what seems so, and does it (for it is characteristic of the good man to work out the good), and does so for his own sake (for he does it for the sake of the intellectual element in him, which is thought to be the man himself); and he wishes himself to live and be preserved, and especially the element by virtue of which he thinks. For existence is good to the virtuous man, and each man wishes himself what is good, while no one chooses to possess the whole world if he has first to become some one else (for that matter, even now God possesses the good); he wishes for this only on condition of being whatever he is; and the element that thinks would seem to be the individual man, or to be so more than any other element in him. And such a man wishes to live with himself; for he does so with pleasure, since the memories of his past acts are delightful and his hopes for the future are good, and therefore pleasant. His mind is well stored too with subjects of contemplation. And he grieves and rejoices, more than any other, with himself; for the same thing is always painful, and the same thing always pleasant, and not one thing at one time and another at another; he has, so to speak, nothing to repent of.

Therefore, since each of these characteristics belongs to the good man in relation to himself, and he is related to his friend as to himself (for his friend is another self), friendship too is thought to be one of these attributes, and those who have these attributes to be friends. Whether there is or is not friendship between a man and himself is a question we may dismiss for the present; there would seem to be friendship in so far as he is two or more, to judge from the afore-mentioned attributes of friendship, and from the fact that the extreme of friendship is likened to one's love for oneself.

But the attributes named seem to belong even to the majority of men, poor creatures though they may be. Are we to say then that in so far as they are satisfied with themselves and think they are good, they share in these attributes? Certainly no one who is thoroughly bad and impious has these attributes, or even seems to do so. They hardly belong even to inferior people; for they are at variance with themselves, and have appetites for some things and rational desires for others. This is true, for instance, of incontinent people; for they choose, instead of the things they themselves think good, things that are pleasant but hurtful; while others again, through cowardice and laziness, shrink from doing what they think best for themselves. And those who have done many terrible deeds and are hated for their wickedness even shrink from life and destroy themselves. And wicked men seek for people with whom to spend their days, and shun themselves; for they remember many a grievous deed, and anticipate others like them, when they are by themselves, but when they are with others they forget. And having nothing lovable in them they have no feeling of love to themselves. Therefore also such men do not rejoice or grieve with themselves; for their soul is rent by faction, and one element in it by reason of its wickedness grieves when it abstains from certain acts, while the other part is pleased, and one draws them this way and the other that, as if they were pulling them in pieces. If a man cannot at the same time be pained and pleased, at all events after a short time he is pained because he was pleased, and he could have wished that these things had not been pleasant to him; for bad men are laden with repentance.

Therefore the bad man does not seem to be amicably disposed even to himself, because there is nothing in him to love; so that if to be thus is the height of wretchedness, we should strain every nerve to avoid wickedness and should endeavour to be good; for so and only so can one be either friendly to oneself or a friend to another.

Chapter 5: Relation of friendship to goodwill.

Goodwill is a friendly sort of relation, but is not identical with friendship; for one may have goodwill both towards people whom one does not know, and without their knowing it, but not friendship. This has indeed been said already.' But goodwill is not even friendly feeling. For it does not involve intensity or desire, whereas these accompany friendly feeling; and friendly feeling implies intimacy while goodwill may arise of a sudden, as it does towards competitors in a contest; we come to feel goodwill for them and to share in their wishes, but we would not do anything with them; for, as we said, we feel goodwill suddenly and love them only superficially.

Goodwill seems, then, to be a beginning of friendship, as the pleasure of the eye is the beginning of love. For no one loves if he has not first been delighted by the form of the beloved, but he who delights in the form of another does not, for all that, love him, but only does so when he also longs for him when absent and craves for his presence; so too it is not possible for people to be friends if they have not come to feel goodwill for each other, but those who feel goodwill are not for all that friends; for they only wish well to those for whom they feel goodwill, and would not do anything with them nor take trouble for them. And so one might by an extension of the term friendship say that goodwill is inactive friendship, though when it is prolonged and reaches the point of intimacy it becomes friendship—not the friendship based on utility nor that based on pleasure; for goodwill too does not arise on those terms. The man who has received a benefit bestows goodwill in return for what has been done to him, but in doing so is only doing what is just; while he who wishes some one to prosper because he hopes for enrichment through him seems to have goodwill not to him but rather to himself, just as a man is not a friend to another if he cherishes him for the sake of some use to be made of him. In general, goodwill arises on account of some excellence and worth, when one man seems to another beautiful or brave or something of the sort, as we pointed out in the case of competitors in a contest.

Chapter 6: Relation of friendship to unanimity.

Unanimity also seems to be a friendly relation. For this reason it is not identity of opinion; for that might occur even with people who do not know each other; nor do we say that people who have the same views on any and every subject are unanimous, e.g. those who agree about the heavenly bodies (for unanimity about these is not a friendly relation), but we do say that a city is unanimous when men have the same opinion about what is to their interest, and choose the same actions, and do what they have resolved in common. It is about things to be done, therefore, that people are said to be unanimous, and, among these, about matters of consequence and in which it is possible for both or all parties to get what they want; e.g. a city is unanimous when all its citizens think that the offices in it should be elective, or that they should form an alliance with Sparta, or that Pittacus should be their ruler—at a time when he himself was also willing to rule. But when each of two people wishes himself to have the thing in question, like the captains in the Phoenissae, they are in a state of faction; for it is not unanimity when each of two parties thinks of the same thing, whatever that may be, but only when they think of the same thing in the same hands, e.g. when both the common people and those of the better class wish the best men to rule; for thus and thus alone do all get what they aim at. Unanimity seems, then, to be political friendship, as indeed it is commonly said to be; for it is concerned with things that are to our interest and have an influence on our life.

Now such unanimity is found among good men; for they are unanimous both in themselves and with one another, being, so to say, of one mind (for the wishes of such men are constant and not at the mercy of opposing currents like a strait of the sea), and they wish for what is just and what is advantageous, and these are the objects of their common endeavour as well. But bad men cannot be unanimous except to a small extent, any more than they can be friends, since they aim at getting more than their share of advantages, while in labour and public service they fall short of their share; and each man wishing for advantage to himself criticizes his neighbour and stands in his way; for if people do not watch it carefully the common weal is soon destroyed. The result is that they are in a state of faction, putting compulsion on each other but unwilling themselves to do what is just.

Chapter 7: The pleasure of beneficence.

Benefactors are thought to love those they have benefited, more than those who have been well treated love those that have treated them well, and this is discussed as though it were paradoxical. Most people think it is because the latter are in the position of debtors and the former of creditors; and therefore as, in the case of loans, debtors wish their creditors did not exist, while creditors actually take care of the safety of their debtors, so it is thought that benefactors wish the objects of their action to exist since they will then get their gratitude, while the beneficiaries take no interest in making this return. Epicharmus would perhaps declare that they say this because they 'look at things on their bad side', but it is quite like human nature; for most people are forgetful, and are more anxious to be well treated than to treat others well. But the cause would seem to be more deeply rooted in the nature of things; the case of those who have lent money is not even analogous. For they have no friendly feeling to their debtors, but only a wish that they may kept safe with a view to what is to be got from them; while those who have done a service to others feel friendship and love for those they have served even if these are not of any use to them and never will be. This is what happens with craftsmen too; every man loves his own handiwork better than he would be loved by it if it came alive; and this happens perhaps most of all with poets; for they have an excessive love for their own poems, doting on them as if they were their children. This is what the position of benefactors is like; for that which they have treated well is their handiwork, and therefore they love this more than the handiwork does its maker. The cause of this is that existence is to all men a thing to be chosen and loved, and that we exist by virtue of activity (i.e. by living and acting), and that the handiwork is in a sense, the producer in activity; he loves his handiwork, therefore, because he loves existence. And this is rooted in the nature of things; for what he is in potentiality, his handiwork manifests in activity.

At the same time to the benefactor that is noble which depends on his action, so that he delights in the object of his action, whereas to the patient there is nothing noble in the agent, but at most something advantageous, and this is less pleasant and lovable. What is pleasant is the activity of the present, the hope of the future, the memory of the past; but most pleasant is that which depends on activity, and similarly this is most lovable. Now for a man who has made something his work remains (for the noble is lasting), but for the person acted on the utility passes away. And the memory of noble things is pleasant, but that of useful things is not likely to be pleasant, or is less so; though the reverse seems true of expectation.

Further, love is like activity, being loved like passivity; and loving and its concomitants are attributes of those who are the more active.

Again, all men love more what they have won by labour; e.g. those who have made their money love it more than those who have inherited it; and to be well treated seems to involve no labour, while to treat others well is a laborious task. These are the reasons, too, why mothers are fonder of their children than fathers; bringing them into the world costs them more pains, and they know better that the children are their own. This last point, too, would seem to apply to benefactors.

Chapter 8: The nature of true self-love.

The question is also debated, whether a man should love himself most, or some one else. People criticize those who love themselves most, and call them self-lovers, using this as an epithet of disgrace, and a bad man seems to do everything for his own sake, and the more so the more wicked he is—and so men reproach him, for instance, with doing nothing of his own accord—while the good man acts for honour's sake, and the more so the better he is, and acts for his friend's sake, and sacrifices his own interest.

But the facts clash with these arguments, and this is not surprising. For men say that one ought to love best one's best friend, and man's best friend is one who wishes well to the object of his wish for his sake, even if no one is to know of it; and these attributes are found most of all in a man's attitude towards himself, and so are all the other attributes by which a friend is defined; for, as we have said, it is from this relation that all the characteristics of friendship have extended to our neighbours. All the proverbs, too, agree with this, e.g. 'a single soul', and 'what friends have is common property', and 'friendship is equality', and 'charity begins at home'; for all these marks will be found most in a man's relation to himself; he is his own best friend and therefore ought to love himself best. It is therefore a reasonable question, which of the two views we should follow; for both are plausible.

Perhaps we ought to mark off such arguments from each other and determine how far and in what respects each view is right. Now if we grasp the sense in which each school uses the phrase 'lover of self', the truth may become evident. Those who use the term as one of reproach ascribe self-love to people who assign to themselves the greater share of wealth, honours, and bodily pleasures; for these are what most people desire, and busy themselves about as though they were the best of all things, which is the reason, too, why they become objects of competition. So those who are grasping with regard to these things gratify their appetites and in general their feelings and the irrational element of the soul; and most men are of this nature (which is the reason why the epithet has come to be used as it is—it takes its meaning from the prevailing type of self-love, which is a bad one); it is just, therefore, that men who are lovers of self in this way are reproached for being so. That it is those who give themselves the preference in regard to objects of this sort that most people usually call lovers of self is plain; for if a man were always anxious that he himself, above all things, should act justly, temperately, or in accordance with any other of the virtues, and in general were always to try to secure for himself the honourable course, no one will call such a man a lover of self or blame him.

But such a man would seem more than the other a lover of self; at all events he assigns to himself the things that are noblest and best, and gratifies the most authoritative element in and in all things obeys this; and just as a city or any other systematic whole is most properly identified with the most authoritative element in it, so is a man; and therefore the man who loves this and gratifies it is most of all a lover of self. Besides, a man is said to have or not to have self-control according as his reason has or has not the control, on the assumption that this is the man himself; and the things men have done on a rational principle are thought most properly their own acts and voluntary acts. That this is the man himself, then, or is so more than anything else, is plain, and also that the good man loves most this part of him. Whence it follows that he is most truly a lover of self, of another type than that which is a matter of reproach, and as different from that as living according to a rational principle is from living as passion dictates, and desiring what is noble from desiring what seems advantageous. Those, then, who busy themselves in an exceptional degree with noble actions all men approve and praise; and if all were to strive towards what is noble and strain every nerve to do the noblest deeds, everything would be as it should be for the common weal, and every one would secure for himself the goods that are greatest, since virtue is the greatest of goods.

Therefore the good man should be a lover of self (for he will both himself profit by doing noble acts, and will benefit his fellows), but the wicked man should not; for he will hurt both himself and his neighbours, following as he does evil passions. For the wicked man, what he does clashes with what he ought to do, but what the good man ought to do he does; for reason in each of its possessors chooses what is best for itself, and the good man obeys his reason. It is true of the good man too that he does many acts for the sake of his friends and his country, and if necessary dies for them; for he will throw away both wealth and honours and in general the goods that are objects of competition, gaining for himself nobility; since he would prefer a short period of intense pleasure to a long one of mild enjoyment, a twelvemonth of noble life to many years of humdrum existence, and one great and noble action to many trivial ones. Now those who die for others doubtless attain this result; it is therefore a great prize that they choose for themselves. They will throw away wealth too on condition that their friends will gain more; for while a man's friend gains wealth he himself achieves nobility; he is therefore assigning the greater good to himself. The same too is true of honour and office; all these things he will sacrifice to his friend; for this is noble and laudable for himself. Rightly then is he thought to be good, since he chooses nobility before all else. But he may even give up actions to his friend; it may be nobler to become the cause of his friend's acting than to act himself. In all the actions, therefore, that men are praised for, the good man is seen to assign to himself the greater share in what is noble. In this sense, then, as has been said, a man should be a lover of self; but in the sense in which most men are so, he ought not.

Chapter 9: Why does the happy man need friends?

It is also disputed whether the happy man will need friends or not. It is said that those who are supremely happy and self-sufficient have no need of friends; for they have the things that are good, and therefore being self-sufficient they need nothing further, while a friend, being another self, furnishes what a man cannot provide by his own effort; whence the saying 'when fortune is kind, what need of friends?' But it seems strange, when one assigns all good things to the happy man, not to assign friends, who are thought the greatest of external goods. And if it is more characteristic of a friend to do well by another than to be well done by, and to confer benefits is characteristic of the good man and of virtue, and it is nobler to do well by friends than by strangers, the good man will need people to do well by. This is why the question is asked whether we need friends more in prosperity or in adversity, on the assumption that not only does a man in adversity need people to confer benefits on him, but also those who are prospering need people to do well by. Surely it is strange, too, to make the supremely happy man a solitary; for no one would choose the whole world on condition of being alone, since man is a political creature and one whose nature is to live with others. Therefore even the happy man lives with others; for he has the things that are by nature good. And plainly it is better to spend his days with friends and good men than with strangers or any chance persons. Therefore the happy man needs friends.

What then is it that the first school means, and in what respect is it right? Is it that most identify friends with useful people? Of such friends indeed the supremely happy man will have no need, since he already has the things that are good; nor will he need those whom one makes one's friends because of their pleasantness, or he will need them only to a small extent (for his life, being pleasant, has no need of adventitious pleasure); and because he does not need such friends he is thought not to need friends.

But that is surely not true. For we have said at the outset that happiness is an activity; and activity plainly comes into being and is not present at the start like a piece of property. If (1) happiness lies in living and being active, and the good man's activity is virtuous and pleasant in itself, as we have said at the outset, and (2) a thing's being one's own is one of the attributes that make it pleasant, and (3) we can contemplate our neighbours better than ourselves and their actions better than our own, and if the actions of virtuous men who are their friends are pleasant to good men (since these have both the attributes that are naturally pleasant),—if this be so, the supremely happy man will need friends of this sort, since his purpose is to contemplate worthy actions and actions that are his own, and the actions of a good man who is his friend have both these qualities.

Further, men think that the happy man ought to live pleasantly. Now if he were a solitary, life would be hard for him; for by oneself it is not easy to be continuously active; but with others and towards others it is easier. With others therefore his activity will be more continuous, and it is in itself pleasant, as it ought to be for the man who is supremely happy; for a good man qua good delights in virtuous actions and is vexed at vicious ones, as a musical man enjoys beautiful tunes but is pained at bad ones. A certain training in virtue arises also from the company of the good, as Theognis has said before us.

If we look deeper into the nature of things, a virtuous friend seems to be naturally desirable for a virtuous man. For that which is good by nature, we have said, is for the virtuous man good and pleasant in itself. Now life is defined in the case of animals by the power of perception in that of man by the power of perception or thought; and a power is defined by reference to the corresponding activity, which is the essential thing; therefore life seems to be essentially the act of perceiving or thinking. And life is among the things that are good and pleasant in themselves, since it is determinate and the determinate is of the nature of the good; and that which is good by nature is also good for the virtuous man (which is the reason why life seems pleasant to all men); but we must not apply this to a wicked and corrupt life nor to a life spent in pain; for such a life is indeterminate, as are its attributes. The nature of pain will become plainer in what follows. But if life itself is good and pleasant (which it seems to be, from the very fact that all men desire it, and particularly those who are good and supremely happy; for to such men life is most desirable, and their existence is the most supremely happy) and if he who sees perceives that he sees, and he who hears, that he hears, and he who walks, that he walks, and in the case of all other activities similarly there is something which perceives that we are active, so that if we perceive, we perceive that we perceive, and if we think, that we think; and if to perceive that we perceive or think is to perceive that we exist (for existence was defined as perceiving or thinking); and if perceiving that one lives is in itself one of the things that are pleasant (for life is by nature good, and to perceive what is good present in oneself is pleasant); and if life is desirable, and particularly so for good men, because to them existence is good and pleasant for they are pleased at the consciousness of the presence in them of what is in itself good); and if as the virtuous man is to himself, he is to his friend also (for his friend is another self):—if all this be true, as his own being is desirable for each man, so, or almost so, is that of his friend. Now his being was seen to be desirable because he perceived his own goodness, and such perception is pleasant in itself. He needs, therefore, to be conscious of the existence of his friend as well, and this will be realized in their living together and sharing in discussion and thought; for this is what living together would seem to mean in the case of man, and not, as in the case of cattle, feeding in the same place.

If, then, being is in itself desirable for the supremely happy man (since it is by its nature good and pleasant), and that of his friend is very much the same, a friend will be one of the things that are desirable. Now that which is desirable for him he must have, or he will be deficient in this respect. The man who is to be happy will therefore need virtuous friends.

Chapter 10: The limit to the number of friends.

Should we, then, make as many friends as possible, or—as in the case of hospitality it is thought to be suitable advice, that one should be 'neither a man of many guests nor a man with none'—will that apply to friendship as well; should a man neither be friendless nor have an excessive number of friends?

To friends made with a view to utility this saying would seem thoroughly applicable; for to do services to many people in return is a laborious task and life is not long enough for its performance. Therefore friends in excess of those who are sufficient for our own life are superfluous, and hindrances to the noble life; so that we have no need of them. Of friends made with a view to pleasure, also, few are enough, as a little seasoning in food is enough.

But as regards good friends, should we have as many as possible, or is there a limit to the number of one's friends, as there is to the size of a city? You cannot make a city of ten men, and if there are a hundred thousand it is a city no longer. But the proper number is presumably not a single number, but anything that falls between certain fixed points. So for friends too there is a fixed number perhaps the largest number with whom one can live together (for that, we found, thought to be very characteristic of friendship); and that one cannot live with many people and divide oneself up among them is plain. Further, they too must be friends of one another, if they are all to spend their days together; and it is a hard business for this condition to be fulfilled with a large number. It is found difficult, too, to rejoice and to grieve in an intimate way with many people, for it may likely happen that one has at once to be happy with one friend and to mourn with another. Presumably, then, it is well not to seek to have as many friends as possible, but as many as are enough for the purpose of living together; for it would seem actually impossible to be a great friend to many people. This is why one cannot love several people; love is ideally a sort of excess of friendship, and that can only be felt towards one person; therefore great friendship too can only be felt towards a few people. This seems to be confirmed in practice; for we do not find many people who are friends in the comradely way of friendship, and the famous friendships of this sort are always between two people. Those who have many friends and mix intimately with them all are thought to be no one's friend, except in the way proper to fellow-citizens, and such people are also called obsequious. In the way proper to fellow-citizens, indeed, it is possible to be the friend of many and yet not be obsequious but a genuinely good man; but one cannot have with many people the friendship based on virtue and on the character of our friends themselves, and we must be content if we find even a few such.

Chapter 11: Are friends more needed in good or in bad fortune?

Do we need friends more in good fortune or in bad? They are sought after in both; for while men in adversity need help, in prosperity they need people to live with and to make the objects of their beneficence; for they wish to do well by others. Friendship, then, is more necessary in bad fortune, and so it is useful friends that one wants in this case; but it is more noble in good fortune, and so we also seek for good men as our friends, since it is more desirable to confer benefits on these and to live with these. For the very presence of friends is pleasant both in good fortune and also in bad, since grief is lightened when friends sorrow with us. Hence one might ask whether they share as it were our burden, or—without that happening—their presence by its pleasantness, and the thought of their grieving with us, make our pain less. Whether it is for these reasons or for some other that our grief is lightened, is a question that may be dismissed; at all events what we have described appears to take place.

But their presence seems to contain a mixture of various factors. The very seeing of one's friends is pleasant, especially if one is in adversity, and becomes a safeguard against grief (for a friend tends to comfort us both by the sight of him and by his words, if he is tactful, since he knows our character and the things that please or pain us); but to see him pained at our misfortunes is painful; for every one shuns being a cause of pain to his friends. For this reason people of a manly nature guard against making their friends grieve with them, and, unless he be exceptionally insensible to pain, such a man cannot stand the pain that ensues for his friends, and in general does not admit fellow-mourners because he is not himself given to mourning; but women and womanly men enjoy sympathisers in their grief, and love them as friends and companions in sorrow. But in all things one obviously ought to imitate the better type of person.

On the other hand, the presence of friends in our prosperity implies both a pleasant passing of our time and the pleasant thought of their pleasure at our own good fortune. For this cause it would seem that we ought to summon our friends readily to share our good fortunes (for the beneficent character is a noble one), but summon them to our bad fortunes with hesitation; for we ought to give them as little a share as possible in our evils whence the saying 'enough is my misfortune'. We should summon friends to us most of all when they are likely by suffering a few inconveniences to do us a great service.

Conversely, it is fitting to go unasked and readily to the aid of those in adversity (for it is characteristic of a friend to render services, and especially to those who are in need and have not demanded them; such action is nobler and pleasanter for both persons); but when our friends are prosperous we should join readily in their activities (for they need friends for these too), but be tardy in coming forward to be the objects of their kindness; for it is not noble to be keen to receive benefits. Still, we must no doubt avoid getting the reputation of kill—joys by repulsing them; for that sometimes happens.

The presence of friends, then, seems desirable in all circumstances.

Chapter 12: The essence of friendship is living together.

Does it not follow, then, that, as for lovers the sight of the beloved is the thing they love most, and they prefer this sense to the others because on it love depends most for its being and for its origin, so for friends the most desirable thing is living together? For friendship is a partnership, and as a man is to himself, so is he to his friend; now in his own case the consciousness of his being is desirable, and so therefore is the consciousness of his friend's being, and the activity of this consciousness is produced when they live together, so that it is natural that they aim at this. And whatever existence means for each class of men, whatever it is for whose sake they value life, in that they wish to occupy themselves with their friends; and so some drink together, others dice together, others join in athletic exercises and hunting, or in the study of philosophy, each class spending their days together in whatever they love most in life; for since they wish to live with their friends, they do and share in those things which give them the sense of living together. Thus the friendship of bad men turns out an evil thing (for because of their instability they unite in bad pursuits, and besides they become evil by becoming like each other), while the friendship of good men is good, being augmented by their companionship; and they are thought to become better too by their activities and by improving each other; for from each other they take the mould of the characteristics they approve—whence the saying 'noble deeds from noble men'—So much, then, for friendship; our next task must be to discuss pleasure.

www.ingramcontent.com/pod-product-compliance
Lightning Source LLC
Chambersburg PA
CBHW020904100426
42737CB00043B/121